Dennis H Goodwin

ELEMENTS OF DATA PROCESSING MATHEMATICS

SECOND EDITION

ELEMENTS OF DATA PROCESSING MATHEMATICS

SECOND EDITION

WILSON T. PRICE
Merritt College

MERLIN MILLER
Merritt College

RINEHART PRESS
San Francisco, California

Library of Congress Catalog Card Number: 71-140239

ISBN 03-084745-1

Printed in the United States of America

7890 038 09876

PREFACE

With the availability of low-cost computers, data processing curricula have become common in technical institutes and community colleges. One of the many important objectives of these programs is to provide vocational training in the vast area of business data processing. Even though the field is a new one to education, specific objectives for each course are, in most cases, relatively easy to define. For example, a first programming course should provide the student with a firm foundation in stored-program concepts. It is much more difficult, however, to clearly and concisely define the mathematical background needed in this area. Students enrolled in vocational data processing courses have, in general, a broad range of backgrounds in mathematics. This revision, as was the original edition, is written for the student with a minimum of one year of high-school algebra, but it has been designed as well for ready adaptation to a course with intermediate algebra as a prerequisite.

In revising a book, the author generally finds that his original work represents both an asset and a liability. Certainly the experience of using and experimenting with his book, and the advice and criticism of his colleagues provide an excellent basis for improving the quality and utility of the book. On the other hand, there is a natural tendency for the author to treat all changes as minor modifications to the original work with a degree of prejudice toward the original. In this revision, we have carefully attempted to maximize the advantage of our experience and that of other users, and to minimize our natural bias toward the original work.

However, the fundamental philosophy upon which the original work was based is characteristic of this revised edition. That is, we feel it necessary to provide the computer programming/data processing student with a low-key approach to the following four broad mathematical areas: (1) number systems, (2) intermediate algebra, (3) computer related mathematics, and (4) modern and Boolean algebra.

A major effort has been made to relate mathematical concepts to computers and computer languages as extensively as possible. This is especially important to avoid the common student reaction (especially to algebra) of "the same old stuff." Consistent with this philosophy, Chapter 1 sets forth the basic notions of flowcharting and using algorithms for later use throughout the book. This chapter also includes the section on sets (with a careful tie-in to flowcharting) which was introduced in a later chapter of the original edition. Chapters 2 and 3 present the basic concepts of number systems. Such topics as number base conversion techniques and binary numbers as they appear in computers have been greatly expanded. The study of logic has been more carefully related to computer languages, programming, and flowcharting and is covered in Chapter 4 of this revised edition.

Chapters 5-9 are devoted to the study of the basic algebraic techniques found in an intermediate algebra course. Not only does algebra represent a working tool of mathematics, but such computer languages as Fortran and Algol are based on algebra. The basic concepts of algebraic expressions, fractions, and exponents are presented in Chapter 5. In the usual intermediate algebra sequence, these are three distinct chapters. We have combined them into one chapter for two reasons. First, it is not our intention to cover these topics as completely as they would be covered in an intermediate algebra course, and second, it makes a convenient one-chapter package that a class with intermediate algebra as a prerequisite can merely review or omit.

Chapter 6, on equations, is divided into two parts: simple equations and quadratic equations. In this chapter, as in Chapter 5, emphasis is placed on topics related to computers. For example, in considering algebraic expressions, the hierarchy of operations is stressed. In Chapter 7 the idea of a function is introduced, with emphasis on functions as defined by tables and graphs. Graphical interpretation is stressed throughout. Nonlinear functions are covered in Chapter 8, where emphasis is placed on the logarithmic function, using logs, and interpolation of the log table. The principal reason for this emphasis on logarithms is to introduce the student to the concept of approximating a function with a straight-line segment. This concept is carried further in a later chapter on numerical methods. The intermediate algebra portion of this text terminates with a study of simultaneous systems of linear equations.

Many people who use this text will later be working with programmers who have an extensive background in computer-related mathematics. Thus, it is important to provide these students with an awareness of such topics of mathematics as matrices and linear programming, series, and iterative methods. These subject areas are covered in Chapters 10-13. Obviously, since each normally represents an entire course of study, none of them can be studied in detail. Because of the important role they play in modern day computing, however, it certainly is possible and of value to the programmer to have a brief exposure to the nature of these areas of mathematics. Perhaps conspicuous by its absence is statistics. It is not included in this text since many precalculus level courses in statistics are now available to the student. Furthermore, the subject is of sufficient importance, we feel, that the prospective programmer should take a full-semester course.

Chapter 14 covers the topic of Boolean algebra using the intuitive approach of studying more tangible switching circuits rather than by use of a purely mathematical approach.

Overall, the materials in approximately the first half of the book are at a high-school algebra level, whereas the presentation of the second half is nearer to a college algebra level. In preparing these materials, our attempt has been to provide the continuity which is desired of a textbook, yet to organize the subject matter into sections which are somewhat independent. Our hope is for a text which may be used in a variety of ways: for a one-semester course where the students have a strong algebra background (one or two years of high-school algebra); for a one- or two-semester (or two-quarter) course where the students have a minimal one year of high-school algebra background; or in some instances for a one-semester or one-quarter course where the students have taken a course in college algebra.

ACKNOWLEDGEMENTS

In preparing a revision of a successful book, an author has the tremendous advantage of benefiting from the wide and varied experiences of others who have used the original edition. We have attempted to take full advantage of the opinions and constructive comments offered by those who have used the book. Highly valued advice was received from our colleagues and students as well as from many other professionals and students who were kind enough to write to us. In addition, the constructive critiques of approximately 15 reviewers whom the publisher utilized provided insight beyond that for which we had hoped. In this edition, we have taken extra care in attempting to minimize errors. The exceptional patience and perseverance of Miss Amy Ng in checking examples and answers will be apparent, we are certain, to all users of the book.

To all of these individuals, we express our sincere appreciation and the hope that this revision is worthy of their contributions. Finally, and with utmost humility, we must thank our wives Mary Price and Alice Miller for their forebearance and patience during preparation of the manuscript.

Oakland, California — W. T. Price
January 1971 — M. Miller

CONTENTS

Preface v

1 ALGORITHMS AND FLOWCHARTS 1
1.1 Algorithms and Their Use 2
1.2 Using Flowcharts 6
1.3 Basic Concepts of Sets 13
1.4 Venn Diagrams 21
1.5 Sets and Flowcharts 28
1.6 Boolean Properties 32

2 THE DECIMAL NUMBER SYSTEM 43
2.1 The Natural Numbers 44
2.2 Extending the Number System 47
2.3 Characteristics of Real Numbers 52
2.4 Characteristics of Decimal Numbers 59

3 OTHER NUMBER BASES 66
3.1 Introduction 67
3.2 Base 2 71
3.3 Binary and Decimal Number Conversion 74
3.4 Binary Arithmetic 85
3.5 Arithmetic Complements 89
3.6 Octal and Hexadecimal Numbers 95
3.7 Octal and Hexadecimal Arithmetic 105

4 LOGICAL FORMS AND PROGRAMMING 111
4.1 Use of Set Descriptions 112
4.2 AND, OR and NOT in Sequence 119
4.3 Flowcharts 123
4.4 IF-THEN Forms 129
4.5 Truth Values of Conditionals and Equivalences 136
4.6 Special Uses of AND, OR, and EOR 141
4.7 Simplification by Flowchart 148

5 BASIC ALGEBRA 153

Algebraic Expressions 154

5.1 Evaluating Expressions 156
5.2 Arithmetic Operations 160
5.3 Factoring 165

Algebraic Fractions 170

5.4 Relationships Involving Fractions 171
5.5 Reducing Rational Numbers 177
5.6 Multiplication and Division of Rational Numbers 179
5.7 Addition and Subtraction of Rational Numbers 181

Exponents and Radicals 184

5.8 Properties of Exponents 184
5.9 Fractional Exponents and Radicals 191

6 EQUATIONS 195

Simple Equations 196

6.1 Basic Notions of Equations 196
6.2 More on Equations 205
6.3 Equations in Several Variables 211
6.4 Simple Inequalities 216

Polynomials 220

6.5 Solving Quadratics by Factoring 221
6.6 Completing the Square and the Quadratic Formula 223

7 FUNCTIONS 232

7.1 Tables and Graphs 233
7.2 Equations and Formulas 242
7.3 The Linear Function 248
7.4 Linear Inequalities 256

8 NONLINEAR FUNCTIONS 261

8.1 The Quadratic Function 262
8.2 Trigonometric Functions 267
8.3 The Exponential and Logarithmic Functions 273
8.4 Common Logarithms 278
8.5 Linear Interpolation 282
8.6 Logarithm Tables 287

9 SIMULTANEOUS SYSTEMS OF EQUATIONS 293

9.1 Systems in Two Variables 294
9.2 Dependent and Inconsistent Systems 302
9.3 Systems in Three Variables 305
9.4 Applications of Linear Systems 310
9.5 Solution by Determinants 312

10 MATRICES 321
10.1 The Notion of a Matrix 322
10.2 Matrix Multiplication 327
10.3 Solving Simultaneous Systems with Matrices 334
10.4 Finding the Inverse of a Matrix 342
10.5 Matrix Multiplication on the Computer 346

11 LINEAR PROGRAMMING 352
11.1 A Graphical Solution 353
11.2 Finding a Minimum Cost 361
11.3 An Algebraic Method 368
11.4 The Algebraic Method in Matrix Form 375
11.5 The Simplex Method 380

12 SERIES 388
12.1 Sequences 389
12.2 Progressions 395
12.3 Convergences of Sequences 400
12.4 Summations 404
12.5 Sums of Arithmetic and Geometric Progressions 408
12.6 Infinite Geometric Series 412

13 NUMERICAL METHODS 416
13.1 Square Root 417
13.2 Roots by the Bolzano Method 421
13.3 Roots by the Method of False Position 426
13.4 Solution of Linear Systems by the Jacobi Method 429
13.5 Solution of Linear Systems by the Gauss-Seidel Method 437

14 BOOLEAN ALGEBRA 445
14.1 Basic Circuits 446
14.2 Additional Circuits 452
14.3 The Form $AB + A$ and Boolean Properties 457
14.4 Other Boolean Properties 463
14.5 Simplifications 466
14.6 An Application of Boolean Algebra 471

Appendix I HEXADECIMAL-DECIMAL CONVERSION TABLE 477
Appendix II FOUR PLACE LOGARITHM TABLE 486
Appendix III ANSWERS TO SELECTED PROBLEMS 489

Index 525

1 ALGORITHMS AND FLOWCHARTS

1.1 ALGORITHMS AND THEIR USE / **2**
What is an Algorithm? / **2**
Examples of Algorithms / **2**
Exercise 1.1 / **5**

1.2 USING FLOWCHARTS / **6**
Why Flowcharts? / **6**
Flow Direction / **7**
Processing / **7**
Input/Output / **7**
Decision / **8**
Termination / **8**
Connector / **8**
Flowcharting the Examples / **8**
On Flowcharting in General / **11**
Exercise 1.2 / **12**

1.3 BASIC CONCEPTS OF SETS / **13**
Punched Cards / **13**
Set Notation / **15**
Subsets / **16**
Intersection of Two Sets / **16**
Union of Two Sets / **17**
Complements / **17**
Format / **18**
Exercise 1.3 / **18**

1.4 VENN DIAGRAMS / **21**
Intersection and Union / **21**
Set Language / **23**
Combining Three Sets / **23**
Exercise 1.4 / **25**

1.5 SETS AND FLOWCHARTS / **28**
Exercise 1.5 / **31**

1.6 BOOLEAN PROPERTIES / **32**
Basic Characteristics of Operations / **32**
Commutative Property / **33**
Associative Property / **35**
Distributive Property / **38**
Exercise 1.6 / **39**

1.1 ALGORITHMS AND THEIR USE

WHAT IS AN ALGORITHM? In almost all cases, the eventual objective in learning mathematics is to develop the facility for solving problems of one type or another. For the vast majority of users, mathematics is simply a tool which can be used to achieve a desired end; mathematics *itself* is not the end objective. Usually the successful student of mathematics has recognized the importance of a logical, step-by-step approach and uses it throughout his study and application of mathematics. Indeed, we commonly hear comments such as "he has a logical mind, he should do well in mathematics."

The very difficulties which students of mathematics encounter commonly plague the beginner in programming. We often hear such comments as "he is good at mathematics so he should be good at programming." (This is often mistaken as meaning that in order to become proficient in programming, one must be a mathematician.) The important notion to recognize is that the logical, systematic approach essential to mathematics is also essential to programming. In both mathematics and programming, we commonly deal with **algorithms.** An algorithm, broadly speaking, is no more than a set of rules or instructions for doing something. In this sense, a recipe for baking a cake is an algorithm; calculating interest for a bank account involves an algorithm; solving a set of linear equations involves an algorithm; computing a satellite orbit involves an algorithm; and so on. Of foremost importance to the programmer is that programming itself involves the preparation of algorithms (programs) for the computer. Without a systematic approach, programming of any complexity becomes virtually impossible.

EXAMPLES OF ALGORITHMS. To gain insight into algorithms, let us consider the simple task of calculating interest.

ALGORITHM
Add the interest rate (expressed in hundredths) to 1.00; raise this sum to the fifth power and multiply by 200. Repeat this process for each of the interest rates.

Although this description is complete, an individual with no prior knowledge of what is meant by compound interest would probably find it somewhat confusing. Note how much clearer is the language of algebra for expressing this algorithm in the following alternative solution.

EXAMPLE 1.1

Calculate the interest on $200 invested for 5 years compounded annually at each of the four interest rates 4, 6, 8, and 10 percent.

ALGORITHM (alternate)
Calculate interest using the formula

$$I = 200(1.0 + i)^5$$

for values of $i = 0.04$, 0.06, 0.08 and 0.10.

That which is confusing and sometimes even ambiguous when written in ordinary English becomes obvious when written in algebraic form. The use of algebra in expressing mathematical operations to be performed in a computer program is often invaluable. In fact, the Fortran programming language is a language which is designed to appear as similar to the language of algebra as possible.

On the other hand, algorithms are hardly restricted to the field of mathematics, as is illustrated by Example 1.2.

EXAMPLE 1.2

Turn on a gas furnace.

ALGORITHM

First turn the fan on, then check the pilot light. If it is burning, turn the gas valve on; if it is not burning, turn off the pilot light valve, wait five minutes, turn the valve on and light the pilot light. Then turn the gas valve on.

Virtually all of us have encountered such a set of instructions at one time or another and have found it necessary to reread them several times to understand their meaning. Frequently the problem is compounded by ambiguities; for example, what is to be done if the furnace has several valves and the main valve is not clearly obvious. (Of course, it is possible to go overboard on clarity as many fathers will attest after attempting to assemble junior's new toy in which "flange C is overlapped with the interior of column A and connected by the inverted hexagonal cross-bolt.") However, consider the clarity introduced in lighting the furnace when the algorithm is stated as follows:

ALGORITHM (alternate)

1. Turn on the fan.
2. Check the pilot light; if on skip to step 7, otherwise continue to step 3.
3. Turn off pilot light valve (small valve with "P" imprinted on handle).
4. Wait five minutes.
5. Turn on pilot light.
6. Light pilot light.
7. Turn on main gas valve (red handled valve near bottom of furnace).

Obviously this detailed, step-by-step approach is far less likely to be confusing. Very often we can get by with writing vague and sometimes ambiguous instructions to another person because a human being can exercise a degree of judgment in carrying out the operations. However, in writing computer programs, we must be much more precise since the computer has no means for exercising judgment to determine what we think; it can only do as it is explicitly told.

EXAMPLE 1.3

An employer plans to pay a bonus to each employee. Those earning \$6000 or over are to be paid 5 percent of their salary; those earning less than \$6000 are to be paid \$300. Prepare a set of instructions (an algorithm) for performing this operation.

ALGORITHM

1. Obtain next employee folder.
2. Compare salary to \$6000; if less, go to step 6.
3. Calculate bonus as
$$\text{bonus} = 0.05 \times \text{salary}.$$
4. Record bonus on employee ledger.
5. Return to step 1.
6. Set bonus to \$300.
7. Go to step 4.

Note that this sequence differs from that for lighting a furnace in that it is repetitive in nature. That is, the sequence of steps is carried out for one employee and then the same set of operations is performed for the next employee. Of course, the difference between the two involves the numbers which are used in the calculations.

The notion of repetition is also included in the next example in which the boss of a not-too-bright employee leaves a job to be completed while he is away. He leaves chalk, an eraser, a chalk board, and a pigeonhole box with cards in it as shown in Figure 1.1. The boss instructed the employee to start with the card in slot 6, to do whatever it says, then proceed to box 7, and so on.

EXAMPLE 1.4

Keep an employee busy during boss's absence.

ALGORITHM

Card Number	Content
1	200
2	1.06
3	5
4	1
5	Place new cards in holes 1, 2, and 3
6	Erase slate, write number from card 1
7	Multiply by number on card 2
8	Round off answer to two decimal places
9	Replace number on card 1 with number from slate
10	Erase slate, write number from card 3
11	Subtract number on card 4
12	Write number from slate on card 3
13	If result is zero go to card 15
14	Go back to card 6
15	Remove cards 1, 2 and 3 and save for boss
16	Go back to card 5

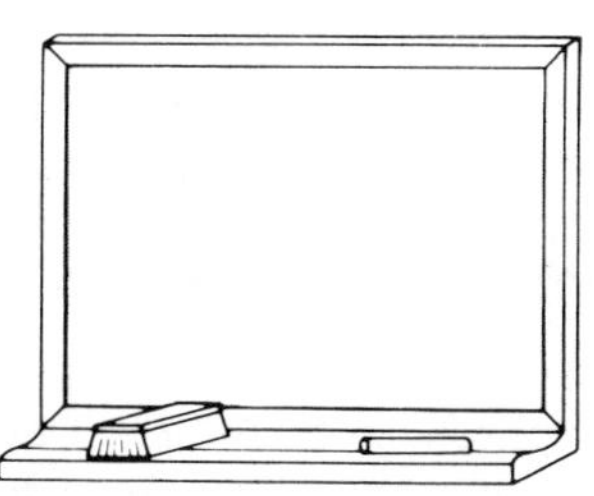

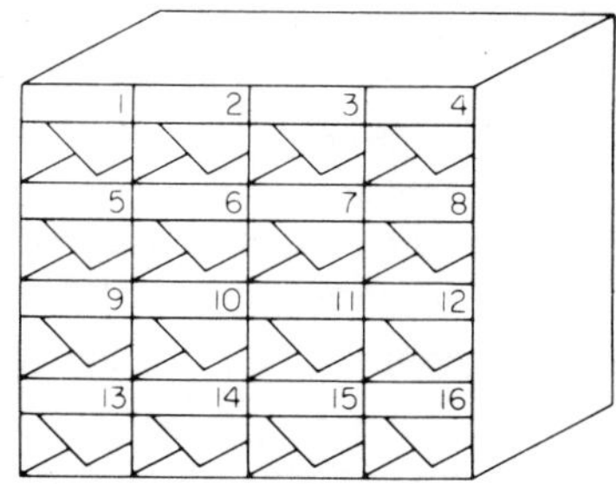

FIGURE 1.1 Pigeonhole example

Upon proceeding through this sequence of instructions, we would find that the contents of cards 1, 2 and 3 (which are to be saved) would be:

Card 1	268.28	final value
Card 2	1.06	unchanged
Card 3	0	final value

It is interesting to note that we can easily perform this sequence of operations without the least idea of what the numbers represent. By taking the instructions in order as directed, it is possible to perform the function required, since each individual operation is relatively simple and independent of other operations. This is an important notion and one which the programmer must *fully* appreciate since it is the manner in which the digital computer operates. However, with a little insight and imagination, we can see that this particular calculation determines the compound interest on $200 invested for 5 years at 6 percent (one of the values used in Example 1.1).

EXERCISE 1.1

The next three questions apply to the algorithm of Example 1.4.

1. Change the instructions as required to begin the set of instructions with card number 1 and put the data on cards 13-16.
2. What would result if the data were inserted in the middle of the sequence of instructions? For example, what if instructions were contained in cards 1-6, data in cards 7-10, and the remainder of the instructions in cards 11-15? (Assume appropriate changes in card number references of the instructions.)
3. Would the algorithm function properly if the first 10 instructions were written on cards 1-10 and the data on cards 11-14? (Assume appropriate changes in card number references of the instructions.) Justify your answer.

Write an algorithm for each of the following.

4. Prepare a cup of instant coffee by first bringing the water to a boil.
5. Fill a water tank to the $\frac{3}{4}$-full mark using a small pail.
6. A paint touch-up job is to be done in a home. The original color was

obtained by adding slight amounts of red and blue paint to a can of white paint. Mix some paint to match the present color.

7. A computer operator obtains a card reader check in loading a program (meaning that the machine has either misread the card in which case it should be reread or the card has been improperly punched in which case it should be repunched). Show the corrective action required.
8. Find the highest score in a list of examination scores.
9. Working as a clerk, you accept $1.00 for purchase of an item. Determine what the change should be in terms of pennies, nickels, dimes, quarters, and half dollars for a given amount of purchase.
10. A person has completed five courses and received his grades. Calculate the grade points for each course and determine his grade point average.
11. Compile a scholarship list of students who are seniors, are carrying more than 15 units, and have grade point averages of 3.0 or greater.
12. Determine if the number 377 is prime, that is, it cannot be divided evenly by any number other than itself and 1. (A simple means is to attempt dividing it by all the integers in the range 2 through 376.)
13. Multiply two numbers together by successive addition. For example, 628 can be multiplied by 27 by successively adding 628 a total of 27 times.

1.2 USING FLOWCHARTS

WHY FLOWCHARTS? The four examples of the preceding section serve to illustrate the basic nature of algorithms. These English and mathematical descriptions appear to be adequate for the relatively simple tasks of the examples. However, as a given problem becomes more complex, the algorithm when written in ordinary English can become very cumbersome. This is commonly due to the fact that the extensive and detailed descriptions together with the ambiguous nature of the English language tend to disguise the basic logical structure of the algorithm. This was especially true of Example 1.2 consisting of descriptions to light a furnace. The problem was partially resolved by listing steps 1,2,3,···. However, a long and confusing sequence of steps can very often be clearly illustrated through use of a pictorial representation (consistent with the notion that "a picture is worth a thousand words"). For example, consider Figure 1.2, which is a picture representation of Example 1.2. This commonly used means for representing an algorithm is called a **flowchart** (sometimes referred to as a **block diagram**). Note that the alternate courses of action are clearly illustrated by this picture whereas they are not so evident in the original algorithm.

Because of the importance of using flowcharts in programming, they are widely used throughout this book. In perusing data processing literature, the reader commonly encounters the same geometric symbol used by different authors to illustrate different operations. In the interest of standardization, all flowcharts in this book will conform to the USASI (USA Standards Institute) flowcharting standards as adopted by IBM.

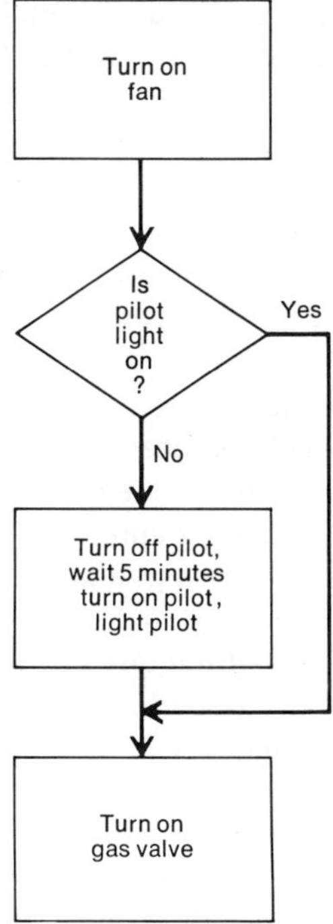

FIGURE 1.2 Flowchart for Algorithm Example 1.2

FLOW DIRECTION. The flow-direction symbol represents the direction of logic flow, which is generally from top to bottom and left to right. To avoid confusion, flow lines are usually drawn with an arrowhead at the point of entry of a symbol. Whenever possible, crossing of flow lines should be avoided.

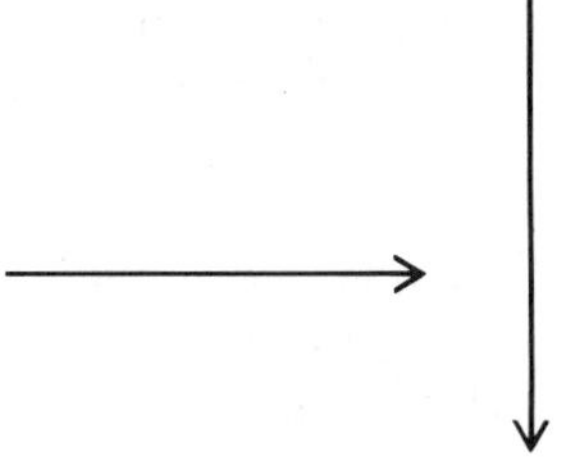

Flow direction

PROCESSING. The processing symbol is used to represent general processing functions not represented by other symbols. These functions are generally those that contain the actual calculating operations of the algorithm (program).

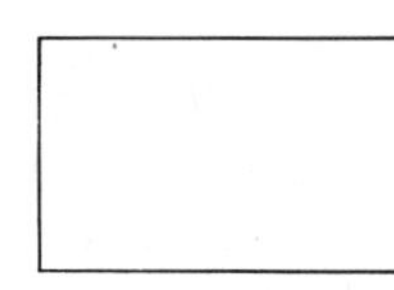

Processing

INPUT/OUTPUT. The input/output symbol is used to denote the operation of obtaining new data for processing (input) or for recording results of computations (output).

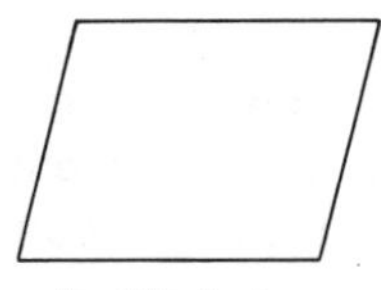

Input/output

DECISION. The decision symbol is used to indicate a point in a program at which a branch to one of two or more alternative points is possible. As shown in Figure 1.2, the criterion for making the choice should be indicated clearly. Also, the condition upon which each of the possible exit paths will be executed should be identified and all the possible paths should be accounted for.

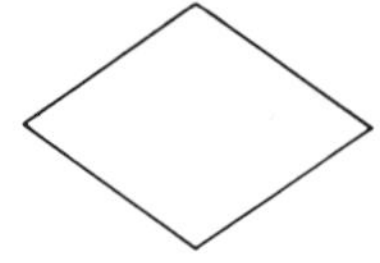

Decision

TERMINATION. The termination symbol represents any point at which processing originates or terminates. With normal program operations, such points will be at the start and completion of the sequence, and sometimes at a terminal point under error conditions. For example, a computer program might require that the quantity a/x be evaluated. However, prior to performing the calculation the programmer might test to ensure that x is not zero. Upon detection of zero, a possible course of action would be to transfer control to an error halt.

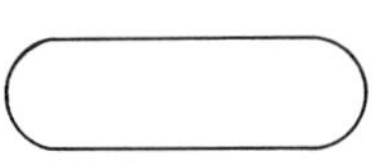

Terminal

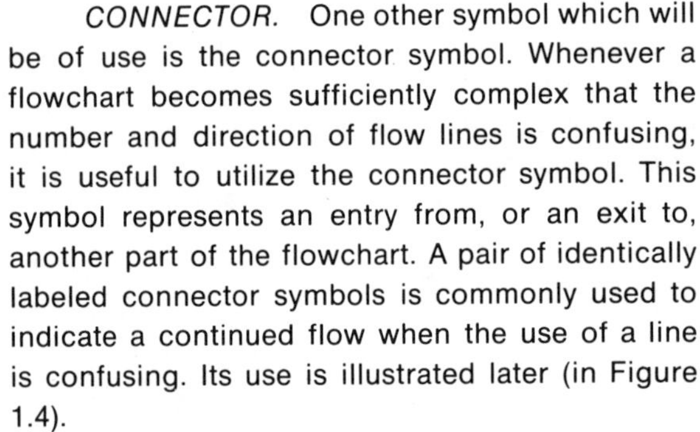

CONNECTOR. One other symbol which will be of use is the connector symbol. Whenever a flowchart becomes sufficiently complex that the number and direction of flow lines is confusing, it is useful to utilize the connector symbol. This symbol represents an entry from, or an exit to, another part of the flowchart. A pair of identically labeled connector symbols is commonly used to indicate a continued flow when the use of a line is confusing. Its use is illustrated later (in Figure 1.4).

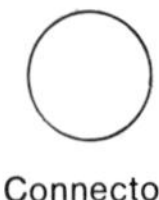

Connector

FLOWCHARTING THE EXAMPLES. As we can see, the flowchart for Example 1.2 (Figure 1.2) utilizes the symbols as defined by the flowcharting standards. It is important to recognize that each individual operation need not be assigned its own block. Several processing functions are commonly indicated by a single block as in Figure 1.2.

The algorithm of Example 1.1 is charted in Figure 1.3 which is a complete flowchart (in contrast to Figure 1.2 which might be better classified as a flowchart segment). Note the manner in which the algorithm is set up in order to provide a convenient representation in flowchart form. This flowchart clearly

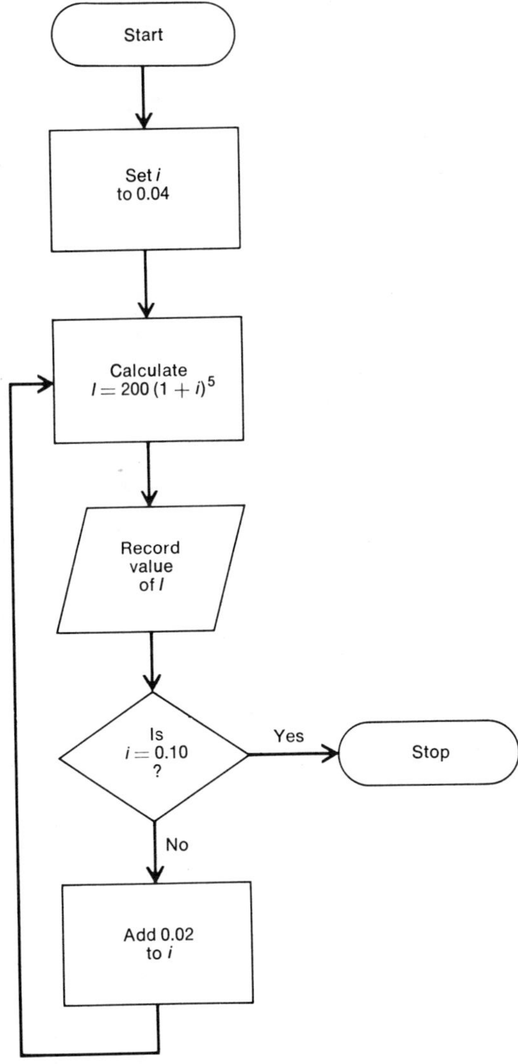

FIGURE 1.3 Flowchart for Algorithm Example 1.1

indicates the procedures to be followed in calculating the interest (I) and that termination will occur after a given number of passes through the loop (commonly termed *iterations*).

Similarly, the algorithm of Example 1.3 is clearly illustrated by the flowchart of Figure 1.4. In (a) we see the usual flow lines causing a return to process the next employee file. An alternate flowchart (b) is included to illustrate use of the connector symbol to avoid drawing the flowline. Although use of the connector symbol provides no advantage in this case, it is commonly used when a large number of lines will appear confusing in a complex flowchart.

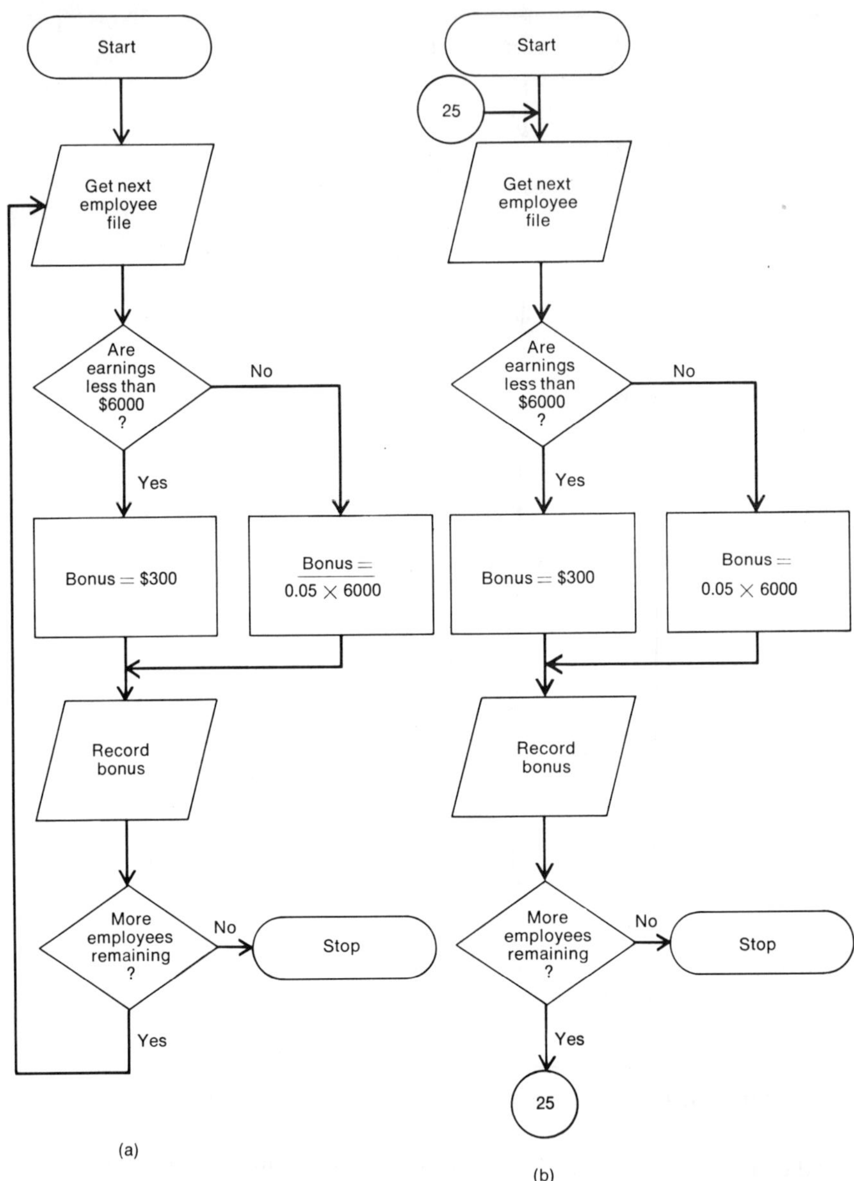

FIGURE 1.4 Flowchart for Algorithm Example 1.3

Finally, the flowchart for Example 1.4 is shown in Figure 1.5 in which the card, or step numbers, of the algorithm are indicated to the above left of each block. This flowchart clearly illustrates the basic nature of the algorithm, that is, one loop is nested within another. After a given number of passes through the inner loop (as controlled by the card in slot 2), control passes to the outer loop which restarts the cycle. Note that the algorithm begins with the block labeled START but it has no end. Of course, knowing the nature of the problem, we see that processing will obviously terminate when slots 1, 2, and 3 contain no more cards. However, this is not represented in the flowchart so the algorithm is said to represent an **infinite loop**; it has no **logical end**. Flowcharts consisting of such infinite loops are commonly encountered but the implication is that processing terminates when no more data is available. Automatic features of the so-called computer software provide for terminating execution of a program when all of the data has been read.

ON FLOWCHARTING IN GENERAL. In drawing flowcharts when programming, the programmer should keep in mind that the primary functions of the flowchart are to provide him with a better insight into his problem while

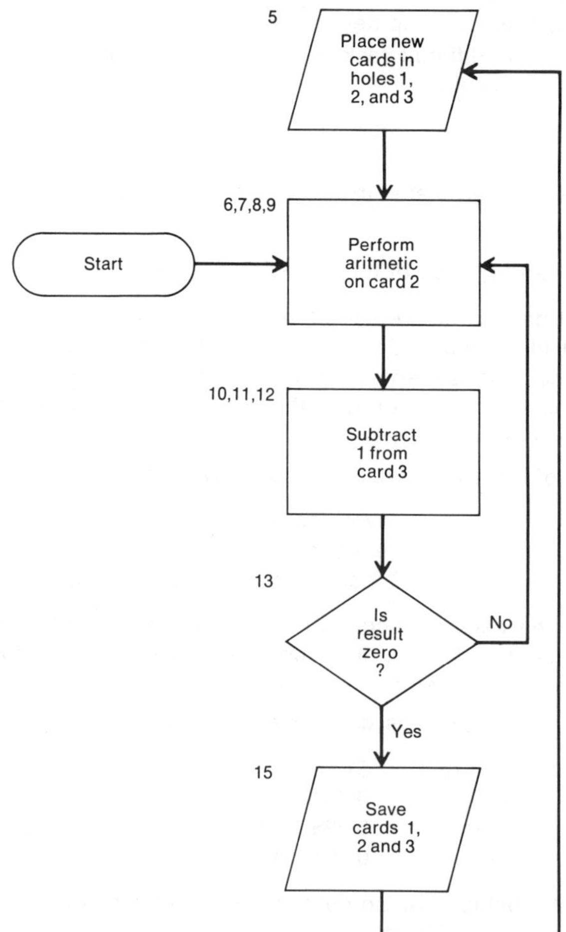

FIGURE 1.5 Flowchart for Algorithm Example 1.4

preparing the program and to document and explain the program to others who might want to use it. The following pointers may be of help in producing useful results:

1. First chart the mainline logic, then incorporate detail.
2. Do not chart every detail or the chart will only be graphic representation, step by step, of the sequence of instructions and "sight of the forest will be lost because of the trees."
3. Put yourself in the position of a reader who might need to refer to your flowchart and try to anticipate his problems in understanding your chart.

The importance of flowcharting cannot be overemphasized. Usually, the most complex part of programming a problem is to determine the basic logic required. Recognizing this, some large business firms employ trained individuals to reduce problems for computer solution to a series of flowcharts. From there, the somewhat less difficult task of coding is performed by someone with a lesser degree of training.

EXERCISE 1.2

Note: Each of the problems from Exercise 1.1 can be used as a flowcharting problem.

Draw a flowchart for each of the following.

1. Student data cards are sorted into an ascending sequence based on the student file number. Check the cards to insure that they are in proper sequence. If a sequencing error is detected, record the number of the card causing the error and the number of the preceding card, then continue checking.
2. A deck of IBM cards includes the following three types of cards:

   Payroll card (code 1)
   Deduction card (code 2)
   Special terminal card (code 3)

   Search the deck to record the employee name and number from each payroll card; pull each deduction card from the file. Upon encountering the terminal card, stop searching.
3. A water company bills customers according to the following schedule:

Flat rate	$1.00
plus	0.20/unit for the first 20 units
plus	0.225/unit for the next 30 units
plus	0.30/unit for all in excess of 50 units

   The water bill is to be computed for each customer.
4. A large number of measurements have been taken which are accurate to the nearest 0.0001 inch. Each number is to be rounded off to the nearest tenth.

5. Commissions for a list of salesmen are to be calculated and recorded as follows:

   (a) If the calculated commission is less than a designated minimum, pay the minimum.
   (b) If the calculated commission is greater than a designated maximum, pay an additional incentive amount.
   (c) Otherwise pay the calculated commission.

6. Examine a list of employees and print the name of each who qualifies as follows:

   1. Has not received a pay increase in the past one year

   and 2. Has a pay rate of

   (a) Less than $6.00 per hour

   or (b) Less than $240 per week

7. Two card sorters have a combined speed of 33 cards per second. One machine is 7 cards per second faster than the other. Determine the speeds of the sorters by trial and error. Begin with a sorting speed of 1 card per second for the slower (and consequently 32 for the faster) and increment by one for each trial.

8. An electronic control circuit includes resistors R_1, R_2 and R_3 and voltages v_a and v_b. The circuit will be activated under the following conditions

$$(R_3 \neq 0 \quad or \quad v_b > 5)$$

and

$$[R_1 < 6 \quad or \quad (R_2 > 10 \quad and \quad v_a \neq 0)]$$

Values for the voltages and resistances to be evaluated are listed on a worksheet. If a circuit will be activated by a given data set, print the message ACTIVE; if not, print NONACTIVE.

1.3 BASIC CONCEPTS OF SETS

PUNCHED CARDS. In determining the weekly payroll of a firm, a payroll clerk must take into account several varying factors for each employee who is paid on an hourly basis. Two principal factors would be the number of hours worked and the hourly wage. This information might be entered on a card as shown in Figure 1.6.

For convenience, assume a company employing 20 people has only two hourly wage classifications, $2.00 and $3.00. Furthermore, any employee working fewer than 20 hours per week is a part-time employee, and any working 20 or more hours per week is a full-time employee. Full-time employees working more than 40 hours per week receive a time-and-a-half rate for all hours over 40. If this information is punched into a separate card for each employee, the complete deck of cards, shown in Figure 1.7, would contain payroll information for the entire set of employees of the company.

A high-speed sorter can very rapidly group these cards in a number of different ways. If desired, the sorter operator can process the deck of cards according to wage classification (resulting in two *subdecks*), or work-week

No. Name Hourly Rate Hours Worked

002 RICHARDSON RALPH R 200 30

FIGURE 1.6

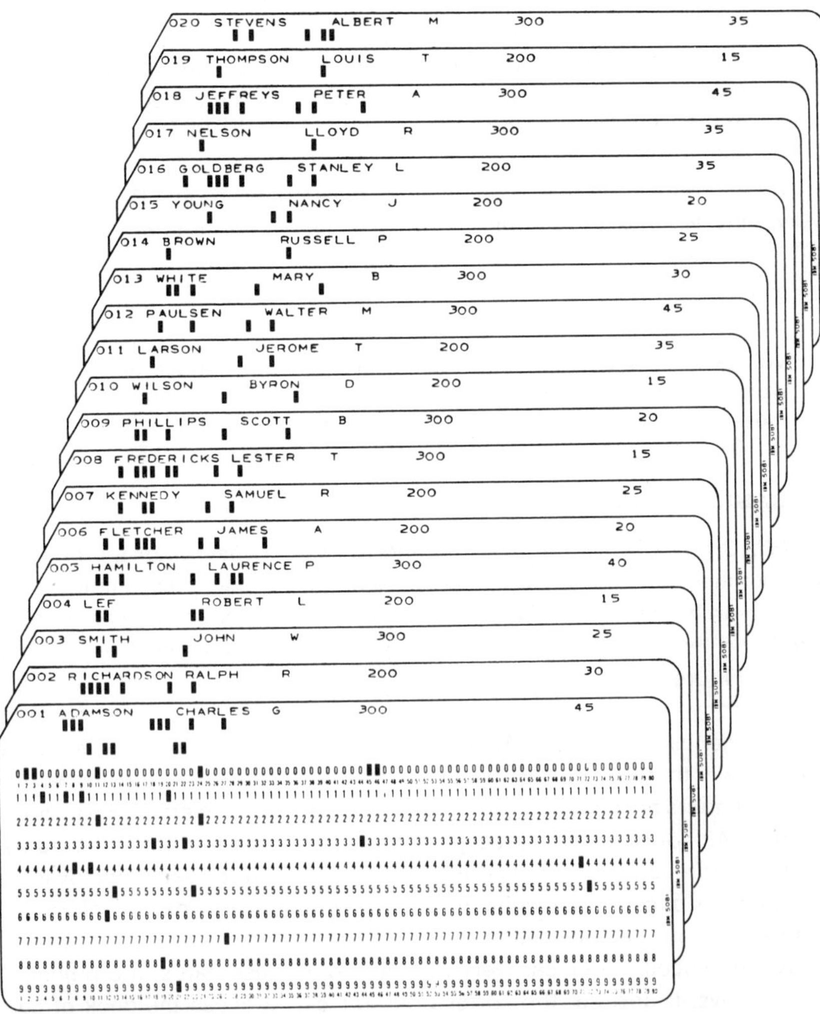

FIGURE 1.7

classification (three subdecks). This is shown schematically in Figure 1.8(a) and (b).

The main deck of cards in Figure 1.7 contains cards of employees earning \$2.00 or \$3.00 per hour and working 10-45 hours for a given week. After the cards have been subdivided according to wage classifications, the cards for all employees earning \$2.00 an hour are in one subdeck, and for those earning \$3.00 an hour in another subdeck. In Figure 1.8(a), these subdecks have been labeled H_2 and H_3 for the \$2.00 and \$3.00 hourly wage classifications, respectively.

The subdivision shown in Figure 1.8(b) results in three subdecks. The letter P represents the subdeck of cards for part-time employees working less than 20 hours per week, F for those full-time employees working 20-40 hours per week, and V for employees working over 40 hours per week. In each subdeck the employees share a common property, although they may not be earning the same amount per hour. For example, in the subdeck labeled P, all employees have worked less than 20 hours but both hourly rates occur.

SET NOTATION. These decks of cards provide examples of the mathematical concept of **sets**. In general, a set is any collection of objects. The objects in a set, called elements or members of the set, can be anything from numbers, to space guns, to IBM cards on which is recorded certain information.

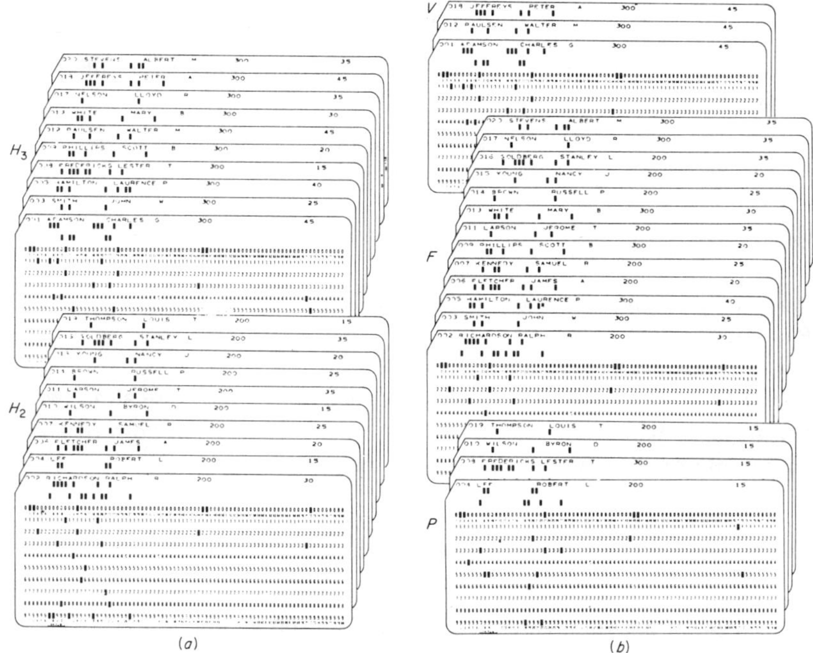

FIGURE 1.8 (a) Cards subdivided into two hourly wage classifications; (b) cards subdivided into three work week classifications

Normally in a particular situation we limit our considerations to some arbitrary set called the **universal** set. In the preceding example, the universal set would be the set of all employees of the company. For this reason, we let *U* represent the main deck of cards of Figure 1.7.

SUBSETS. After the cards have been grouped by the sorter according to the number of hours worked, the universal set *U* has been divided into three distinct sets, depicted in Figure 1.8(b) as subdecks *P, F,* and *V.* Since each of these subdecks is a portion of the main deck or universal set, they are said to be **subsets** of the larger set. The mathematical notation,

$$P \subset U$$

is used to indicate that *P* is a subset of *U*, or that every member of *P* is a member of *U* (and similarly for *F* and *V*).

The subset H_2 can also be indicated by listing the file numbers (or names). If one examines Figure 1.8 and notes those cards with a \$2.00 rate, he can make the following table:

002	006	010	014	016
004	007	011	015	019

This table then is a *representation* of the set H_2 just as is a portion of Figure 1.8. However, to stress the set concept, we write

$$H_2 = \{002,004,006,007,010,011,014,015,016,019\}.$$

The brace ({ . . . }) notation suggests a *boundary* of the set (or subset). In a similar manner,

$$P = \{004,008,010,019\}.$$

Note that the question

Is *P* a subset of H_2? or Is $P \subset H_2$ true?

can be readily answered using the brace notation. Since 008 is in *P* but not in H_2, *P* is not a subset of H_2.

INTERSECTION OF TWO SETS. A second method of relating two sets results in the formation of a new set. The set of elements in *both* H_2 and *P* is called the **intersection** of H_2 and *P* and is indicated by

$$H_2 \cap P \qquad (H_2 \text{ intersect } P).$$

Thus

$$H_2 \cap P = \{004,010,019\}.$$

This is the set of part-time employees earning \$2.00 an hour.

When two sets have no common members, their intersection is the empty set (also called the null set). This is represented by { } or more commonly by ϕ.

Then the fact that H_2 and H_3 share no common members can be indicated by

$$H_2 \cap H_3 = \phi.$$

Examples:

$$\{1,2,3,4\} \cap \{2,4,5\} = \{2,4\}$$
$$\{1,4,7\} \cap \{2,3,5\} = \phi$$
$$\{2,3,4\} \cap \{1,2,3,4,5\} = \{2,3,4\}$$
$$\text{also note } \{2,3,4\} \subset \{1,2,3,4,5\}$$

UNION OF TWO SETS. In Figure 1.8, we might place the deck of cards labeled H_2 on the deck H_3. This combination of sets H_2 and H_3 is called the union of two sets and is indicated by

$$H_2 \cup H_3 \qquad (H_2 \text{ union } H_3).$$

In this case,

$$H_2 \cup H_3 = U,$$

or the union of H_2 and H_3 is the universal set.

For

$$H_2 = \{002,004,006,007,010,011,014,015,016,019\}$$

and

$$P = \{004,008,010,019\},$$

$$H_2 \cup P = \{002,004,006,007,008,010,011,014,015,016,019\}.$$

Note that those members of H_2 also in P (or in $H_2 \cap P$) are included only once in the union. H_2 has 10 members, P has 4 members but $H_2 \cup P$ has 11 (not 14) members. (See problem 15, Exercise 1.3.)

Stated in general terms, the **union** of A and B is the set of elements in A, or in B or in both A and B.

Examples:

$$\{1,2\} \cup \{3,4\} = \{1,2,3,4\}$$
$$\{1,2\} \cup \{2,3\} = \{1,2,3\}$$
$$\{1,2\} \cup \{1,2,3\} = \{1,2,3\}$$

COMPLEMENTS. In finding a union or an intersection, two sets are combined to form a new set. A single set can also lead to a new set by considering those elements which are *not* in a given set but still are within the boundary of the universal set. For example,

the set of employees not in P
is the set of employees who work 20 hours or more.

This set, called the **complement** of P is indicated by

$$P' \text{ (read } P \text{ complement).}$$

In this case

$$P' = F \cup V.$$

Since those employees *not* earning \$2.00 per hour receive \$3.00 per hour (and vice versa)

$$H_2' = H_3 \qquad \text{and} \qquad H_3' = H_2,$$

or

$$H_2 \text{ complement} = H_3 \qquad \text{and} \qquad H_3 \text{ complement} = H_2.$$

Example:

$$\text{If } A = \{1,2,3\}; \qquad B = \{3,4,5\}$$
$$\text{and } U = \text{universal set} = \{1,2,3,4,5\},$$
$$\text{then } A' = \{4,5\} \text{ and } B' = \{1,2\}.$$

FORMAT. The format used in arranging facts can be very significant n how one visualizes a problem. For example, consider

$$\begin{aligned} A &= \{1, \quad 3, \quad 5\}, \\ B &= \quad \{2, 3, 4\}, \\ U &= \{1,2, 3, 4,5\}. \end{aligned}$$

lot only are the complements of A and of B easily determined from this ar- angement but also the members of $A \cap B'$, $A' \cap B$, or $A \cup B$ become apparent. or example, any member of $A \cap B'$ must be in A but *not* in B. Hence,

$$A \cap B' = \{1,5\}.$$

ɿ a similar manner,

$$A' \cap B = \{2,4\}$$

nd

$$A \cup B' = \{1,3,5\}.$$

ɩlso, since

$$A \cap B = \{3\}$$
$$(A \cap B)' = \{1,2,4,5\},$$

ɩnd since

$$A \cup B = \{1,2,3,4,5\} = U$$
$$(A \cup B)' = \phi.$$

EXERCISE 1.3

1. Using the brace notation and Figure 1.8, indicate the file numbers of the following sets.

   (a) H_2 (b) H_3 (c) P (d) F (e) V

2. List by file number, the members of each set.

   (a) $H_2 \cap F$ (c) $H_2 \cup P$ (e) $P \cup F$ (g) $H_2 \cap V$
   (b) $H_3 \cap V$ (d) $H_3 \cup F$ (f) $F \cup V$ (h) $H_3 \cap P$

3. Indicate the complement of each set using the letters P, F and V.

   (a) F (b) V (c) $P \cup F$ (d) $P \cup V$ (e) $F \cup V$

4. True or false?

   (a) $(H_2 \cap F) \subset F$ (c) $(H_2 \cup F) \subset F$ (e) $P \cap F = \phi$
   (b) $(H_2 \cap F) \subset H_2$ (d) $F \subset (H_2 \cup F)$ (f) $H_2 \cap V = \phi$

5. Indicate the set representation (using letters) for the group of employees:
   (a) earning \$2.00 hourly and working 20-40 hours per week.
   (b) who are part time and earn \$3.00 hourly.
   (c) who are not part time and earn \$2.00 hourly.
   (d) working 40 hours or less and earning \$2.00 hourly.
   (e) who are not part time and do not work over 40 hours.

6. Each of the following sets will be the empty set ϕ or the universal set U. Indicate which is correct.

   (a) $H_2 \cap H_3$ (d) $F' \cap F$ (g) $V \cap V'$
   (b) $H_2 \cup H_3$ (e) $F' \cup F$ (h) $V' \cup V$
   (c) $P \cap V$ (f) $H_2' \cap H_3'$ (i) ϕ'
   (j) U'

7. If $A = \{1,2,\ \ 4\}$
   $B = \ \ \{2,3,\ \ 5\}$
   $U = \{1,2,3,4,5\}$

   List the members of the following sets.

   (a) $A \cap B$ (c) A' (e) $A' \cap B'$ (g) $(A \cap B)'$
   (b) $A \cup B$ (d) B' (f) $A' \cup B'$ (h) $(A \cup B)'$

8. List the members of all sets indicated in problem 7, if

$$A = \ \ \{2,\ \ 4\}$$
$$B = \{1,\ \ 3\}$$
$$U = \{1,2,3,4,5\}$$

9. List the members of all sets indicated in problem 7, if

$$A = \{a,b,\ \ d\}$$
$$B = \ \ \{b,c,\ \ \ \ f\}$$
$$U = \{a,b,c,d,e,f\}$$

In problems 10-12, let

$U =$ set of employees of a company,
$A =$ set of men who are employees,
$B =$ set of women who are employees,
$C =$ set of employees who work in the data processing department,
$D =$ set of employees who are under 21,
$E =$ set of employees who are over 60.

10. A single letter or ϕ can represent each of the following sets.
    (a) $A \cup B$ (b) $A \cap B$ (c) A' (d) B' (e) $D \cap E$

11. Describe (verbally) each of the following sets:
    (a) $A \cap D$ (c) $B \cap C$ (e) $(D \cup E)'$ (g) $(C \cap D)'$
    (b) $B \cap E$ (d) $A' \cap D$ (f) $C \cap D$ (h) $A \cap D'$

12. Find a set representation for the set of:
    (a) male employees who work in data processing.
    (b) employees in data processing who are 21 or over.
    (c) women employees who are 60 or under.

In problems 13 and 14

$A - B =$ set of members of A which are *not* in B.

Example:

$A = \{1,2,3\}$
$B = \{3,4\}$

then $A - B = \{1,2\}$

13. If $A = \{2, 4, 6\}$ $C = \{1,2 \quad 6\}$
    $B = \{1, \quad 4,5\}$ $U = \{1,2,3,4,5,6\}$

    List the members of each set.

    (a) $A - B$ (c) $B - C$ (e) $C - A$ (g) $U - A$ (i) $U - B$
    (b) $A - C$ (d) $B - A$ (f) $C - B$ (h) A' (j) $C - U$

14. If $A = \{1,3,4\}$, $B = \{3,4\}$, $C = \{1,4\}$, and $U = \{1,2,3,4\}$, list the members of the sets in (a)-(h).

    (a) $A - B$ (c) $B - C$ (e) A' (g) $U - B$
    (b) $A - C$ (d) $B - A$ (f) $U - A$ (h) B'

    True or false?

    (i) $B \subset A$ (k) $A - B = B - A$ (m) $B' = U - B$
    (j) $C \subset A$ (l) $A' = U - A$ (n) $B \subset C$

15. (a) If A has 7 members, B has 6 members, and $A \cap B$ has 4 members, how many elements are in $A \cup B$?

    Consider: $\{1,2,3,4,5,6,7\}$
    $\{4,5,6,7,8,9\}$

    (b) A—14 members; B—17 members; $A \cap B$—7 members; $A \cup B$ has how many members?
    (c) A—7 members; $A \cap B$—3 members; $A \cup B$—13 members; B has how many members?
    (d) B—9 members; $A \cup B$—27 members; $A \cap B$—0 members; A has how many members?
    (e) A—s members; B—t members; $A \cap B$—u members; then does $A \cup B$ have $(s + t - u)$ members?

16. Let X represent a set whose members are not listed. If we know that $A = \{1,2,3\}$, $U = \{1,2,3,4\}$, $A \cap X = \{2,3\}$, and $A \cup X = U$; list the members of X.

17. True or false (the members of X are unknown)?

    (a) If $A = \{1,2\}$ and $A \cap X = \{2\}$, then 2 is a member of X.
    (b) $A \cap B$ is always a subset of A.
    (c) $A \cup B$ is a subset of A.
    (d) If $A = \{1,2,3\}$, $A \cup X$ could be $\{2,3,4\}$ for some set X.
    (e) If $A = \{1,2,3\}$ and $A \cup X = \{1,2,3,4\}$, then 4 is a member of set X.
    (f) If $A = \{1,2,3\}$, $A \cup X = \{1,2,3,4\}$ then 4 is in $A \cap X$.
    (g) If $A = \{1,2,3\}$, $A \cup X = \{1,2,3,4\}$ then 2 will always be in $A \cap X$.
    (h) If A has 4 members, then $A \cap X$ can have from 0 to 4 members?
    (i) If $A \cap X$ has 4 members and X has 6 members, then A must have 2 members.

18. If $A = \{1,2,\ \ 4\}$
$B = \qquad \{3,4,5\}$
$U = \{1,2,3,4,5\}$

List those members of:

(a) A which are not in B. (d) U which are not in B.
(b) B which are not in A. (e) U which are not in $A \cup B$.
(c) U which are not in A. (f) U which are not in $(A \cap B)$.
(g) Find set representations (such as $A \cap B'$) for a through f.

1.4 VENN DIAGRAMS

INTERSECTION AND UNION. As an example of a typical data processing type of problem, consider the following.

EXAMPLE 1.5

The director of personnel of a company with 5000 employees wishes to notify all employees whose last names begin with A through H and are covered under a new group insurance that they must have a physical examination by April 1.

Since this involves repeated testing of a particular set of conditions, it is a task which is appropriate for programming on a computer. In effect, the computer would be programmed to process a universal set of all employees and select out the required subset. However, the programmer (who writes a sequence of instructions which can be executed by the computer) must first visualize how to fit the parts together. As a possible aid, he might use a pictorial representation called a **Venn diagram**.

In Figure 1.9(a) let the interior of the rectangle represent the universal set of all employees. One could imagine 5000 dots inside the rectangle, each dot corresponding to an employee. However, the dots are specially arranged so that each dot inside the closed curve in Figure 1.9(b) corresponds to an employee whose last name begins with A through H (we label this set H). It follows that the shaded portion of Figure 1.9(c) is H' (H complement).

In Figure 1.10(a), the interior of the second closed figure represents the set of employees (I) covered under the new group insurance. The "overlap" of these two sets is shaded in Figure 1.10(b) and can be represented by $H \cap I$.

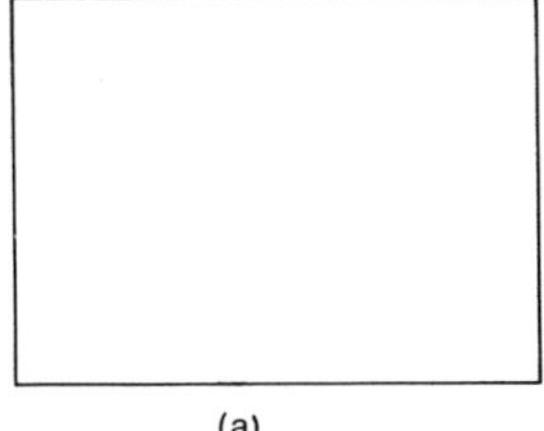

(a)

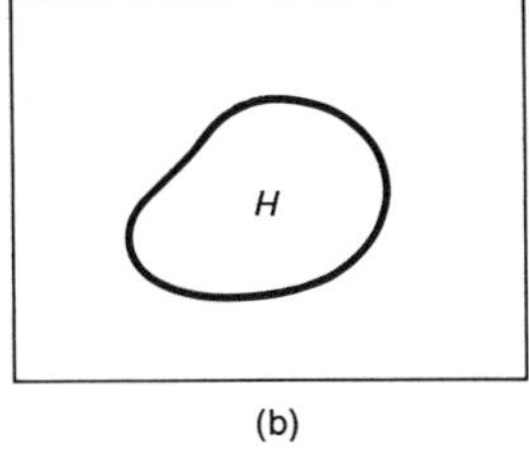

(b)

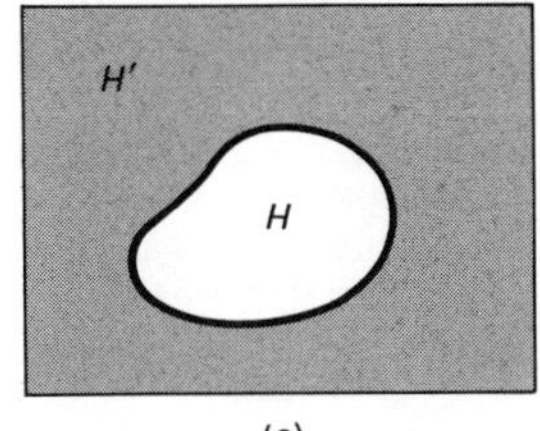

(c)

FIGURE 1.9 (a) Universal set; (b) H'

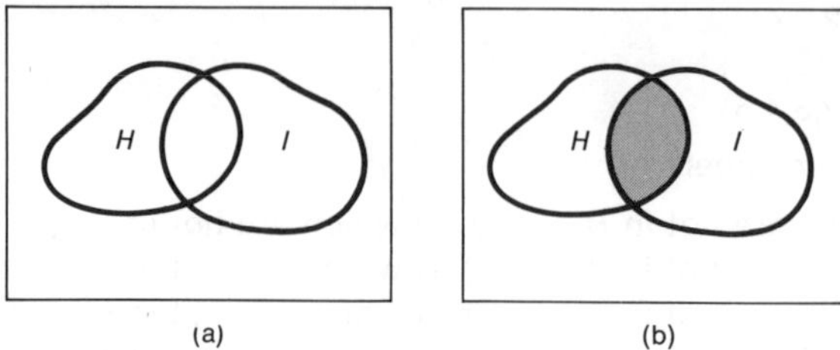

FIGURE 1.10 (a) Sets H and I; (b) $H \cap I$

This is the set of employees who must take the physical examination. The union of two sets can also be indicated in a Venn diagram. In Figure 1.11, the shaded portion represents those employees in H or in I or in both.

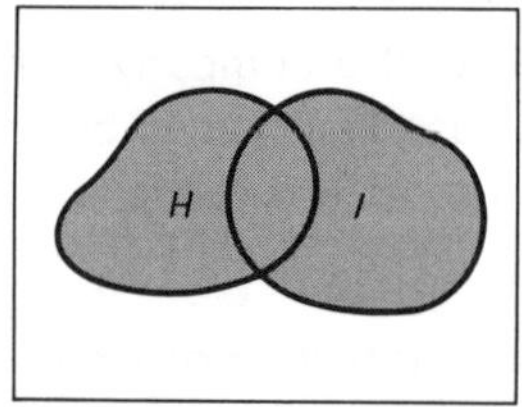

FIGURE 1.11 $A \cup B$

As a second illustration of the use of Venn diagrams, consider the sets

$$\begin{aligned} A &= \{1, 3, 5, 7\}, \\ B &= \{2, 4, 6\}, \\ C &= \{1,2, 7,8,9\}, \\ \text{Universal set} &= \{1,2,3,4,5,6,7,8,9,10\} \end{aligned}$$

Before a Venn diagram can be drawn accurately, one must note that A and B share no common elements or

$$A \cap B = \phi \qquad (A \text{ intersect } B \text{ is empty})$$

also

$$A \cap C \quad \text{and} \quad B \cap C \quad \text{are not empty.}$$

Then we begin as in Figure 1.12(a). Since $A \cap C = \{1,7\}$ and $B \cap C = \{2\}$, we have the Venn diagram in Figure 1.12(b). The remainder of the universal set can be inserted as shown in Figure 1.12(c). We see that the final Venn diagram

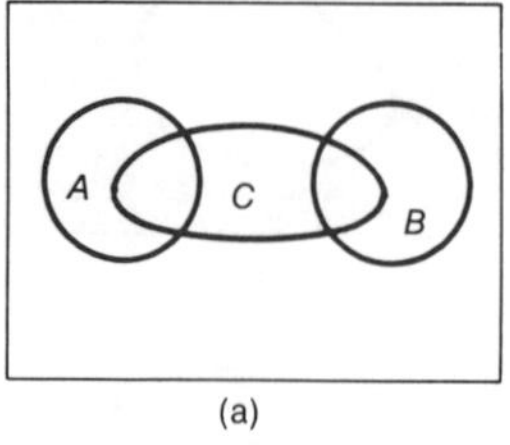

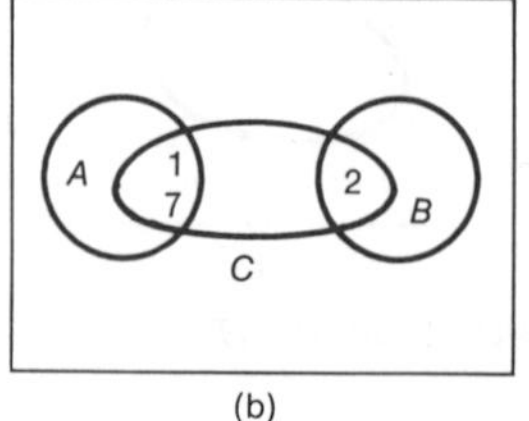

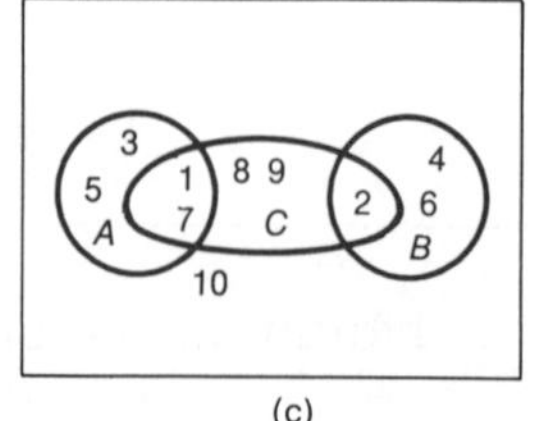

FIGURE 1.12

offers a clear picture of various set relationships. The reader should use Figure 1.12(c) to verify each of the following.

$A \cap C$ is a subset of C or $(A \cap C) \subset C$
also $(B \cap C) \subset C$

$A \cup C = \{3,5,1,7,8,9,2\}$ or $\{1,2,3,5,7,8,9\}$
$A' = \{2,4,6,8,9,10\}$
B is a subset of A' or $B \subset A'$
since $\{2,4,6\} \subset \{2,4,6,8,9,10\}$
$[(A \cup B) \cup C]' = \{10\}$

SET LANGUAGE. A major difficulty for most students in basic algebra is the translation of a problem from verbal to equation form; this basic notion was stressed earlier in this chapter and will receive continued emphasis. On one hand, persistent effort can often improve one's ability for problem solving, but the effective use of mathematics, as well as the computer, requires that we understand the structure of the problem which is to be solved. As we learned in the preceding section, the Venn diagram provides a visualization which clearly shows set relationships (or the structure of the problem). Now let us compare a verbal description, a set representation, and the corresponding Venn diagram (Figure 1.13). As we shall see, the verbal form is closely related to the other two forms and hence something is missing if a student feels that he understands one form but not another.

COMBINING THREE SETS. Finding a union (or intersection) is a **binary** operation in that **two** sets are combined to form a single set. When three sets are involved, parentheses are used (as in basic algebra) to indicate which union or intersection is to be found first. Thus

$$(A \cup B) \cap C$$

indicates that the union is determined first and then the intersection.

EXAMPLE 1.6

$$A = \{1,2\}$$
$$B = \{1, \quad 3\}$$
$$C = \quad \{2,3,4\}$$
$$U = \{1,2,3,4\}$$

$$A \cup B = \{1,2,3\}$$
$$(A \cup B) \cap C = \{1,2,3\} \cap \{2,3,4\} = \{2,3\}$$
$$A' \cap B = \{3\}$$
$$(A' \cap B) \cup C = \{3\} \cup \{2,3,4\} = \{2,3,4\}$$

EXAMPLE 1.7
Find a set representation for each shaded area in Figure 1.14.

The vertically shaded area is in A, not in B, and not in C. Thus its representation is $A \cap B' \cap C'$.

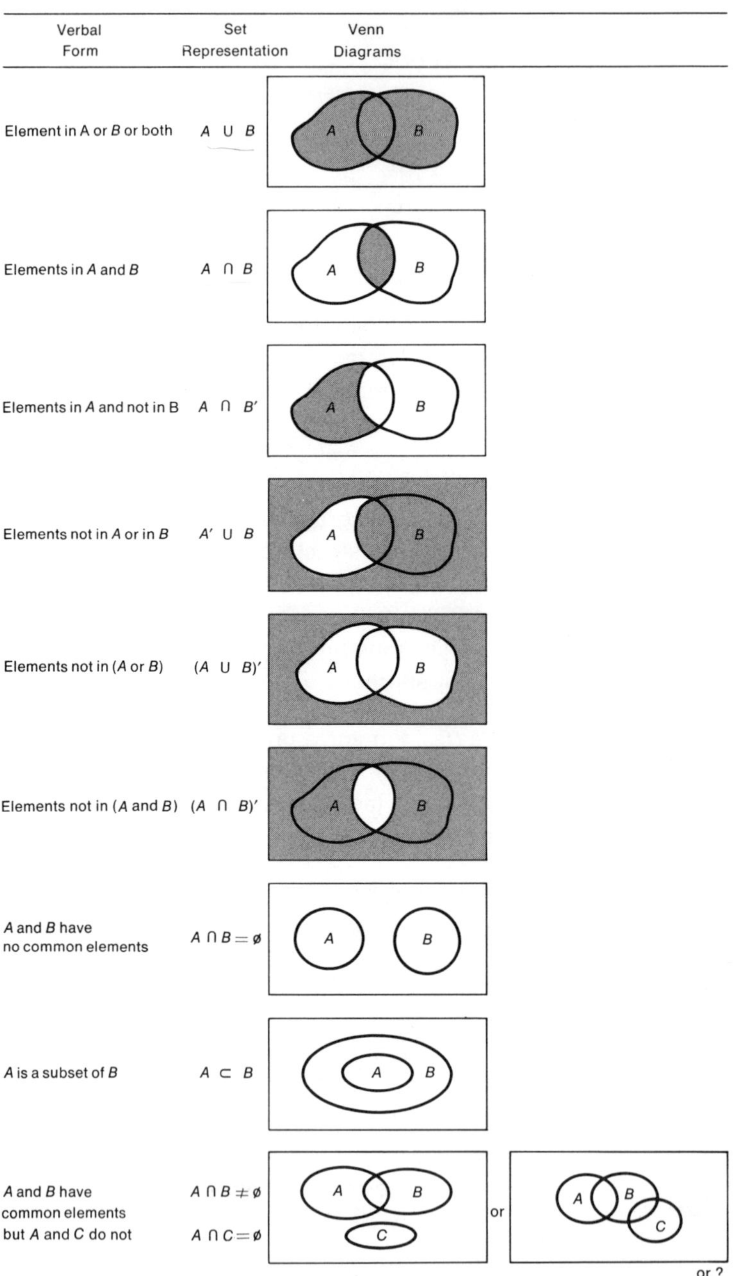
Verbal Form
Set Representation
Venn Diagrams
Element in A or B or both
A ∪ B
A
B
Elements in A and B
A ∩ B
A
B
Elements in A and not in B
A ∩ B′
A
B
Elements not in A or in B
A′ ∪ B
A
B
Elements not in (A or B)
(A ∪ B)′
A
B
Elements not in (A and B)
(A ∩ B)′
A
B
A and B have no common elements
A ∩ B = ∅
A
B
A is a subset of B
A ⊂ B
A
B
A and B have common elements but A and C do not
A ∩ B ≠ ∅
A ∩ C = ∅
A
B
C
or
A
B
C
or ?

FIGURE 1.13

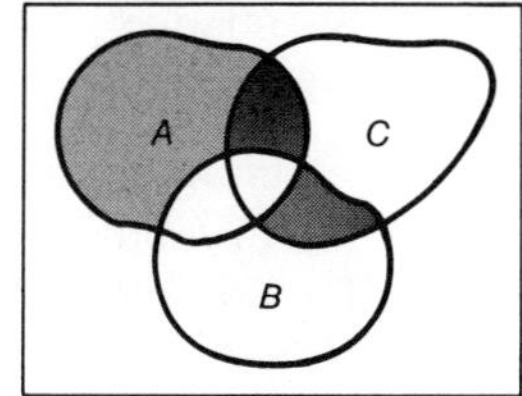

FIGURE 1.14

The horizontally shaded area consists of 2 parts:

The top part is represented by $A \cap B' \cap C$.
The bottom part is represented by $A' \cap B \cap C$.
Both parts could be represented by $(A \cap B' \cap C) \cup (A' \cap B \cap C)$.

EXERCISE 1.4

1. Reproduce Figure 1.15 and shade the following sets. Use a different diagram for each set.

   (a) $A' \cap B$ (c) $(A \cup B) \cap B'$ (e) $(A \cup B) \cap A$
   (b) $A \cup B'$ (d) $(A \cup B) \cup A$ (f) $(A \cap B) \cap A$

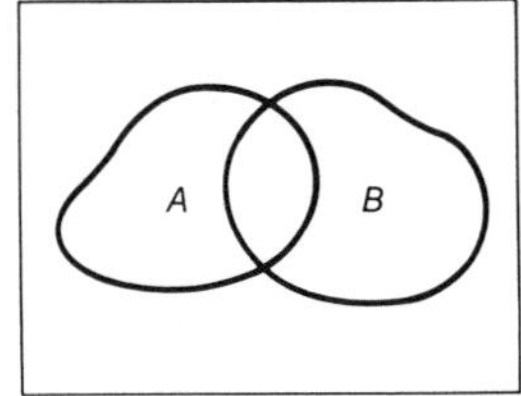

FIGURE 1.15

2. Using six Venn diagrams where $A \cap B = \phi$, shade each of the sets listed in problem 1.
3. In Figure 1.16 find a set representation for each of the nonoverlapping numbered areas.

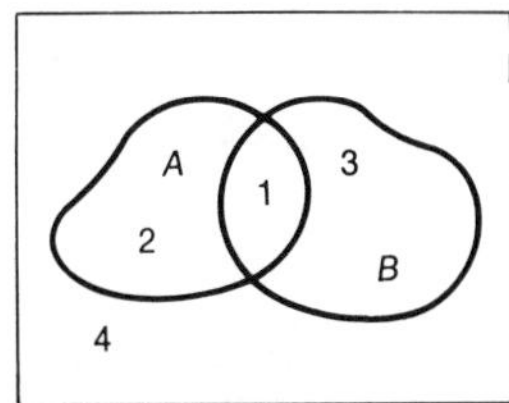

FIGURE 1.16

4. Find a set representation for the shaded area in Figure 1.17.
5. Find a set representation for the elements:

(a) in A or not in B.
(b) not in A and in B.
(c) not in A and not in B.
(d) in A, not in B, but in C.
(e) in A but not in (B or C).
(f) not in B but in (A and C).

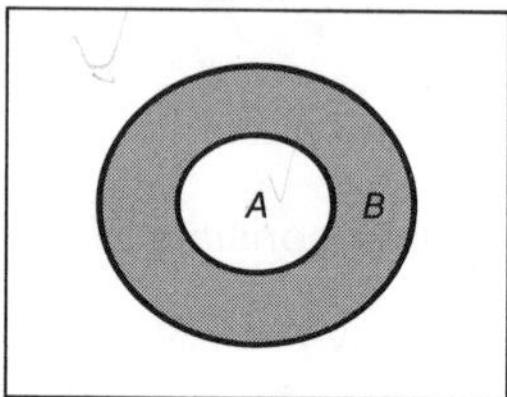

FIGURE 1.17

6. Find a set representation for the shaded area in Figure 1.18.
7. Find a set representation for those elements in A but not in B, or in B but not in A. (Use Figure 1.18.)

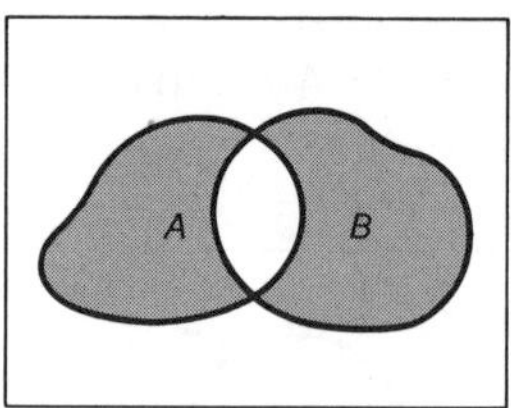

FIGURE 1.18

8. If $A = \{1,2,\ \ 4\}$
$B = \{2,3\}$
$C = \{3,4\}$
$U = \{1,2,3,4\}$

list the members of each set:

(a) $A' \cap B$
(b) $A' \cup C$
(c) $(A \cap B') \cap C$
(d) $A' \cap (B \cap C)$
(e) $(A \cap B) \cap C'$
(f) $A \cap (B \cup C)$
(g) $A' \cup (B \cap C)$
(h) $(A \cap B') \cup C$
(i) $(A' \cup B') \cap C$

9. Let M = set of male employees,
T = set of employees who are 30 years of age or under,
F = set of employees who are 40 or over,
A = set of employees earning \$125 per week or less,
B = set of employees earning \$200 per week or more.

Find a set representation for each set.

(a) All male employees who are 40 or over.
(b) All female employees under 30 who earn \$125 per week or less.
(c) All male employees between 30 and 40 years of age.

(d) All female employees earning between \$125 and \$200 per week.
(e) All female employees, 40 or over, who earn \$200 per week or more.
(f) All male employees between 30 and 40 earning between \$125 and \$200 per week.

In Figure 1.19, the numbers refer to the eight nonoverlapping areas. For example, the area numbered 3 can be represented by $(B \cap C) \cap A'$ since it is part of B, part of C, but outside of A.

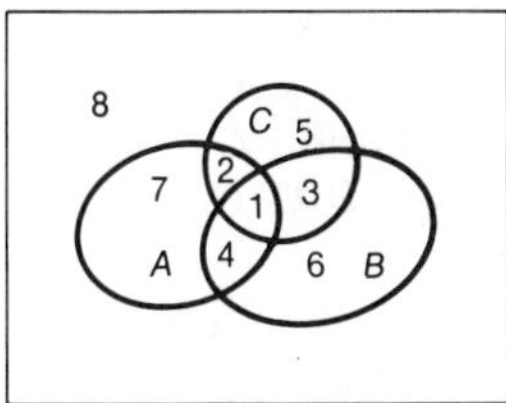

FIGURE 1.19

10. Indicate the set representation for each numbered area in Figure 1.19.

11. In Figure 1.19 find a set representation for the elements in each of the following areas:

(a) 2 or 3 (c) 2 or 5 or 7 (e) (1 or 2) or 4
(b) 3 or 6 (d) (1 or 3) or (4 or 6) (f) 5 or 7

12. Given the sets $A = \{1,2,3,5\}$, $B = \{3,4,5,7\}$, $C = \{2,3,6,7,8\}$, $U = \{1,2,3,4,5,6,7,8,9\}$, draw a Venn diagram relating all three sets. Then indicate whether each of the following is true or false.

(a) $\{3,7\}$ is a subset of B. (d) $\{1\}$ is a subset of $(B \cup C)'$.
(b) $\{5,7\}$ is a subset of B. (e) $(A \cup B) \cap (A \cup B') = A$.
(c) $\{5\}$ is a subset of $B \cap C$. (f) $(A \cap B) \cup (A \cap B') = A$.

13. Draw a Venn diagram which satisfies each set of conditions.

(a) $A \cap B = \phi$, $A \cap C \neq \phi$, $B \cap C = \phi$, no set is a subset of another.
(b) $A \subset B$, $A \cap C = \phi$, $B \cap C \neq \phi$.
(c) $B \subset C$, $A \subset C'$.
(d) $A \cap B = \phi$, $A \subset C$, $B \subset C$.
(e) $(A \cup B) \subset C$.
(f) $(A \cup B) \subset C'$.

14. Sketch two identical Venn diagrams where $(A \cap B) \cap C \neq \phi$ and shade the sets in (a) and (b) below:

(a) $A \cap (B \cup C)$.
(b) $(A \cap B) \cup (A \cap C)$.
(c) Are the shaded areas in (a) and (b) the same?
(d) Is this similar to the commutative, associative, or distributive property of ordinary mathematics?

15. The sequence in which unions, intersections, and complements are formed is, of course, very significant. This sequence for $(A' \cup B) \cap C$

could be indicated as *C-U-I* (for *complement-union-intersection*) while for $(A \cap B)' \cup C$ it would be *I-C-U*.

Indicate the correct sequence for each of the following:

(a) $(B \cap C) \cup A'$
(b) $(A \cup C)' \cap B'$
(c) $(A \cup B) \cap (B \cup C)$
(d) $(A' \cup B)'$
(e) $(A' \cap B) \cup (A \cap B')$
(f) $[(A' \cup B)' \cap C]'$

16. State whether each of the following is true or false.

(a) If x is in $(A \cap B)$, then x is in A.
(b) If x is in $(A \cup B)$, then x is in B.
(c) If $A \subset B$ and x is in A, then x is in B.
(d) If $A \subset B$ and x is in B', then x is in A'.
(e) If $A \cap B \neq \phi$ and $B \subset C$, then $A \cap C \neq \phi$.
(f) If $A \subset B$ and $A \subset C$, then $B \subset C$.
(g) If $A \subset B$, then $B' \subset A'$.
(h) If $A \cap B = \phi$ and x is in A, then x is in B'.
(i) If $A \cap B = \phi$ and $C \subset A$, then $B \cap C = \phi$.
(j) The complement of a union equals the union of the complements.
(k) The complement of an intersection equals the union of the complements.

1.5 SETS AND FLOWCHARTS

We have seen in preceding sections how Venn diagrams provide a convenient visualization of set relationships; now let us consider how flowcharts go beyond Venn diagrams and show the sequence in which the components of a problem fit together. As an illustration, consider the following situation.

EXAMPLE 1.8

Any student who has completed 40 units or more and has at least a 3.0 gradepoint average can apply for a scholarship.

A computer can be programmed to furnish a list of eligible students if it has access to the required information. In effect, it will be finding the intersection of two sets. Let

C = set of students who have completed 40 or more units,
G = set of students who have at least a 3.0 grade point average (GPA).
U = set of all students.

Then $C \cap G$ is the set of eligible students. The computer will be programmed to determine which students are in set C and also in set G. The flowchart symbolism of Figure 1.20(a) illustrates the initial phase of checking the number of units completed. Now it is important to recall that if a student is not in C then he cannot be in $C \cap G$. In this case, the execution of the computer program should continue to the next student, as shown in Figure 1.20(b). However, when a student has completed 40 units or more, a check is made of his grade point average. If his GPA is 3.0 or better then he is eligible; otherwise, execution should continue to the next student as shown in Figure 1.21.

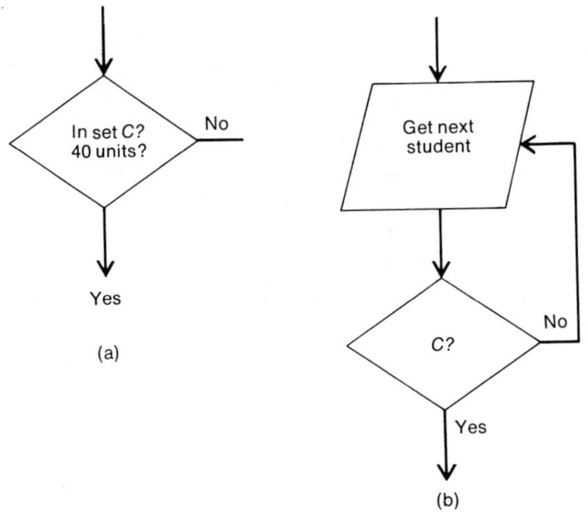

In set C?
40 units?
No
Yes

(a)

Get next student
C?
No
Yes

(b)

FIGURE 1.20

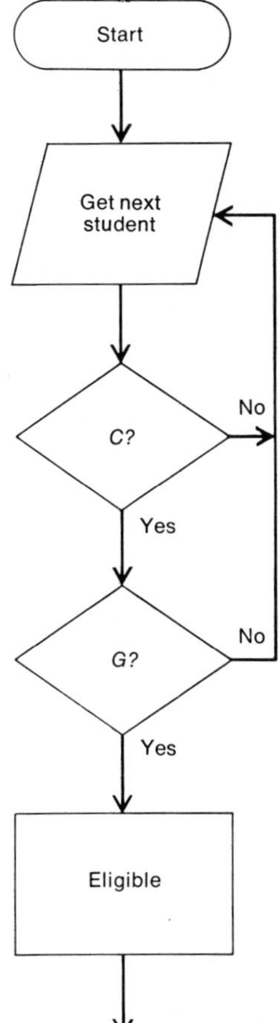

FIGURE 1.21 Flowchart for Example 1.8

EXAMPLE 1.9

Draw a flowchart which illustrates printing a list of employees for all members of the shaded set in Figure 1.22.

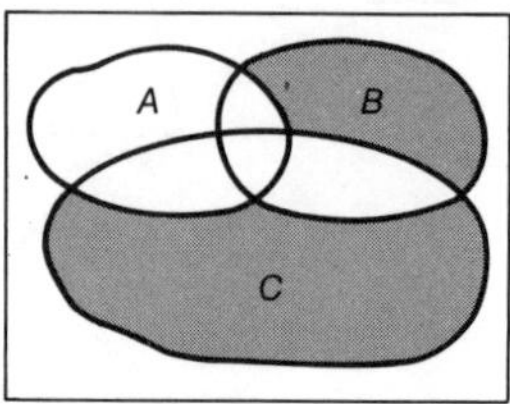

FIGURE 1.22 Example 1.9—Venn diagram

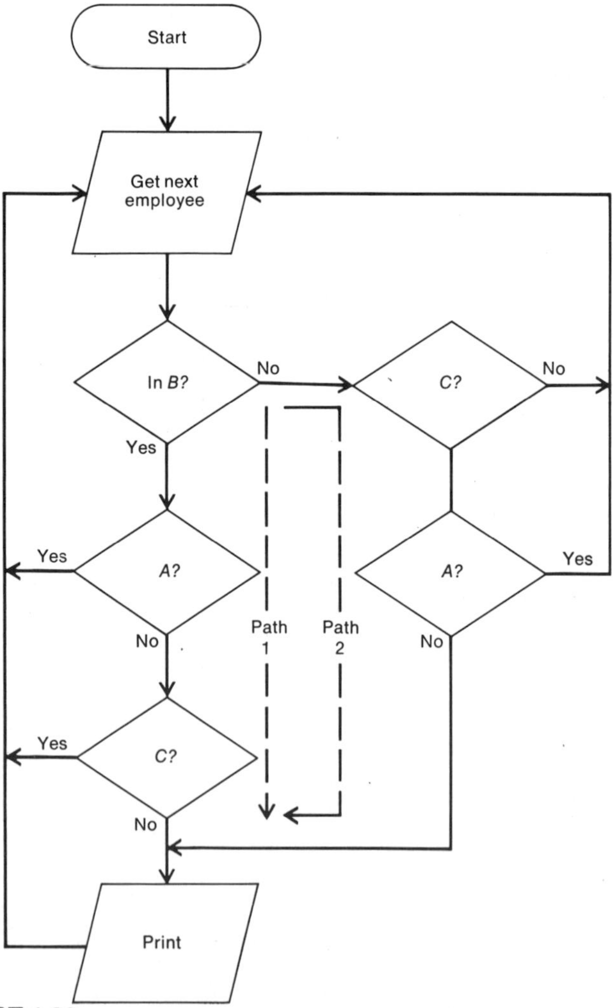

FIGURE 1.23

In examining Figure 1.22 we might note that the shaded area represents the union of two sets. To be in the set illustrated by the upper portion of the diagram, an employee must be in B but not in A and not in C. In the flowchart of Figure 1.23, this set corresponds to path 1. (Note that if an employee is in A, he cannot be in the shaded area of the Venn diagram.) The set in the lower portion of Figure 1.22 consists of those employees in C but not in A and not in B; this corresponds to "the not in B," "in C" and "not in A" sequence of path 2 in Figure 1.23.

A good programmer, however, will note that this flowchart can be improved. Either from the flowchart or the Venn diagram, we should note that if an employee is not in A, then he will not be on the list (see problem 23). But is there a more systematic method of finding a simpler flowchart? For this type of problem, some general properties of Boolean algebra which can be useful will be discussed in the next section.

EXERCISE 1.5

Draw a flowchart which describes each of the following situations. A list is to be printed containing the names of those employees in the following sets. It may be convenient to draw a Venn diagram and shade the appropriate set.

1. $A \cup B$
2. $A \cup B'$
3. $A' \cup B'$
4. $A' \cap B$
5. $A' \cap B'$
6. $(A \cup B) \cup C$
7. $(A \cup B) \cup C'$
8. $(A \cup B) \cap C$
9. $(A \cap B) \cup C$
10. $(A \cap B) \cap C'$

Draw a flowchart indicating a list is to be printed which includes all members of the shaded set.

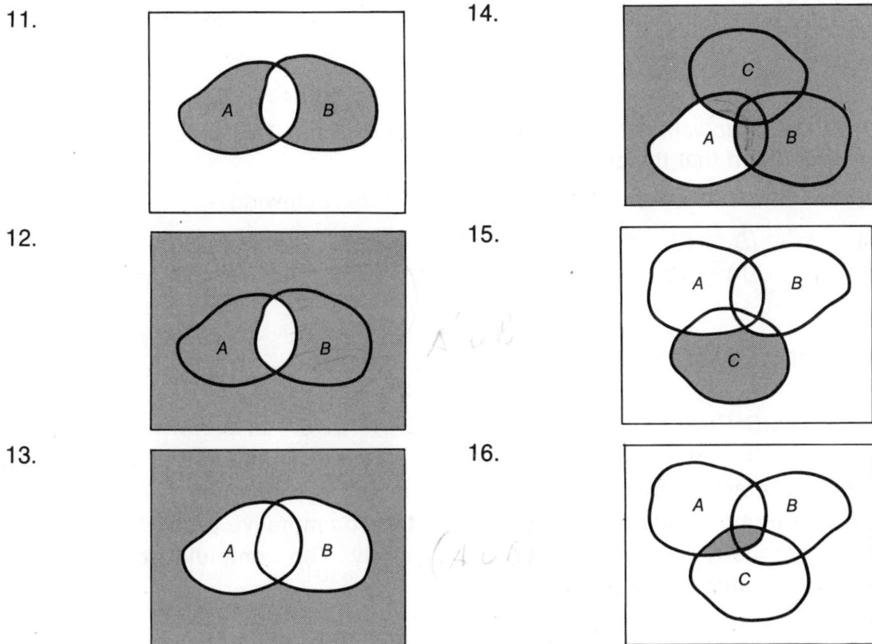

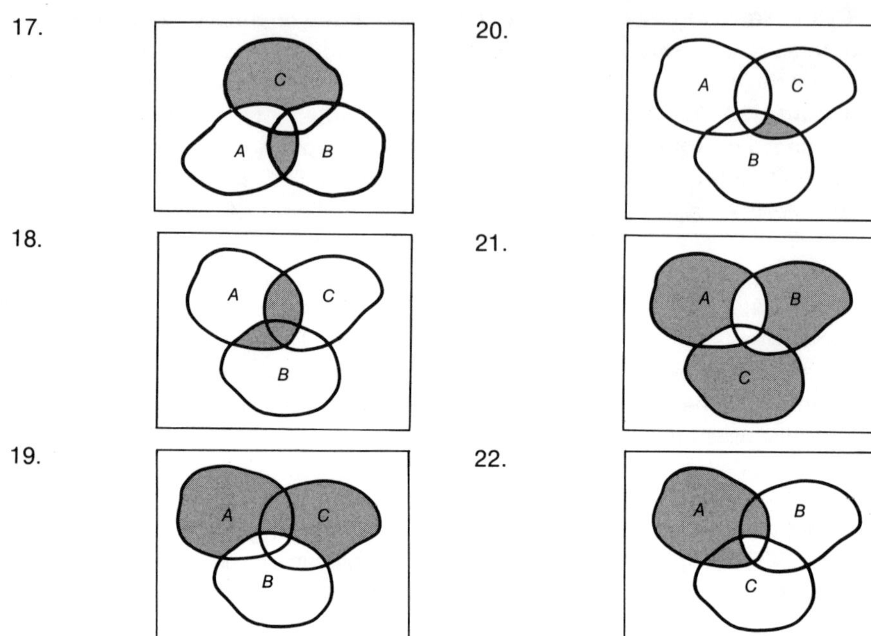

23. In Figure 1.22 the top set is $B \cap (A' \cap C')$ and the bottom set is $C \cap (A' \cap B')$. Their union

$$[B \cap (A' \cap C')] \cup [C \cap (A' \cap B')]$$

corresponds to the flowchart in Figure 1.23. This union can be simplified to

$$A' \cap [(B \cap C') \cup (C \cap B')].$$

Draw a flowchart for this representation. (*Note:* This union imposes the condition that the employee not be in A.)

Draw a flowchart (to print a list) for each of the following:

24. $A' \cup (B \cap C')$
25. $(A \cup B') \cap C'$
26. $(A \cup B) \cap (A \cup C)$
27. $(A' \cap B) \cup (A' \cap C')$
28. $(A \cap B') \cup (A' \cap B')$
29. $(A \cup B) \cap (A \cup C')$

1.6 BOOLEAN PROPERTIES

BASIC CHARACTERISTICS OF OPERATIONS. In performing arithmetic operations on numbers, we seldom give conscious thought to some of the very important properties involved. Three basic properties (which will be covered in more detail in Chapter 2) are the commutative property, the associative property, and the distributive property. The commutative law states that for any numbers a and b

$$a + b = b + a \quad \text{for addition,}$$
$$a \cdot b = b \cdot a \quad \text{for multiplication.}$$

This is obvious for our number system; we easily recognize, for example, that $5+7=7+5$ and $3 \cdot 27 = 27 \cdot 3$.

The associative law indicates that the order in which addition and multiplication operations are performed has no effect on the result; that is, for any numbers *a, b* and *c*

$$(a+b)+c=a+(b+c) \qquad \text{for addition,}$$
$$(a \cdot b) \cdot c = a \cdot (b \cdot c) \qquad \text{for multiplication.}$$

In other words, $5+17+26$ can be treated as $5+17=22$ then $22+26=48$, or else as $17+26=43$ then $5+43=48$ with the same result. (The principle also applies to multiplication.)

Finally, the distributive law (of multiplication over addition) actually represents a basic tool used in algebra (called factoring); for any numbers *a, b,* and *c,* its general form becomes

$$a(b+c)=ab+ac.$$

For example, $13(9+12)=13(21)=273$ and $13 \cdot 9+13 \cdot 12=117+156=273$; that is, the results are the same. Although the commutative and associative properties for real numbers each represent a pair of relationships, no corresponding relationship for the distributive property (that is, a distributive property of addition over multiplication) exists for our number system. However, this second distributive property *does* apply to set relationships as we shall see in the following paragraphs.

COMMUTATIVE PROPERTY. It is obvious that arithmetic operations of our number system conform to the commutative property; now let us consider whether or not the commutative property holds for unions and intersections of sets. In other words, for sets *A* and *B*, are the following equations true?

$$A \cap B = B \cap A \qquad \text{for intersection,}$$
$$A \cup B = B \cup A \qquad \text{for union.}$$

It is quite apparent by referring to the Venn diagrams in Figure 1.24, but, we might ask, how is this significant, especially relative to computers? Suppose a computer is used to determine which employees, covered under pension plan *A*, are over 50. The list will include all members of $A \cap B$ where

$A=$ set of employees covered under pension plan *A*,
$B=$ set of employees over 50.

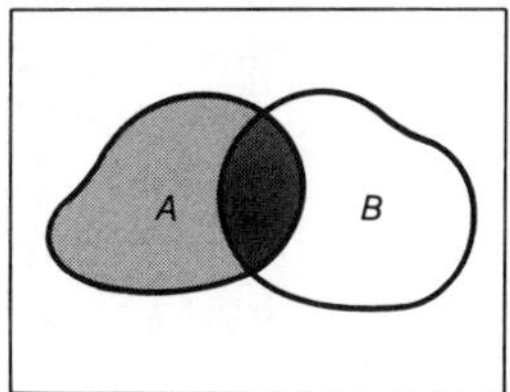

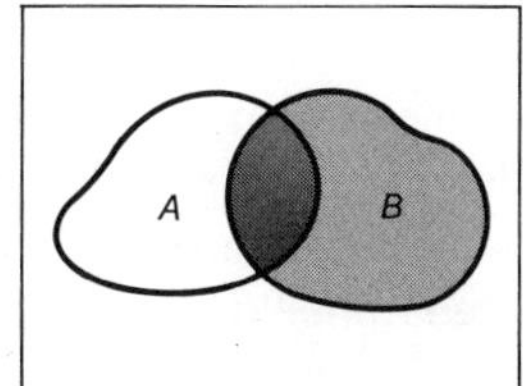

FIGURE 1.24 (a) $A \cap B$; (b) $B \cap A$

The commutative property asserts that the list is the same if $A \cap B$ or $B \cap A$ is used. However, suppose that roughly $\frac{1}{2}$ the employees are covered under pension plan A but only 20 percent are over 50. Which intersection is better? The flowcharts for the two possibilities are shown in Figure 1.25. The question

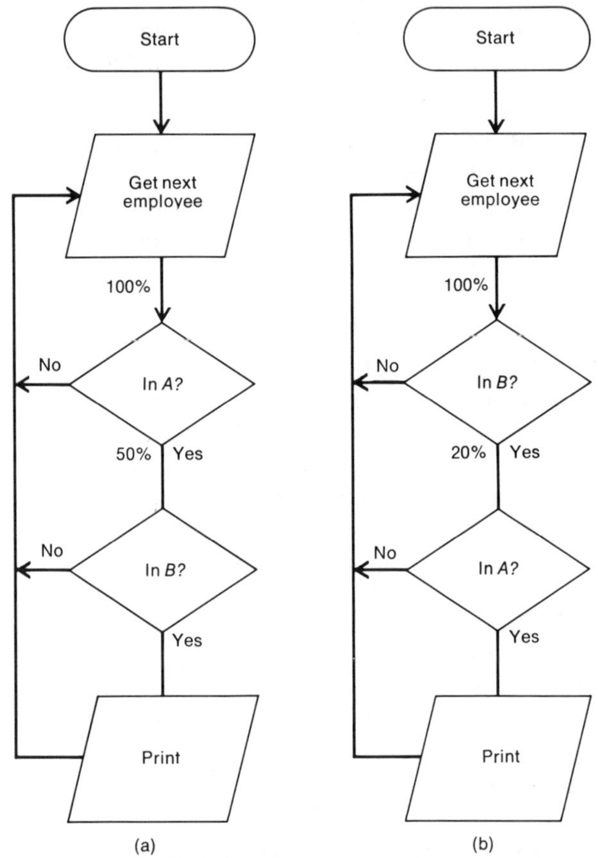

FIGURE 1.25 (a) Flowchart—$A \cap B$; (b) flowchart—$B \cap A$

"Is the employee in set B?" will be false in 8 out of 10 cases. Furthermore, if an employee is not over 50 then he will not be on the list. In Figure 1.25(a) roughly 50 percent would be tested twice (condition A and condition B) while in Figure 1.25(b) only 20 percent would be so tested. This indicates that the flowchart in Figure 1.25(b) describes the more efficient program since fewer tests (less execution time) are required. The reader should also examine Figure 1.26, to determine if it provides a better description of the number of employees checked (see problem 1). The "route" followed (in terms of numbers of employees checked) is through the universal set and then A (or B).

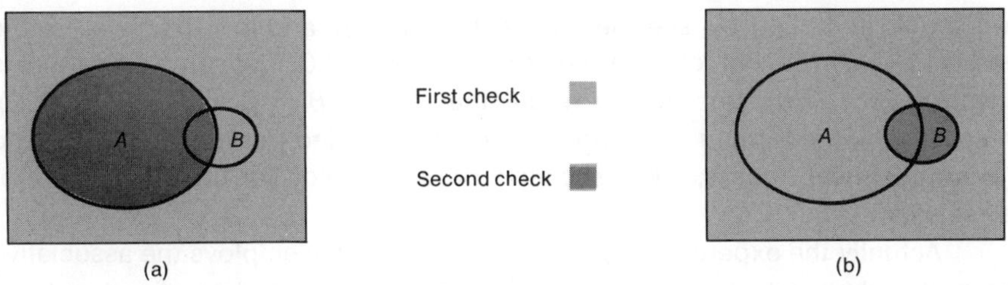

FIGURE 1.26 (a) $(U \cap A)$; (b) $(U \cap B)$

ASSOCIATIVE PROPERTY. In set notation, the associative law for the sets *A, B,* and *C* would appear as

$$(A \cap B) \cap C = A \cap (B \cap C) \qquad \text{for intersection,}$$
$$(A \cup B) \cup C = A \cup (B \cup C) \qquad \text{for union.}$$

Now we ask the question: "Does this property hold for sets?" We readily arrive at the conclusion that these relationships are indeed true if we refer to a step-by-step consideration using Venn diagrams in Figure 1.27. In 1.27(a)

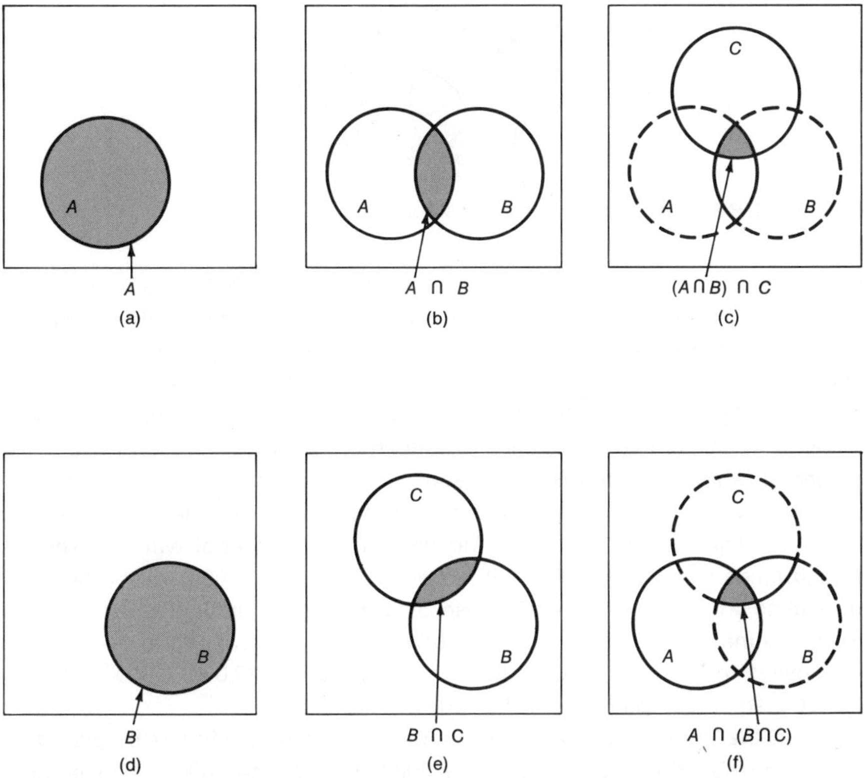

FIGURE 1.27 The associative property

we see A, in 1.27(b) we see the intersection $A \cap B$, and in 1.27(c) we see the result $(A \cap B) \cap C$, which is formed from $A \cap B$ and C. The same procedure is involved in 1.27(d), (e), and (f) where the sets B, $B \cap C$ and $A \cap (B \cap C)$, respectively, are formed. It is apparent that the shaded portions in 1.27(a) and (e) are identical; that is, the associative property holds for the intersection of three sets.

Actually the experienced programmer commonly employs the associative property when selecting data records according to predetermined criteria. For example, assume that we are to examine a list of 20,000 licensed drivers to find those satisfying the following conditions:

$A =$ set of female drivers,
$B =$ set of drivers under 21 years of age,
$C =$ set of drivers with driver training in high school.

Furthermore, we will assume that, having performed statistical studies previously, we have an idea of approximately how many will fall into each group as illustrated in the Venn diagram of Figure 1.28. (Note that numbers given are

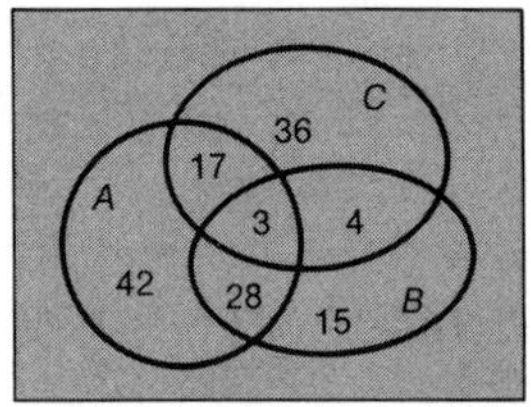

FIGURE 1.28 Categories of drivers (in hundreds)

in hundreds.) Thus we will be searching the list of 20,000 drivers for the approximate 300 who satisfy the required conditions; that is, those that fall in $A \cap B \cap C$. Now we must ask ourselves if the sequence in which the testing is performed makes any difference. Obviously whether we use $(A \cap B) \cap C$ or $A \cap (B \cap C)$ (Figures 1.27(b), (c), (d) and 1.27(e), (f), (g), respectively) is immaterial as far as the final result is concerned. However, efficient use of the computer is an important factor to keep in mind. This is clearly illustrated by the flowchart of Figure 1.29(a) (where the numbers are obtained from the Venn diagram). The test indicated by the first decision symbol will be executed 20,000 times but since 11,000 will be eliminated, only 9000 will be tested to determine if they are in set B. Of these, 5900 will be disqualified and 3100 tested to determine if they are in set C. In other words, determination of the set of employees in $(A \cap B) \cap C$ would require a total of $20{,}000 + 9000 + 3100 = 32{,}100$ tests in compiling the list of 300 drivers.

On the other hand, careful examination of this data indicates that of the 20,000 drivers, 9000 are in A but only 5000 are in B and 6000 in C. Therefore, we should test B first, eliminating 75 percent of the drivers from further testing,

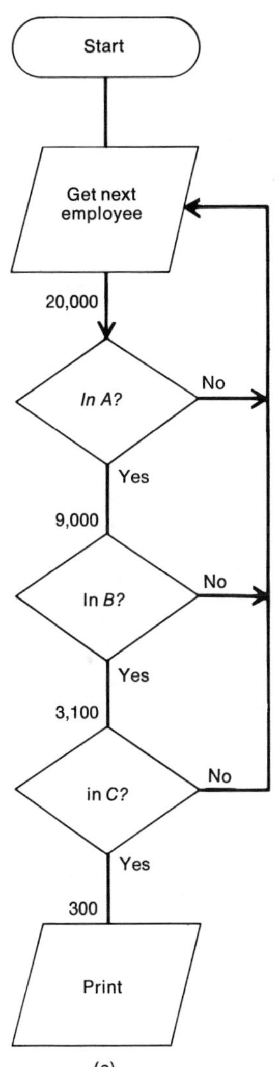

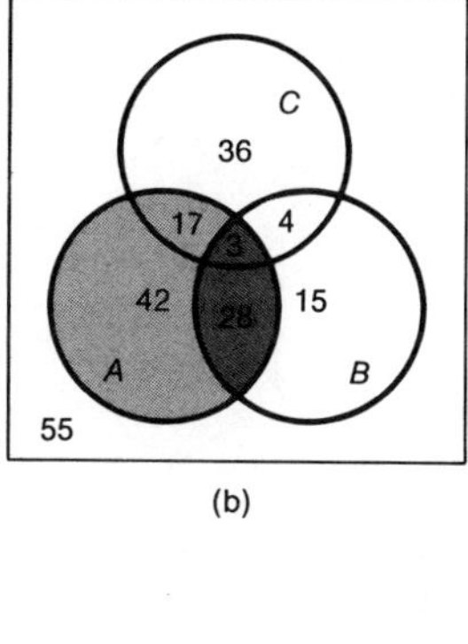

FIGURE 1.29 (a) Flowchart–$(A \cap B) \cap C$; (b) $(A \cap B) \cap C$

then test C and finally A; in set notation, we are interested in $(B \cap C) \cap A$. The flowchart, Venn diagram, and associated numbers are shown in Figure 1.30. We can see that the total number of tests required is $20{,}000 + 5000 + 700 = 25{,}700$, which is approximately 25 percent less than the result in Figure 1.29.

On the one hand, these techniques can be valuable to the programmer in preparing a more efficient program. On the other hand, the reader must recognize that data sets are often such that they do not necessarily yield a "most efficient" approach or else insufficient knowledge of the data is available to make such estimates regarding relative quantities.

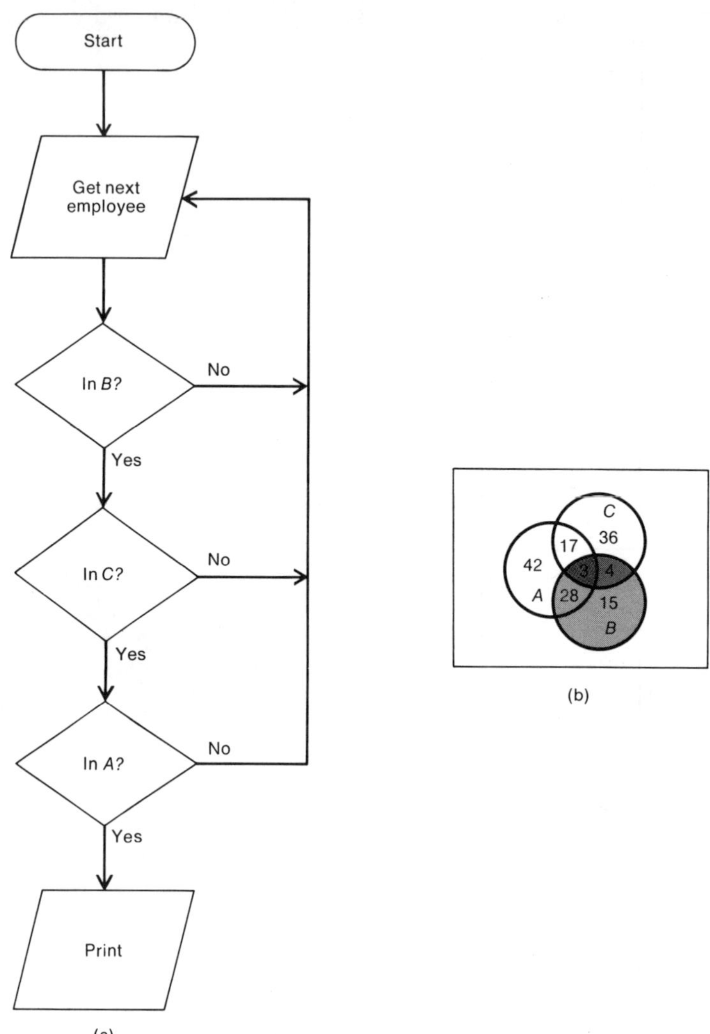

FIGURE 1.30 (a) Flowchart—$(B \cap C) \cap A$; (b) $(B \cap C) \cap A$

DISTRIBUTIVE PROPERTY. In an earlier paragraph the distributive property of multiplication over addition was illustrated and it was emphasized that a corresponding distributive property of addition over multiplication is nonexistent for our number system. However, set operations allow greater flexibility. The two possibilities are

$$A \cap (B \cup C) = (A \cap B) \cup (A \cap C),$$
$$A \cup (B \cap C) = (A \cup B) \cap (A \cup C).$$

The fact that *both* are true indicates a departure from ordinary algebra. Sets with the operations, union and intersection, provide an example of what is called **Boolean algebra**. (See Chapters 4 and 14.) As indicated below, the "rules" in Boolean algebra differ from those in arithmetic or basic algebra.

The two distributive properties can be more easily remembered if one thinks of A and $\cap$ (or A and $\cup$) as being "fastened together." Note that $A \cap$ occurs three times in the first equation. The same is true for $A \cup$ in the second equation. But this does not indicate why they are true. To illustrate the left side of

$$A \cap (B \cup C) = (A \cap B) \cup (A \cap C),$$

the union of B and C is shaded in Figure 1.31(a). That portion of $B \cup C$ which is also part of A is crosshatched in 1.31(b). The right side, $(A \cap B) \cup (A \cap C)$,

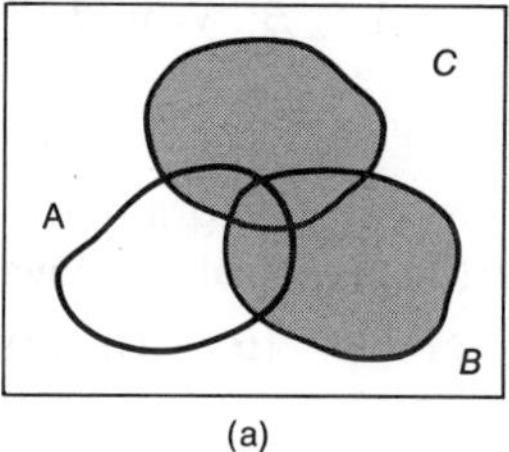

(a)

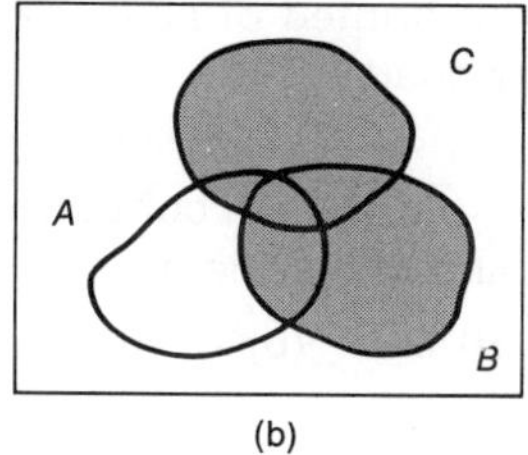

(b)

FIGURE 1.31 (a) $B \cup C$; (b) $A \cap (B \cup C)$

involves two intersections and then the union of these intersections which is illustrated in Figure 1.32. It is apparent that the final result in 1.32(c) is identical to that of 1.31(b).

Justification of the second distributive property is left as an exercise for the reader (problem 6, Exercise 1.6).

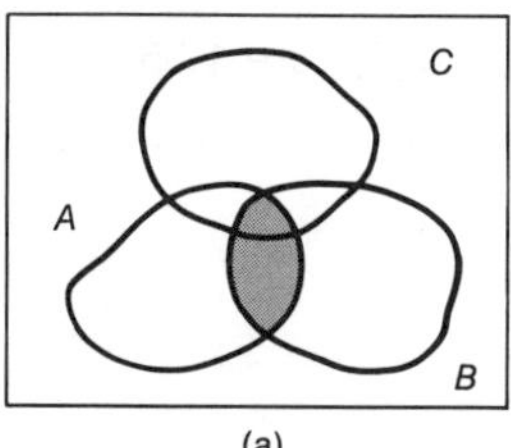

(a)

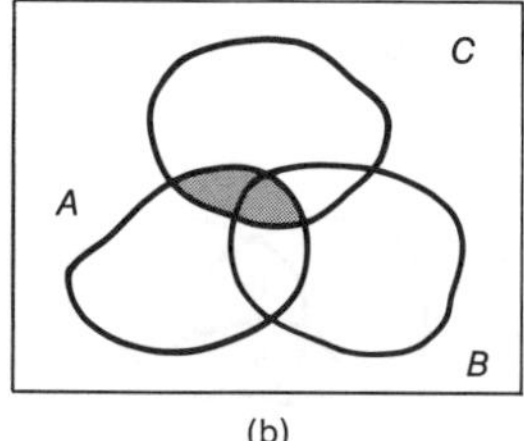

(b)

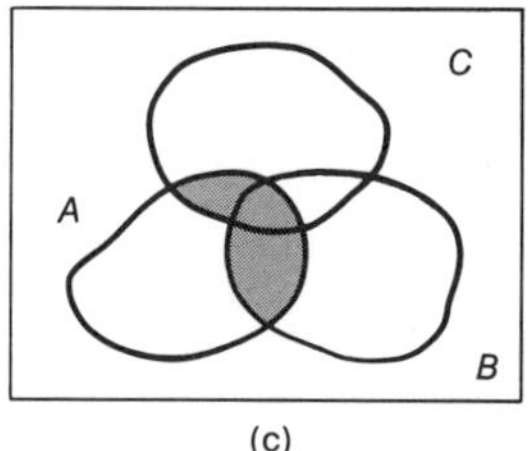

(c)

FIGURE 1.32 (a) $A \cap B$; (b) $A \cap C$; (c) $(A \cap B) \cup (A \cap C)$

EXERCISE 1.6

1. In Figure 1.33, A contains 74 members, B, 97 members, $A \cap B$, 31 members and $A' \cap B'$, 100 members.
   (a) Should the flowchart in Figure 1.25(a) or 1.25(b) be used?
   (b) How many checks would be made in each case?
   (c) How many members are in $A' \cap B$? In $A \cap B'$?

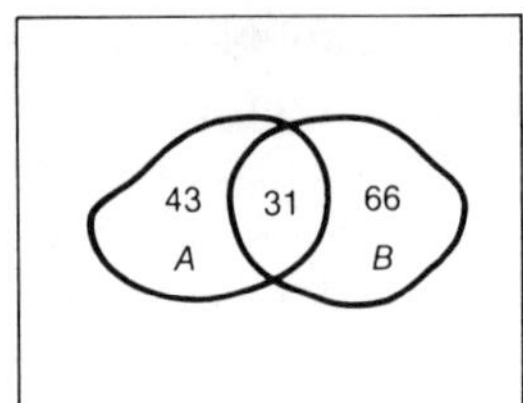

FIGURE 1.33

2. In problem 1, suppose A contains 123 members, B, 97 members, $A \cap B$, 43 members, and $A' \cap B'$, 200 members. Answer parts (a), (b), and (c) in problem 1.
3. (a) Using the information from problem 1, draw a flowchart indicating a list is to be printed of all employees in $A \cap B'$. Would $B' \cap A$ lead to a better flowchart?
   (b) Repeat part (a) of this problem for $A' \cap B$ (or $B \cap A'$).
4. Using the information contained in the Venn diagram in Figure 1.34, determine (and draw) the best flowchart for a list of employees in each set:
   (a) $(A \cap B) \cap C$ (b) $(A \cup B) \cup C$ (c) $A' \cap (B \cup C)$ (d) $(A \cup C) \cap B'$

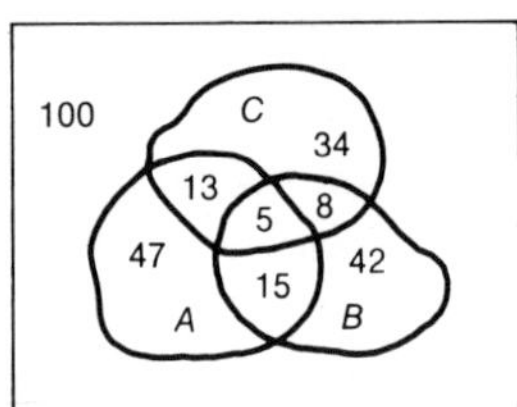

FIGURE 1.34

5. Repeat problem 4 using the information contained in Figure 1.35.

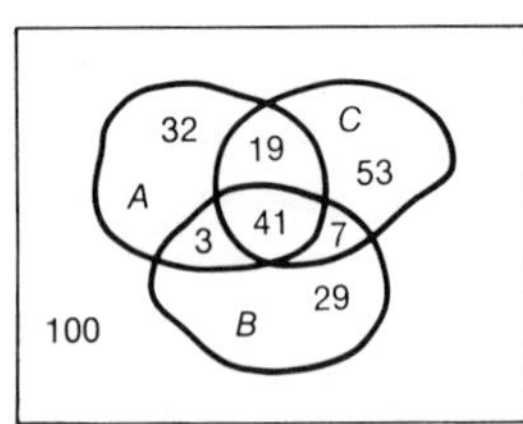

FIGURE 1.35

6. The second distributive property is

$$A \cup (B \cap C) = (A \cup B) \cap (A \cup C).$$

   (a) Reproduce Figure 1.36 and shade each set according to the instructions:

$B \cap C$ vertical shading
$A \cup (B \cap C)$ horizontal shading

(b) In a second copy of Figure 1.36, shade the following sets:

$A \cup B$ vertical shading
$A \cup C$ horizontal shading

(c) Do the results of parts (a) and (b) indicate that the second distributive property holds?

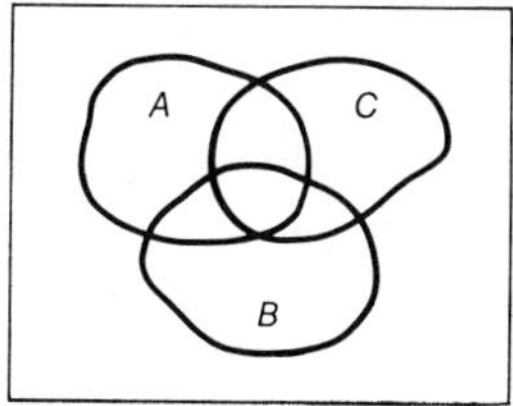

FIGURE 1.36

7. If $A = \{1,2,4\}$, $B = \{1,3,4\}$, $C = \{2,3,5\}$, $U = \{1,2,3,4,5\}$, verify both distributive properties by listing the members of the following sets:
   (a) $A \cap (B \cup C)$ (c) $A \cup (B \cap C)$
   (b) $(A \cap B) \cup (A \cap C)$ (d) $(A \cup B) \cap (A \cup C)$
8. Repeat problem 7, if $A = \{1,3,5\}$, $B = \{2,4,5\}$, $C = \{1,4\}$, $U = \{1,2,3,4,5\}$.
9. The flowchart for $A' \cap (B \cup C)$ is shown in Figure 1.37.

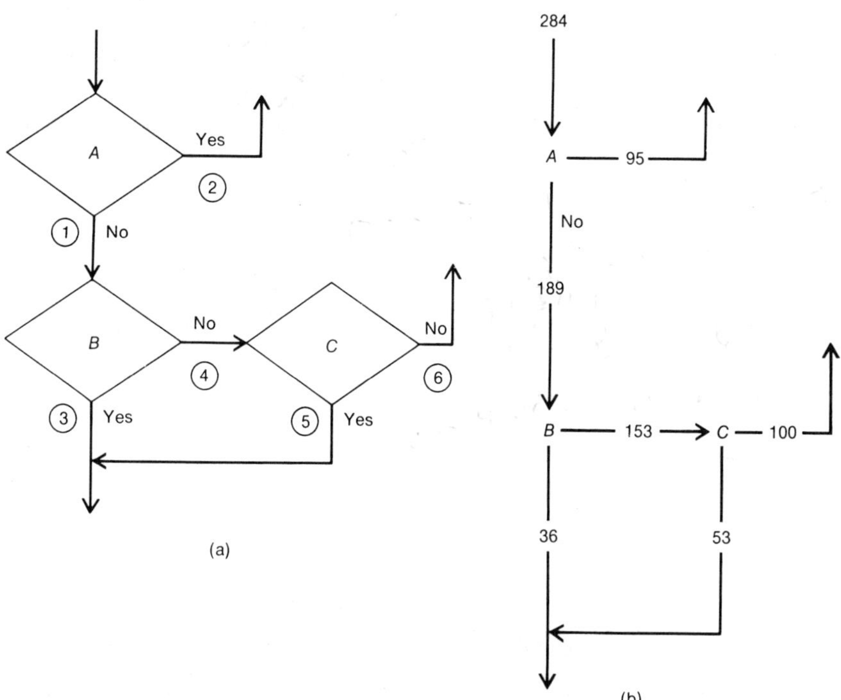

FIGURE 1.37 (a) Flowchart—$A' \cap (B \cup C)$; (b) Number of tests

(a) Sets can be assigned to the circled numbers as follows:

(1) A' (3) $A' \cap B$ (5) $(A' \cap B') \cap C$
(2) A (4) $A' \cap B'$ (6) $(A' \cap B') \cap C'$

(b) Using Figure 1.35, indicate the number of members in each set listed in part (a).

(c) Figure 1.37(b) provides a convenient schematic to show the total number of tests for inclusion in each set. Verify that the total number of tests is 626.

10. Using Figure 1.35, repeat problem 9 for

(a) $A' \cap (C \cup B)$ (b) $(B \cup C) \cap A'$ (c) $(C \cup B) \cap A'$

11. Using Figure 1.35, repeat problem 9 for

(a) $(A \cup C) \cap B'$ (b) $(C \cup A) \cap B'$ (c) $B' \cap (A \cup C)$ (d) $B' \cap (C \cup A)$

12. Using Figure 1.34, repeat problem 11.

REFERENCES

Dinkines, Flora, *Elementary Concepts of Modern Mathematics.* New York, Appleton-Century-Crofts, 1964.

Nahikian, Howard M., *Topics in Modern Mathematics.* New York, The Macmillan Company, 1966.

Rice, J. K. and Rice, J. R., *Introduction to Computer Science.* New York, Holt, Rinehart and Winston, Inc., 1969.

Schriber, T. J., *Fundamentals of Flowcharting.* New York, John Wiley & Sons, Inc., 1969.

Youse, *An Introduction to Mathematics.* Boston, Allyn and Bacon, Inc., 1970.

2 THE DECIMAL NUMBER SYSTEM

2.1 THE NATURAL NUMBERS / **44**
The Evolution of Numbers / **44**
Correspondence / **44**
Addition and Multiplication of Natural Numbers / **45**
Symbols Representing Numbers / **45**
Closure and the Computer / **46**
Exercise 2.1 / **46**

2.2 EXTENDING THE NUMBER SYSTEM / **47**
The Integers / **47**
Rational Numbers / **48**
Irrational Numbers / **49**
The Real Numbers / **50**
Sorting / **50**
Exercise 2.2 / **51**

2.3 CHARACTERISTICS OF REAL NUMBERS / **52**
Laws Relating to Real Numbers / **52**
Other Principles / **53**
Addition and Subtraction of Signed Numbers / **54**
Multiplication and Division of Signed Numbers / **55**
The Notion of Inverses / **56**
Integer Arithmetic Within the Computer / **56**
Exercise 2.3 / **58**

2.4 CHARACTERISTICS OF DECIMAL NUMBERS / **59**
Exponents / **59**
Place Value and Cipherization / **60**
Scientific Notation / **61**
Repeating and Non-repeating Decimals / **62**
Approximations / **63**
Significant Digits / **63**
Exercise 2.4 / **64**

Printed characters such as 13, −7, 6.293, and 9/17 are immediately recognized as numbers and are as much a part of our everyday language as the words on this page. It is interesting that we use them, together with the basic operations of arithmetic, in all phases of our life, yet the average person is completely unaware of their truly abstract nature. The newcomer to algebra, and indeed many advanced students of mathematics, frequently are puzzled by the apparent triviality of rules relating to the study of the number system. However, an understanding of these rules is basic to the study of mathematics in general and is of special importance in the study of computer-related mathematics.

2.1 THE NATURAL NUMBERS

THE EVOLUTION OF NUMBERS. The earliest attempts of man to evolve a counting system were undoubtedly based on the principle of correspondence. For instance, the need of a shepherd to convey the idea of four sheep to another shepherd could have been satisfied by holding up four fingers or making four marks in the sand, a *one-to-one correspondence.* (It is interesting to note that in practically every civilization, the symbol for the number 1 is a single mark, usually a vertical stroke.)

Counting by the use of a one-to-one correspondence was simple enough but certainly limited in scope. Several thousand years ago, when man first began living in large groups, systems began to evolve for representing large quantities. One such system involved the Hindu-Arabic *numerals* (or symbols) 1, 2, 3, 4, 5, 6, 7, 8, 9, 0, which we use today. The versatility of these *digits* was tremendously increased by the principles of place value and cipherization (topics we shall study later). Thus, to represent the total number of one man's fingers and toes, it was not necessary to invent ten additional symbols but only to use combinations of the ten basic digits.

Although we speak of this development briefly, the Hindu-Arabic numerals did not become part of European civilization until the twelfth century. Even up to the seventeenth century most Northern Europeans performed their calculations on the abacus and recorded results in Roman numerals.

CORRESPONDENCE. In thinking of the *natural* numbers (1, 2, 3, 4, ⋯, 15, 16, ⋯) we can arrange them in convenient order. Since 8 is smaller than 15, we say that 8 precedes 15 or 8 is less than 15, which may be written $8 < 15$. Similarly, 15 is greater than 8, which may be written $15 > 8$. We can further compare the whole numbers to one another and find a definite ordering of 1, 2, 3, 4, ...

This ordering property suggests a useful graphical method for representing numbers. Consider a straight line marked by a series of points equally spaced and beginning with the point labeled *A*, as shown in Figure 2.1. Now we can do as the shepherd did and set up a correspondence between the numbers we have been discussing and the points on this straight line, as shown

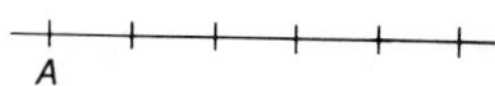

FIGURE 2.1

in Figure 2.2. With this association we can readily see the order of the natural numbers: if one number is smaller than another, it lies toward the left on this axis; if larger, it lies toward the right.

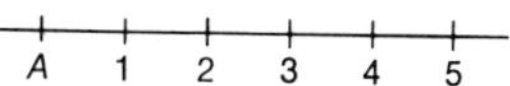

FIGURE 2.2

ADDITION AND MULTIPLICATION OF NATURAL NUMBERS. We can associate the operations of arithmetic with the representation of Figure 2.2 by considering 1 equivalent to the distance from A to the number 1, 2 equivalent to the distance from A to the number 2, and so on. Thus $2+3=5$ can be represented as shown in Figure 2.3.

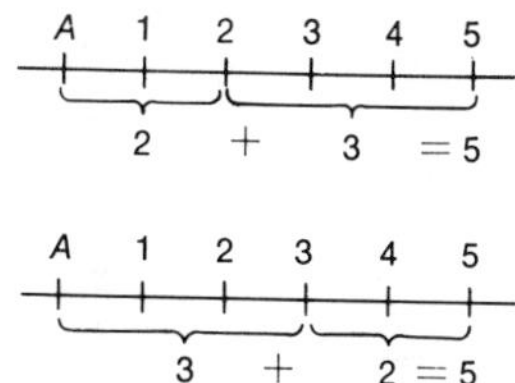

FIGURE 2.3

Note that under the operation of addition, the sum of two natural numbers is a third natural number. This property is called *closure*, and we say that the natural numbers are *closed* under the operation of addition. Further, it is quite significant that the order of representing the numbers is of no consequence; that is, $2+3=3+2$. This is called the *commutative law* for addition, and we say that the natural numbers *commute* under the operation of addition.

In multiplying two numbers it is always possible to approach the operation as one of successive additions. This is the basis of many calculating and computing devices. For example,

$$3\times 2=2+2+2=6 \qquad \text{or} \qquad 2\times 3=3+3=6.$$

It can be seen that the natural numbers are closed and commutative under the operation of multiplication also. Another way to express this is to consider the natural numbers as a collection or a *set*, where each number is an *element* of that set. Then we can say that *the set of natural numbers is closed and commutative under the operations of addition and multiplication.*

SYMBOLS REPRESENTING NUMBERS. In algebra, relationships which involve numbers are studied. Often a particular number is not known and must be deduced from existing facts. It is also often convenient to state rules in equation form so that they apply to all numbers. For convenience, letters and other symbols are used. As an example, the closure laws for addition and multiplication of natural numbers may be defined in the following way: If a and b are natural numbers, c and d are also natural numbers, where

$$c = a + b \qquad \text{and} \qquad d = a \times b.$$

Similarly, the commutative law may be summarized as follows: If a and b are natural numbers,

$$a + b = b + a \qquad \text{and} \qquad a \times b = b \times a.$$

In each case the letters are being used to represent arbitrary choices of natural numbers.

CLOSURE AND THE COMPUTER. Both of these concepts, closure and commutivity, appear so obvious that it hardly seems necessary to point them out. However, these basic relationships are important to many other assumptions and relationships in mathematics. Furthermore, within the computer, neither of these properties necessarily holds true. Although these notions will be discussed more extensively later in this chapter, consider a desk calculator with an accumulator register capable of storing any eight-digit positive number as shown in Figure 2.4. The smallest possible number is stored in (a) and the

FIGURE 2.4 Storage registers

largest is stored in (b). Unlike the set of natural numbers, this set (which, technically speaking, is a *subset* of the natural numbers) consists of a finite number of elements—100,000,000 to be exact. Now if we consider arithmetic operations, we find that difficulties arise, as is evident by the following:

$$75{,}123{,}400 + 50{,}000{,}000 = 125{,}123{,}400.$$
$$75{,}123{,}400 \times 10 = 751{,}234{,}000.$$

Since both of these are beyond the machine capacity, we see that this particular subset of the natural numbers is not closed under either addition or multiplication.

EXERCISE 2.1

In each of the following questions justify your answer, either by words or by an example, as to which fails to satisfy the requirements.

1. Are the even natural numbers closed (a) under the operation of addition (that is, does the addition of two even natural numbers always give another even natural number)? (b) under the operation of multiplication?

2. Are the odd natural numbers closed under the operation of (a) addition? (b) multiplication?
3. Are the odd natural numbers commutative under the operation of (a) addition? (b) multiplication?
4. Is the set of digits consisting only of 0 and 1 closed under the operation of (a) addition? (b) multiplication?
5. Is the number set consisting of the positive powers of 10—that is, 1, 10, 100, 1000, and so on—closed under the operation of (a) addition? (b) multiplication?
6. Is the number set of problem 5 commutative under (a) addition? (b) multiplication?
7. Using the symbols m and n, state the closure law for (a) addition (b) multiplication.
8. Using the symbols x and y, state the commutative law for (a) addition (b) multiplication.

2.2 EXTENDING THE NUMBER SYSTEM

Historically speaking, the concept of fractions evolved very early in the history of mathematics. Negative numbers, on the other hand, were slow to be accepted, even by mathematicians, and it was not until the seventeenth century that the negative number concept became an established part of the number system. In our consideration here of the number system we will depart from its historic evolution and consider the negative numbers and zero as a logical addition to the natural numbers.

THE INTEGERS. Within the system of natural numbers it is always possible, given any two of them (say a and b), to find a third (say x), such that $x = a + b$ merely by adding a and b. For instance, if a is 2 and b is 7, x is $2 + 7 = 9$ (closure).

On the other hand, consideration of the operation of subtraction can introduce problems in dealing strictly with natural numbers. Given natural numbers a and b, it is not always possible to find a third natural number x such that $x = b - a$. Although $x = 4$ if $b = 7$ and $a = 3$, no natural number can be found for x if $b = 3$ and $a = 7$ (or, for that matter, if $a = b = 0$). Generally speaking, x will be a natural number only if $b > a$.

To overcome this dilemma, a new set of numbers was created to satisfy equations such as $x = 3 - 7$. These are called the negative integers and are obtained from the natural numbers by the introduction of a minus sign (for example, -4, -16). The negative integers, together with zero and the positive integers (natural numbers), extend the number line infinitely in both directions, as shown in Figure 2.5.

Note that the interpretation of "greater than" and "less than" is consistent with this representation. That is, if a number b lies to the right of a number c on the axis, then $b > c$ (for example, $4 > 1$, $4 > -2$, $1 > -6$).

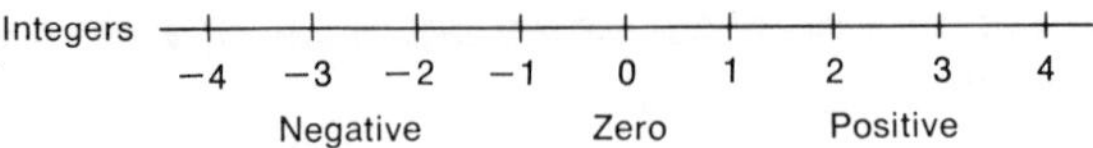

FIGURE 2.5

In a rigorous mathematical development of the number system, a clear distinction is often made between the natural numbers and the positive integers. More specifically, the natural numbers have no sign, whereas the positive integers are considered *signed* numbers and are represented as +1, +2, +3, and so on. However, we will make no such distinction and will assume that the absence of a sign implies a positive signed number.

RATIONAL NUMBERS. In extending the number system to include the negative integers and zero we considered the arithmetic operation of subtraction, the *inverse* operation of addition. As we have seen, multiplication of integers is closed. In other words, when a and b are integers, their product, which may be represented as $x = ab$, is an integer. If instead we consider the quotient of b and a ($x = b/a$), we can find many values of a and b for which no integer exists for x (for example, $b = 3$ and $a = 2$).

Just as the inverse operation to addition dictated a need to expand the natural numbers, the inverse of multiplication (division) required the definition of additional numbers. Obviously these numbers are of the form b/a, where a and b are integers; in other words, they are the fractions. Technically speaking they are called the *rational numbers*, and are defined: *Rational numbers are pairs of integers a and b in the form b/a, with the restriction that* $a \neq 0$ *(a is not equal to 0)*.

The term rational number is easily remembered by noting that a rational number is formed by the ratio of two integers. However, we might question whether or not the integers themselves are rational numbers. The answer is that any integer may be put in the form b/a simply by letting $a = 1$; for example, $3 = 3/1$. Of course, any number of forms could be used to satisfy the definition, since

$$3 = \frac{3}{1} = \frac{6}{2} = \frac{9}{3} = \cdots.$$

On the number line, each fraction can be shown to correspond to a point on the line, as shown in Figure 2.6. It is interesting to note that in any interval on the line, such as 0-1, we can find an infinite number of rational numbers; by the same reasoning we can find an infinite number of points on the line. From this it would appear that the rational numbers represent all of the numbers

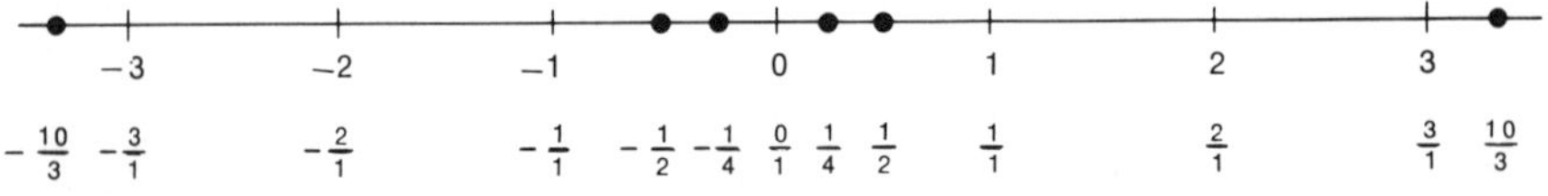

FIGURE 2.6

and that each and every point on the line has been used. However, this conclusion is false.

IRRATIONAL NUMBERS. The need for a further expansion of the number system becomes apparent when we attempt to obtain square roots of integers such as 2. It is known that the square root of two is 1.414, but this value is only an approximation. The value of $\sqrt{2}$ has been computed to many hundreds of places, but without conclusion, because $\sqrt{2}$ cannot be put in the form b/a. Numbers that cannot be put in this form are called *irrational numbers.* However, if we concede that there must be some number which when squared gives 2, we can argue that it must correspond to a point on the number line. Obtaining a distance equal to $\sqrt{2}$ is a relatively simple geometric task if we recall the Pythagorean theorem. In Figure 2.7, a right triangle has been constructed with the length of each leg equal to one unit.

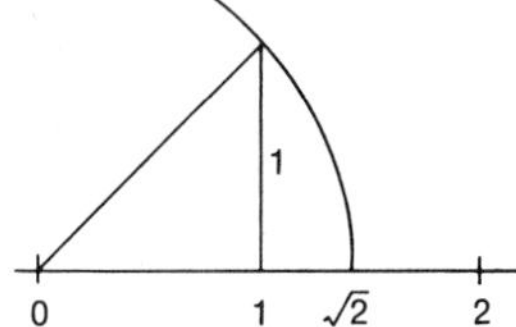

FIGURE 2.7

The length of the diagonal, which is equal to $\sqrt{2}$ by the theorem, has been transferred to the axis by drawing an arc. The point of intersection then represents the point on the number line corresponding to $\sqrt{2}$. As with the rational numbers, we can find an infinite number of square roots, third roots, and so on, which are irrational.

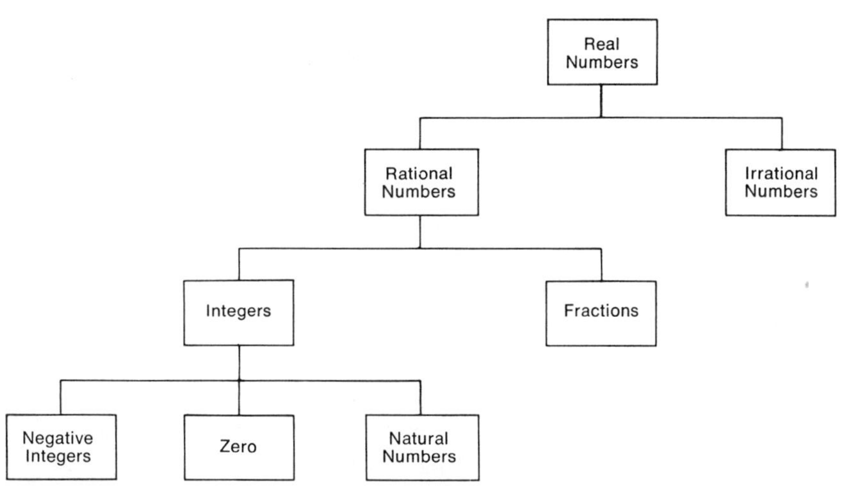

FIGURE 2.8 The real number system

THE REAL NUMBERS. The rational and irrational numbers together represent the *real number* system, and the number line which we have been discussing is commonly called the *number line* or the *real axis*. In arriving at this system, we first discussed the natural numbers, then the integers, the rational numbers, and finally the irrational numbers. It is interesting to note that the natural numbers are a subset of the integers, the integers a subset of the rational numbers, and both the rational and irrational numbers are subsets of the real number system. This is represented schematically in Figure 2.8.

SORTING. A common need in programming is to arrange a set of data values into a desired sequence, usually an ascending sequence. Whether the quantities are experimental measurements consisting of fractional parts (rational numbers) or simply employee file numbers (natural numbers), the need is the same. That is, they must be ordered according to the relationships "greater than" and "less than" as illustrated by the number line of Figure 2.6. The process of "ordering" a hand of playing cards actually involves a sorting process. Most card players search for the smallest card (of a given suit), place it to the left, search for the next larger card, place it after the smallest, and so on. (Problem 8, Exercise 1.1 involves developing an algorithm for finding the *largest* value in a data set.) Although this technique is convenient for playing cards, it is not especially compatible with the computer.

A means somewhat more versatile consists of comparing and reordering adjacent pairs. For instance, let us arrange the following digits in ascending order using this method:

5 4 3 2 1

We first compare the 5 and the 4; since the former is larger we change the order, giving

4 5 3 2 1

Comparing the next pair, 5 and 3, we see that the 5 is once again larger, so we exchange these two, giving

4 3 5 2 1

Continuing in this manner, we would eventually obtain

4 3 2 1 5

Having completed this first pass, the largest of the digits would be last.

After four such passes have been completed the digits would be reordered as required:

5 4 3 2 1	original number
4 3 2 1 5	completion of first pass
3 2 1 4 5	completion of second pass
2 1 3 4 5	completion of third pass
1 2 3 4 5	completion of fourth pass

For the sake of simplicity, we compared each of the four pairs of digits during each of the passes. Note, however, that after the first pass we need not compare the last pair, 1 and 5, since 5 is already in its proper position. Similarly, after the third pass it is only necessary to compare the first two digits, not each of the four pairs. Although we would undoubtedly use this more efficient method if we were manually performing the sorting operation, special consideration of this factor is necessary in writing a computer program.

The above example required four passes to complete the sequencing. However, if the numbers are not completely out of sequence, the task is shorter. For instance, consider the following example:

3 2 5 1 4	original number
2 3 1 4 5	completion of first pass
2 1 3 4 5	completion of second pass
1 2 3 4 5	completion of third pass

In this case, it was not necessary to complete four passes since the desired sequencing was obtained in three. Once again this would probably be automatic if we were arranging the digits ourselves, but special provisions are required in preparing a program in order to avoid the maximum number of passes when they are not necessary.

EXERCISE 2.2

1. Insert the proper symbol $(<, =, >)$ between the following pairs of real numbers (assume that *a, b,* and *c* are positive integers):

(a)	1	2	(i)	0	0
(b)	3	27	(j)	1/5	1/4
(c)	100	87	(k)	1/6	7/48
(d)	-5	12	(l)	$-1/100$	0
(e)	4	-4	(m)	$a+b$	a
(f)	-4	-5	(n)	$a+b$	$b+a$
(g)	9	1	(o)	$a+b+c$	$c+a+b$
(h)	-27	0	(q)	0	$-1/b$

2. Arrange the following numbers in ascending order separated by the $<$ symbol $(1 < 2 < 3 \cdots)$: (a) 19, -6, 4/3, 0, -9, 12. (b) $3a$, $2a$, 0, $-a/5$, $a/5$, $-6a$ (a is a positive integer).
3. Arrange the following numbers in descending order separated by the $>$ symbol $(8 > 7 > 6 \cdots)$: (a) 16, -5, 8, -1, 0, $-15/16$. (b) $-4c/5$, 0, $6c$, $-6c$, $9c/5$ (c is a positive integer).
4. (a) In 2(b), would the required ordering of $2a$ and $3a$ be possible if a could be any integer value? (b) Why?
5. Are the negative numbers closed under the operations of (a) addition? (b) multiplication? (c) If not, give counter-examples.
6. Are the rational numbers closed under the operation of (a) addition? (b) multiplication? (c) If not, give counter-examples.
7. The product of two irrational numbers (which are taken up in later chap-

ters) is frequently a third irrational number. For instance, $\sqrt{2} \times \sqrt{3} = \sqrt{6}$. However, the irrationals are not, in general, closed under multiplication. Give an obvious case where closure is not satisfied.

8. Prepare a flowchart to illustrate the sorting algorithm. Assume that 10 numbers (commonly called *fields* in data processing) are to be sorted and that nine complete passes are to be carried out as illustrated by the first example.

9. Prepare a flowchart to illustrate the sorting algorithm assuming that the number of fields is not known (refer to this count as n). Make successively fewer comparisons for each pass and provide for termination if the fields are in sequence prior to the maximum required number of passes. *Food for thought*: What is the implication of no interchanges occurring during a given pass?

2.3 CHARACTERISTICS OF REAL NUMBERS

LAWS RELATING TO REAL NUMBERS. The properties of closure and commutivity, although not technically applying to numbers within computers, are fundamental to our number system. Two other properties fall in the same category; they are the *associative* law and the *distributive* law. In adding a group of numbers such as $7 + 5 + 8$ we usually pay little attention to the sequence of combining the individual pairs. For instance, we can add 7 to 5 to get 12, and then add 12 to 8 to get 20. On the other hand, we can first determine $5 + 8$ and then combine this with 7. In other words $(7 + 5) + 8 = 7 + (5 + 8) = 20$. This operation for addition and a comparable one for multiplication may be formulated as follows:

1. The Associative Law for Addition
   If a, b, and c are any real numbers

$$(a + b) + c = a + (b + c).$$

2. The Associative Law for Multiplication
   If a, b, and c are any real numbers

$$(ab)c = a(bc).$$

The *distributive law for multiplication with respect to addition* has the following formal definition:

If a, b, and c are real numbers

$$a(b + c) = ab + ac.$$

An example of this is

$$\begin{aligned} 2(3 + 9) &= 2(3) + 2(9) \\ 2(12) &= 6 + 18 \\ 24 &= 24. \end{aligned}$$

The misuse of this property in algebraic manipulations is very common, and frequently causes basic errors in setting up algebraic expressions. Within the computer, numbers are commonly stored in several different formats. One of them, to be discussed later in this chapter, renders these laws invalid for certain applications; however, for normal processing, resulting inaccuracies are not important.

In retrospect, note that these laws come in pairs. For example, the commutative law for addition and the commutative law for multiplication. The distributive law of multiplication with respect to addition, on the other hand, appears not to have a partner. However, it is interesting to note that a distributive law of addition with respect to multiplication is an extremely important and useful component of abstract algebraic systems such as Boolean algebra (see Chapter 14).

OTHER PRINCIPLES. In the addition process, zero plays a unique role since it is the only real number which, when added to another real number, will not cause a change. For instance, $3 + 0 = 3$. For this reason zero is commonly called the *identity* element for addition or the *additive identity.* More generally speaking, we call zero the additive identity because, for any real number a,

$$0 + a = a + 0 = a.$$

The number 1 plays a somewhat similar role in the operation of multiplication of real numbers, for example, $23 \times 1 = 23$. Thus 1 is called the identity for multiplication or the *multiplicative identity.* In general, we call 1 the multiplicative identity because, for any real number b,

$$b \times 1 = 1 \times b = b.$$

In dealing with signed numbers it is frequently useful to speak of the *magnitude* or *absolute value* of a signed number. The absolute value of any number is that number without its sign. Relating this to the real axis, it may be interpreted as the number of units from zero to the number point or simply as the distance from zero. The two quantities -20 and $+20$ are equidistant from the zero point and correspondingly have the same absolute value of 20. Absolute value is commonly indicated by two vertical lines on each side of the number:

$$\begin{aligned} |20| &= 20 & |10-7| &= |7-10| = 3 \\ |-20| &= 20 & |5-3| &= |3-5| = 2 \\ |-10/3| &= 10/3 & |a-b| &= |b-a| \\ |10/3| &= 10/3. & |0| &= 0. \end{aligned}$$

In ordering a set of absolute values, the procedure is somewhat different than with signed numbers, for example, $2 > -7$ but $|2| < |-7|$ since $|2| = 2$ and $|-7| = 7$. This can be generalized by saying, if $a > b$ (a and b are real numbers) it does not necessarily follow that $|a| > |b|$. Similarly, if $c < d$ it does not necessarily follow that $|c| < |d|$. These relationships are obvious when dealing with numbers themselves but frequently cause difficulty when letters (representing numbers) are substituted, as in the study of algebra. The absolute value of a real number a is frequently defined as

$$|a| = a \quad \text{if} \quad a \geq 0$$
$$|a| = -a \quad \text{if} \quad a < 0$$

ADDITION AND SUBTRACTION OF SIGNED NUMBERS. The notion of adding natural numbers is introduced in Figure 2.3 where the addition processing is related to a counting process on the number line. For signed numbers, where a positive number implies counting to the right, a negative number can logically be interpreted as implying counting to the left. Using this basic notion, the following rules for adding signed numbers can easily be evolved.

1. If the numbers have *like* signs, add the absolute values and prefix their common sign.
2. If the numbers have *opposite* signs, take the difference of their absolute values and prefix the sign of the number that has the larger absolute value.

It is interesting that when defined in this manner, subtraction becomes simply a special case of addition. As we shall learn in Chapter 3, the following rule for subtraction is effectively designed into some computers.

The difference *of two signed numbers is obtained by changing the sign of the subtrahend and proceeding with the rule for addition.*

Examples: $(+16) - (+8) = (+16) + (-8) = +8$
$(+16) - (-8) = (+16) + (+8) = +24$

The rules for addition of signed numbers (and subsequently subtraction) can be simplified for design of calculating machines by using a different representation for negative numbers. The method of storing a negative number in its *complement* form and of changing sign of a number by *complementing* it is commonly used in computers. For example, if we have a computer capable of storing four-digit numbers (that is, designed with four-digit registers), we might require that the numbers 5716 and −2463 be added. However, in our computer the negative quantity could be stored in its complement form which is obtained by subtracting its absolute value from 10000 yielding

$$\begin{array}{r} 10000 \\ \underline{-2463} \\ 7537 \end{array}$$

The arithmetic process then becomes one of addition as shown in Figure 2.9. Since the register capacity is only four digits, the high-order carry is lost yield-

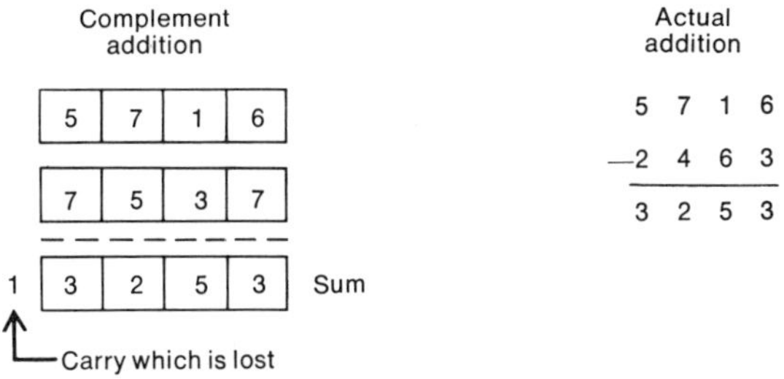

FIGURE 2.9 Using complements

ing the correct result. This is no coincidence and is the basis for problem 10, Exercise 2.3. By slightly expanding this technique, it is possible to arrive at a completely consistent set of rules for performing arithmetic in this fashion. The principles of complementary arithmetic in general and binary complements in particular are extensively developed in Chapter 3 because of their wide use in computers.

In the design of computers, a number of different techniques have been employed to provide arithmetic capabilities. An interesting technique used in the IBM 1620 computer involves storing an addition table in storage registers 100 through 199. Any pair of digits to be added (say 4 and 3) are used to form the register address 143 which is the table storage location containing the resulting sum digit, 7. Use of this table was eliminated in later models of the 1620 because of the relatively slow speed of this method.

MULTIPLICATION AND DIVISION OF SIGNED NUMBERS. Although most modern computers provide individual instructions for performing multiplication and division operations, in some of the earlier machines multiply and divide instructions were available only as optional features. However, this did not preclude the possibility of multiplying and dividing since multiplication can be performed by successive addition and division by successive subtraction. For multiplication, the technique suggested in problem 13, Exercise 1.1 involves adding 628 a total of 27 times in multiplying 628 by 27. Although simple, this procedure can be very time-consuming in the machine. For example, multiplying 989,898 by 987,654 would require almost one million additions.

A more efficient approach is to apply the successive addition process taking into account positional significance of the multiplier digits. For example:

```
    1384
  × 312
  ------
    1384  }
    1384  }   that is, 1384 × 2
   1384       that is, 1384 × 10
  1384    }
  1384    }   that is, 1384 × 300
  1384    }
  ------
  431808
```

Regarding signs, it is necessary to employ the conventional rules for multiplying (or dividing) signed numbers, that is,

> *The product (or quotient) of two signed numbers is the product (or quotient) of the absolute values with the sign determined by the signs of the original numbers as follows:* (a) *positive if the original signs are the same* (b) *negative if the original signs are different*

Whenever both numbers have positive signs, the result is obviously positive; the product of two negative numbers being positive is not obvious but is actually dictated by the requirement that signed numbers satisfy the distributive rule (see problem 12, Exercise 2.3).

THE NOTION OF INVERSES. By inspection of the number line, we can see that it is symmetric. For every positive number to the right of zero, there is a corresponding negative number to the left. Whenever we add such a pair, the result is the identity element for addition, 0; for example,

$$2+(-2)=0 \text{ and } \frac{10}{3}+\left(-\frac{10}{3}\right)=0.$$

The term *inverse* is commonly used to describe this relationship between pairs of numbers. More generally speaking, the sum of a number and its additive inverse (any given number has only *one* inverse) is the additive identity.

We might expect a parallel concept for multiplication to exist; that is, *the product of a number and its inverse for multiplication is the identity for multiplication.* That such quantities exist is obvious; for instance, 7 has the inverse $\frac{1}{7}$ since $7 \times \frac{1}{7} = 1$. In general, $1/a$ is the multiplicative inverse of a whenever $a \neq 0$, since

$$a \times \frac{1}{a} = \frac{1}{a} \times a = 1.$$

The need for excluding zero is apparent if we recognize that division by zero is undefined. (For implications of this, refer to problem 7, Exercise 1.3.)

INTEGER ARITHMETIC WITHIN THE COMPUTER. In most digital computers, arithmetic is performed on integer quantities. Decimal point positioning is either accounted for by the programmer or by special programs rather than by the machine itself. However, many modern computers also have capabilities for automatically handling decimal positioning. Other than the difficulties presented earlier regarding possible overflow, no problems would appear to exist. However, we must give great care to what is meant by *integer* arithmetic, especially when performing division. For example, when dividing 13 by 5 we have

$$\frac{13}{5} = 2.6.$$

But 2.6 is *not* an integer so we must think of the result as

$$\frac{13}{5} = 2 \text{ with a remainder of 3.}$$

This is what occurs within the computer. When programming in the computers language, both the quotient and remainder are available to the program but when programming in Fortran, the remainder is lost. As a result, in Fortran

$$\frac{0}{100} \rightarrow 0$$

$$\frac{99}{100} \rightarrow 0$$

$$\frac{100}{100} \rightarrow 1$$

Although this integer arithmetic (also referred to as *fixed point* arithmetic) tends to be confusing and very distracting to the beginner, it provides the programmer with a very useful tool. However, great care must be exercised because of peculiar appearing results which occur, as evidenced by the following examples.

Example:

$$\frac{10 \cdot 2}{10} = 2 \qquad \text{can be evaluated as follows:}$$

$$10 \cdot 2/10 = 20/10 = 2,$$

or

$$(2/10) \cdot 10 \rightarrow (0)/10 = 0.$$

Example:

$$\frac{5 \cdot 3 \cdot 9}{4 \cdot 7} \simeq 4.82:$$

$$5 \cdot 3 \cdot 9/(4 \cdot 7) = 135/28 \rightarrow 4,$$

or

$$(5/4) \cdot 3 \cdot 9/7 \rightarrow 1 \cdot 3 \cdot 9/7 = 27/7 \rightarrow 3,$$

or

$$(5/7) \cdot 3 \cdot 9/4 \rightarrow 0 \cdot 3 \cdot 9/4 = 0.$$

Example:

$$\frac{1}{5} + \frac{7}{5} + \frac{2}{5} = 2:$$

$$\frac{1}{5} + \frac{7}{5} + \frac{2}{5} \rightarrow 0 + 1 + 0 = 1,$$

or

$$(1 + 7 + 2)/5 = 10/5 = 2.$$

EXERCISE 2.3

1. Consider a computer with four-digit registers. Assuming that high-order carries beyond the register capacity are lost for intermediate results as well as final results in arithmetic computations, show that the associative law does not hold for

$$(+8773) + (5891) + (-4992).$$

2. Using the assumptions of problem 1, show that the distributive law does not hold for

$$49[(+500) + (-400)].$$

3. Illustrate, by substituting several values for m and n, that

$$|m - n| = |n - m|.$$

(Note that this is *not* a proof of the relationship)

4. Demonstrate, by appropriate choice of values for a and b, that

$$|a| + |b| \neq |a + b|$$

In general, what *can* be said of the relationship between

$$|a| + |b| \text{ and } |a + b|?$$

5. Within the computer a number is stored in an area which we shall refer to as NUM. Define an algorithm in the form of a flowchart to store the absolute value of NUM in another area called ANUM.

6. Perform the indicated arithmetic operations.

(a) $25 - (+13)$	(e) $(-a) - (+a)$	(i) $(25)(13)$	(m) $(-a)(-b)$
(b) $25 - (-13)$	(f) $(-a) + (-a)$	(j) $(25)(-13)$	(n) $(0)(-25)$
(c) $-25 - (-13)$	(g) $(-a) - (-a)$	(k) $(-25)(13)$	(o) $(-1)(x)$
(d) $-25 - (+13)$	(h) $0 - (-x)$	(l) $(-25)(-13)$	(p) $(-2)(-5)(-10)$

7. With reference to problem 6, what can be said (a) of the product of an odd number of negative numbers and (b) of an even number of negative numbers?

8. Show that division by zero leads to contradictions by solving for

$$x \text{ if } a/0 = x \ (a \neq 0).$$

9. Given two real numbers a and b, what can be said of them if

(a) $ab < 0$? (b) $ab = 0$? (c) $ab > 0$?

10. The notion of adding numbers with unlike signs (that is, 5716 and -2463) was illustrated using complements in this section. Describe why the high-order carry can be ignored to yield the correct result.

11. Using the complement, add the two numbers 2563 and -5716. What is the significance of the result relative to the true answer? Experiment with other numbers using unlike signs to determine the key. Explain.

12. Demonstrate that the product of two negative numbers must be positive by applying the distributive law to

$$(-3)[(+2) + (-2)] \quad \text{(which equals 0).}$$

13. Define an algorithm in the form of a flowchart to perform the multiplication operation by successive addition using the technique illustrated in this section. Assume eight-digit numbers which may yield a 16-digit product.
14. Define a division algorithm using successive subtraction of the divisor from the dividend until the dividend has been reduced to zero. What occurs if the divisor is zero?
15. Using integer arithmetic perform the following operations. (For successive multiplications and divisions, perform them in order from left to right.)

(a) $5/3$	(c) $14/5 - 10/5$	(e) $2 \times (3/4) \times 2$	(g) $5 \times 6 \times 9/4/3$
(b) $7/4 + 3/4$	(d) $3/4 + 1/4 + 5/4$	(f) $2 \times 2 \times 3/4$	(h) $5 \times 6/4 \times 9/3$

2.4 CHARACTERISTICS OF DECIMAL NUMBERS

In Chapter 1 we discussed the evolution of our number system, its properties, and its abstractness but made only fleeting reference to it as a base 10 system. This section will be devoted to a study of its decimal nature. However, the first discussion will be devoted to an initial exploration of exponents since exponential forms are basic to the decimal system.

EXPONENTS. Exponents are a commonly used shorthand form of mathematics. For instance, the exponent form as illustrated below can be used to save considerable writing:

$$a \times a \times a \times a \times a = a^5,$$

or

$$10 \times 10 \times 10 \times 10 \times 10 \times 10 = 10^6.$$

Generalized, the relationship is

$$\underbrace{a \times a \times a \times \cdots \times a}_{n \text{ times}} = a^n$$

Where a is any real number, called the *base*, and n is any natural number, called the *exponent*. (We shall discover later that n can be any real number.) The use here is obvious, the exponent n merely indicating how many times the base a should be included as a factor in the expression. We should recognize that this is a symbolism arrived at by definition, and that the product of eight values of a can be defined as a^8. This is much the same as defining the symbol \$ to represent dollars, or # to represent pounds. Although detailed properties of exponents are covered in Chapter 5, a few of the basic relationships necessary for this chapter and Chapter 3 are summarized as follows. (The student with a weak algebra background might find it useful to study Chapter 5 at this point.)

Relationship	Examples
$x^a \times x^b = x^{a+b}$	$x^3 \times x^4 = x^7$
	$10^5 \times 10^3 = 10^8$
$x^{a+b} = x^a \times x^b$	$x^9 = x^{2+7} = x^2 \times x^7$
	$10^6 = 10^{1+5} = 10 \times 10^5$
	$2^5 = 2^{2+3} = 2^2 \times 2^3$
$x^0 = 1$ (by definition)	$10^0 = 1$
	$2^0 = 1$
$x^{-a} = \frac{1}{x^a}$	$x^{-5} = \frac{1}{x^5}$
	$10^{-3} = \frac{1}{10^3}$ (= 0.001)

PLACE VALUE AND CIPHERIZATION. Two very important characteristics that are fundamental to the decimal system are (a) the principle of *place* or *positional* value, and (b) the principle of *cipherization.* Since the decimal system includes only nine distinct digits and zero, it is necessary to use the principle of place value in counting above nine. A multiple-digit number such as 4378 uses four of the ten available digits, yet each of them has a different *significance.* What is actually implied by the symbolism 4378 is:

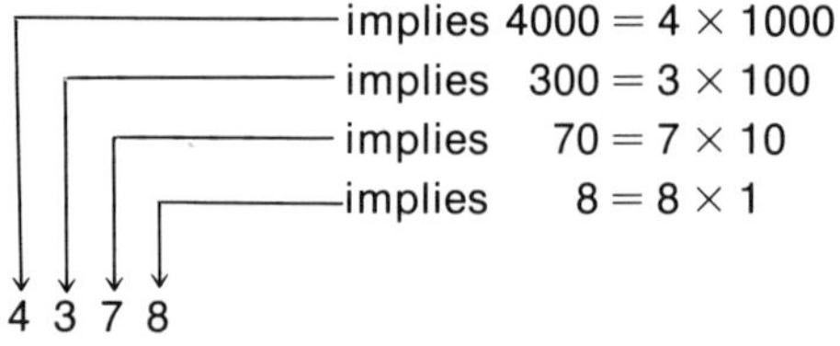

$$4378 = (4 \times 1000) + (3 \times 100) + (7 \times 10) + (8 \times 1).$$

Thus, the value of each digit is represented not only by the digit itself but also by its place or position. The inclusion of zero in the number system is important here since it serves the purpose of a *place holder.* Obviously, there is a considerable difference between the numbers 352 and 3520. Were it not for the use of place value it might be necessary to incorporate many hundreds of different symbols into our number system. However, the principle of cipherization is probably the most important single characteristic of the decimal system.

Using exponential forms, numbers such as 4378 and 13905 may be expressed as

$$\begin{aligned} 4378 &= 4 \times 1000 + 3 \times 100 + 7 \times 10 + 8 \\ &= 4 \times 10^3 + 3 \times 10^2 + 7 \times 10^1 + 8 \times 10^0; \\ 13905 &= 1 \times 10000 + 3 \times 1000 + 9 \times 100 + 0 \times 10 + 5 \\ &= 1 \times 10^4 + 3 \times 10^3 + 9 \times 10^2 + 0 \times 10^1 + 5 \times 10^0. \end{aligned}$$

Both of these representations resemble the manner of writing algebraic polynomials (see Chapter 6); for example, $x^4 + 2x^3 + 9x^2 + 4x^1 + x^0$. This arrangement into an ordered sequence of consecutive powers of a given base is the principle of cipherization. It should be noted that this principle also permits the representation of numbers with values less than 1 in the required form:

$$21.1 = 2 \times 10^1 + 1 \times 10^0 + 1 \times 10^{-1}$$

and

$$443.762 = 4 \times 10^2 + 4 \times 10^1 + 3 \times 10^0 + 7 \times 10^{-1} + 6 \times 10^{-2} + 2 \times 10^{-3}.$$

Although our decimal system uses a base of 10, the choice of a number system base is arbitrary and, as we will find in Chapter 3, base 10 leaves much to be desired in computer applications.

SCIENTIFIC NOTATION. The decimal characteristics of our number system provide the very convenient method for expressing large and small numbers called *scientific notation.* For instance, in computer usage, we speak of operating times in terms of microseconds (millionths of a second), where one microsecond is

$$\frac{1}{1{,}000{,}000} \text{ sec} = \frac{1}{10^6} \text{ sec} = 1 \times 10^{-6} \text{ sec} = 1\ \mu\text{sec}.$$

The cost of computers is jokingly referred to in terms of "kilobucks" or "megabucks," which are 10^3 and 10^6 dollars, respectively.

In chemistry and physics there are many large and small quantities, such as the speed of light:

$$\begin{aligned} 186{,}000 \text{ miles per sec} &= 1.86 \times 100{,}000 \text{ miles per sec} \\ &= 1.86 \times 10^5 \quad \text{miles per sec.} \end{aligned}$$

Other commonly used quantities are the Avogadro number, 6.02×10^{26}, and the mass of a proton, 1.67×10^{-27}. (Try writing these out.)

The standard practice in using scientific notation is to transform all numbers to a form with one digit to the left of the decimal multiplied by the appropriate power of 10. This is illustrated by the following examples:

$$\begin{aligned} 186{,}000 &= 1.86 \times 10^5 \\ 0.186 \times 10^6 &= 1.86 \times 10^5 \\ 0.123 &= 1.23 \times 10^{-1} \\ 0.00123 &= 1.23 \times 10^{-3} \\ 0.000567 \times 10^5 &= 5.67 \times 10^{-4} \times 10^5 = 5.67 \times 10. \end{aligned}$$

In performing arithmetic with large and small numbers, scientific notation and rules for exponents are commonly employed. The product of two numbers such as 6,250,000 and 0.00012 is easily found by multiplying, but care must be taken to insure proper decimal placement. Using scientific notation, this can be accomplished as shown below:

$$\begin{aligned} 6{,}250{,}000 \times 0.00012 &= 6.25 \times 10^6 \times 1.2 \times 10^{-4} \\ &= 6.25 \times 1.2 \times 10^6 \times 10^{-4} \\ &= 7.5 \times 10^2 \\ &= 750. \end{aligned}$$

Although intuitively obvious, the rearrangement of terms in progressing from the first equation to the second is possible because of the commutative law. By proper grouping of the exponential forms, the laws for operations with exponents may be applied. Other examples are:

$$\frac{45{,}000 \times 2{,}000{,}000}{30{,}000} = \frac{4.5 \times 10^4 \times 2 \times 10^6}{3 \times 10^4}$$

$$= \frac{4.5 \times 2}{3} \times 10^4 \times 10^6 \times 10^{-4}$$

$$= 3 \times 10^6.$$

$$\frac{65{,}000 \times 120{,}000}{0.00000050} = \frac{6.5 \times 10^4 \times 1.2 \times 10^5}{5.0 \times 10^{-7}}$$

$$= \frac{6.5 \times 1.2}{5.0} \times 10^4 \times 10^5 \times 10^7$$

$$= 1.56 \times 10^{16}.$$

$$\frac{8{,}000{,}000 \times 8000 \times 0.0006}{12{,}000{,}000 \times 160{,}000} = \frac{8 \times 10^6 \times 8 \times 10^3 \times 6 \times 10^{-4}}{1.2 \times 10^7 \times 1.6 \times 10^5}$$

$$= \frac{8 \times 8 \times 6}{1.2 \times 1.6} \times 10^6 \times 10^3 \times 10^{-4} \times 10^{-7} \times 10^{-5}$$

$$= 200 \times 10^{-7}$$

$$= 2.0 \times 10^{-5}.$$

These are operations with which the scientist and engineer become very familiar in performing multiplication and division on a slide rule.

REPEATING AND NONREPEATING DECIMALS. In the discussion of the real number system, it was emphasized that rational numbers are numbers that can be put in the form b/a, where b and a are integers; irrational numbers cannot be put in this form. The difference between rational and irrational real numbers in decimal form is frequently very subtle. The rational numbers 1/2, 1/10, 3/4, and 9/8 have equivalent decimal forms of 0.5, 0.1, 0.75, and 1.125, respectively. On the other hand, such rational numbers as 1/3 and 1/7 have no such concise decimal form, although they are approximately 0.333333 and 0.142857, respectively. One property that rational numbers have in common is that their decimal forms always exhibit a repeating pattern. For example, $\frac{1}{2} = 0.5000 \cdots$ repeats in 0 after the initial five, $\frac{1}{3} = 0.3333 \cdots$ repeats 3 continuously, and $1/7 = 0.142857142857$ repeats the combination 142857 every six digits.

It is an interesting puzzle that every such fraction repeats itself when written in decimal form, and the number of digits in the cycle is always less than the value of the denominator. Irrational numbers do not exhibit this property. For instance, $\sqrt{2}$ has been computed to hundreds of decimal places and no such pattern emerges. Actually it is quite useless to attempt to find such patterns. Euler, by a relatively simple proof, showed that $\sqrt{2}$ cannot be expressed in the form b/a and therefore is not expressible in repeating decimal form.

APPROXIMATIONS. In performing practical calculations, whether by hand or with a calculator or computer, it is usually necessary to make approximations for irrationals and frequently for rationals as well. For instance, commonly used decimal approximations for several numbers are shown below (the symbol $\simeq$ means *approximately equal to*):

$$\sqrt{2} \simeq 1.414 \qquad \pi \simeq 3.1416$$
$$\pi \simeq 3.14 \qquad \tfrac{1}{7} \simeq 0.143$$

The value for $\sqrt{2}$ is correct to three decimal places, the two values of π are correct to two and four decimal places, respectively, and the value of $\frac{1}{7}$ is correct to three decimal places. However it is possible to express any of these numbers to as many places as is required. That is, a 10- or 20-digit approximation of $\sqrt{2}$ may be used in a calculation if necessary.

SIGNIFICANT DIGITS. The approximation $\sqrt{2} \simeq 1.414$ is said to consist of four significant figures, that is, the number of digits in the number. Similarly, the two approximations for π consist of three and five significant figures, respectively. However, the approximation for $\frac{1}{7}$ consists of three significant figures (the leading zero is merely for convenience). Furthermore, the numbers 0.0032, 30.02, and 3.1020 consist of two, four, and five significant figures, respectively. Note that both zeros in 3.1020 and the zeros in 30.02 are considered to be significant figures, whereas the leading zeros in 0.0032 are not; they serve only as place holders.

These examples serve to illustrate some of the following principles relative to significant digits:

1. *All nonzero digits and zeros occurring between nonzero digits are significant.* By this criterion, all of the following numbers have five significant figures.

   23671 29.731 0.61152 23002 9.5701

2. *Terminal zeros following the decimal point are always considered significant.* Each of the following numbers has four significant figures.

   23.10 471.0 2.000 650.0

   Just as the number 471.3 is correct to the nearest tenth, so is 471.0. To write 471.3 as 471.30 would imply that the number is correct to the nearest hundredth, which might not be true; the same can be said for writing 471.0 and 471.00.

3. *In a number less than 1, zeros immediately following the decimal but preceding any nonzero digit are not significant.* Each of the following numbers has three significant figures.

   0.0123 0.957 0.230 0.0000501

   In the first and last of these illustrations, the zeros immediately following the decimal serve only as place holders.

4. *Terminal zeros in an integer usually are not significant.* For instance, the speed of light is normally stated as 186,000 miles per second. This commonly used value is an approximation correct only to the nearest 1000 miles per second, so it consists of but three significant figures.

The last situation may cause a certain amount of confusion since the general tendency is to state approximate values. For example, a firm may have a yearly operating budget of $27,000. In all probability a detailed budget accounting might give some amount such as $27,121.63, so the approximation of $27,000 consists of only two significant figures. However, there does exist a possibility that the amount is exactly $27,000, in which case all five digits would be significant. In a situation such as that, the sum would probably be written as $27,000.00, which implies that this number consists of seven significant figures.

For amounts which are not monetary, the use of scientific notation can serve to clarify a given situation. For instance, if it is known that the numbers 26,130,000, 194,000, and 1,000,000 each contain four significant figures, they could be represented as

$$26{,}130{,}000 = 2.613 \times 10^7,$$
$$194{,}000 = 1.940 \times 10^5,$$
$$1{,}000{,}000 = 1.000 \times 10^6.$$

EXERCISE 2.4

Express each of the following numbers in the form illustrated by the example

$$203 = 2 \times 10^2 + 0 \times 10^1 + 3 \times 10^0.$$

1. 62	8. 37.65	15. 0.156
2. 359	9. 92.10	16. 0.9215
3. 421	10. 11.113	17. 0.9006
4. 67001	11. 200.001	18. 0.0156
5. 92105	12. 40.0045	19. 0.0203
6. 84110	13. 501.687	20. 0.0067
7. 22101	14. 2.563	21. 0.00909

Express each of the following in scientific notation, for example:

$$43.62 \times 10^2 = 4.362 \times 10^3$$

22. 43798
23. 22706
24. 217
25. 4801
26. 32.76
27. 59.62
28. 9087.6
29. 643.9
30. 0.591
31. 0.653
32. 0.00541
33. 0.00493
34. 0.0000472
35. 0.00000802
36. 64.7×10^2
37. 975.2×10^4
38. 75.4×10^3
39. 801×10^{-4}
40. 6201×10^{-7}
41. 0.0827×10^{-3}
42. 0.0000065×10^6
43. 0.000043×10^2
44. 0.0061×10^{-4}
45. 0.00901×10^{-6}

Perform the indicated operations. Express the final answer in scientific notation.

46. 6000×400000
47. $30000 \times 40000 \times 60000$
48. 3000×0.0000031
49. 20000×0.00015
50. 0.00023×0.000053
51. 0.006×0.00002
52. $\dfrac{50000 \times 300000}{60000000}$
53. $\dfrac{60000 \times 24000000}{12000000}$
54. $\dfrac{900000 \times 7000000}{0.0006}$
55. $\dfrac{0.00007 \times 0.0003}{60000}$
56. $\dfrac{0.00002 \times 0.00074}{14000}$
57. $\dfrac{0.00004 \times 60000}{0.003}$
58. $\dfrac{0.00003 \times 0.0005}{0.0000006}$
59. $\dfrac{0.0069}{0.006 \times 0.023}$

What is the degree of accuracy (number of significant figures) of each of the following?

60. 321
61. 507
62. 40.31
63. 41.30
64. 576.00
65. 65,000,000
66. 386,000
67. 902.00
68. 4,000,000.0
69. 4000.2
70. 0.650
71. 0.00631
72. 0.0494
73. 0.00060
74. 0.0003

REFERENCES

Dodes, Irving Allen and Greitzer, S. L., *Numerical Analysis.* New York, Hayden Book Company, Inc., 1964.

Keedy, M. L., and Bittinger, M. L., *Mathematics: A Modern Introduction.* Reading, Mass.: Addison-Wesley Publishing Company, Inc., 1970.

Richardson, Moses, *Fundamentals of Mathematics,* 3d ed. New York, The Macmillan Company, 1966.

Schaaf, William L., *Basic Concepts of Elementary Mathematics,* 2d ed. New York, John Wiley & Sons, Inc., 1965.

3 OTHER NUMBER BASES

3.1 INTRODUCTION / **67**
Early Computational Devices / **67**
Electrical Devices / **68**
Coding Methods / **69**
Exercise 3.1 / **71**

3.2 BASE 2 / **71**
Binary Numbers / **71**
Counting in Binary / **73**
Binary Fractions / **74**

3.3 BINARY AND DECIMAL NUMBER CONVERSION / **74**
Binary to Decimal Conversion—Method 1 / **74**
Decimal to Binary Conversion—Method 1 / **75**
Exercise 3.3 / **76**
Binary to Decimal Conversion—Method 2 / **76**
Decimal to Binary Conversion—Method 2 / **79**
Exercise 3.3 (Continued) / **81**
Conversion of Fractions / **81**
Exercise 3.3 (Continued) / **85**

3.4 BINARY ARITHMETIC / **85**
Addition / **85**
Subtraction / **87**
Multiplication / **88**
Exercise 3.4 / **89**

3.5 ARITHMETIC COMPLEMENTS / **89**
The Notion of a Complement / **89**
Binary Complements / **91**
Negative Numbers / **92**
Register Arithmetic / **93**
Words and Halfwords / **94**
Exercise 3.5 / **95**

3.6 OCTAL AND HEXADECIMAL NUMBERS / **95**
Correspondence with Binary and Decimal / **95**
Octal to Binary Conversion / **98**
Hexadecimal Conversion / **98**
Conversion to Decimal / **99**
Conversion From Decimal / **101**
Using a Conversion Table / **101**
Fractions / **101**
Exercise 3.6 / **104**

3.7 OCTAL AND HEXADECIMAL ARITHMETIC / **105**
Addition / **105**
Subtraction / **107**
Multiplication and Division / **108**
Exercise 3.7 **109**

3.1 INTRODUCTION

EARLY COMPUTATIONAL DEVICES. For centuries man has constructed devices to assist him with his computations. Undoubtedly one of the earliest of these is the abacus, which has been in use some 2000-3000 years yet is still commonly encountered today. Figure 3.1(a) is an illustration of a Japanese Soroban in which the number 2571851 has been stored. Each row of beads is used to represent one digit of a number; having eight such rows, this abacus can hold an eight digit number. Note the coding scheme which is used. Referring to any given string, each of the four lower beads can be used to represent 1 whereas the upper bead may represent 5. Beads which are positioned toward the inner bar are counted and those positioned away are not counted. The operation of adding another number, say 425, is easily accomplished by moving the required beads in each string with the result shown in Figure 3.1(b). Note that in the third string from the right (hundreds position), a carry resulted so beads were returned and the carry of one is propagated to the next string. This is a manual operation and must be carried out by the operator.

The actual mechanical carry did not come into being until 1642 when a 19-year-old French boy constructed the first true adding machine in hopes of alleviating the tiresome work of his accountant father. Blaise Pascal, later to become widely known in mathematics and physics, marked the digits 0 through

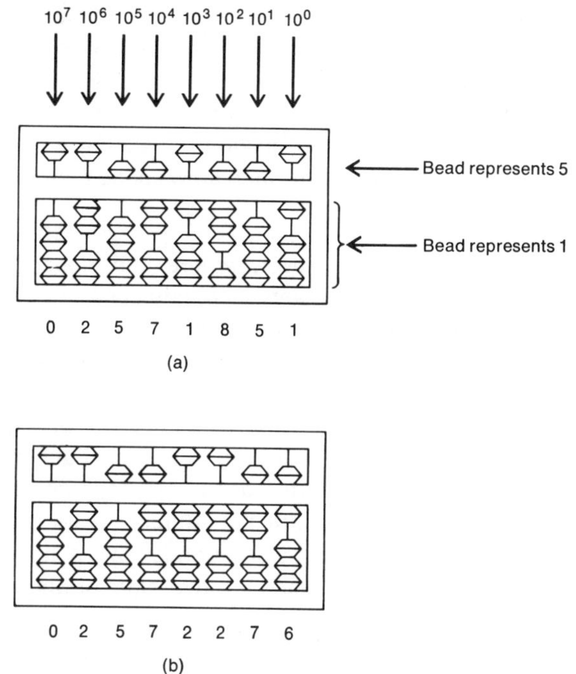

FIGURE 3.1 The Abacus

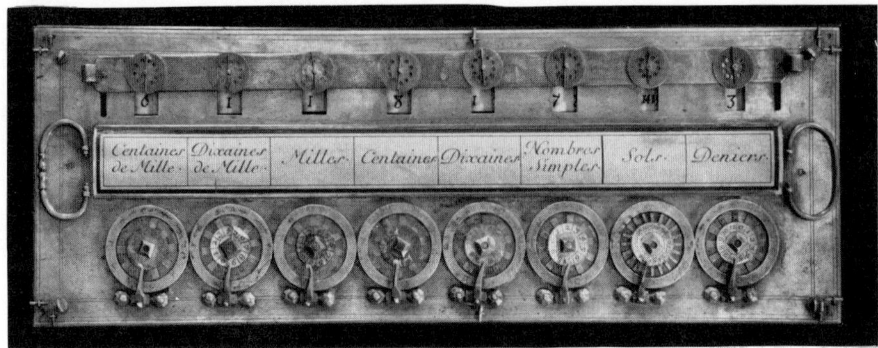

FIGURE 3.2 Pascal's adding machine (Photograph, Bettman Archive)

9 around the edges of circular dials. He then connected the dials by gears so that at the end of each revolution of a dial, a tooth would mesh with the gear connected to the next dial on the left and advance it $\frac{1}{10}$ revolution. Figure 3.2 shows Pascal's adding machine. Note that each wheel can be positioned at any of ten positions, thus representing one decimal digit. These same principles are used in such modern devices as adding machines and gas and electric meters; anyone who has driven an automobile has noted the odometer (mileage gage) which is fundamentally a Pascal type adding machine. Although these principles have been developed to a high degree of efficiency in present day electromechanical calculating devices, any mechanical device is severely restricted by its relatively low speed. For example, mechanical operations which take place in hundredths or thousandths of a second might appear to be very fast. But when considered in views of millions of such operations required in common data processing applications, such slow speed methods became quite impractical.

ELECTRICAL DEVICES. However, where the speed of mechanical devices is severely restricted by their very mechanical nature, electrical devices can be operated at much higher speeds.* Prior to and during World War II, much research and design was concentrated on the use of electronic devices for computations. Let us consider two means by which numeric information could be stored electronically. In Figure 3.3, an array of eight columns of indicators can be used to store an eight-digit number. Each column consists of nine elements, one to represent each of the nine nonzero digits. Note that we can think of these elements as electric light bulbs, electric switches, vacuum tubes, or whatever. The important consideration is that each has what we term an *on* state and an *off* state; that is, each is a *two-state* or **binary** device. As we

**Technically speaking, mechanical devices are partially limited by the inertia of the mechanical components. A similar "electrical-magnetic inertia" also limits operating speeds in electronic systems. In fact, beginning with the third generation of computers, the "slow" speed of electricity through a conductor became a design consideration.*

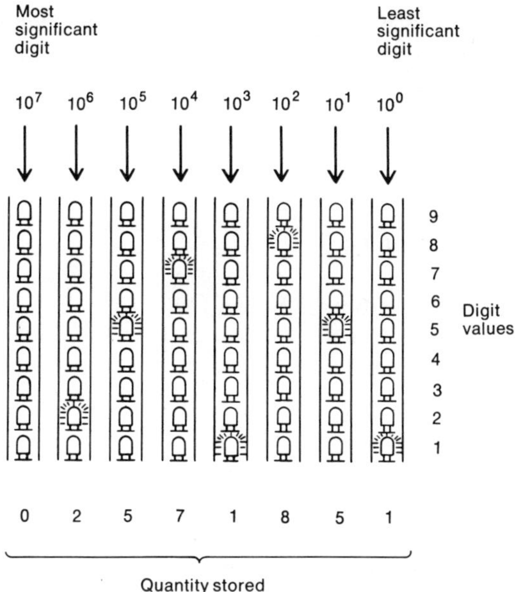

FIGURE 3.3 A storage device

can see in Figure 3.3, a decimal digit in any given digit position is represented by that element being energized or in the *on* state. The absence of an *on* element in a column indicates the digit zero. With this representation, it is apparent that the group of indicators in Figure 3.3 contains the number 2571851 (the same number as is stored in Figure 3.1(a)). By comparing Figures 3.1 and 3.3, we see that the two methods for storing information are very similar.

Although this method might appear simple from a conceptual point of view, the storage of large amounts of information with this method would be cumbersome. We can see that each eight-digit storage unit would require at least 72 electronic elements. The obvious solution to a person untrained in electronics is to use but one vacuum tube (or other electronic device) for each decimal digit and use different levels of electrical or magnetic state to represent the digits. Thus, for example, an electronic component when activated by one volt would store the digit 1, when activated by two volts, the digit 2, and so on. Although this might sound like a good idea, the problems of achieving a simple electronic device for use in computers which has ten separate stable states is practically insurmountable. As a result the two-state or binary element, such as that used in Figure 3.3, is currently in use and will continue to be used for the foreseeable future.

CODING METHODS. However, different schemes can be employed to more efficiently utilize the capacity of circuit components. For example, let us consider Figure 3.4 in which four elements are used to represent each digit. To illustrate this scheme, the number 12346789 is stored; the reader will note that in each digit position, the sum of the *on* elements is taken. Actually this is very similar to the technique used in the abacus (Figure 3.1) where the beads on one part of the abacus represent 5 and those on the other part each represent 1; the number stored is determined by totalling the beads in each string. The choice of 8, 4, 2 and 1 for row values is a logical one which can be justified as:

1. 1 and 2 are required
2. $1 + 2 = 3$
3. 4 is required to progress beyond 3
4. $1 + 4 = 5$
   $2 + 4 = 6$
   $1 + 2 + 4 = 7$
5. 8 is required to progress beyond 7
6. $1 + 8 = 9$

Obviously the 8 row could have been assigned any value in the range 5-8 while still providing the required capacity. However, as we shall see, the values 1, 2, 4, and 8 are the first four powers of two and are far more compatible with the binary nature of the computer. The coding method illustrated in Figure 3.4 is commonly used in modern computers and is called **binary coded decimal** (commonly referred to as **BCD**). The principle advantage of BCD is that numbers are effectively stored in the familiar decimal form—very convenient for human beings but inconvenient for design of computing machines. For example, consider an operation in which 32000 is added to the storage unit of Figure 3.4. Design of hardware to affect this function is relatively simple; the result, of course, would be 12378789. However, propagation of a carry becomes a problem since the capacity of each digit position is $8 + 4 + 2 + 1 = 15$. Thus the circuitry will involve special complexity to propagate a carry when any given position exceeds 9. This is analogous to the manual carry operation required on the abacus.

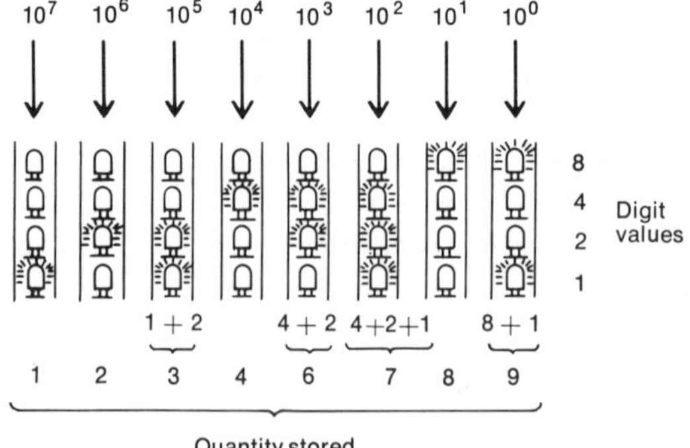

FIGURE 3.4 An improved storage device

We have now reached a logical point from which to begin a study of binary numbers, for which we will use the BCD illustrated by Figure 3.4 as a reference. Later in the chapter, our studies will be expanded to base 8 (octal) and to base 16 (hexadecimal). A base 16 number system is actually suggested by the preceding example. Remember that a base 10 number system exhibits 10 different states (that is, digits); each digit position in Figure 3.4 is capable of exhibiting 16 different states (values from 0-15). Hence it can conceivably be used to illustrate a base 16 number system.

EXERCISE 3.1

1. What is the advantage of using a two-state electronic device for storing information as opposed to a device capable of several states (such as 10)?
2. A variation of the binary coded decimal method for storing decimal information is the *excess-three* system. The four-bit code for each decimal digit is formed by adding three, then converting the result to binary. In other words 2 and 8 would be stored as $2 + 3 = 5 \rightarrow 0101$ and $8 + 3 = 11 \rightarrow 1011$. Make a table of the coding for the decimal digits 0 through 9. Compare the codings for 0 and 9, 1 and 8, 2 and 7, and so on. These are termed *complements*, and have important applications in computers.
3. Another code is the so-called *two-out-of-five* system which utilizes five binary type indicators to store each decimal digit. In this coding technique each decimal digit 1-9 is represented by two *on* bits. If three of the five bit values are 0, 1 and 2, what will be the other two bit values to achieve a two-out-of-five method? (That is, complete the following table through the digit 9.)

Decimal Digit	Two-of-five code				
	b	*a*	2	1	0
1	0	0	0	1	1
2	0	0	1	0	1
3	0	0	1	1	0

If there is more than one set of values which can be used for *a* and *b*, show all versions of the table.

3.2 BASE 2

BINARY NUMBERS. The rapid evolution of the computer field has involved the use of several number systems in computing equipment. For instance, the IBM 650 display panel uses a biquinary (basically a base 5 system), and the Univac uses an octal or base 8 system. However, arithmetic operations in these computers are done in a binary or base 2 system. Nearly all computers operate in the binary mode because of its ability to describe an electrical circuit as either being on or off. However, we should not get the idea that a base 2 number system is something new and unique to the computer. There is evidence that a binary system was recognized 5000 years ago in China and again 3000 years ago in Egypt. But it was only within the last 300 years that Baron Gottfried von Leibniz delved into the system and documented his work.

In our intuitive consideration of the binary coded decimal method for coding numbers, the term binary was used because of the two-state nature of the device. Actually the particular digit values chosen in Figure 3.4 have far greater implications than the preceding discussion tends to indicate. To illustrate, we will consider one column (digit position) "turned on its side" as shown in Figure 3.5. By the convention adopted in the preceding section, an *on* lamp

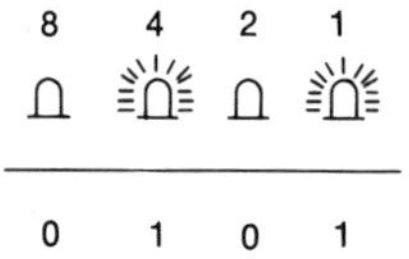

8	4	2	1
0	1	0	1

FIGURE 3.5 Binary coding

indicates that the value is to be included whereas an *off* lamp indicates that it should be excluded. Thus the number shown is $4 + 1 = 5$. Obviously, the same message can be conveyed by using the digit 1 to represent the *on* condition and 0 the *off*. Now it is possible to represent this value as 0101, but we must remember the *positional significance* of each 0 or 1. Mathematically this actually represents

$$\begin{aligned} 0 \times 8 + 1 \times 4 + 0 \times 2 + 1 \times 1 &= 0 + 4 + 0 + 1 \\ &= 5. \end{aligned}$$

One more careful observation and we have "arrived." That is, the digit values used here and in Figures 3.4 and 3.5 are actually the first four powers of 2, or

$$\begin{aligned} 1 &= 2^0, \\ 2 &= 2^1, \\ 4 &= 2^2, \\ 8 &= 2^3. \end{aligned}$$

Then the representation becomes

$$0 \times 8 + 1 \times 4 + 0 \times 2 + 1 \times 1 = 0 \times 2^3 + 1 \times 2^2 + 0 \times 2^1 + 1 \times 2^0.$$

Note that this is precisely the polynomial type form which our familiar base 10 number system exhibits. Thus we have established an intuitive basis for further consideration of a base 2 number system. That is, the system incorporates the two distinct digits 0 and 1, and utilizes the principle of positional value which is based on powers of the number system base 2. However, we must take care to distinguish between a binary and a decimal number. For example, is 101 the decimal quantity "one hundred one" or the binary quantity (without its leading 0) which is equivalent to the decimal quantity 5? The standard practice is to use subscripts whenever this is not clear by context. Then the binary quantity 101_2 is easily distinguished from the decimal quantity 101_{10}.

COUNTING IN BINARY. To gain more insight into the nature of binary numbers, consider Table 3.1 which shows binary quantities and their decimal equivalents.

TABLE 3.1 BASE 2 AND BASE 10 NUMBERS

Base 2					Base 10
2^3	2^2	2^1	2^0		
8	4	2	1		
0	0	0	0	→ 0000	0
0	0	0	1	→ 0001	1
0	0	1	0	→ 0010	2
0	0	1	1	→ 0011	3
0	1	0	0	→ 0100	4
0	1	0	1	→ 0101	5
0	1	1	0	→ 0110	6
0	1	1	1	→ 0111	7
1	0	0	0	→ 1000	8
1	0	0	1	→ 1001	9
1	0	1	0	→ 1010	10
1	0	1	1	→ 1011	11
1	1	0	0	→ 1100	12
1	1	0	1	→ 1101	13
1	1	1	0	→ 1110	14
1	1	1	1	→ 1111	15

We can see that these both represent "counting sequences" (from 0 through 15). Of course, when counting with decimal numbers we "use up" all of the digits after reaching 9 so we start the familiar cycle again, that is, 0, 1, 2, 3, 4, 5, 6, 7, 8, 9, 10, 11, 12, ···. In counting with a binary system, a similar situation arises except that we use up all of the digits much more rapidly. With the digits 0 and 1 we can see in Table 3.1 that the repetition begins with the binary equivalent of 2_{10}. A careful study of this table will show that, effectively, the same technique is involved in binary counting as in decimal counting.

For convenience, this exploration of binary numbers has focused on quantities consisting of four digits. Since they are binary as opposed to decimal, they are commonly referred to as **binary digits** or **bits**. However, we must not get the impression that binary quantities are always limited to four binary digits. For instance, the quantities 10110_2 and 10011101_2 actually represent

$$\begin{aligned}10110_2 &= 1 \times 2^4 + 0 \times 2^3 + 1 \times 2^2 + 1 \times 2^1 + 0 \times 2^0\\ &= 1 \times 16 + 0 \times 8 + 1 \times 4 + 1 \times 2 + 0 \times 1\\ &= 16 + 4 + 2\\ &= 22_{10}.\end{aligned}$$

$$\begin{aligned}10011101_2 &= 1 \times 2^7 + 0 \times 2^6 + 0 \times 2^5 + 1 \times 2^4 + 1 \times 2^3 + 1 \times 2^2 + 0 \times 2^1 + 1 \times 2^0\\ &= 1 \times 128 + 0 \times 64 + 0 \times 32 + 1 \times 16 + 1 \times 8 + 1 \times 4 + 0 \times 2 + 1 \times 1\\ &= 128 + 16 + 8 + 4 + 1\\ &= 157_{10}.\end{aligned}$$

Because large binary numbers with their strings of 1s and 0s are often confusing to read, they are commonly grouped in threes or in fours as follows:

number	in fours	in threes
10110	1 0110 or 0001 0110 or	10 110 010 110
10011101	1001 1101	010 011 101

As we shall learn later in this chapter, these groupings are related to the base 8 and base 16 number systems, respectively, in a very important way and essentially provide the basis for computers being classified as base 8 or base 16 machines.

BINARY FRACTIONS. In Chapter 2 we studied decimal numbers of the form 23.752. Furthermore we noted that such numbers could be expressed in a pseudopolynomial form by making use of negative exponents, that is,

$$23.752 = 2 \times 10^1 + 3 \times 10^0 + 7 \times 10^{-1} + 5 \times 10^{-2} + 2 \times 10^{-3}$$

Similarly, a binary fraction form is usually written in the form 101.1101. This number may be expanded and converted to its decimal form by the usual method, which yields

$$\begin{aligned} 101.1101_2 &= 1 \times 2^2 + 0 \times 2^1 + 1 \times 2^0 + 1 \times 2^{-1} + 1 \times 2^{-2} + 0 \times 2^{-3} + 1 \times 2^{-4} \\ &= 1 \times 4 + 0 \times 2 + 1 \times 1 + 1 \times \frac{1}{2} + 1 \times \frac{1}{4} + 0 \times \frac{1}{8} + 1 \times \frac{1}{16} \\ &= 4 + 1 + \frac{1}{2} + \frac{1}{4} + \frac{1}{16} \\ &= 4 + 1 + 0.5 + 0.25 + 0.0625 \\ &= 5.8125_{10}. \end{aligned}$$

In other words,

$$101.1101_2 = 5.8125_{10}.$$

The so called *binary point* used to indicate fractional quantities in binary is analogous to the decimal point which we commonly use with decimal numbers.

3.3 BINARY AND DECIMAL NUMBER CONVERSION

BINARY TO DECIMAL CONVERSION—METHOD 1. Occasionally the programmer will find it necessary to convert binary representations from the machine to decimal, or conversely, to convert a decimal quantity to binary. From the preceding section, we recall that conversion from binary to decimal can easily be performed by expanding the binary number in the powers of its base. That is,

$$101101_2 = 1 \times 2^5 + 0 \times 2^4 + 1 \times 2^3 + 1 \times 2^2 + 0 \times 2^1 + 1 \times 2^0$$

$$= 32 + 8 + 4 + 1$$

$$= 45_{10}.$$

$$11.011_2 = 1 \times 2^1 + 1 \times 2^0 + 0 \times 2^{-1} + 1 \times 2^{-2} + 1 \times 2^{-3}$$

$$= 2 + 1 + \frac{1}{4} + \frac{1}{8}$$

$$= 3.375_{10}.$$

This is a convenient method to remember and to use since it relies on the basic characteristic of the number systems we have studied and it involves no "gimmicks." When converting whole numbers, it is usually a simple matter to scan the binary number from right to left, writing down appropriate powers in the process. Furthermore, this method has the advantage that it works equally well for binary integers and fractions. Another technique which we shall consider requires separate treatment of the integer and fractional parts of a number.

DECIMAL TO BINARY CONVERSION—METHOD 1. To perform the inverse operation, that is, to convert from base 10 to base 2, it is convenient to write out a table of powers of two. For instance converting 93_{10} is readily converted by reference to the following table (from Table 3.2):

$$\begin{array}{ll} 2^0 = 1 & 2^4 = 16 \\ 2^1 = 2 & 2^5 = 32 \\ 2^2 = 4 & 2^6 = 64 \\ 2^3 = 8 & 2^7 = 128 \end{array}$$

Since the number to be converted is between 2^6 (64) and 2^7 (128), six will be the largest power of base 2 required. The process now consists merely of breaking the number 93 into a sum of powers of two with the largest being 64. This gives

$$\begin{array}{rl} 93 & \\ \underline{-64} & 2^6 \\ 29 & \\ \underline{-16} & 2^4 \\ 13 & \\ \underline{-8} & 2^3 \\ 5 & \\ \underline{-4} & 2^2 \\ 1 & 2^0 \end{array}$$

or

$$93 = 64 + 16 + 8 + 4 + 1 = 2^6 + 2^4 + 2^3 + 2^2 + 2^0.$$

In order to include all powers of two from zero through six, this would be written

$$93 = 1 \times 2^6 + 0 \times 2^5 + 1 \times 2^4 + 1 \times 2^3 + 1 \times 2^2 + 0 \times 2^1 + 1 \times 2^0.$$

Thus,

$$93_{10} = 1011101_2.$$

Once we understand the basic nature of the number systems with which we are dealing, conversion from one system to another is a straightforward, although somewhat cumbersome, task. To facilitate conversion, both to and from decimal, a number of techniques have been developed. The two which we shall study (one to convert from binary to decimal and the other from decimal to binary) are commonly used by programmers.

EXERCISE 3.3

Using the method of the preceding sections, convert the following decimal numbers to binary.

1. 0, 4, 9, 27, 63, 64, 91, 593.
2. 1, 8, 13, 36, 127, 128, 129, 427.

Using the method of the preceding sections, convert the following binary numbers to decimal.

3. 1, 10, 101, 1111, 10000, 10001, 1011010.
4. 0, 11, 111, 11101, 100000, 11111, 100001, 1010101.
5. Why is it meaningless to speak of the binary number 102_2?

BINARY TO DECIMAL CONVERSION—METHOD 2. As we have seen, the decimal equivalent of a binary number is easily obtained by direct expansion in powers of 2. A shortcut method which usually can be performed mentally consists of the procedure illustrated by the following example.

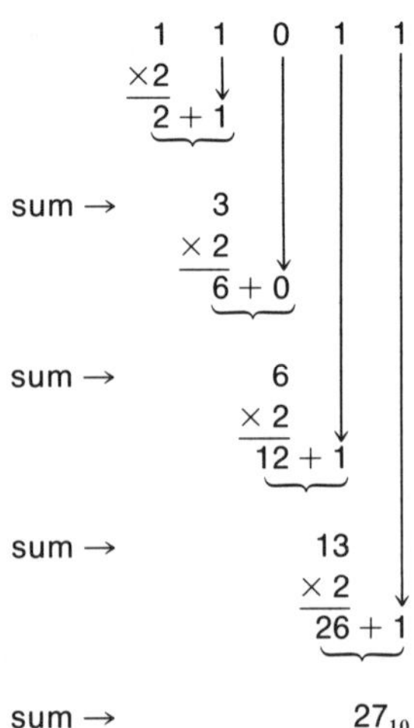

Note that the number 11011_2 (equivalent to 27_{10}) is converted to decimal according to the following rule.

Double the leftmost binary digit and add it to the next digit (progressing right). Double this sum and add it to the next digit. Repeat this process until the rightmost digit has been added. The resulting sum is the decimal equivalent of the binary number.

The general procedure is further illustrated by the flowchart of Figure 3.6. The "artwork" of the preceding example can be dispensed with as shown with the following conversions of 110101_2 and 11101_2.

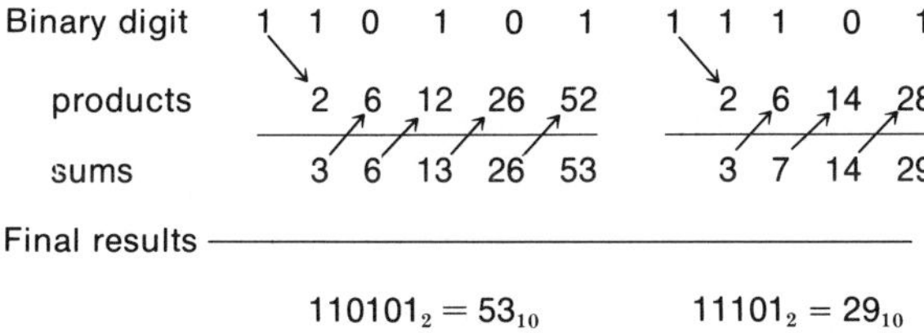

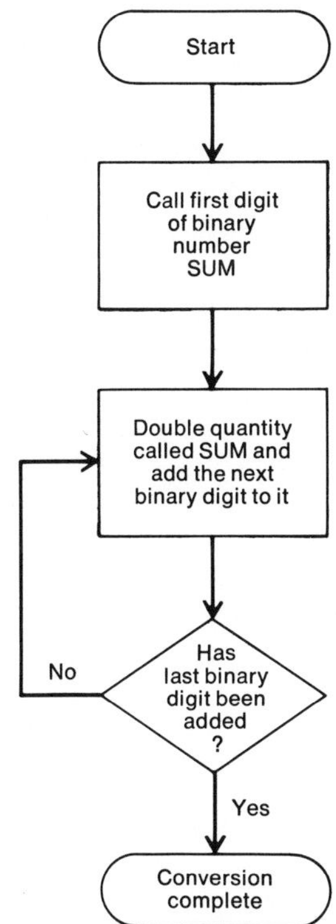

FIGURE 3.6 Conversion from binary to decimal

With a little practice, the average person can convert from binary to decimal in his head with little problem. However, the beginner is often confused and sometimes cannot remember whether to start the operation with the leftmost or rightmost digit. An intuitive aid is to remember that the leftmost digit represents the greatest power of two and thus should logically be "involved in" more doubling than any of the other digits. Actually, this method is easily justified as follows:

> *The basic rules for multiplying quantities with exponents,* $(x^a \cdot x^b = x^{a+b})$ *as described in Chapter 2 can be used in reverse to factor such quantities.**

For example,

(a)

$$2^4 = 2^1 \cdot 2^3 = 2 \cdot 2^3.$$

(b)

$$2^2 + 2^1 = 2^1(2^1 + 2^0) = 2(2+1).$$

(c)

$$\begin{aligned} 2^3 + 2^2 + 2^1 + 2^0 &= (2^3 + 2^2 + 2^1) + 2^0 \\ &= 2^1[2^2 + 2^1 + 2^0] + 2^0 \\ &= 2[2^2 + 2^1 + 1] + 1. \end{aligned}$$

Within the brackets of the last example, the $2^2 + 2^1$ terms may further be factored with the following result:

$$\begin{aligned} &= 2[2^1(2^1 + 2^0) + 1] + 1 \\ &= 2[2(2+1) + 1] + 1. \end{aligned}$$

Applying this principle to the expansion of 110101_2 yields the following interesting results.

$110101_2 = 1 \times 2^5 + 1 \times 2^4 + 0 \times 2^3 + 1 \times 2^2 + 0 \times 2^1 + 1 \times 2^0$	factor 2 from shaded portion
$= 2(1 \times 2^4 + 1 \times 2^3 + 0 \times 2^2 + 1 \times 2^1 + 0 \times 2^0) + 1$	factor 2 from shaded portion
$= 2(2(1 \times 2^3 + 1 \times 2^2 + 0 \times 2^1 + 1 \times 2^0) + 0) + 1$	factor 2 from shaded portion
$= 2(2(2(1 \times 2^2 + 1 \times 2^1 + 0 \times 2^0) + 1) + 0) + 1$	factor 2 from shaded portion
$= 2(2(2(2(1 \times 2^1 + 1 \times 2^0) + 0) + 1) + 0) + 1$	

Now we can evaluate the expression by beginning the arithmetic operations within the innermost parentheses.

$$= 2(2(2(2\underbrace{(1 \times 2 + 1)}_{3} + 0) + 1) + 0) + 1.$$

Note the digits of the original binary number double and add.

$$= 2(2(2\underbrace{(2 \times 3 + 0)}_{6} + 1) + 0) + 1$$

double and add

$$= 2(2\underbrace{(2 \times 6 + 1)}_{13} + 0) + 1$$

double and add

$$= 2\underbrace{(2 \times 13 + 0)}_{26} + 1$$

*For the reader with a weak background in basic algebra, it may be desirable to review Chapter 5 before considering the following derivations.

$$= 2 \times 26 + 1 \qquad \text{double and add}$$
$$= 53_{10}$$

Although it would be incorrect to consider this a mathematical proof of the double-and-add conversion method, it certainly tends to substantiate its validity. (Remember that in mathematics a general proof cannot be made using a particular example.) The method used, however, is general in nature and can be used to derive equivalent techniques for converting from a number in any given base to base 10 (or other base, if necessary).

DECIMAL TO BINARY CONVERSION—METHOD 2. The following alternate conversion method from decimal to binary involves successive division by 2 and is essentially the inverse of the preceding binary to decimal conversion method.

> *Repeatedly divide the decimal number by two, saving the remainders which may be 0 or 1, until a quotient of 0 is obtained. The equivalent binary number is composed of the remainders, the first remainder being the rightmost digit and the last remainder being the leftmost digit.*

For example, the numbers 55_{10} and 216_{10} are converted as follows:

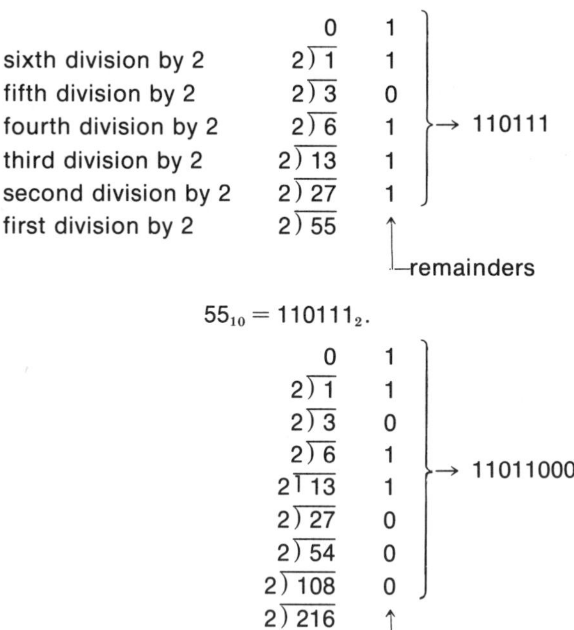

$$55_{10} = 110111_2.$$

$$216_{10} = 11011000_2.$$

Here again this simple technique is often confusing to the beginner who simply memorizes the steps then forgets whether the first remainder is the leftmost or rightmost significant digit of the resulting binary number. A branch of mathematics called number theory provides a simple method of proof for this technique which we shall apply to the conversion of 55_{10}. Although we might not know the binary equivalent of 55_{10}, we do know that it can be expressed as follows:

$$55_{10} = a \times 2^5 + b \times 2^4 + c \times 2^3 + d \times 2^2 + e \times 2^1 + f \times 2^0.$$

Here a, b, c, d, e, and f represent the bits, which may be 0 or 1, of the binary number. Even though we do not know the values of these unknowns, we can still convert this representation to the factored form used earlier.

$$55_{10} = 2(2(2(2(a \times 2 + b) + c) + d) + e) + f.$$

Now remember, our algorithm for performing the conversion involves dividing the decimal number by two and saving the remainder (as the rightmost digit of the equivalent binary number).

$$\frac{55}{2} = 27;\ \text{remainder of } 1.$$

Since, in the preceding general equation, the right hand side of the equation is simply "another way of expressing" 55_{10}, we shall be able to divide it by 2 and achieve the same result; that is,

$$\frac{2(2(2(2(a \times 2 + b) + c) + d) + e) + f}{2} = 2(2(2(a \times 2 + b) + c) + d) + e;\ \text{remainder of } f.$$

But by number theory we can conclude that

$$\begin{aligned} 27 &= 2(2(2(a \times 2 + b) + c) + d) + e \quad &&\text{(whole number portion)},\\ 1 &= f &&\text{(remainder)} \end{aligned}$$

In other words f, which represents the rightmost bit, is the equal to the remainder resulting from division of the original decimal number. Continuing, we find

$$\frac{27}{2} = 13;\ \text{remainder of } 1.$$

$$\frac{2(2(2(a \times 2 + b) + c) + d) + e}{2} = 2(2(a \times 2 + b) + c) + d;\ \text{remainder of } e,$$

thus $e = 1$.

$$\frac{13}{2} = 6;\ \text{remainder of } 1.$$

$$\frac{2(2(a \times 2 + b) + c) + d}{2} = 2(a \times 2 + b) + c;\ \text{remainder of } d,$$

thus $d = 1$.

$$\frac{6}{2} = 3;\ \text{remainder of } 0.$$

$$\frac{2(a \times 2 + b) + c}{2} = a \times 2 + b;\ \text{remainder of } c,$$

thus $c = 0$.

$$\frac{3}{2} = 1; \text{ remainder of } 1.$$

$$\frac{a \times 2 + b}{2} = a; \text{ remainder of } b,$$

thus $b = 1$.

$$\frac{1}{2} = 0; \text{ remainder of } 1.$$

$$\frac{a}{2} = 0; \text{ remainder of } a,$$

thus $a = 1$.

Summarizing, we have

$$\begin{aligned} a &= 1, \\ b &= 1, \\ c &= 0, \\ d &= 1, \\ e &= 1, \\ f &= 1, \end{aligned}$$

or

$$\begin{aligned} 55_{10} &= 1 \times 2^5 + 1 \times 2^4 + 0 \times 2^3 + 1 \times 2^2 + 1 \times 2^1 + 1 \times 2^0 \\ &= 110111_2. \end{aligned}$$

Although this justification of the remainder conversion technique uses a specific value (55), it is easily generalized to prove the conversion method for any two number bases, as well as base 10 or base 2.

EXERCISE 3.3 (Continued)

Using the method of the preceding sections, convert the following decimal numbers to binary.

6. 0, 4, 9, 27, 63, 64, 91, 593.
7. 1, 8, 13, 36, 127, 128, 129, 427.

Using the method of the preceding sections, convert the following binary numbers to decimal.

8. 1, 10, 101, 1111, 10000, 10001, 1011010.
9. 0, 11, 111, 11101, 100000, 11111, 100001, 1010101.
10. Draw a flowchart to illustrate the necessary steps in converting from decimal to binary using the remainder technique.

CONVERSION OF FRACTIONS. In converting binary fractions to their decimal equivalents, effectively the same principle which is used with integers is involved. However, its actual implementation is much more cumbersome.

Usually the expansion of the powers of two will give the results much more easily, for example,

$$\begin{aligned} .1011_2 &= 1 \times 2^{-1} + 0 \times 2^{-2} + 1 \times 2^{-3} + 1 \times 2^{-4} \\ &= 1 \times \frac{1}{2} + 0 \times \frac{1}{4} + 1 \times \frac{1}{8} + 1 \times \frac{1}{16} \\ &= 0.5 + 0.125 + 0.0625 \\ &= .6875_{10}. \end{aligned}$$

If more than a few significant bits are involved, then a table of the decimal equivalents of negative powers of 2, such as that shown in Table 3.2, is useful.

TABLE 3.2 POWERS OF TWO

2^n	n	2^{-n}
1	0	1.0
2	1	0.5
4	2	0.25
8	3	0.125
16	4	0.062 5
32	5	0.031 25
64	6	0.015 625
128	7	0.007 812 5
256	8	0.003 906 25
512	9	0.001 953 125
1 024	10	0.000 976 562 5
2 048	11	0.000 488 281 25
4 096	12	0.000 244 140 625
8 192	13	0.000 122 070 312 5
16 384	14	0.000 061 035 156 25
32 768	15	0.000 030 517 578 125
65 536	16	0.000 015 258 789 062 5
131 072	17	0.000 007 629 394 531 25
262 144	18	0.000 003 814 697 265 625
524 288	19	0.000 001 907 348 632 812 5
1 048 576	20	0.000 000 953 674 316 406 25
2 097 152	21	0.000 000 476 837 158 203 125
4 194 304	22	0.000 000 238 418 579 101 562 5
8 388 608	23	0.000 000 119 209 289 550 781 25
16 777 216	24	0.000 000 059 604 644 775 390 625
33 554 432	25	0.000 000 029 802 322 387 695 312 5
67 108 864	26	0.000 000 014 901 161 193 847 656 25
134 217 728	27	0.000 000 007 450 580 596 923 828 125
268 435 456	28	0.000 000 003 725 290 298 461 914 062 5
536 870 912	29	0.000 000 001 862 645 149 230 957 031 25
1 073 741 824	30	0.000 000 000 931 322 574 615 478 515 625
2 147 483 648	31	0.000 000 000 465 661 287 307 739 257 812 5

Using this, we can quickly convert a binary number such as 010110111.

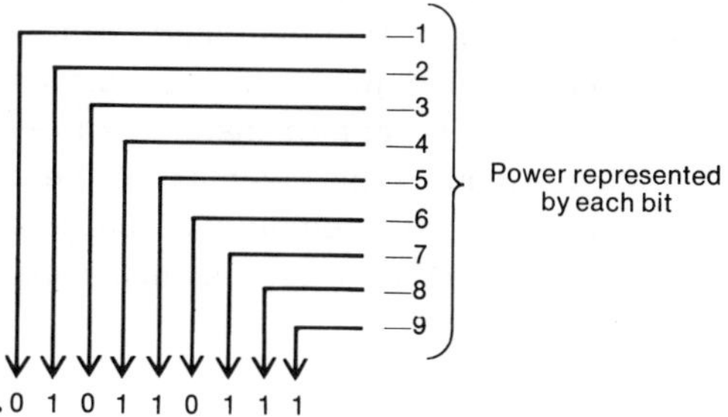

Therefore

$$
\begin{aligned}
0.010110111_2 = \;& \,0.25 \\
&+ 0.0625 \\
&+ 0.03125 \\
&+ 0.0078125 \\
&+ 0.00390625 \\
&+ 0.001953125 \\
\hline
& \,0.357421875_{10}
\end{aligned}
$$

That is,

$$0.010110111_2 = 0.357421875_{10}.$$

The mathematical theory used to evolve the successive division technique for converting decimal integers to binary can also be applied to yield the successive multiplication method for converting decimal fractions to binary. Although a justification for the technique is not included in this book, the method is as follows.

> *Multiply the decimal fraction by the number system base (2 in this case). The resulting integer part forms the leftmost bit of the desired binary result. Repeat the multiplication using the fractional result from the preceding multiplication; the integer part of the process is the next binary digit. Continue this process until the desired accuracy is obtained.*

For example, consider converting 0.828125_{10} to binary.

Decimal fraction	× 2 = Product	Integer part	
0.828125	× 2 = 1.65625	1	resulting binary fraction
0.65625	× 2 = 1.3125	1	
0.3125	× 2 = 0.625	0	
0.625	× 2 = 1.25	1	
0.25	× 2 = 0.50	0	
0.50	× 2 = 1.00	1	

therefore

$$0.828125_{10} = 0.110101_2.$$

The reader will note that this decimal fraction is conveniently converted to an exact binary equivalent. This occurred because the decimal quantity .828125 was "constructed" (by the authors) from the binary form in order to simplify the first example. In actual practice, this seldom occurs and the binary form resulting from the conversion is an approximation expressed to a predetermined number of significant digits. (This is similar to the problem discussed in Chapter 2 of representing 1/3 in the computer as .33333333—an exact decimal representation is impossible.) To illustrate this, consider the innocent appearing decimal fraction 0.1_{10}.

Decimal fraction	× 2 =	Product	Integer part	
0.1	× 2	0.2	0	
0.2	× 2	0.4	0	Repeating sequence
0.4	× 2	0.8	0	
0.8	× 2	1.6	1	
0.6	× 2	1.2	1	
0.2	× 2	0.4	0	Repeating sequence
0.4	× 2	0.8	0	
0.8	× 2	1.6	1	
0.6	× 2	1.2	1	
0.2	× 2	0.4	0	Repeating sequence
0.4	× 2	0.8	0	
0.8	× 2	1.6	1	
0.6	× 2	1.2	1	

Therefore

$$0.1_{10} \simeq 0.0001100110011_2.$$

That the binary equivalent of 0.1_{10} is a never ending binary decimal is obvious by referring to the conversion where successive multiplications are shown grouped in fours to illustrate the repeating nature of the conversion. This illustrates a basic problem within the computer when storing decimal fractions in binary fraction form. That is, let us assume that our machine has internal capacity for storing the number of digits shown in this example. Then the actual decimal quantity stored is not 0.1_{10}, but less, as determined using Table 3.2.

$$\begin{aligned} 0.0001100110011_2 = {} & \ 0.0625 \\ & + 0.03125 \\ & + 0.00390625 \\ & + 0.001953125 \\ & + 0.000244140625 \\ & \underline{+ 0.0001220703125} \\ & \ 0.0999755859375_{10} \end{aligned}$$

Although this discrepancy is usually of little concern to the engineer (who uses a safety factor of 3 or 4 times in his design), consider the accountant who must determine the total cost of 1000 items each costing \$0.10. Using this binary approximation, he comes up with a total of $99.97 (or 99.98). When ledgers must balance to the nearest penny, this minor error is of major significance.* As a result, prior to the third generation of computers, machines were generally designed as binary computers for math-science applications or decimal (binary coded decimal) for business applications. However, most third generation computers are designed to accommodate both binary arithmetic operations for scientific calculations and decimal arithmetic operations for business calculations. For instance, the IBM System/360 includes a special set of binary floating-point instructions and a separate set of decimal instructions.

EXERCISE 3.3 (Continued)

Convert the following binary quantities to decimal:

11. 0.11, 0.10011, 0.10101, 0.11001111, 0.00011101
12. 0.1, 0.1111, 0.0011011, 0.10101110, 0.0001010101
13. 11.01, 101.101, 11100.00111, 1.01101, 1101101.101101
14. Draw a flowchart to illustrate the necessary steps in converting a decimal fraction to a binary fraction using the multiplication technique of this chapter.

Convert the following decimal quantities to binary. (Carry 12 bits to the right of the binary point for approximations.)

15. 0.5, 0.3125, 0.661875, 0.2, 0.124, 0.007165
16. 0.75, 0.59375, 0.1015625, 0.6, 0.444, 0.000673
17. 24.375, 573.5078125, 682.138, 63.23
18. 0.75, 3.0, 13.78125, 0.7, 22.3, 19.833, 6.26
19. 0.375, 5.0, 15.2578125, 0.9, 22.6, 17.314, 9.38

3.4 BINARY ARITHMETIC

ADDITION. Familiar relationships for adding base 10 numbers which are learned in elementary school are summarized in Table 3.3. The sum of any

TABLE 3.3 ADDITION TABLE FOR BASE 10

+	0	1	2	3	4	5	6	7	8	9
0	0	1	2	3	4	5	6	7	8	9
1	1	2	3	4	5	6	7	8	9	$^{1}0$
2	2	3	4	5	6	7	8	9	$^{1}0$	$^{1}1$
3	3	4	5	6	7	8	9	$^{1}0$	$^{1}1$	$^{1}2$
4	4	5	6	7	8	9	$^{1}0$	$^{1}1$	$^{1}2$	$^{1}3$
5	5	6	7	8	9	$^{1}0$	$^{1}1$	$^{1}2$	$^{1}3$	$^{1}4$
6	6	7	8	9	$^{1}0$	$^{1}1$	$^{1}2$	$^{1}3$	$^{1}4$	$^{1}5$
7	7	8	9	$^{1}0$	$^{1}1$	$^{1}2$	$^{1}3$	$^{1}4$	$^{1}5$	$^{1}6$
8	8	9	$^{1}0$	$^{1}1$	$^{1}2$	$^{1}3$	$^{1}4$	$^{1}5$	$^{1}6$	$^{1}7$
9	9	$^{1}0$	$^{1}1$	$^{1}2$	$^{1}3$	$^{1}4$	$^{1}5$	$^{1}6$	$^{1}7$	$^{1}8$

** To avoid this problem when using a binary machine, binary fractions are not used. Thus the monetary amount $1.25 would be stored as 125 cents or, in some cases as 1250 mils, where a mil is 1/10 of a cent. It then becomes the responsibility of the programmer to take into account decimal point placement.*

two digits, for example 7 and 6, is readily found by locating 7 in the extreme left column and 6 in the top row and then reading down and across, as indicated by the shaded area. The sum (3 with a carry of 1) is located at the intersection of the indicated row and column.

Memorizing such an addition table would have been considerably easier had the number system been base 2, as is evident by inspecting Table 3.4.

TABLE 3.4 BINARY ADDITION TABLE

+	0	1
0	0	1
1	1	$^{1}0$

The operation of adding two binary numbers is much the same as adding two base 10 numbers. For example, consider the addition of the two decimal numbers 392 and 469.

```
 1 1    carries
 392    addend
 469    augend
 ---
 861    sum
```

Whenever the sum of a pair of digits exceeds 9, only the units digit is recorded and a high-order carry of 1 is propagated to the next position. The addition process in binary arithmetic is essentially the same, as is illustrated by the binary numbers 11001 and 1011.

```
 11  11     carries
 011001     addend
   1011     augend
 ------
 100100     sum
```

Converting each of these to base 10 gives the following check:

$$11001_2 = 25 \qquad 25 + 11 = 36.$$
$$1011_2 = 11$$
$$100100_2 = 36.$$

Frequently, in adding binary numbers, carries will be propagated to nearly every position, as is evident in the following addition.

```
 11111111       carries
 1001100111     addend
   10111011     augend
 ----------
 1100100010     sum
```

Two simple rules of thumb are apparent from this example. First, if a column including the carry consists of an odd number of 1 digits, a 1 should be placed in the sum position of that column; otherwise a 0 should be recorded. Second, if there are two or three 1 digits (including a carry) in a column, a carry should be placed in the next column. Although binary addition is sufficiently simple that memorization of such rules is of little or no value, techniques similar to this are used in computer arithmetic units to simplify electronic equipment.

SUBTRACTION. The process of subtraction in binary arithmetic, although not actually difficult, often causes the beginner more difficulty than addition. First, let us consider Table 3.4 in the following form:

$$\begin{aligned} 0+0&=0\\ 1+0&=1\\ 0+1&=1\\ 1+1&=10 \end{aligned}$$

Recalling that subtraction is the inverse operation of addition leads to the following subtraction rules:

$$\begin{aligned} 0-0&=0\\ 1-0&=1\\ 1-1&=0\\ 10-1&=1. \end{aligned}$$

The difference $0-1$ is normally handled by the familiar process of borrowing from the next column to the left, giving $10-1=1$. Although the term "borrowing" is commonly used, we will use it reluctantly only for a lack of a better word. We will be borrowing with no intention of paying back.

To illustrate subtraction, let us consider the following examples:

$$\begin{array}{rlr} & \text{borrowing} & {}^{1\,12} \\ 10110 & \text{minuend} & \not{2}\,\not{2} \\ -\quad 100 & \text{subtrahend} & -\ 4 \\ \hline 10010 & \text{difference} & 18 \end{array}$$

Here the binary subtraction is straightforward; it is not necessary to borrow from an adjacent column. In subtracting the equivalent decimal numbers it is necessary to borrow in the units position, where 4 is subtracted from 12, thus reducing 2 in the tens position to 1. Another example is

$$\begin{array}{rlr} {}^{0\ 10}\qquad & \text{borrowing} & {}^{1\,12} \\ \not{1}\,\not{0}\,1\,1\,0 & \text{minuend} & \not{2}\,\not{2} \\ -\ 1000 & \text{subtrahend} & -\ 8 \\ \hline 1110 & \text{difference} & 14 \end{array}$$

In both the binary and base 10 subtractions, it is necessary to borrow from the column on the left when the subtrahend digit exceeds the minuend digit.

An example of even greater complexity is

$$\begin{array}{rlr} {}^{0\ 10\ 0\ 1010} & \text{borrowing} & \\ \not{1}\,\not{0}\,\not{1}\,\not{1}\,\not{0} & \text{minuend} & 22 \\ -1011 & \text{subtrahend} & -11 \\ \hline 1011 & \text{difference} & 11 \end{array}$$

The procedure becomes more complex in the second and third binary columns, since, after borrowing from the second column, the digit becomes zero and a borrow from the third column is also required. The extreme case is shown below, where it is necessary to borrow from each position:

```
 0 1 1 1 10   borrowing
 1̸ 0̸ 0̸ 0̸ 0̸    minuend         16
 − 1 0 1 1    subtrahend     −11
 ---------                   ---
     1 0 1    difference       5
```

MULTIPLICATION. The operations of multiplication, division, and obtaining square root are considerably easier in the binary than in the decimal system. Although we will not dwell upon these operations, it will be useful to consider multiplication very briefly. For many of us, the process of learning multiplication tables in our grade school studies was an unpleasant one. The use of a binary system in place of decimal would have made this task much easier because of the limited number of combinations. With only 1 and 0, the multiplication table appears as shown in Table 3.5.

TABLE 3.5 BINARY MULTIPLICATION TABLE

×	0	1
0	0	0
1	0	1

The multiplication operation then becomes a yes or no situation, which is nothing more than a series of successive additions, as shown in the following example:

```
   101101       45
      101        5
   ------      ---
   101101      225
  000000
 101101
 --------
 11100001
```

If the binary multiplier is large, it is usually convenient to perform the multiplication by summing partial products in order to avoid excessive propagation of carries. For example,

```
     1010111
       11011
     -------
     1010111
    1010111
   ---------
   100000101   first partial product
  10101110
  ----------
  1110111101   second partial product
 1010111
 ------------
 100100101101  final product
```

EXERCISE 3.4

Add the following binary numbers:

1. 101
10
2. 1011
110
3. 1101
101
4. 1101
1011
5. 11101
11
6. 100111
111100
7. 11111
1
8. 1111111
1
9. 101101
011101
10. 100010
011101
11. 100
10
11
12. 101
110
101

Perform the following binary subtractions:

13. 111
−10
14. 1110
−101
15. 11011
−1111
16. 11111
−10101
17. 10100
−1011
18. 10000
−11

Find the product of the following binary numbers:

19. 11011
101
20. 10110
11
21. 11011
111
22. 111001
1011
23. 101011
11011
24. 110001
1111

By closely examining the method of long division with decimal numbers, devise a means for dividing binary numbers and perform the following divisions. (Note the simplicity of long division with binary numbers.)

25. 10) 100110
26. 11) 11011
27. 101) 1000001
28. 111) 1001101
29. 10) 110011
30. 101) 1000010
31. 111) 1001111
32. 110) 1000111

3.5 ARITHMETIC COMPLEMENTS

THE NOTION OF A COMPLEMENT. When writing signed decimal numbers, we distinguish between a negative quantity and a positive quantity by using the minus sign (−) or the plus sign (+). Although this is a convenient device for people, it is not very desirable for use with machines. In fact, negative quantities are normally stored in their *complement* form. Furthermore, the operation of subtraction normally involves the use of complements. Although a given number can have many different types of complements, one of them is the difference between the number and the next higher power of the base. For example, in the decimal system the complement of 159 is 1000 − 159, which equals 841. Another type of complement is obtained by subtracting the number from any higher power of the base. The complement of 159 might then be

$$1000 - 159 = 841,$$

or

$$10{,}000 - 159 = 9841,$$

or

$$100{,}000 - 159 = 99{,}841,$$

and so on. Obtaining this type of complement involves borrowing from each position, an operation that can be cumbersome from a mechanical point of view. To simplify the mechanics, the following operations can be performed:

$$\begin{aligned} 1{,}000{,}000 - 159 &= 1{,}000{,}000 - 1 - 159 + 1 \\ &= 999{,}999 - 159 + 1 \\ &= 999{,}840 + 1 \\ &= 999{,}841. \end{aligned}$$

Although this form appears much more complex, the technique is a simple one because it involves the operation of subtracting the number whose complement is desired from a series of nines, thus requiring no borrowing. Once the operation is complete, 1 is added as a "correction factor," giving the correct result.

To illustrate subtraction using this complement, consider

$$\begin{aligned} 130 - 109 &= 130 - 109 + 1{,}000{,}000 - 1{,}000{,}000 \\ &= 130 + (100{,}000 - 109) - 1{,}000{,}000 \\ &= 130 + 999{,}891 - 1{,}000{,}000. \end{aligned}$$

As a result of this manipulation, the second term in the final expression is the complement of the original subtrahend, reducing the problem of subtraction to one of addition and a very simple subtraction. Continuing from above, we have

$$\begin{aligned} 130 - 109 &= 1{,}000{,}021 - 1{,}000{,}000 \\ &= 21. \end{aligned}$$

In other words, if the power of the base is large enough, the subtraction is accomplished merely by discarding the high-order 1 in the term 1,000,021. To appreciate the significance of this, let us assume that we are designing a mechanical desk calculator with registers capable of holding 6-digit numbers. Furthermore, we shall assume that whenever two numbers are added whose sum contains more than 6 digits, the high-order carry will be lost. Two numbers to be added might then appear in the registers as shown in Figure 3.7. Register contents during the process of subtracting using the complement (as performed in the preceding example) are shown in Figure 3.8.

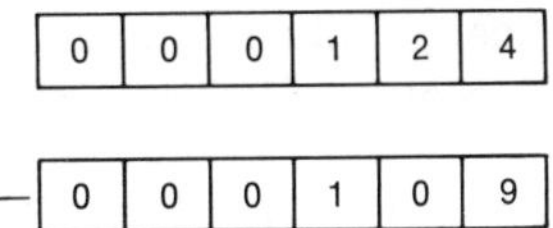

FIGURE 3.7 Register contents

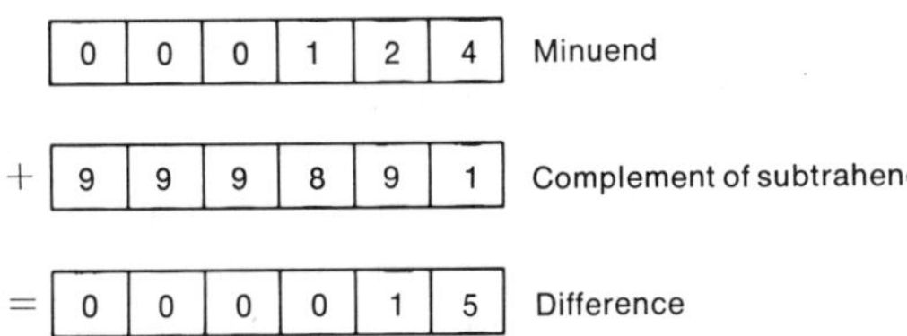

FIGURE 3.8 Complementary arithmetic

BINARY COMPLEMENTS. Let us now apply these same principles to binary numbers and note the inherent simplicity of the process. The complement of a binary number may be obtained in much the same way as that of a decimal number. For example, let us obtain the complement of 1101101 (109_{10}) using these methods:

$$\begin{array}{r} 100000000 \\ -\quad 1101101 \\ \hline 10010011 \end{array} \quad \text{is equivalent to} \quad \begin{array}{r} 11111111 \\ -\quad 1101101 \\ \hline 10010010 \\ +\qquad\quad 1 \\ \hline 10010011 \end{array}$$

The equivalent operation is possible because $10000000 = 1111111 + 1$.

The most significant point here is that this type of complement, which is often called the *twos* (2s) *complement*, is simply the *inverse* (of the original binary number) plus 1. Thus, if we consider a computer with 8-bit registers we can see how useful this complement is in performing subtractions by referring to Figure 3.9. Again, note that the complement of the subtrahend was obtained by inverting each bit, then adding 1 to the result. Also, the "correction factor"

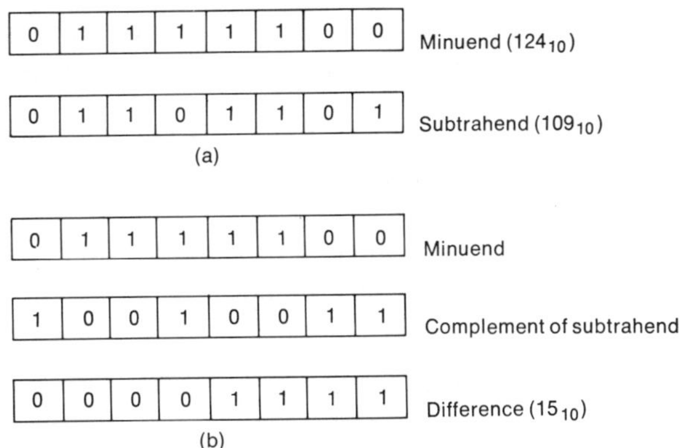

FIGURE 3.9 Subtraction with complements

of 100000000 was automatically deducted, because the high-order carry was beyond the capacity of the register and was thus lost.

NEGATIVE NUMBERS. Let us consider the binary equivalent of the subtraction $23 - 92 = -69$, that is, a subtraction that produces a negative result. Figure 3.10 illustrates the results as they would appear in 8-bit registers.

A cursory inspection of this result and conversion to decimal appears to yield 187, which we know is incorrect. However, careful study of the result and the process involved in arriving at this result will show that this is the complement of the correct difference; that is, the complement of 10111011 is $01000101 = 69_{10}$. As we know, $23 - 92 = -69$. In fact, whenever the subtrahend is larger than the minuend, producing a negative result, the difference will be in a 2s complement form. In general, it is convenient always to represent negative numbers in this form. In binary computers it is common practice to use the first bit position to indicate sign, with a 0 bit indicating a positive number and a 1 bit indicating a negative number. This is consistent with the result of Figure 3.10. In Figure 3.10(a), the positive binary equivalents of 23_{10} and 92_{10} have zeros in the first bit positions. In (b) the subtrahend has been

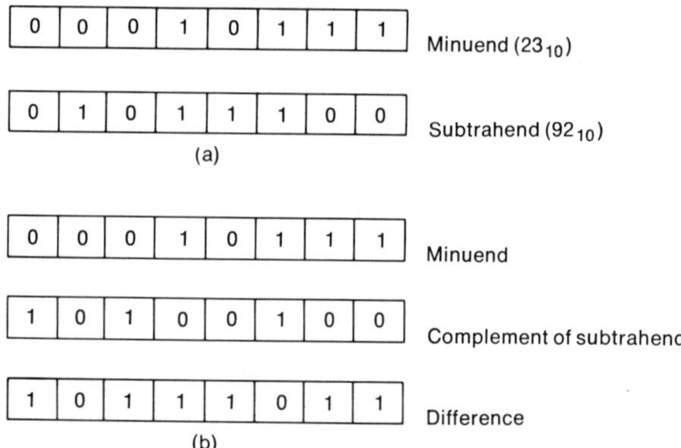

FIGURE 3.10 Negative result

converted to its complement, which becomes negative, because the first bit changes to 1. Note that in using the complement, we change the operation from subtraction to addition. In other words, this illustrates a basic elementary-school rule for subtracting signed numbers; that is, change the sign of the subtrahend and add. The difference, as we have already seen, is negative and also has a 1 bit in the first position.

Thus, using an 8-bit register (first bit reserved for sign), the smallest and largest positive numbers we could have would be

$$00000000 = 0_{10},$$
$$01111111 = 127_{10} \qquad (2^7 - 1).$$

Similarly, the smallest and largest (in magnitude) negative numbers we could have would be

$$\begin{aligned} 11111111 &= -00000001 \\ &= -1_{10}, \\ 10000000 &= -10000000 \\ &= -128_{10} \qquad (-2^7). \end{aligned}$$

This apparent lack of "symmetry" is often distracting and confusing to the beginner. As an intuitive guide, it is sometimes helpful to note that there are as many positive numbers (including zero) as negative numbers, because minus zero does not exist.

REGISTER ARITHMETIC. It is of utmost significance that in performing arithmetic operations, *we need not concern ourselves with the signs of the numbers upon which we are operating.* The sign is significant only when it is necessary to interpret the meaning of the number. For example, consider the following set of examples (the usual "artwork" has been dispensed with but 8 bits are carried in each instance). In each of the first four examples, the binary result coincides with the decimal result, even in Examples 2 and 4 where high-order carries resulted. In Examples 5 and 6, the results are invalid be-

		Base 2	
	Base 10	Basic Form	Complementing Subtrahend
Example 1:	+22 −+31 −9	00010110 −00011111	00010110 +11100001 11110111 (= -9_{10})
Example 2:	−22 −+31 −53	11101010 −00011111	11101010 +11100001 11001011 (= -53_{10})
Example 3:	+22 −−31 53	00010110 −11100001	00010110 +00011111 00110101 (= 53_{10})
Example 4:	−22 −−31 9	11101010 −11100001	11101010 +00011111 00001001 (= 9_{10})
Example 5:	−80 −+65 −145	10110000 −01000001	10110000 +10111111 01101111 (=111_{10})
Example 6:	+80 −−65 145	01010000 −10111111	01010000 +01000001 10010001 (=111_{10})

cause the numbers that are being added form sums whose magnitudes are too large for the 8 bits. Each of these is said to cause an *overflow condition.*

An overflow condition is recognized by the computer by observing the carries into and out of the sign bit (left-most bit). If the carries in and out of the sign bit are the same, the arithmetic operation is valid; if they are different, an overflow condition exists. The four possibilities are

Carry In	Carry Out	Overflow
No	No	No
Yes	Yes	No
No	Yes	Yes
Yes	No	Yes

Thus, of the preceding six examples, the carry and overflow conditions would be as follows:

Example	Carry in	Carry out	Overflow
1	No	No	No
2	Yes	Yes	No
3	No	No	No
4	Yes	Yes	No
5	No	Yes	Yes
6	Yes	No	Yes

As can be seen by careful inspection of these six examples, it is not necessary to include additional examples to illustrate addition since employment of complementary arithmetic results in addition in each instance.

WORDS AND HALFWORDS. Although the preceding examples utilize the notion of an 8-bit register to illustrate the characteristics of binary numbers, these principles apply regardless of the register size. The number of bits comprising the register in any binary type of machine is a basic characteristic of the machine. For example, both the Honeywell Series 15 and the IBM 1130 computers have a register size of 16 bits. This is referred to as a **word** in both machines. On the other hand, the IBM System/360 word consists of 32 bits although **halfwords** consisting of 16 bits are also used. However, it is important to recognize that the preceding discussion applies to 16- and 32-bit words just as readily as it does to 8-bit words. The only significant difference lies in the magnitude of numbers which the respective registers can hold. The ranges of positive and negative 16- and 32-bit numbers are:

16 bits—word for Honeywell Series 15 and IBM 1130;
halfword for IBM S/360

0111 1111 1111 1111 $= 2^{15} - 1 = 32{,}767_{10}$ (largest positive number).
0000 0000 0000 0000 $= 0$ (smallest positive number).
1111 1111 1111 1111 $= -0000\ 0000\ 0000\ 0001_2$ (complementing)
$= -1_{10}$ (smallest, in magnitude, negative number).
1000 0000 0000 0000 $= -1000\ 0000\ 0000\ 0000_2$ (complementing)
$= -2^{15} = -32{,}768_{10}$ (largest, in magnitude, negative number).

32 bits—word for IBM S/360

$$0111\ 1111\ 1111\ 1111\ 1111\ 1111\ 1111\ 1111 = 2^{31} - 1$$
$$= 2{,}147{,}483{,}647 \quad \text{(largest).}$$

$$1000\ 0000\ 0000\ 0000\ 0000\ 0000\ 0000\ 0000$$
$$= -1000\ 0000\ 0000\ 0000\ 0000\ 0000\ 0000\ 0000_2 \text{ (complementing)}$$
$$= -2{,}147{,}483{,}648 \text{ (smallest).}$$

EXERCISE 3.5

Express the following decimal numbers in 8-bit binary form (where the left-most bit is used to indicate sign); negative numbers should be in 2s complement form.

1. 38, −38, 127, −128, −1, −73, 128, −129.
2. 27, −27, −26, 63, −64, −91, 132, −132.

Perform the following subtractions using complementary arithmetic. Check your results by converting to decimal.

3. 11010 − 1011	5. 10000 − 1111	7. 101001 − 11100
4. 10010 − 1101	6. 10001 − 1001	8. 100001 − 11111

Convert the following decimal numbers to binary (in 8-bit registers with the first bit representing sign), then perform the indicated arithmetic operations. Indicate whether or not carries into and out of the sign bit occur and whether or not an overflow condition occurs.

9. (+47) + (+52)	14. (−47) + (−52)	19. (−83) − (−57)
10. (+47) − (+52)	15. (−128) + (+127)	20. (−1) + (+1)
11. (+47) + (−52)	16. (−128) − (+127)	21. (−1) − (−1)
12. (+47) − (−52)	17. (−83) − (+127)	22. (+64) − (32)
13. (−47) + (+52)	18. (−83) − (+57)	23. (+25) − (+38)

24. What would be the largest and smallest (positive and negative) numbers which could be stored (using the first bit for sign) with a 6-bit register? With a 12-bit register?

3.6 OCTAL AND HEXADECIMAL NUMBERS

CORRESPONDENCE WITH BINARY AND DECIMAL. Although all digital computers operate in a binary mode in one manner or another, the binary number system is not the only one of importance in the study of computers. Where the binary system, with 0 and 1 corresponding to the on and off states of electronic components, is ideal for use in the computer, the large binary numbers are often cumbersome for the programmer. For example, communication from one programmer to another is hardly enhanced when the topic is a binary number within the computer such as 1001010110101110. Communication would be improved by converting to base 10, but the binary-decimal conversion is a cumbersome process. As a result, two other number bases, 8 and 16, are commonly used with computers because of their simple relationship to binary. (That is, both 8 and 16 are powers of two.)

Whereas the base or radix in our decimal system is 10 and in the binary system is 2, the base in an *octal* (base 8) system is 8 and in a *hexadecimal* (base 16) system is 16. Since an octal system requires eight distinct digits, the digits 0 through 7 are used. On the other hand, a hexadecimal system requires 16 unique digits, thus the 10 digits used in decimal are insufficient. Rather than generate new and strange "hieroglyphics," the ten digits and the first six letters of the alphabet are most commonly used. The correspondence between decimal, hexadecimal, binary, and octal is shown in Table 3.6. In the counting sequence, when the digit 7 is reached in octal and F (15_{10}) in hexadecimal, all of the digits have "been used" in the respective systems. At this point, in each instance, a high-order carry to the next digit is necessary, exactly as in decimal and binary counting.

TABLE 3.6 OCTAL AND HEXADECIMAL NUMBERS

Decimal	Hexa-decimal	Binary	Octal	Decimal	Hexa-decimal	Binary	Octal
0	0	0000	0	16	10	1 0000	20
1	1	0001	1	17	11	1 0001	21
2	2	0010	2	18	12	1 0010	22
3	3	0011	3	19	13	1 0011	23
4	4	0100	4	20	14	1 0100	24
5	5	0101	5	21	15	1 0101	25
6	6	0110	6	22	16	1 0110	26
7	7	0111	7	23	17	1 0111	27
8	8	1000	10	24	18	1 1000	30
9	9	1001	11	25	19	1 1001	31
10	A	1010	12	26	1A	1 1010	32
11	B	1011	13	27	1B	1 1011	33
12	C	1100	14	28	1C	1 1100	34
13	D	1101	15	29	1D	1 1101	35
14	E	1110	16	30	1E	1 1110	36
15	F	1111	17	31	1F	1 1111	37

In their structures, all four of these numbers systems are the same in that they all use the principle of ciphering, that is, place values representing powers of the number system base. For example, the quantity 101 would normally be interpreted as the decimal quantity *one hundred one*. However, consider its meaning in each of the four number systems which we have studied.

$$101_2 = 1 \times 2^2 + 0 \times 2^1 + 1 \times 2^0 \rightarrow 4 + 1 = 5_{10}.$$
$$101_8 = 1 \times 8^2 + 0 \times 8^1 + 1 \times 8^0 \rightarrow 64 + 1 = 65_{10}.$$
$$101_{10} = 1 \times 10^2 + 0 \times 10^1 + 1 \times 10^0 \rightarrow 100 + 1 = 101_{10}.$$
$$101_{16} = 1 \times 16^2 + 1 \times 16^1 + 1 \times 16^0 \rightarrow 256 + 1 = 257_{10}.$$

Perhaps the curious reader wondered about the comment earlier that base 8 and 16 are commonly used in computers because of their important relationship to binary. To illustrate this, we shall consider the binary number 101101001. (Technically speaking, we cannot consider the following a proof with the choice of a particular value; however, the binary quantity could easily be generalized to yield a mathematical proof.

$$\begin{aligned} 101101001 = {} & 1 \times 2^8 + 0 \times 2^7 + 1 \times 2^6 + 1 \times 2^5 + 0 \times 2^4 \\ & + 1 \times 2^3 + 0 \times 2^2 + 0 \times 2^1 + 1 \times 2^0. \end{aligned} \tag{3.1}$$

Grouping the terms in threes and using the rules for exponents from the previous chapter gives

$$\begin{aligned} 101101001 = {} & (1 \times 2^2 + 0 \times 2^1 + 1 \times 2^0)2^6 \\ & + (1 \times 2^2 + 0 \times 2^1 + 1 \times 2^0)2^3 \\ & + (0 \times 2^2 + 0 \times 2^1 + 1 \times 2^0)2^0. \end{aligned} \tag{3.2}$$

But $2^6 = 8^2$; $2^3 = 8^1$; and $2^0 = 8^0$; therefore,

$$\begin{aligned} 101101001 = {} & (1 \times 2^2 + 0 \times 2^1 + 1 \times 2^0)8^2 \\ & + (1 \times 2^2 + 0 \times 2^1 + 1 \times 2^0)8^1 \\ & + (0 \times 2^2 + 0 \times 2^1 + 1 \times 2^0)8^0 \end{aligned} \tag{3.3}$$

$$= (4 + 0 + 1)8^2 + (4 + 0 + 1)8^1 + (0 + 0 + 1)8^0$$

$$= 5 \times 8^2 + 5 \times 8^1 + 1 \times 8^0 \tag{3.4}$$

The final expression is precisely the required form for an octal number, so we have converted the binary number 101101001_2 to its octal equivalent 551_8.

Note the form of Eq. 3.3. Within each group of parentheses, base 2 appears with exponents of 2, 1, and 0. Thus, if we consider each set of parentheses as containing a three-digit binary number, we may rewrite Eq. 3.3 as

$$101101001 = (101)8^2 + (101)8^1 + (001)8^0.$$

As can readily be seen, the three sets of binary numbers within the parentheses make up the pure binary number on the left. This is no coincidence, as an inspection of the preceding equations will show; rather it is a direct consequence of the relationship between the binary and octal systems. Thus, in order to convert from binary to octal, it is only necessary to group the binary digits in threes and perform the conversion directly. For example,

$$\begin{aligned} 101110_2 &= \underbrace{101}\ \underbrace{110} \\ &= \ \ 5 \quad\ \ 6 \\ &= 56_8. \\ 111011 &= \underbrace{111}\ \underbrace{011} \\ &= \ \ 7 \quad\ \ 3 \\ &= 73_8. \\ 1011011101_2 &= \underbrace{001}\ \underbrace{011}\ \underbrace{011}\ \underbrace{101} \\ &= \ \ 1 \quad\ \ 3 \quad\ \ 3 \quad\ \ 5 \\ &= 1335_8. \end{aligned}$$

OCTAL TO BINARY CONVERSION. The inverse operation, that of converting octal numbers to binary numbers, uses the same principle and is equally as simple. This becomes evident from an examination of the following examples. Since

$$110_2 = 1 \times 2^2 + 1 \times 2^1 + 0 \times 2^0 = 4 + 2 + 0,$$

$$\begin{aligned} 6_8 &= 110_2. \\ 66_8 &= 110\ 110 \\ &= 110110_2. \\ 34_8 &= 011\ 100 \\ &= 11100_2. \\ 7654_8 &= 111\ 110\ 101\ 100 \\ &= 111110101100_2. \end{aligned}$$

HEXADECIMAL CONVERSION. This convenient relationship between binary and octal should not be considered unique to the octal system; indeed, a similar relationship can be shown between binary and any number system with a power of 2 as a base. The difference is, of course, in the number of binary digits within each grouping. Since the hexadecimal digit range is $0\text{-}F_{16}$ ($0000\text{-}1111_2$) we might anticipate from the derivation for octal that the grouping for hexadecimal is in fours. That is, for example

$$\begin{aligned} 0010_2 &= 2_{16} \\ 0100_2 &= 4_{16} \\ 0111_2 &= 7_{16} \\ 1011_2 &= B_{16}. \end{aligned}$$

Therefore,

$$\begin{aligned} 10010001111011_2 &= \underbrace{0010}_{2}\ \underbrace{0100}_{4}\ \underbrace{0111}_{7}\ \underbrace{1011}_{B} \\ &= 247B_{16}. \end{aligned}$$

Conversion to binary from hexadecimal is every bit as simple, it is only necessary to expand each hexadecimal digit into its four-bit binary form as follows.

$$\begin{aligned} 29CA_{16} &= 0010\ 1001\ 1100\ 1010 \\ &= 10100111001010_2 \\ FC2B_{16} &= 1111\ 1100\ 0010\ 1011 \\ &= 1111110000101011_2 \end{aligned}$$

With this convenient relationship between binary and hexadecimal, a programmer or operator can easily look at a row of lights on a computer console which shows 1s and 0s, for example 0001100111000010; group them in fours and record the value as the hexadecimal number 19C2.

In the System/360 as in many other computers, the basic unit of storage is the *byte*, which consists of eight bits (thus a 360 word consists of four bytes). Because of the versatility of the eight-bit storage unit which has virtually become a standard, hexadecimal is commonly used. The contents of an eight-bit unit can easily be represented by two hexadecimal digits.

Although many computers are classified as octal or as hexadecimal machines, it is important not to be mislead by this terminology. The octal or hexadecimal numbers are used simply as a convenient device to represent binary quantities (usually storage addresses). Within the computer, operations take place in binary (except where BCD is used, then the operations are not purely binary in nature), and storage addresses appear in binary form. For example, the Honeywell Series 15 computer uses an ingenious addressing method. The storage of this machine is divided into so-called *sectors* of 512 words each. Each word has a unique address in the range $000\text{-}511_{10}$, which is $000000000\text{-}111111111_2$; thus any word in a given sector can be addressed by use of nine bits. Represented in octal, these addresses range from 000_8 to 777_8. A machine with a storage capacity of eight sectors (numbered 0 through 7) would have a storage consisting of 4096 words. Then the address of any storage location would be specified by a three-bit sector number and a nine-bit address within the sector. Represented in octal, a typical location such as 512_8 in sector 4 would appear as the four-digit octal number

$$4\underbrace{512}$$

Sector ↑ ↑ address within the sector

These features are used to advantage in design of Series 15 instructions.

CONVERSION TO DECIMAL. Whenever a programmer finds it necessary to delve into his program to the extent that he must work with octal or hexadecimal quantities, he usually finds it convenient to work entirely within the number system of his machine. However, it is sometimes necessary to convert to or from decimal. Obviously, the conversion to decimal simply involves expanding the octal or hexadecimal number in its polynomial form, then performing the indicated operations. For example:

$$\begin{aligned} 3762_8 &= 3 \times 8^3 + 7 \times 8^2 + 6 \times 8^1 + 2 \times 8^0 \\ &= 3 \times 512 + 7 \times 64 + 6 \times 8 + 2 \times 1 \\ &= 2034_{10}. \\ 13B5_{16} &= 1 \times 16^3 + 3 \times 16^2 + 11 \times 16^1 + 5 \times 16^0 \\ &= 1 \times 4096 + 3 \times 256 + 11 \times 16 + 5 \times 1 \\ &= 5045_{10}. \end{aligned}$$

If this appears cumbersome to the reader, then a few attempts at converting from decimal to octal or hexadecimal is usually sufficient to convince one of the value of using the special conversion techniques described in the sections on binary. Better still, the use of conversion tables (when available) greatly simplifies the process.

The principle of the double-and-add technique for converting from binary to decimal arose from a conveniently factored form of the binary expansion.

This same technique can be applied to octal or hexadecimal; that is:

$$
\begin{aligned}
23762_8 &= 2 \times 8^4 + 3 \times 8^3 + 7 \times 8^2 + 6 \times 8^1 + 2 \times 8^0 \\
&= 8(8(8\underbrace{(2 \times 8 + 3)}_{19} + 7) + 6) + 2 \\
&= 8(8\underbrace{(8 \times 19 + 7)}_{159} + 6) + 2 \\
&= 8(8 \times 159 + 6) + 2 \\
&= 8 \times 1278 + 2 \\
&= 10226_{10}.
\end{aligned}
$$

Obviously the same procedure can be applied to hexadecimal. However, it is usually somewhat simpler to apply the following algorithm for either octal or hexadecimal:

OCTAL TO DECIMAL

Multiply the left-most octal digit by the number system base (8); add the next high-order octal digit to this product. Multiply the sum by 8 and add the next octal digit. Continue this process until all octal digits are exhausted.

HEXADECIMAL TO DECIMAL

Multiply the decimal equivalent of the left-most hexadecimal digit by the number system base (16); add the decimal equivalent of the next hexadecimal digit to this product. Multiply the sum by 16 and add the decimal equivalent of the next hexadecimal digit. Continue this process until all hexadecimal digits are exhausted.

For example, consider converting 3762_8 and $24BC_{16}$ to decimal.

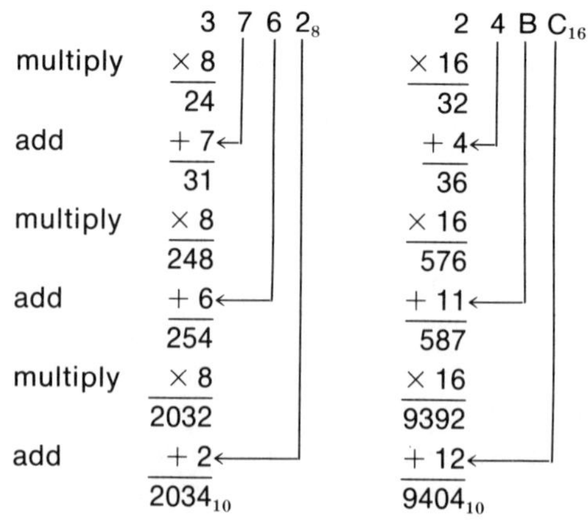

Hence

$$3762_8 = 2034_{10} \qquad 24BC_{16} = 9404_{10}$$

CONVERSION FROM DECIMAL. The division-remainder method used with binary numbers is equally applicable to octal and hexadecimal.

> *Divide the decimal number repeatedly by the number system base, until a zero quotient is obtained. The first remainder represents the least significant digit of the result (octal or hexadecimal as the case may be); the last remainder is the most significant digit.*

For example, consider converting 1710_{10} to octal and to hexadecimal.

	quotient / division	remainder
	0	3
fourth division	8) 3	2
third division	8) 26	5
second division	8) 213	6
first division	8) 1710	↑ remainders

Remainders read upward: 3256

$$1710_{10} = 3256_8$$

	quotient / division	remainder
	0	6
third division	16) 6	10 → A
second division	16) 106	14 → E
first division	16) 1710	↑ remainders

Remainders read upward: 6AE

$$1710_{10} = 6AE_{16}$$

USING A CONVERSION TABLE. It is obvious from the preceding examples that number base conversion is a tedious and error-prone job. If such conversion is frequently necessary then the programmer usually has a set of conversion tables close at hand. Table 3.7 is a conversion table indicating the correspondence between the numbers 0000_8 through 1777_8 and 0000_{10} through 1023_{10}; similarly Table 3.8 indicates the correspondence between the numbers 000_{16} through $1FF_{16}$ and 0000_{10} through 0511_{10}. Appendix 2 includes a set of eight such tables (for hexadecimal) covering the range 000_{16} through FFF_{16}. The use of these tables is illustrated with the example 0751_8 for Table 3.7 and $1E9_{16}$ for Table 3.8. Note that the first three digits (two for hexadecimal) determine the proper row and the last digit determines the proper column as shown by the shaded areas. The number at the intersection, in each case, is the decimal equivalent.

$$0751_8 = 489_{10}$$
$$1E9_{16} = 489_{10}$$

FRACTIONS. Fractional quantities in octal and hexadecimal exist in exactly the same form as in decimal and binary. Thus we have, for example:

$$
\begin{aligned}
36.624_8 &= 3 \times 8^1 + 6 \times 8^0 + 6 \times 8^{-1} + 2 \times 8^{-2} + 4 \times 8^{-3} \\
&= 3 \times 8 + 6 \times 1 + 6 \times \frac{1}{8} + 2 \times \frac{1}{64} + 4 \times \frac{1}{512} \\
&= 24 + 6 + 0.75 + 0.03125 + 0.0078125 \\
&= 30.7890625_{10};
\end{aligned}
$$

$$
\begin{aligned}
2B.C8_{16} &= 2 \times 16^1 + 11 \times 16^0 + 12 \times 16^{-1} + 8 \times 16^{-2} \\
&= 2 \times 16 + 11 \times 1 + 12 \times \frac{1}{16} + 8 \times \frac{1}{256} \\
&= 32 + 11 + 0.75 + 0.03125 \\
&= 43.78125_{10}.
\end{aligned}
$$

TABLE 3.7 OCTAL-DECIMAL CONVERSION TABLE

OCTAL 0000 to 0777 — DECIMAL 0000 to 0511

	0	1	2	3	4	5	6	7
0000	0000	0001	0002	0003	0004	0005	0006	0007
0010	0008	0009	0010	0011	0012	0013	0014	0015
0020	0016	0017	0018	0019	0020	0021	0022	0023
0030	0024	0025	0026	0027	0028	0029	0030	0031
0040	0032	0033	0034	0035	0036	0037	0038	0039
0050	0040	0041	0042	0043	0044	0045	0046	0047
0060	0048	0049	0050	0051	0052	0053	0054	0055
0070	0056	0057	0058	0059	0060	0061	0062	0063
0100	0064	0065	0066	0067	0068	0069	0070	0071
0110	0072	0073	0074	0075	0076	0077	0078	0079
0120	0080	0081	0082	0083	0084	0085	0086	0087
0130	0088	0089	0090	0091	0092	0093	0094	0095
0140	0096	0097	0098	0099	0100	0101	0102	0103
0150	0104	0105	0106	0107	0108	0109	0110	0111
0160	0112	0113	0114	0115	0116	0117	0118	0119
0170	0120	0121	0122	0123	0124	0125	0126	0127
0200	0128	0129	0130	0131	0132	0133	0134	0135
0210	0136	0137	0138	0139	0140	0141	0142	0143
0220	0144	0145	0146	0147	0148	0149	0150	0151
0230	0152	0153	0154	0155	0156	0157	0158	0159
0240	0160	0161	0162	0163	0164	0165	0166	0167
0250	0168	0169	0170	0171	0172	0173	0174	0175
0260	0176	0177	0178	0179	0180	0181	0182	0183
0270	0184	0185	0186	0187	0188	0189	0190	0191
0300	0192	0193	0194	0195	0196	0197	0198	0199
0310	0200	0201	0202	0203	0204	0205	0206	0207
0320	0208	0209	0210	0211	0212	0213	0214	0215
0330	0216	0217	0218	0219	0220	0221	0222	0223
0340	0224	0225	0226	0227	0228	0229	0230	0231
0350	0232	0233	0234	0235	0236	0237	0238	0239
0360	0240	0241	0242	0243	0244	0245	0246	0247
0370	0248	0249	0250	0251	0252	0253	0254	0255
0400	0256	0257	0258	0259	0260	0261	0262	0263
0410	0264	0265	0266	0267	0268	0269	0270	0271
0420	0272	0273	0274	0275	0276	0277	0278	0279
0430	0280	0281	0282	0283	0284	0285	0286	0287
0440	0288	0289	0290	0291	0292	0293	0294	0295
0450	0296	0297	0298	0299	0300	0301	0302	0303
0460	0304	0305	0306	0307	0308	0309	0310	0311
0470	0312	0313	0314	0315	0316	0317	0318	0319
0500	0320	0321	0322	0323	0324	0325	0326	0327
0510	0328	0329	0330	0331	0332	0333	0334	0335
0520	0336	0337	0338	0339	0340	0341	0342	0343
0530	0344	0345	0346	0347	0348	0349	0350	0351
0540	0352	0353	0354	0355	0356	0357	0358	0359
0550	0360	0361	0362	0363	0364	0365	0366	0367
0560	0368	0369	0370	0371	0372	0373	0374	0375
0570	0376	0377	0378	0379	0380	0381	0382	0383
0600	0384	0385	0386	0387	0388	0389	0390	0391
0610	0392	0393	0394	0395	0396	0397	0398	0399
0620	0400	0401	0402	0403	0404	0405	0406	0407
0630	0408	0409	0410	0411	0412	0413	0414	0415
0640	0416	0417	0418	0419	0420	0421	0422	0423
0650	0424	0425	0426	0427	0428	0429	0430	0431
0660	0432	0433	0434	0435	0436	0437	0438	0439
0670	0440	0441	0442	0443	0444	0445	0446	0447
0700	0448	0449	0450	0451	0452	0453	0454	0455
0710	0456	0457	0458	0459	0460	0461	0462	0463
0720	0464	0465	0466	0467	0468	0469	0470	0471
0730	0472	0473	0474	0475	0476	0477	0478	0479
0740	0480	0481	0482	0483	0484	0485	0486	0487
0750	0488	0489	0490	0491	0492	0493	0494	0495
0760	0496	0497	0498	0499	0500	0501	0502	0503
0770	0504	0505	0506	0507	0508	0509	0510	0511

OCTAL 1000 to 1777 — DECIMAL 0512 to 1023

	0	1	2	3	4	5	6	7
1000	0512	0513	0514	0515	0516	0517	0518	0519
1010	0520	0521	0522	0523	0524	0525	0526	0527
1020	0528	0529	0530	0531	0532	0533	0534	0535
1030	0536	0537	0538	0539	0540	0541	0542	0543
1040	0544	0545	0546	0547	0548	0549	0550	0551
1050	0552	0553	0554	0555	0556	0557	0558	0559
1060	0560	0561	0562	0563	0564	0565	0566	0567
1070	0568	0569	0570	0571	0572	0573	0574	0575
1100	0576	0577	0578	0579	0580	0581	0582	0583
1110	0584	0585	0586	0587	0588	0589	0590	0591
1120	0592	0593	0594	0595	0596	0597	0598	0599
1130	0600	0601	0602	0603	0604	0605	0606	0607
1140	0608	0609	0610	0611	0612	0613	0614	0615
1150	0616	0617	0618	0619	0620	0621	0622	0623
1160	0624	0625	0626	0627	0628	0629	0630	0631
1170	0632	0633	0634	0635	0636	0637	0638	0639
1200	0640	0641	0642	0643	0644	0645	0646	0647
1210	0648	0649	0650	0651	0652	0653	0654	0655
1220	0656	0657	0658	0659	0660	0661	0662	0663
1230	0664	0665	0666	0667	0668	0669	0670	0671
1240	0672	0673	0674	0675	0676	0677	0678	0679
1250	0680	0681	0682	0683	0684	0685	0686	0687
1260	0688	0689	0690	0691	0692	0693	0694	0695
1270	0696	0697	0698	0699	0700	0701	0702	0703
1300	0704	0705	0706	0707	0708	0709	0710	0711
1310	0712	0713	0714	0715	0716	0717	0718	0719
1320	0720	0721	0722	0723	0724	0725	0726	0727
1330	0728	0729	0730	0731	0732	0733	0734	0735
1340	0736	0737	0738	0739	0740	0741	0742	0743
1350	0744	0745	0746	0747	0748	0749	0750	0751
1360	0752	0753	0754	0755	0756	0757	0758	0759
1370	0760	0761	0762	0763	0764	0765	0766	0767
1400	0768	0769	0770	0771	0772	0773	0774	0775
1410	0776	0777	0778	0779	0780	0781	0782	0783
1420	0784	0785	0786	0787	0788	0789	0790	0791
1430	0792	0793	0794	0795	0796	0797	0798	0799
1440	0800	0801	0802	0803	0804	0805	0806	0807
1450	0808	0809	0810	0811	0812	0813	0814	0815
1460	0816	0817	0818	0819	0820	0821	0822	0823
1470	0824	0825	0826	0827	0828	0829	0830	0831
1500	0832	0833	0834	0835	0836	0837	0838	0839
1510	0840	0841	0842	0843	0844	0845	0846	0847
1520	0848	0849	0850	0851	0852	0853	0854	0855
1530	0856	0857	0858	0859	0860	0861	0862	0863
1540	0864	0865	0866	0867	0868	0869	0870	0871
1550	0872	0873	0874	0875	0876	0877	0878	0879
1560	0880	0881	0882	0883	0884	0885	0886	0887
1570	0888	0889	0890	0891	0892	0893	0894	0895
1600	0896	0897	0898	0899	0900	0901	0902	0903
1610	0904	0905	0906	0907	0908	0909	0910	0911
1620	0912	0913	0914	0915	0916	0917	0918	0919
1630	0920	0921	0922	0923	0924	0925	0926	0927
1640	0928	0929	0930	0931	0932	0933	0934	0935
1650	0936	0937	0938	0939	0940	0941	0942	0943
1660	0944	0945	0946	0947	0948	0949	0950	0951
1670	0952	0953	0954	0955	0956	0957	0958	0959
1700	0960	0961	0962	0963	0964	0965	0966	0967
1710	0968	0969	0970	0971	0972	0973	0974	0975
1720	0976	0977	0978	0979	0980	0981	0982	0983
1730	0984	0985	0986	0987	0988	0989	0990	0991
1740	0992	0993	0994	0995	0996	0997	0998	0999
1750	1000	1001	1002	1003	1004	1005	1006	1007
1760	1008	1009	1010	1011	1012	1013	1014	1015
1770	1016	1017	1018	1019	1020	1021	1022	1023

TABLE 3.8 HEXADECIMAL-DECIMAL CONVERSION*

	0	1	2	3	4	5	6	7	8	9	A	B	C	D	E	F
000	0000	0001	0002	0003	0004	0005	0006	0007	0008	0009	0010	0011	0012	0013	0014	0015
010	0016	0017	0018	0019	0020	0021	0022	0023	0024	0025	0026	0027	0028	0029	0030	0031
020	0032	0033	0034	0035	0036	0037	0038	0039	0040	0041	0042	0043	0044	0045	0046	0047
030	0048	0049	0050	0051	0052	0053	0054	0055	0056	0057	0058	0059	0060	0061	0062	0063
040	0064	0065	0066	0067	0068	0069	0070	0071	0072	0073	0074	0075	0076	0077	0078	0079
050	0080	0081	0082	0083	0084	0085	0086	0087	0088	0089	0090	0091	0092	0093	0094	0095
060	0096	0097	0098	0099	0100	0101	0102	0103	0104	0105	0106	0107	0108	0109	0110	0111
070	0112	0113	0114	0115	0116	0117	0118	0119	0120	0121	0122	0123	0124	0125	0126	0127
080	0128	0129	0130	0131	0132	0133	0134	0135	0136	0137	0138	0139	0140	0141	0142	0143
090	0144	0145	0146	0147	0148	0149	0150	0151	0152	0153	0154	0155	0156	0157	0158	0159
0A0	0160	0161	0162	0163	0164	0165	0166	0167	0168	0169	0170	0171	0172	0173	0174	0175
0B0	0176	0177	0178	0179	0180	0181	0182	0183	0184	0185	0186	0187	0188	0189	0190	0191
0C0	0192	0193	0194	0195	0196	0197	0198	0199	0200	0201	0202	0203	0204	0205	0206	0207
0D0	0208	0209	0210	0211	0212	0213	0214	0215	0216	0217	0218	0219	0220	0221	0222	0223
0E0	0224	0225	0226	0227	0228	0229	0230	0231	0232	0233	0234	0235	0236	0237	0238	0239
0F0	0240	0241	0242	0243	0244	0245	0246	0247	0248	0249	0250	0251	0252	0253	0254	0255
100	0256	0257	0258	0259	0260	0261	0262	0263	0264	0265	0266	0267	0268	0269	0270	0271
110	0272	0273	0274	0275	0276	0277	0278	0279	0280	0281	0282	0283	0284	0285	0286	0287
120	0288	0289	0290	0291	0292	0293	0294	0295	0296	0297	0298	0299	0300	0301	0302	0303
130	0304	0305	0306	0307	0308	0309	0310	0311	0312	0313	0314	0315	0316	0317	0318	0319
140	0320	0321	0322	0323	0324	0325	0326	0327	0328	0329	0330	0331	0332	0333	0334	0335
150	0336	0337	0338	0339	0340	0341	0342	0343	0344	0345	0346	0347	0348	0349	0350	0351
160	0352	0353	0354	0355	0356	0357	0358	0359	0360	0361	0362	0363	0364	0365	0366	0367
170	0368	0369	0370	0371	0372	0373	0374	0375	0376	0377	0378	0379	0380	0381	0382	0383
180	0384	0385	0386	0387	0388	0389	0390	0391	0392	0393	0394	0395	0396	0397	0398	0399
190	0400	0401	0402	0403	0404	0405	0406	0407	0408	0409	0410	0411	0412	0413	0414	0415
1A0	0416	0417	0418	0419	0420	0421	0422	0423	0424	0425	0426	0427	0428	0429	0430	0431
1B0	0432	0433	0434	0435	0436	0437	0438	0439	0440	0441	0442	0443	0444	0445	0446	0447
1C0	0448	0449	0450	0451	0452	0453	0454	0455	0456	0457	0458	0459	0460	0461	0462	0463
1D0	0464	0465	0466	0467	0468	0469	0470	0471	0472	0473	0474	0475	0476	0477	0478	0479
1E0	0480	0481	0482	0483	0484	0485	0486	0487	0488	0489	0490	0491	0492	0493	0494	0495
1F0	0496	0497	0498	0499	0500	0501	0502	0503	0504	0505	0506	0507	0508	0509	0510	0511

*From IBM Systems Reference Library manual C26-5927-0.

Conversion from octal or hexadecimal to binary remains the simple task of expanding as with integer quantities; for example:

$$36.624_8 = 011\ 110.110\ 010\ 100$$
$$= 11110.1100101_2;$$

$$2B.C8_{16} = 0010\ 1011.1100\ 1000$$
$$= 101011.11001_2.$$

With the inverse process, that is, converting from binary to octal or hexadecimal, it is only necessary to remember that the grouping begins with the point. For example,

$$11010.0010111_2 = 011\ 010.001\ 011\ 100$$
$$= 32.134_8;$$

$$11010.0010111_2 = 0001\ 1010.0010\ 1110$$
$$= 1A.2E_{16}.$$

As described in Chapter 2, most computers have provisions for handling numbers in *floating-point* format. In the IBM 1620, a binary-coded decimal machine, floating-point decimal numbers appear very similar to the scientific notation form. However, in a binary machine, the quantities will be in a binary floating-point format. Where we might think of the decimal quantity 3672 in its floating-point form as

$$3672 = 0.3672 \times 10^4$$

it would appear in binary floating point as

$$3672_{10} = 111001011000_2 = 0.111001011 \times 2^{12};$$

similarly

$$1011.001 = 0.1011001 \times 2^4;$$

and

$$0.000110111 = 0.110111 \times 2^{-3}.$$

This is basically the form, with minor variations, which is used within the computer. However, within the computer, the decimal point is not included but is understood to be at the left of the most significant digit. Furthermore, only the fractional part, or so-called *mantissa*, and the exponent are stored in the computer, both in binary form. Thus the quantity 1101100.100110011 essentially would be stored as

$$1101100.100110011 = 0.1101100100110011 \times 2^7,$$

but 2^7 is 10000000

$$\underbrace{1101100100110011}_{\text{Mantissa}} \qquad \underbrace{10000000}_{\text{exponent}}.$$

In hexadecimal, these would appear as

D933 80

There are other considerations that will change the appearance of the number in storage but these will not be discussed in this book.

EXERCISE 3.6

Without using a table, find the decimal equivalents of the following octal numbers. Check your results against the table whenever possible. Also convert them to binary.

1. 7	4. 162	7. 7172	10. 16.64
2. 17	5. 77	8. 6012	11. 23.16
3. 161	6. 777	9. 7116	12. 21.124

Without using a conversion table find the decimal equivalents of the following hexadecimal numbers. Check your results against the table. Also, convert them to binary.

13. 9	16. 1B1	19. 1FF	22. 1F.C
14. C	17. 199	20. 1A0	23. 30.88
15. 1D	18. BB	21. 200	24. 24.C4

Using Tables 3.7 and 3.8 in reverse, find the octal and hexadecimal equivalents of the following decimal numbers.

25. 15
26. 16
27. 17
28. 256
29. 64
30. 449
31. 271
32. 511
33. 489
34. 170
35. 273
36. 424

The hexadecimal number 1000_{16} may be converted to decimal form by expressing it in polynomial form, or $1000_{16} = 1 \times 16^3 = 4096_{10}$. Similarly, $2000_{16} = 8192_{10}$, $3000_{16} = 12{,}288_{10}$, and so on. Thus the conversion table of Appendix 1 may be used to convert four-digit hexadecimal numbers to decimal, that is

$$21C5_{16} = 2000_{16} + 1C5_{16},$$

but

$$\begin{aligned} 2000_{16} &= 8192_{10} \\ 1C5_{16} &= 453_{10} \qquad \text{(from the table).} \end{aligned}$$

Therefore,

$$\begin{aligned} 21C5_{16} &= 8192_{10} + 453_{10} \\ &= 8645_{10}. \end{aligned}$$

Using this principle, find the decimal equivalent of the following hexadecimal numbers:

37. 1110
38. 31CC
39. 21D0
40. 5050
41. A01A
42. B066

Using the above principle, find the hexadecimal equivalent of the following decimal numbers:

43. 4097
44. 8200
45. 4296
46. 4605
47. 8300
48. 12,300

49. In this section the binary-octal relationship was developed using a particular binary number. Using the binary number 10010110101, develop the same relationship for binary and hexadecimal numbers.
50. Develop a table (similar to Table 3.6) for hexadecimal, binary and base 4 numbers using the digits 0, 1, 2 and 3. Develop a technique for converting from base 4 to base 16 and reverse.
51. Using the letters *a, b, c, d, e* and *f* (to represent the digits 0, 1, 2, 3, 4 and 5) make a table similar to Table 3.6 relating this number system to the decimal numbers for 1_{10} through 25_{10}. What is the base of this system?
52. Why is it meaningless to speak of the octal number 238_8 with reference to the usage in this chapter?

3.7 OCTAL AND HEXADECIMAL ARITHMETIC

ADDITION. Although most calculated results are printed by the computer in decimal form, it is often convenient for the programmer to work directly with hexadecimal numbers when attempting to detect errors in his program. Thus, an ability to perform addition and, occasionally, subtraction on hexadecimal numbers can be a great time-saver. In section 3.4 we studied the concept of the decimal addition table (Table 3.3) and related the addition process to binary numbers. The addition of a pair of decimal numbers is so automatic that we give little attention to the details of the operation. However,

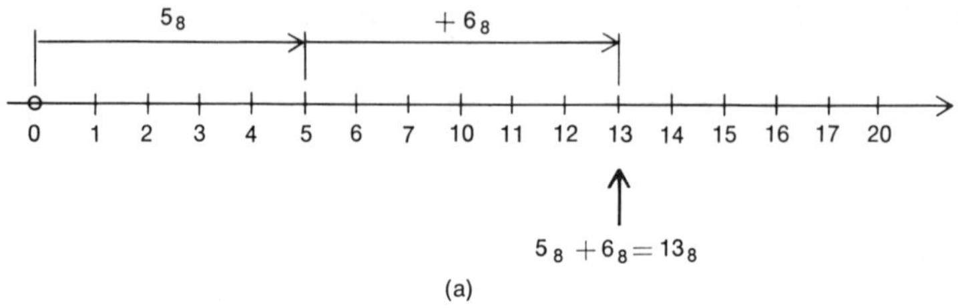

(a)

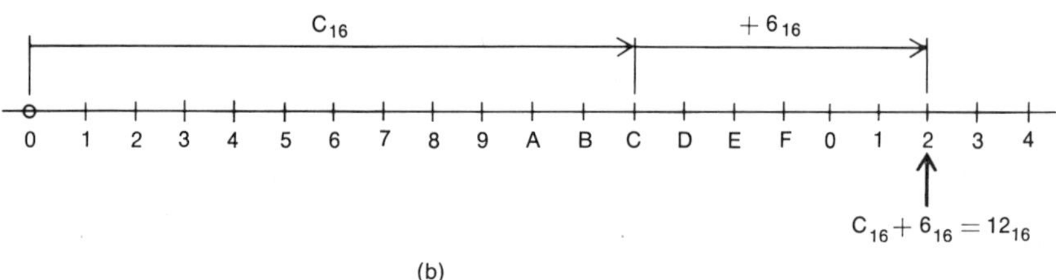

(b)

FIGURE 3.11 Addition using the number line (a) for octal numbers; (b) for hexadecimal numbers

such operations in a base with which we are not so familiar requires considerably more attention to the step-by-step detail. On the other hand, the basic principle of addition in other bases, such as octal or hexadecimal, is identical to that of decimal. A base 8 or 16 addition table similar to Table 3.3 can easily be constructed by "finger counting" or by drawing an octal or hexadecimal number line and then referring to the geometric interpretation of the addition process (as is done for real numbers in Chapter 2). Referring to Figure 3.11(a) and 3.11(b), we see the usual correspondence of equally spaced points on a line to consecutive integers, but this time the numeric quantities are in octal and hexadecimal, respectively. Now the addition, $5_8 + 6_8$ is easily represented in (a) by counting off 6 additional units beyond 5 which yields 13_8, or 3_8 with a carry of 1; in hexadecimal (b), 6 additional units are counted beyond C_{16} yielding 12_{16}, or 2_{16} with a carry of 1.

This technique for adding octal and hexadecimal numbers is effectively "finger counting" and is commonly used by the beginner until he becomes sufficiently familiar with addition that it becomes as automatic as decimal addition. The following additional examples, which can be verified by finger counting, further illustrate the process.

Octal	Hexadecimal
4 + 3 = 7	4 + 3 = 7
4 + 4 = 10 (0 with 1 carry)	4 + 6 = A
4 + 6 = 12 (2 with 1 carry)	4 + C = 10 (0 with 1 carry)
7 + 7 = 16 (6 with 1 carry)	F + F = 1E (E with 1 carry)

Addition of numbers consisting of several digits involves exactly the same operations enumerated for decimal addition as is illustrated by the following examples.

Octal				Hexadecimal		
1	1 1	1 1 1	carries	1 1	1 1	1 1 1
4216	4716	172.613		1C2	13E4	1FF.9C2
2545	3525	427.32		9F	24F	23.715
6763	10443	622.133	sum	261	1633	223.0D7

In other words,

$$4216_8 + 2545_8 = 6763_8 \qquad 1C2_{16} + 9F_{16} = 261_{16}$$
$$4716_8 + 3525_8 = 10443_8 \qquad 13E4_{16} + 24F_{16} = 1633_{16}$$
$$172.613_8 + 427.32_8 = 622.133_8 \qquad 1FF.9C2_{16} + 23.715_{16} = 223.0D7_{16}$$

Note that carries are propagated and that the points are aligned just as in decimal operations.

SUBTRACTION. Subtraction of octal or hexadecimal numbers follows the same rules as decimal subtraction. In subtracting 6_8 from 15_8, we can either use the number line approach from Chapter 2, finger count, or use an addition table such as Table 3.3 in reverse. The number line approach (which is effectively finger counting) is used in Figure 3.12 to illustrate subtracting 6_8 from 15_8 (in (a)) and 6_{16} from 15_{16} (in (b)). The processing of subtracting when both numbers consist of several digits is carried out exactly as with decimal

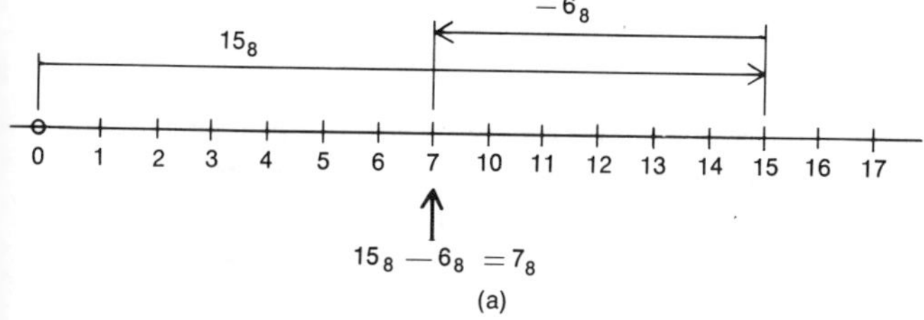

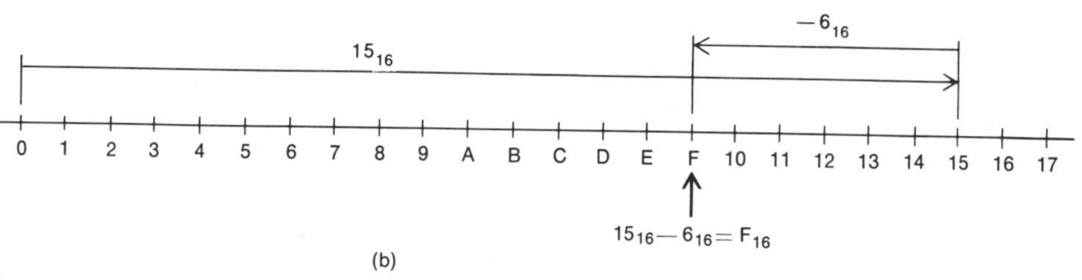

FIGURE 3.12 Subtraction using the number line (a) for octal numbers; (b) for hexadecimal numbers

numbers, even to the extent of borrowing when the subtrahend digit exceeds the minuend digit. Although the steps involved are straightforward, the operation of subtraction appears to be considerably more prone to error than addition, especially for the beginner. The reader may discover that these operations (especially when performed with hexadecimal numbers) are simpler if he converts each pair of digits to their decimal equivalents, performs the subtraction in decimal, then converts back to hexadecimal. However, it is imperative to remember that the borrow represents 16_{10}, not 10_{10}. For instance:

$$\begin{aligned} 19_{16} - C_{16} &= 25_{10} - 12_{10} \\ &= 13_{10} \\ &= D_{16}; \end{aligned}$$

check:

$$C_{16} + D_{16} = 19_{16}.$$

These principles are illustrated by the following examples:

Octal				Hexadecimal		
	6 14	16 4 ~~6~~ 14	Borrows		D 1B	1A D ~~A~~ 19
5 7 4 3	5 ~~7~~ ~~4~~ 3	~~5~~ ~~7~~ ~~4~~ 3	Minuend	4 E B 9	4 ~~E~~ ~~B~~ 9	4 ~~E~~ ~~B~~ ~~9~~
2 5 0 1	1 6 6 2	1 7 6 2	Subtrahend	1 1 A 7	1 1 E 8	2 5 B B
3 2 4 2	4 0 6 1	3 7 6 1	Difference	3 D 1 2	3 C D 1	2 8 F E

That is,

$$\begin{aligned} 5743_8 - 2501_8 &= 3242_8 & \qquad 4EB9_{16} - 11A7_{16} &= 3D12_{16} \\ 5743_8 - 1662_8 &= 4061_8 & \qquad 4EB9_{16} - 11E8_{16} &= 3CD1_{16} \\ 5743_8 - 1762_8 &= 3761_8 & \qquad 4EB9_{16} - 25BB_{16} &= 28FE_{16} \end{aligned}$$

Note that borrows, although appearing somewhat more awkward, are handled exactly as in decimal.

MULTIPLICATION AND DIVISION. The development of octal and hexadecimal numbers in this chapter has progressed from the basic ability to count to the performing of addition and subtraction using the counting ability. Since multiplication actually involves successive addition, the construction of a multiplication table would be a simple matter. For instance,

Octal	*Hexadecimal*
$6_8 \times 1_8 = 6_8$	$6_{16} \times 1_{16} = 6_{16}$
$6_8 \times 2_8 = 6_8 + 6_8 = 14_8$	$6_{16} \times 2_{16} = 6_{16} + 6_{16} = C_{16}$
$6_8 \times 3_8 = (6_8 \times 2_8) + 6_8 = 22_8$	$6_{16} \times 3_{16} = (6_{16} \times 2_{16}) + 6_{16} = 12_{16}$
$6_8 \times 4_8 = (6_8 \times 3_8) + 6_8 = 30_8$	$6_{16} \times 4_{16} = (6_{16} \times 3_{16}) + 6_{16} = 18_{16}$
and so on	and so on

Then multiplication of two octal or of two hexadecimal numbers proceeds in exactly the same way as multiplication of two decimal numbers. Similarly, division uses the method of successive subtractions and functions in the same manner as long division with decimal numbers.

Further consideration of these operations is left to the reader.

EXERCISE 3.7

1. Construct an addition table for the octal digits.
2. Construct an addition table for the hexadecimal digits.
3. Construct a multiplication table for the octal digits.
4. Construct a multiplication table for the hexadecimal digits.

Add the following octal numbers:

5. 1425
6346
6. 2571
2443
7. 7176
1513
8. 25713
747
9. 77625
264
10. 512.673
74.772
11. 6010.72
541.3
12. 6267
1315
4225

Add the following hexadecimal numbers:

13. 164A
3B42
14. 2596
8815
15. 9A6C
2824
16. 10079
11BA
17. BBF6
4412
18. 59A.6C
917.1C
19. 9A.11B
31.196
20. 1572
3874
5762

Perform the indicated subtractions on the following octal numbers:

21. 5713
−1252
22. 52136
−4251
23. 20000
−5416
24. 71662
−62663
25. 17600
−7151
26. 51556
−1772
27. 47711
−16777
28. 23.716
−2.467
29. 5100.27
−301.17

Perform the indicated subtractions on the following hexadecimal numbers:

30. 5A67
−1955
31. 6FB3
−4A41
32. 59A1
−2689
33. 563A
−46CA
34. 6831
−65A
35. 5AC52
−EF38
36. 20000
−3A5F
37. 68.C1D
−2.D8A
38. 971.83
−28.AB

Using the table constructed in problem 3, multiply the following pairs of octal numbers:

39. 6
    7

40. 15
    4

41. 122
    13

42. 471
    25

43. 4771
    23

44. 6113
    212

Using the table constructed in problem 4, multiply the following pairs of hexadecimal numbers:

45. 6
    7

46. C
    6

47. 14
    8

48. 2AA
    21

49. 4AC1
    8A

50. 51A2
    193

51. Do the octal numbers form a "set" as do the real numbers described in Chapter 2? Would you expect the octal numbers to satisfy the same rules as decimal numbers such as closure, the associative property, and the distributive property? Illustrate with examples. Answer these same questions for hexadecimal numbers.
52. What are the octal and hexadecimal identity elements for addition and multiplication?
53. Would it be possible to use the digits 1, 2, 3, 4, 5, 6, 7, and 8 to represent the digits 0, 1, 2, 3, 4, 5, 6, and 7, respectively, in an octal system? If so, what would be the sums $1 + 1$, $1 + 4$, and $4 + 3$ in this seemingly "odd" representation?
54. Problem 51, Exercise 3.6 involves using the letters *a, b, c, d, e,* and *f* to represent the digits of a base 6 number system. Construct an addition table for this system. Construct a multiplication table.

REFERENCES

Brightman, R. W., Luskin, B., and Tilton, T., *Data Processing for Decision-Making.* New York, the Macmillan Company, 1968.

Capettini, Heigho, and Jeltema, *Computing Concepts in Mathematics.* Chicago, Science Research Associates, Inc., 1968.

DeAngelo, S., and Jorgensen, P., *Mathematics for Data Processing.* New York, McGraw-Hill, Inc., 1970.

Keedy, M. L., and Bittinger, M. L., *Mathematics: A Modern Introduction.* Reading, Mass., Addison-Wesley Publishing Company, Inc., 1970.

Number Systems (Student Text), IBM Form Number C20-1618-3. White Plains, N.Y., International Business Machines Corporation, 1968.

4 LOGICAL FORMS AND PROGRAMMING

4.1 USE OF SET DESCRIPTIONS / **112**
Truth Values / **113**
Basic Components / **114**
AND *Statements* / **114**
OR *Statements* / **115**
Negations / **116**
Exercise 4.1 / **118**

4.2 AND, OR, AND NOT IN SEQUENCE / **119**
Truth Tables / **120**
Exercise 4.2 / **122**

4.3 FLOWCHARTS / **123**
AND, OR *Forms* / **124**
Truth Tables and Flowcharts / **125**
A More Realistic Problem / **126**
Exercise 4.3 / **129**

4.4 IF-THEN FORMS / **129**
Conditionals / **129**
Computer Statements / **130**
Other Conditional Forms / **134**
Exercise 4.4 / **135**

4.5 TRUTH VALUES OF CONDITIONALS AND EQUIVALENCES (OPTIONAL) / **136**
Statements and Conclusions / **138**
Equivalence / **139**
Exercise 4.5 / **140**

4.6 SPECIAL USES OF AND, OR AND EOR / **141**
The Inclusive OR / **142**
The AND / **143**
The Exclusive OR / **144**
Testing Binary Data / **144**
Exercise 4.6 / **146**

4.7 SIMPLIFICATION BY FLOWCHART / **148**
Exercise 4.7 / **150**

As computer applications become more widespread, knowledge of the fact that computers use base 2 instead of base 10 also becomes more prevalent. But how can electronic components within a computer respond to the symbols man uses for communication with himself and others? A partial answer relates to the simplicity of base 2 in the use of only two symbols, 0 and 1. A correspondence to an on-off (switch) situation in a computer and to the true-false property of statements provides a link between man-made symbols and electronic components.

4.1 USE OF SET DESCRIPTIONS

The key person in "communicating" with a computer is the programmer who writes a sequence of instructions which can be executed by the computer. Before this sequence can be written, however, the programmer must be able to "visualize" the problem and relate the various components. As an illustration consider the following problem.

EXAMPLE 4.1

A department store executive wishes to have a list of customers (for advertising purposes) in the Lotworth district who last year charged $200 or more or purchased a new television set.

We have noted that "set language" can provide a means of visualizing the problem (Chapter 1).

Let $U =$ the set of all customers of the company,
LOTworth = the set of all customers who live in the Lotworth district,
CHArge = the set of all customers who last year charged $200 or more,
Tv = the set of all customers who purchased a new television set.

To be on the list, a customer must

1. live in the Lotworth district and have charged $200 or more, or
2. live in the Lotworth district and have purchased a new television set.

Thus, a customer on the list is in

1. set *LOTworth* and also in set *CHArge*, or
2. set *LOTworth* and also in set *Tv*.

A mathematician (being rather frugal with symbols) would indicate that *CUSTOMER* is in

1. $LOTworth \cap CHArge$, or
2. $LOTworth \cap Tv$

which can be combined to

$$(LOTworth \cap CHArge) \cup (LOTworth \cap Tv)$$

Using a Venn diagram (assuming that $LOTworth \cap CHArge \cap Tv \neq \emptyset$), the shaded area in Figure 4.1 represents those customers on the list. Now, on one

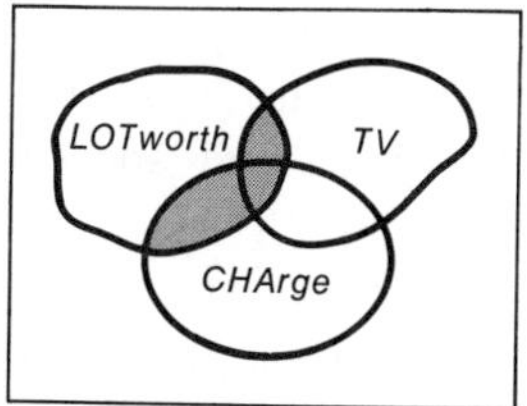

FIGURE 4.1 $(LOTworth \cap CHarge) \cup (LOTworth \cap Tv)$

hand, the programmer has a vivid picture of the problem, but on the other, a computer cannot read Venn diagrams.

TRUTH VALUES. Fortunately, a computer can be programmed to determine if the statement,

CUSTOMER is in the Lotworth district,

is true or false. Of course, it needs some help which could be provided by indicating 0 or 1 in an appropriate column on an IBM card for each customer. Whether the statement,

CUSTOMER charged \$200 or more,

is true or false (for a particular customer) could be determined by considering $(C - 200)$ where C equals the amount charged. If $C - 200$ is positive or zero then the statement is true or its **truth value** is true.

For convenience in writing: "*CUSTOMER* is in the Lotworth district and charged \$200 or more" is abbreviated to

LOTWORTH and *CHARGE*.

Now we must take great care with our symbolism to avoid becoming hopelessly confused. In the preceding paragraph, *LOTworth* and *CHArge* are used to represent the *set* of residents of Lotworth and the *set* of residents who have charged \$200 or more, respectively. Now on the other hand, *LOTWORTH* and *CHARGE* are being used in an entirely different sense. That is, *LOTWORTH* represents the *statement*,

CUSTOMER is in the Lotworth district,

which may be either true or false. Similarly, *CHARGE* represents the following statement which also may be either true or false.

CUSTOMER has charged \$200 or more.

Relative to the computer, a statement will normally appear as part of a program whereas the *set* of residents (or their names) may appear as a list printed according to instructions from a computer.

Thus a computer (and a programmer) must be able to determine not only if *LOTWORTH* is true or false but also if "*LOTWORTH* and *CHARGE*" is true or false and what to do in each of four possible situations. However, before proceeding, it is worth noting that this type of problem is considered in a branch of mathematics called **symbolic logic**. By looking over a logician's shoulder we will see a structuring of language which can be very useful to the programmer.

BASIC COMPONENTS. The basic component in symbolic logic is a statement (or assertion) which has a truth value of true or false (but not both). The statement, represented by *LOTWORTH,*

CUSTOMER lives in the Lotworth district,

is an example of a **simple** statement while

LOTWORTH and *CHARGE*

is a **compound** statement. Examples of other compound statements which form the underlying structure of symbolic logic include:

LOTWORTH OR *CHARGE*
IF *CHARGE,* THEN *LOTWORTH*
NOT *LOTWORTH*
CHARGE IF AND ONLY IF *LOTWORTH*

Each of these compound statements has a truth value which depends on the truth values of its components. In a list of instructions to a computer relating to a business application, one may see the following: (COBOL statements)

```
1.    MULTIPLY SALARY BY 0.2 GIVING TAX.

2.     ... TAX-DEDUCTION IS GREATER THAN ZERO ...

3.     ... QUANTITY-ON-HAND IS EQUAL TO 400 ...

4.    DISPLAY TAKE-HOME-PAY.
```

Statements 1 and 4 do not have a truth value but indicate to a computer an operation which will be performed. Statements 2 and 3 will be true or false depending on the values given (by the computer) to TAX-DEDUCTION and QUANTITY-ON-HAND. Actually statements 2 and 3 will appear in the form (Section 4.4).

```
2.    IF TAX-DEDUCTION IS GREATER THAN ZERO THEN
          SUBTRACT TAX-DEDUCTION FROM SALARY.

3.    IF QUANTITY-ON-HAND IS LESS THAN 400 THEN
          WRITE MESSAGE-RECORD FROM REORDER-LIST.
```

AND *STATEMENTS.* Now we consider the truth values of

```
TAX-DEDUCTION IS GREATER THAN ZERO AND
    RATE IS EQUAL TO 4.00
```

For convenience, abbreviate this to

T AND *R*

Logicians assert that this AND statement is true only when *T* is true and also *R* is true. Otherwise it is false. In table form, we have

T	*R*	*T* AND *R*
T	T	T
T	F	F
F	T	F
F	F	F

If the tax-deduction equals $30 and the rate equals 3.00, then *T* is true and *R* is false. This corresponds to line 2 of the table and *T* AND *R* is false. In summary, *T* AND *R* is false if one or both components are false which agrees with common usage.

OR *STATEMENTS.* However, when considering the compound statement represented by

T OR *R*,

logicians point out that the word OR is used in two ways. One use (exclusive sense) asserts that

T or *R* means *T* is true, or *R* is true, but both are *not* true.

The second (inclusive sense) asserts that

T OR *R* means *T* is true, or *R* is true, or both are true.

Hence the programmer must be careful when his supervisor uses the word *or* since his list of instructions should differ in the two cases to match Table 4.1.

TABLE 4.1

T	*R*	*T* OR *R* Exclusive	*T* OR *R* Inclusive
T	T	F	T
T	F	T	T
F	T	T	T
F	F	F	F

Borrowing from Fortran IV, .OR. will be used to represent the inclusive OR and .AND. to represent AND. Thus the inclusive OR of Table 4.1 would appear as *T*.OR.*R*. (The AND case would be represented as *T*.AND.*R*). In addition, the negation NOT which is described in the next paragraph will be represented by .NOT. which is the Fortran form. Although Fortran does not provide the direct capability for the exclusive OR case, many machine languages do. However, for consistency here, .EOR. will be used to represent the exclusive OR. The reader should carefully note that this is *not* part of the Fortran language, however.

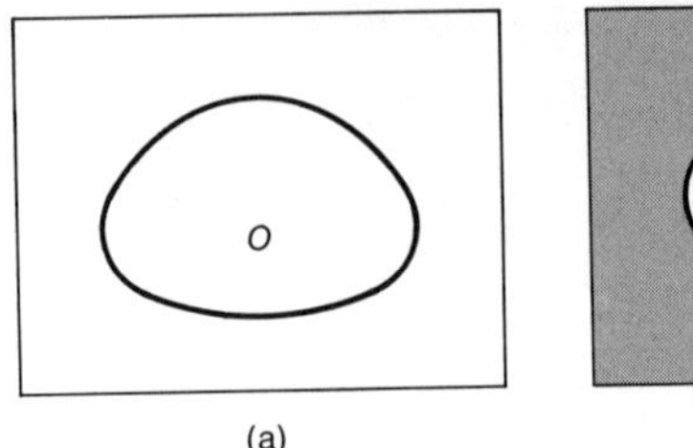

(a)

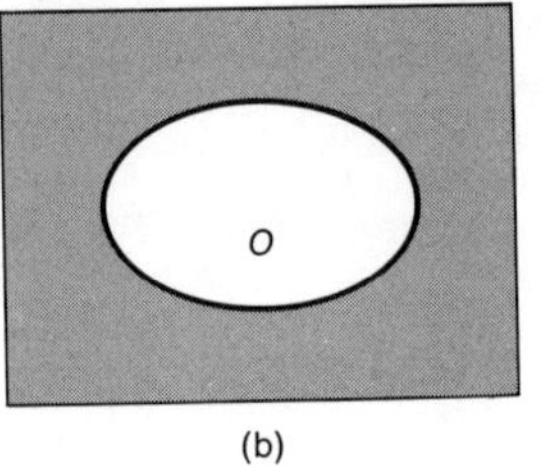

(b)

FIGURE 4.2 (a) Set *O*; (b) Set *O′*

NEGATIONS. In set terminology, the set of a company's employees over 45 years of age (*O*) can be indicated in Figure 4.2 where the rectangle represents the set of all employees. Then *O′* (*O* complement) is the shaded part of Figure 4.2(b) Next, consider

O: EMPLOYEE is over 45.

Looking over the logician's shoulder again, we would see

.NOT.*O* representing *EMPLOYEE* is *not* over 45,
or *EMPLOYEE* is 45 or under.

Logicians assert that .NOT.*O*, called the **negation** of *O*, is a statement such that

(.NOT.*O*) is false when *O* is true,
.AND.(.NOT.*O*) is true when *O* is false.

The definition of .NOT.*O* must be followed carefully to determine the negation of a compound statement. Suppose

M: *EMPLOYEE* is male.

Then the truth values of the negation of (*O*.AND.*M*) or of .NOT.(*O*.AND.*M*), can easily be determined by a truth table (Table 4.2).

TABLE 4.2

O	*M*	*O*.AND.*M*	.NOT.(*O*.AND.*M*)
T	T	T	F
T	F	F	T
F	T	F	T
F	F	F	T

The last column in Table 4.2 is just a reversal of column 3. Since *O*.AND.*M* is true only when both *O* and *M* are true, this is the only situation where .NOT.(*O*.AND.*M*) is false. Otherwise, it is true.

However, if a programmer needs to print a list of employees who are NOT (over 45 and male), is he limited to translating Table 4.2 or is there another possibility? If he were to ask a logician, the reply would be that

(.NOT.*O*).OR.(.NOT.*M*)

could be used in place of the negation of *O*.AND.*M*. Why? Because

(.NOT.*O*).OR.(.NOT.*M*) and .NOT.(*O*.AND.*M*)

have the same truth values or, as the logician would say, they are **logically equivalent.**

To find the truth values of (.NOT.*O*).OR.(.NOT.*M*), we begin as shown in Table 4.3. The third column is a reversal of the first column and the fourth column is a reversal of the second column.

TABLE 4.3

O	*M*	.NOT.*O*	.NOT.*M*
T	T	F	F
T	F	F	T
F	T	T	F
F	F	T	T

Then recall that

(.NOT.*O*).OR.(.NOT.*M*)

is an inclusive OR statement and is true when one or both components are true. Table 4.4 indicates that this OR statement is false only on the first line when both components are false. A comparison with Table 4.2 indicates that the truth values of

.NOT.(*O*.AND.*M*) and (.NOT.*O*).OR.(.NOT.*M*)

are the same and hence they are logically equivalent.

TABLE 4.4

O	*M*	.NOT.*O*	.NOT.*M*	(NOT.*O*).OR.(.NOT.*M*)
T	T	F	F	F
T	F	F	T	T
F	T	T	F	T
F	F	T	T	T

Now the programmer knows that his list can be printed by considering

EMPLOYEE is not over 45 .OR. *EMPLOYEE* is not male

or

EMPLOYEE is 45 or under .OR. female.

These are, however, only possibilities provided by the logician; the programmer must decide which to use. In Section 4.6 we will see how flowcharts can be used to select the best form.

EXERCISE 4.1

Construct truth tables for each of the following:

1. *P*.AND..NOT.*Q*
2. .NOT.*P*.OR.*Q*
3. .NOT.*P*.AND..NOT.*Q*
4. .NOT.(*P*.AND.*Q*)

Translate to a symbolic form, using the letters indicated at the end of each statement:

5. The Giants will be in first place and the Dodgers in second place, or the Giants will be in second place and the Reds in first place. (*G, D, S, R*)
6. The Cubs beat the Dodgers or the Giants lost to the Braves, and the Mets beat the Reds. (*C, G, M*)
7. Stock prices will rise, or we should sell bonds and buy stocks. (*P, B, S*)
8. We can send the painting by ship and have it in London by Thursday, or wait a week and send it by air express. (*P, L, W, A*)
9. The employee is under 50 but (and) over 30, and has been employed for 17 years. (*U, O, W*)
10. ($x < 7$ or $x = 7$) and $x > 3$ (*L, E, G*)
11. $x < 9$ and ($x > 4$ or $x = 4$) (*L, G, E*)
12. The first enrollment period began on September 1, 1970, and was scheduled to end on March 31, 1971, but (and) Congress extended the deadline. (*S, M, E*)
13. No deduction is allowable for losses from sales or exchanges of property between (a) members of a family, (b) a corporation and an individual owning more than 50 percent of the corporation's stock, (c) a grantor and fiduciary of any trust, (d) a fiduciary and a beneficiary of the same trust. (*M, C, I, G, F, B*)
14. A reasonable allowance for the exhaustion, wear, tear, and obsolescence of property shall be allowed as a depreciation deduction. (*E, W, O*)
15. The allowance does not apply to inventories or stock-in-trade, nor to land apart from improvements. (*I, S, L*)
16. Expenditures such as taxes, interest, repairs, insurance, agent's commissions, maintenance, and similar items can be deducted. (*T, I, R, N, A, M, S*)
17. You must see dramatic results or pick up the phone and send us a collect telegram and your money will be wired back immediately. (*R, P, S, W*)

Determine the truth value of the following statements if

EMPLOYEE-NUMBER = 500	TAX = 60
SALARY = 700	INSUR-DEDUCTION = 40
X-CODE = 1	

18. EMPLOYEE-NUMBER IS LESS THAN 400 OR X-CODE = 1
19. SALARY = 700 AND TAX = 50
20. SALARY − TAX = 640 AND X-CODE = 1
21. TAX + INSUR-DEDUCTION = SALARY − 500 OR X-CODE = 0

Using Venn diagrams similar to Figure 4.1 shade the area corresponding to:

22. *LOTworth* ∪ (*CHArge* ∩ *Tv*)
23. *LOTworth* ∩ (*CHArge* ∪ *Tv*)
24. *CHArge* ∪ *Tv*
25. *LOTworth* ∩ *Tv*

If *CUSTOMER* is any member of the set *LOTworth* ∪ (*CHArge* ∩ *Tv*) of problem 22, determine the truth value of each of the following statements.

26. *LOTWORTH*.OR.*CHARGE*
27. *LOTWORTH*.OR.*TV*
28. *LOTWORTH*.AND.*CHARGE*
29. *TV*.OR.*CHARGE*

If *CUSTOMER* is any member of the set (*LOTworth* ∩ *Tv*) of problem 25, determine the truth value of each of the following statements.

30. *LOTWORTH*.OR.*TV*
31. LOTWORTH.AND.*TV*
32. *CHARGE*.OR.*TV*
33. *CHARGE*.AND.*TV*

Using Table 4.1 as a guide, construct truth tables for each of the following:

34. (.NOT.*P*).EOR.*Q*
35. *P*.EOR.(.NOT.*Q*)
36. .NOT.(*P*.EOR.*Q*)
37. (.NOT.*P*).EOR.(.NOT.*Q*)
38. Show by truth tables that .NOT.(*P*.OR.*Q*) is logically equivalent to (.NOT.*P*) .AND.(.NOT.*Q*).

The logical equivalences,

.NOT.(*P*.OR.*Q*) and (.NOT.*P*).AND.(.NOT.*Q*)
.NOT.(*P*.AND.*Q*) and (.NOT.*P*).OR.(.NOT.*Q*)

are called DeMorgan's Laws. Use these to find a form which is logically equivalent to:

39. .NOT.(*P*.OR.(.NOT.*Q*))
40. .NOT.(.NOT.*P*.AND.*Q*)
41. .NOT.((.NOT.*P*).OR.*Q*)
42. .NOT.((.NOT.*P*).AND.(.NOT.*Q*))

Use DeMorgan's Laws to find compound statements that have the same meaning as each of the following.

43. It is not the case that the employee is over 50 and has worked more than 5 years.
44. It is not the case that the employee is over 50 or has worked more than 5 years.

4.2 AND, OR, AND NOT IN SEQUENCE

It is probably apparent that many problems will not have the simple logical forms of

```
P.AND.Q          P.OR.Q
```

A capability of handling more complex logical forms such as

```
(P.AND.Q).OR.R
```

is included in Fortran IV and Cobol. Parentheses are interpreted in the same way as in algebra, when included, but

```
P.AND.Q.OR.R
```

would determine the same truth values as the preceding form. This follows from the sequence in which logical connectives are considered (that is, hierarchy of operations). When no parentheses indicate otherwise, .NOT. is considered first, then .AND., and finally .OR., with operations on connectives of the same hierarchy level proceeding from left to right.

```
P.OR..NOT.Q.AND.R
```

would be interpreted as

```
P.OR.((.NOT.Q).AND.R)
```

If we assume that *P* is false, *Q* is false, and *R* is true, then .NOT.*Q* is true and ((.NOT.*Q*).AND.*R*) is true. Therefore, we see that

```
P.OR..NOT.Q.AND.R
```

is true.

TRUTH TABLES. The truth table for a compound statement involving three different simple statements such as

```
(.NOT.P.OR.Q).AND.R
```

requires eight lines. The four possible combinations of truth values for *Q* and *R* are matched with *P* true and then with *P* false (Table 4.5).

TABLE 4.5

P	*Q*	*R*
T	T	T
T	T	F
T	F	T
T	F	F
F	T	T
F	T	F
F	F	T
F	F	F

Then to determine the truth values of this compound statement in the most efficient manner, we note that the overall form is an .AND. statement. Thus when *R* is false, the .AND. statement is false (Table 4.6). In effect, we are considering

```
R.AND.(.NOT.P.OR.Q)
```

TABLE 4.6

P	Q	R	(.NOT.P	.OR.	Q)	.AND.	R
T	T	T					
T	T	F				F	F
T	F	T					
T	F	F				F	F
F	T	T					
F	T	F				F	F
F	F	T					
F	F	F				F	F

In the remaining four cases, the .OR. statement is true except on line 3. This results in the final set of truth values as shown in Table 4.7.

TABLE 4.7

P	Q	R	(.NOT.P	.OR.	Q)	.AND.	R
T	T	T	F	T	T	T	
T	T	F				F	F
T	F	T	F	F	F	F	
T	F	F				F	F
F	T	T	T	T		T	
F	T	F				F	F
F	F	T	T	T		T	
F	F	F				F	F

As a second illustration, consider

```
(P.AND..NOT.Q).OR..NOT.R
```

Since the overall form is an .OR. statement, it will be true when .NOT.R is true (or when R is false). This leads immediately to 1/2 of the desired truth values (Table 4.8). The .AND. statement will be false when P is false (lines 5 and 7) and in turn the .OR. statement is false (since .NOT.R is also false). In the two remaining cases the .AND. statement is false on line 1 and true on line 3.

TABLE 4.8

P	Q	R	(P	.AND.	.NOT.Q)	.OR.	.NOT.R
T	T	T	T	F	F	F	
T	T	F				T	T
T	F	T	T	T	T	T	
T	F	F				T	T
F	T	T	F	F		F	
F	T	F				T	T
F	F	T	F	F		F	
F	F	F				T	T

In considering these truth tables it should be carefully noted that they are *not* ends in themselves. Rather the construction of a truth table enables one to better understand the general structure of the logical form. In the next section, flowcharts are related to logical forms. Then the shortcuts we have just demonstrated will be essential in drawing a flowchart.

EXERCISE 4.2

Consider each of the following a Fortran IV statement. Insert parentheses to correspond to a computer's interpretation of each.

1. *P*.AND..NOT.*Q*.OR.*R*
2. *P*.AND.*Q*.OR..NOT.*R*
3. *P*.OR.*Q*.OR.*R*
4. *P*.AND.*Q*.AND..NOT.*R*
5. *P*.AND.*Q*.OR.*R*.AND.*S*
6. *P*.OR.*Q*.AND.*R*.AND.*S*
7. *P*.OR.*Q*.OR.*R*.AND.*S*

Construct truth tables for each of the following:

8. (.NOT.*P*.AND.*Q*).OR.*R*
9. .NOT.(*P*.AND.*Q*).OR.*R*
10. (*P*.OR..NOT.*Q*).AND..NOT.*R*
11. *P*.AND.(.NOT.*Q*.OR.*R*)

Construct four-line truth tables for each of the following:

12. *P*.OR.(.NOT.*P*.AND.*Q*)
13. *P*.AND.(.NOT.*P*.OR.*Q*)
14. (*P*.AND.*Q*).OR.(*P*.AND..NOT.*Q*)
15. (*P*.OR.*Q*).AND.(*P*.OR..NOT.*Q*)

Determine if the following pairs are logically equivalent:

16.	*P*.EOR.*Q*	(*P*.AND..NOT.*Q*).OR.(.NOT.*P*.AND.*Q*)
17.	.NOT.*P*.OR.*Q*	.NOT.(*P*.AND.*Q*)
18.	*P*.OR.(*Q*.OR.*P*)	*P*.OR.*Q*
19.	*P*.OR..NOT.*P*	.NOT.*P*.AND.*P*

Determine whether each of the following is true or false.

20. If *P* is true, then *P*.OR.(*Q*.AND.*R*) is true.
21. If *P* is false, then *P*.OR.(*Q*.AND.*R*) is false.
22. If *P* is true, then *P*.AND.*Q* is true.
23. If *Q* is false, then *P*.AND.*Q* is false.
24. If *P*.AND.*Q* is true, then *P* is true.
25. If *P*.OR.*Q* is false, then *Q* is false.
26. If *P*.EOR.*Q* is false, then *Q* is false.
27. If *P*.EOR.*Q* is true, then *P* is true.
28. If .NOT.(*P*.OR.*Q*) is true, then *P* is false.

Construct truth tables for each of the following:

29. *P*.AND.(*Q*.EOR..NOT.*R*)
30. .NOT.*P*.EOR.(*Q*.AND.*R*)

4.3 FLOWCHARTS

We now relate the logical structure provided by AND, OR, and NOT statements to flowcharts. As an illustration, consider the following situation.

EXAMPLE 4.2

A list is to be printed of all employees under 25 who are single or eligible for the draft.

Let

UNDER represent *EMPLOYEE* is under 25,
SINGLE represent *EMPLOYEE* is single,
ELIG represent *EMPLOYEE* is eligible for the draft.

The logical structure of the problem becomes evident by noting that employees who are single or eligible for the draft can be represented by

```
SINGLE.OR.ELIG
```

Inclusion of the third statement results in

```
UNDER.AND.(SINGLE.OR.ELIG)
```

To be on the list, an employee must be under 25 and also single or eligible for the draft. Note the translator must determine where parentheses are placed to match the symbolic with the verbal form. If the reader is uncertain of the match, he may find it helpful to consider a different placement of parentheses or to change a logical connective. Then a check of the match better reveals the structure of the verbal form.

We begin the flowchart by considering the .OR. statement by itself. The truth table for

```
SINGLE.OR.ELIG
```

is given in Table 4.9.

TABLE 4.9

SINGLE	*ELIG*	*SINGLE*.OR.*ELIG*
T	T	T
T	F	T
F	T	T
F	F	F

The flowchart begins with a consideration of the truth value of *SINGLE* as in Figure 4.3(a). The truth table (Table 4.9) indicates that the .OR. statement is true when *SINGLE* is true (Figure 4.3(b)). When *SINGLE* is false, the truth value of *SINGLE*.OR.*ELIG* depends on the truth value of *ELIG* (Figure 4.3(c)). Now returning to the complete statement *UNDER*.AND.(*SINGLE*.OR.*ELIG*) we recall that an .AND. statement is true only when *both* components are true (Table 4.10).

TABLE 4.10

UNDER	*SINGLE*	*ELIG*	*UNDER* .AND. (*SINGLE* .OR. *ELIG*)		
T	T	T		T	T
T	T	F		T	T
T	F	T		T	T
T	F	F		F	F
F	T	T	F	F	
F	T	F	F	F	
F	F	T	F	F	
F	F	F	F	F	

Thus when *UNDER* is false, the AND statement is false (Figure 4.4(a)). However, if *UNDER* is true, the truth value of *SINGLE*.OR.*ELIG* must be checked. Thus we append the .OR. flowchart in Figure 4.3(c) as shown in Figure 4.4(b).

AND, OR *FORMS.* Example 4.2 includes both AND and OR statements. To better illustrate the difference between these two logical forms, consider the flowcharts in Figure 4.5. Here *P* and *Q* represent statements which have a truth value. For *P*.AND.*Q* the flowchart (Figure 4.5(a)) indicates that both components must be true to reach the print instruction while for *P*.OR.*Q* only one component need be true for the print instruction to be executed.

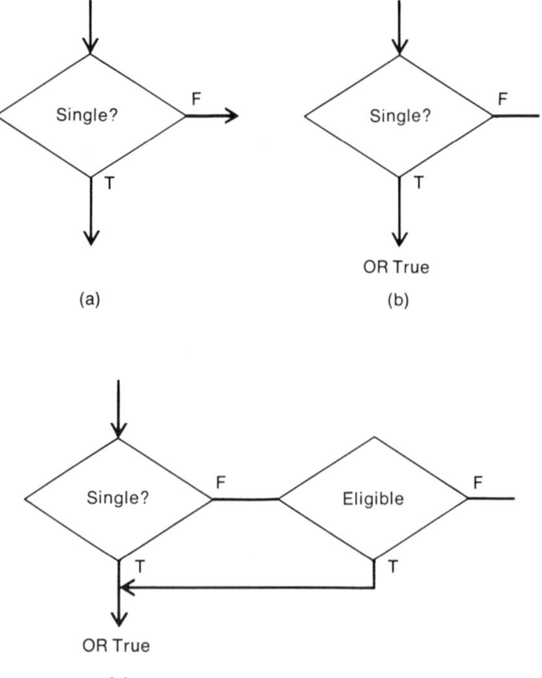

FIGURE 4.3

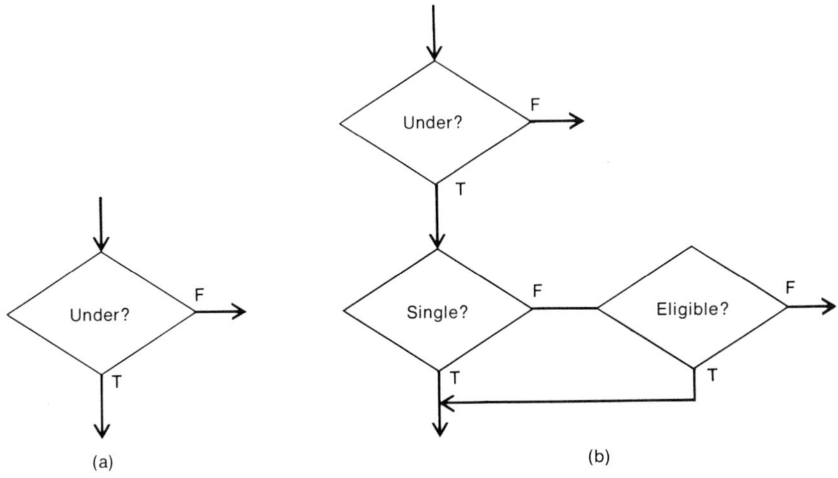

FIGURE 4.4

TRUTH TABLES AND FLOWCHARTS. The reader may wonder—must a truth table be constructed before a flowchart? The programmer will translate the flowchart into a list of instructions to the computer, but is a truth table really necessary?

Actually, constructing a flowchart and a truth table are similar activities. The truth table and flowchart for .NOT.*P*.AND.*Q* are indicated in Figure 4.6 as an aid in comparison.

The small numbers in the flowchart relate to lines of the truth table. The statement .NOT.*P*.AND.*Q* being true corresponds to the print instruction in the flowchart. When .NOT.*P*.AND.*Q* is false, the flowchart indicates that the next card should be read.

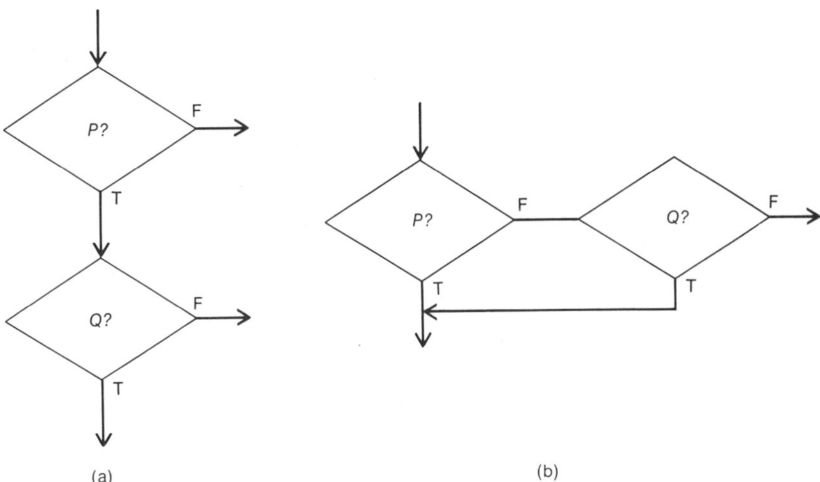

FIGURE 4.5 (a) *P*.AND.*Q*; (b) *P*.OR.*Q*

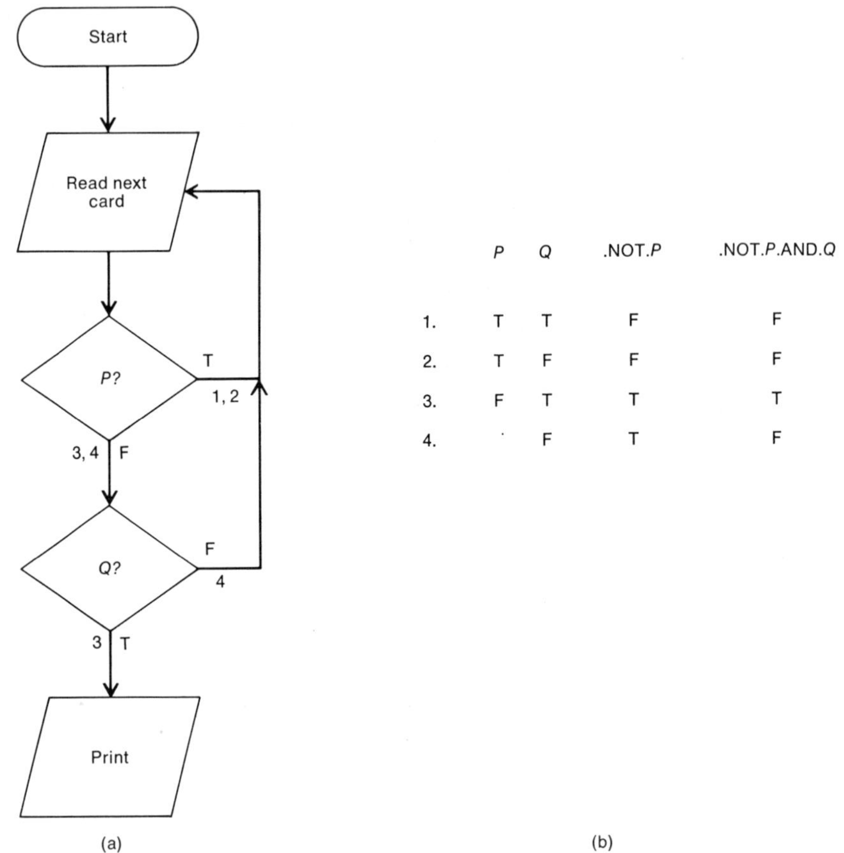

	P	Q	.NOT.P	.NOT.P.AND.Q
1.	T	T	F	F
2.	T	F	F	F
3.	F	T	T	T
4.		F	T	F

(b)

FIGURE 4.6 (a) .NOT.*P*.AND.*Q*; (b) Truth table

The importance of the flowchart to the programmer is indicated by the numbers 1 and 2 in the flowchart, when .NOT.*P* is false. In the truth table there are *two* lines where .NOT.*P* is false, but the programmer should condense this to one check. This results in fewer instructions to the computer and less computer time.

In effect, the programmer, in constructing a flowchart, is not concerned *only* with the truth values of .AND. and .OR. statements (as in a truth table) but also with a sequence of steps. He must determine what to do *next* after determining truth values.

This sequencing of steps is much less difficult when the logical structure of a problem is clear. For convenience this can be indicated in a symbolic form such as

```
(P.OR.Q).AND.(.NOT.P.OR.R)
```

or if one has a feeling for the structure of the problem, it can be translated directly to a flowchart.

A MORE REALISTIC PROBLEM. We now consider a problem with a more complex logical structure.

EXAMPLE 4.3

The homeowners property tax exemption will be granted to the owner of (1) a single family residence, (2) a condominium, (3) a duplex, provided the owner was living in the dwelling on March 1 and files a form by April 15.

It is important to note that even if a programmer does not use a symbolic logical form to represent the problem, he must recognize that parts 1, 2, and 3 relate to an .OR. situation and the last two to an .AND. situation. Then he could indicate this relationship by the form

```
R.OR.C.OR.D.AND.(L.AND.F)
```

On the other hand, he might recognize that this form could be interpreted in several ways; for example, inserting grouping symbols, we have

```
(R.OR.C.OR.D).AND.(L.AND.F)
```

Again the major difficulty in translating the verbal form is recognizing which placement of parentheses provides a match with the symbolic form. The *structure* of the problem is readily apparent after the correct choice is made.

Next, we note that the overall problem is an AND statement which is false if either component is false. This could lead to the flowchart in Figure 4.7 as a general organization of the problem. But the OR statement is true when just one component is true. This leads to the expansion of the basic flowchart as shown in Figure 4.8. Then, since *L*.AND.*F* is true only when both components are true, we have the final flowchart as shown in Figure 4.9.

An alert programmer would recognize (see problem 12, Exercise 4.3) that

```
(R.OR.C.OR.D).AND.(L.AND.F)
```

is the same as (logically equivalent to)

```
(L.AND.F).AND.(R.OR.C.OR.D)
```

and that this is significant (see problem 13, Exercise 4.3). However, a facility in manipulating complex forms (or flowcharts) is not easily attained. It requires

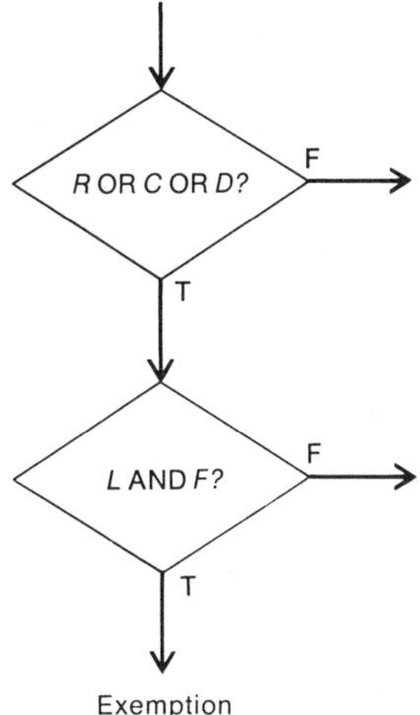

FIGURE 4.7

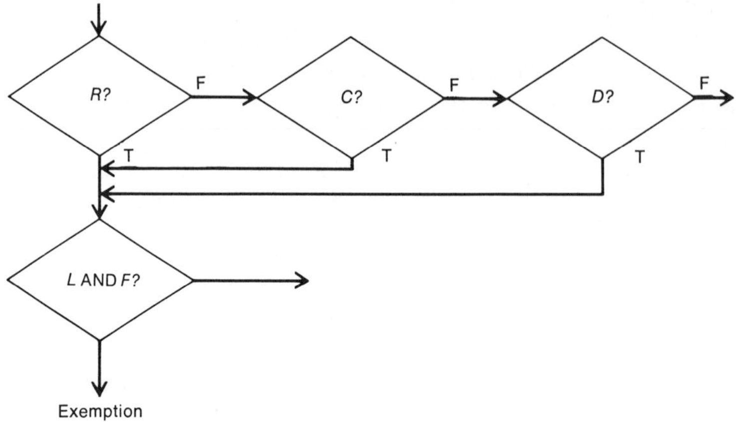

FIGURE 4.8

backing off from the form (or flowcharts, or problem) and noting the general structure. The reader should keep in mind that even though the construction of a flowchart does not require a truth table, this activity does help in becoming familiar with structure.

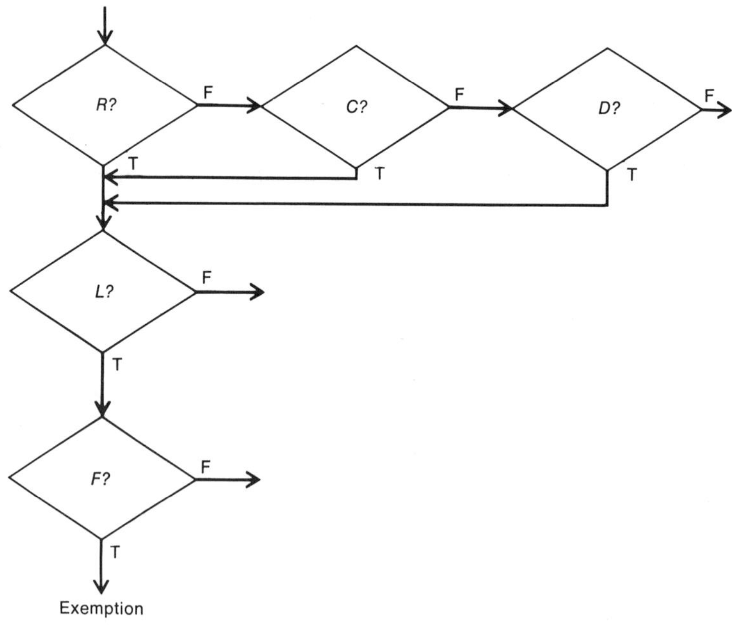

FIGURE 4.9

EXERCISE 4.3

Assume a list is to be printed when each logical form is true. Construct a flowchart for each situation.

1. .NOT.*P*.AND.(*Q*.OR.*R*)
2. *P*.OR.(*Q*.AND.*R*)
3. (*P*.OR.*Q*).AND..NOT.*R*
4. (*P*.AND.*Q*).OR..NOT.*R*
5. .NOT.(*P*.OR.*Q*)
6. .NOT.(*P*.AND..NOT.*Q*)
7. *P*.OR.(.NOT.*Q*.OR.*R*)
8. .NOT.(*P*.OR.*Q*).AND.*R*
9. A bonus is to be paid to all employees who are over 50 years of age and have been employed for 10 years or more, or are over 60 years of age and have been employed for 5 years or more. Draw a flowchart for this situation.

Draw a flowchart which produces a list when each of the following is true. (Problem 16, Exercise 4.2 may be of some help.)

10. *P*.EOR.*Q*
11. .NOT.*P*.EOR.*Q*
12. In Example 4.3, let *P* have the same truth value as *R*.OR.*C*.OR.*D* and *Q* the same truth value as *L*.AND.*F*. Is *P*.AND.*Q* logically equivalent to *Q*.AND.*P*? Why?
13. Draw a flowchart for (*L*.AND.*F*).AND.(*R*.OR.*C*.OR.*D*) in Example 4.3. Explain when and why it would provide a better computer program.

Draw a flowchart which produces a list when each of the following is true.

14. PAY IS GREATER THAN 800 OR RATE = 4.2 AND (TAX = 150 OR X-CODE = 1)
15. NET = 300 OR SUPPLY = 240 AND RATE = 2.7
16. (*P*.OR..NOT.*Q*).AND.(*P*.OR.*Q*)
17. ((*P*.AND.*Q*).OR..NOT.*R*).OR.*S*
18. (*P*.AND.*Q*).OR.(.NOT.*R*.OR.*S*)
19. *P*.AND.(*Q*.OR.(.NOT.*R*.or.*S*))
20. .NOT.*P*.OR..NOT.(*Q*.OR.(*R*.AND..NOT.*S*))
21. .NOT.(*P*.AND.(*Q*.AND.*R*))
22. .NOT.(*P*.OR.(*Q*.OR.*R*))

Construct truth tables for each of the following.

23. .NOT.(*P*.AND.(*Q*.AND.*R*))
24. .NOT.(*P*.OR.(*Q*.OR.*R*))
25. .NOT.(*P*.AND.(*Q*.OR.*R*))

4.4 IF-THEN FORMS

CONDITIONALS. In the preceding sections, AND, OR, and NOT statements were converted to flowcharts which indicated a particular sequence. A key part of each flowchart involved statements which had the property of being true or false. In the list of instructions to the computer, these statements are part of another basic logical form which is called a **conditional** or an implication. These have the form

```
IF PAY IS GREATER THAN 100,
    THEN MULTIPLY PAY BY 0.2 GIVING TAX.                Cobol

IF(A.AND.B.OR.C)  GO TO 73                          Fortran
```

These can be represented by

IF P, THEN Q

or

$$P \rightarrow Q \qquad (P \text{ implies } Q)$$

where P is called the **condition** (or antecedent) and Q the **conclusion**. In logic P *and* Q have truth values which lead to a truth value for the conditional, $P \rightarrow Q$ (Section 4.5). However, as a computer instruction,

IF P, (THEN) Q

the truth value (or sign) of P determines whether Q (which has no truth value) or the following instruction is executed. Thus in the conditional (Cobol) IF PAY IS GREATER THAN 100, MULTIPLY PAY BY 0.2 GIVING TAX, it is important to note that MULTIPLY NET BY 0.2 GIVING TAX describes an operation which can be performed by a computer. But for PAY IS GREATER THAN 100 this is not the case. The computer will check stored information to determine if this is true or false. Then on the basis of this truth value a "path" (indicated schematically in a flowchart) can be selected. This is described now in more detail.

COMPUTER STATEMENTS. As examples of IF-THEN statements that occur in different computer languages, we consider the following:

Fortran (FORmula TRANslation):

```
IF (A-2), 7,8,9              Arithmetic IF
IF (A.GT.B) GO TO 1200       Logical IF
```

Cobol (COmmon Business Oriented Language):

```
IF TOTAL PRICE IS GREATER THAN 40.00 OR CREDIT
IS EQUAL TO 100.00 THEN PERFORM DISCOUNT CALCULATION.

IF NET IS GREATER THAN 200
MULTIPLY NET BY .15 GIVING TAX,
SUBTRACT TAX FROM NET GIVING TAKE-HOME-PAY
ELSE MULTIPLY NET BY .20 GIVING TACKS
SUBTRACT TACKS FROM NET GIVING TAKE-HOME-PAY.
DISPLAY TAKE-HOME-PAY
```

Algol (*ALGO*rithmic *L*anguage):

```
IF I < J ∨ J: = N THEN A(I) = J
```

Mad (*M*ichigan *A*lgorithm *D*ecoder):

```
WHENEVER J.E.O. TRANSFER TO ARRAY
```

Balgol (*B*urroughs *ALGO*rithmic *L*anguage):

```
EITHER IF A EQL B; GO TO R1;
OR IF A LSS B; X = 2
OTHERWISE; X = 2A
```

In Fortran, Mad, and Balgol, the word THEN is omitted with the understanding that the word IF and a comma, semicolon, or parentheses provides sufficient identification. In Cobol the word THEN is not required by the computer but may be used for convenience.

In terms of execution, the Fortran arithmetic IF statement differs considerably from the others. The number represented by the expression (A-2) in the above Fortran statement will be negative, zero, or positive for a particular value of A. The statements labeled 7, 8, or 9 will be executed depending on whether the expression is less than zero, equal to zero, or greater than zero. For example, if A = 2, A − 2 = 0, and statement 8 is executed next. This type of instruction enables a computer to BRANCH or perform a CONDITIONAL TRANSFER. The Fortran statement also differs from the others in that the antecedent (A-2) does not have a truth value. The claim A − 2 = 0 or A − 2 < 0 can be labeled true or false for a particular value of I but this is not the case for the expression, A − 2.

Next we consider the Fortran logical IF statement.

```
IF(A.GT.B) GO TO 1200
```

1. If A.GT.B (A > B) is true, the GO TO 1200 statement is executed.
2. If A.GT.B is false, control is transferred to the next statement in the sequence.

The effect of the first Cobol IF statement is similar to the Fortran logical IF statement except here the condition is an OR statement. The truth values of the OR statement determine whether PERFORM DISCOUNT CALCULATION is executed or skipped (Figure 4.10).

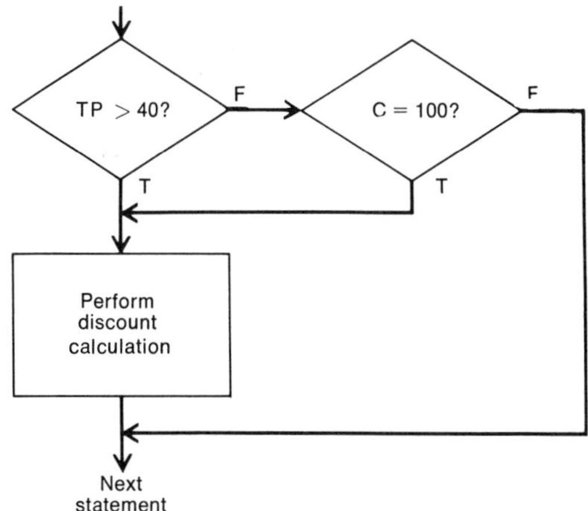

FIGURE 4.10

We now reverse procedures and use the flowchart of Figure 4.11 to illustrate how the second set of Cobol statements (seven lines) is interpreted. When the condition, NET IS GREATER THAN 200 is true, the next two statements (before ELSE) are executed. Then, the statements between ELSE and the *period* are skipped and the final statement, DISPLAY TAKE-HOME-PAY, is performed.

If the condition NET IS GREATER THAN 200 is false, the statements before ELSE are skipped and those between ELSE and the *period* are performed. Then, as before, the last statement is executed.

The general structure of an IF-ELSE Cobol statement is shown in Figure 4.12; note that only the condition P has a truth value. This condition P, the word ELSE, and the *period* divide the statements A through E into two sets. The first set, $\{A,B,C\}$, is performed when P is true while the second set $\{D,E\}$, is performed when P is false. In either case statement F is performed last.

The Balgol example contains a compound conditional. The execution of this sequence of statements is as follows: The first condition, A EQL B, is checked to determine its truth value. If it is true, the GO TO R1 statement is executed. If A EQL B is false, this statement is bypassed and the next statement is considered. Since it also is a conditional statement, the truth value of A LSS B is checked. If A $<$ B, X is set equal to 2, but if A $>$ B, the OTHERWISE statement is executed.

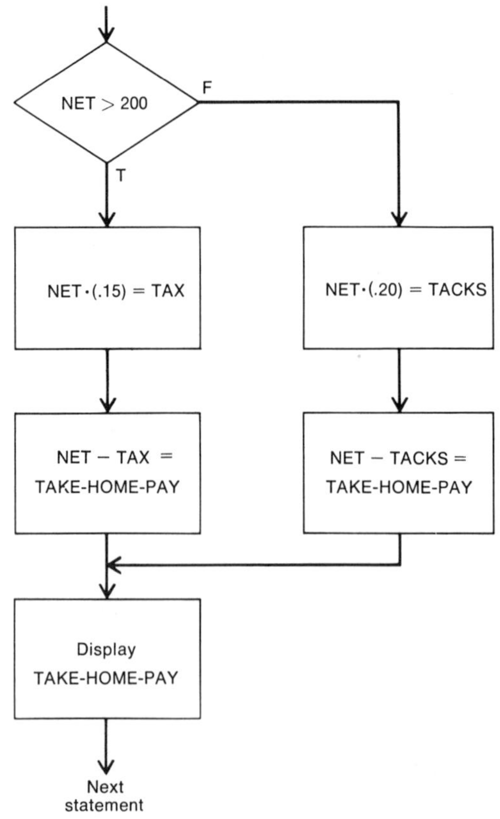

FIGURE 4.11

```
IF P
   A
   B
   C
  Else
   D
   E.
   F
```

(a)

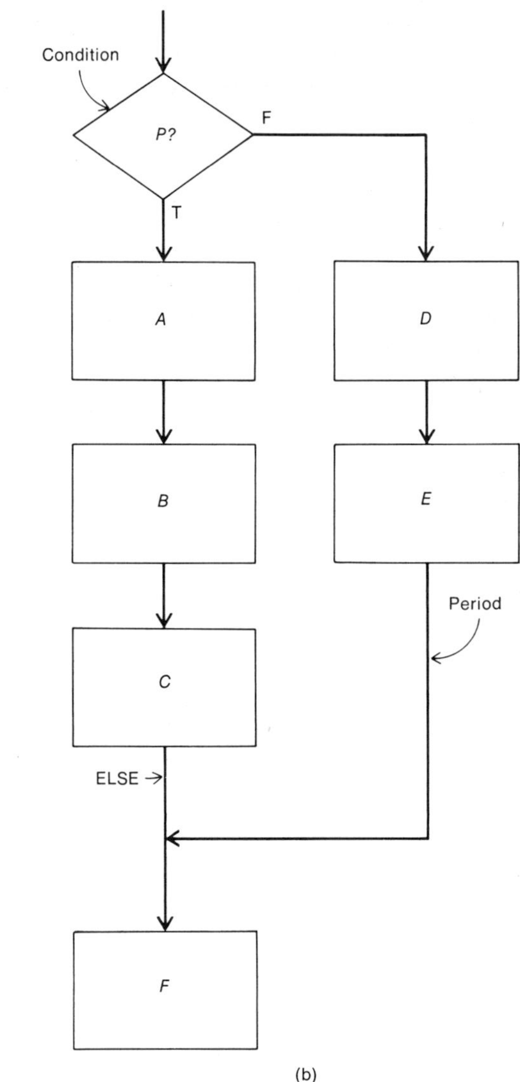

(b)

FIGURE 4.12 (a) Symbolic equivalent Cobol Statement; (b) flowchart

Whenever a condition is true, the corresponding conclusion is executed and the remaining conditionals and the OTHERWISE statement are bypassed. The OTHERWISE statement is executed only when none of the conditions is true.

Similar compound conditional statements can be used in Algol, Cobol, and Mad. For example, Algol would include a statement of the form

```
IF B THEN S1 ELSE S2;
```

If B is true, the statement S1 will be executed and S2 bypassed. When B is false, S1 is skipped and S2 is executed. In both situations, the program proceeds to the statement immediately following the conditional statement.

OTHER CONDITIONAL FORMS. Recognition of conditional forms may also be important to a programmer before a program is written. The following statement refers to a property tax exemption of $750 which was granted to eligible taxpayers (California, 1969).

> *If the parcel number or other legal description of the property and the address of the dwelling are printed on the form when you receive it, check to see that they are printed correctly and correct them if they are not.*

As a taxpayer, one would read the statement and attempt to follow the instructions. But a programmer who has to "translate" this statement must look at it from a different point of view. A check of the logical structure would reveal (in order)

IF, OR, AND, (THEN), AND, IF.

This may lead to the following:

IF ((Parcel *N*..OR.Legal *D*.).AND.Address) (correct),
THEN (Printed *C*? .AND. (IF NOT Printed *C*, THEN Correct)).

However, the flowchart of Figure 4.13 will give a better indication of the general organization. Additional information is needed to tie down the loose ends, but note that

Printed *C* .AND. (IF NOT Printed *C*, THEN Correct)

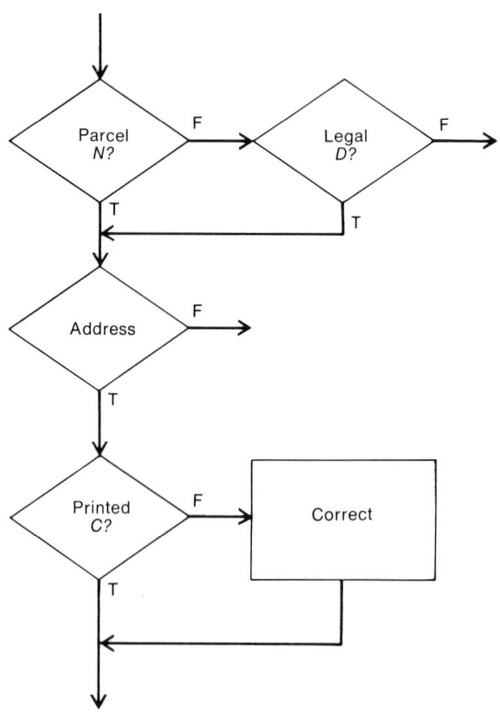

FIGURE 4.13

requires a single check. This type of simplification will be pursued further in Section 4.7.

EXERCISE 4.4

Use the indicated letters to write each of the following in symbolic form.

1. If wages increase, prices will also increase. (*W,P*)
2. If prices increase, pensioners will be hurt the most. (*P,H*)
3. If wages increase, pensioners will be hurt the most. (*W,H*)
4. If the Giants and Dodgers win, the Reds will fall to third place. (*G,D,R*)
5. If an employee is over 50 or he has worked 10 years or more and is under 50, the company will pay him a bonus. (*O,W,B*)
6. The present trend can be halted if (a) slum housing is demolished, (b) industrial employment is increased, (c) civic leaders are more energetic. (*T,S,I,C*)
7. If the major factors leading to senility are uselessness, boredom, and inactivity, education for leisure is important. (*U,B,I,E*)
8. If that means I am not a circus leader, the master of ceremonies of a Senate nightclub, a tamer of political lions, or a wheeler and dealer, I must accept the words. (*C,M,T,W,A*)
9. If he had his choice, he would prefer to go down in history as a foreign affairs expert rather than a great leader. (*C,E,L*)
10. If computers are used in place of credit cards, each individual will select a code of 5 letters that only he and the computer would know. (*C,S,H*)
11. If the card number and secret code do not match, or if the card has been reported lost, the computer will activate a reject light or a police siren. (*M,L,R,P*)

Construct flowcharts for each of the following:

12. IF (A .GT. B) GO TO 1200 (Fortran IV logical IF)
13. IF (A-2), 7, 9, 9 (Fortran arithmetic IF)
14. IF QUOTIENT OVERFLOW 40, 20
15. Problem 5 above
16. Problem 11 above
17. The Balgol example discussed in this section
18. IF QUANTITY-ON-HAND IS LESS THAN 400 OR WAREHOUSE-TWO IS NUMERIC THEN PERFORM REORDER-700-S AND PERFORM REORDER-500-T ELSE PERFORM REORDER-300-N.
19. IF SALARY = 400 AND S-CODE = 1
THEN MOVE 80 TO TAX AND MOVE 30 TO PENSION.
20. IF SALARY = 650 THEN MOVE 120 TO TAX AND MOVE 7 TO DEDUCTION AND MOVE SALARY-TAX-DEDUCTION TO WAGE

21.
```
IF A AND B
    C
  ELSE
    D.
    E
```

22.
```
IF A OR B
    C
    D
  ELSE
    E.
    F
```

23. IF (A AND B) OR C
D
ELSE
E
F.
G

24. IF (A OR B) AND C
D
E
ELSE
F.
G

25. IF (A OR B) OR C
D
ELSE
E.
F

Use the indicated letters to write each of the following in symbolic form.

26. If you have age-revealing brown spots, blotches, or if you want clearer skin, use SKIN-BLOOM. (*A,B,C,S*)
27. If the job is in dispute the association will send men to inspect the work and if they judge it faulty, the original contractor will be instructed to correct it. (*D,S,J,C*)
28. If the consumer is spending too much for goods and thus causing prices to rise, his taxes can be increased or his borrowing discouraged. (*S,P,T,B*)
29. If unions are pressing or other costs are rising, the producing firm can almost always raise prices and pass the added costs along to the public. (*U,C,R,P*)
30. If someone else's name is printed on the form and you were an owner of the property (or a purchaser under contract of sale) and an occupant on March 1, 1970, strike out the printed name and insert your own name, or add your name if you and the one whose name is printed were co-owners. (*N,W,P,O,S,I,A,C*)
31. Draw flowcharts for problems 26 and 27 above.
32. Draw flowcharts for problems 28 and 29 above.
33. Draw a flowchart for problem 30 above.

4.5 TRUTH VALUES OF CONDITIONALS AND EQUIVALENCES

We now take a brief look at the truth values of IF-THEN statements. It may be of help to note that a logician is concerned with a set of statements and the possibility of drawing a conclusion from these statements. For example, imagine a situation where the following statements are true.

A card is bent or the power is off.

and later

The power is on.

Does the conclusion,

therefore a card is not bent?

follow from the first two statements? In this case the answer is yes. A second example is:

If a card is not bent, then the power is off.

and later

The power is off.

Does it follow that

therefore a card is not bent.

Here, logicians assert that the conclusion is incorrect. The reader will note that this example could occur in everyday conversation. However, to fit this into a bigger picture, we need the truth values of an IF-THEN statement. A logician asserts that

IF P, THEN Q

is false *only* when P is true and Q is false as shown in Table 4.11.

TABLE 4.11

P	Q	IF P, THEN Q
T	T	T
T	F	F
F	T	T
F	F	T

The first two lines of this table are easy to accept in terms of common usage but the reader should not expect the *same* conviction for lines 3 and 4. Normally when condition P is false, an IF-THEN statement is ignored and not used. We can, however, point out a reasonable expectation in these last two cases.

Consider

IF A is true, THEN (A.OR.B) is true.

On line three of Table 4.11

the condition P is false .AND. the conclusion Q is true

or

A is false .AND.(A.OR.B) is true.

This is certainly possible (when B is true) so this does not conflict with the total implication being true. On line four of Table 4.11

the condition P is false .AND. the conclusion is false

or

A is false .AND.(A.OR.B) is false.

This again does not conflict with the implication being true. From this point of view, Table 4.11 is reasonable.

EXAMPLE 4.4

Construct a truth table for .NOT.$P \to Q$

P	*Q*	.NOT.*P*	.NOT.*P* → *Q*
T	T	F	T
T	F	F	T
F	T	T	T
F	F	T	F

.NOT.*P* → *Q* is false only when .NOT.*P* is true .AND.*Q* is false.

Again it should be noted that the truth values of an IF-THEN statement were not needed in showing how IF statements are used in a program. They are offered here to complete the picture of the five basic components of symbolic logic.

STATEMENTS AND CONCLUSIONS. As a brief glimpse of the use of the truth values of an IF-THEN statement, consider the second example in this section.

If a card is not bent, then the power is off.
The power is off. } both true
Is *a card is bent* necessarily true?

Let

BENT represent *A card is bent,*
POWER represent *the power is off.*

Then, *both* statements,

IF NOT *BENT,* THEN *POWER*
POWER

are true on lines 1 and 3 in Table 4.12.

TABLE 4.12

BENT	*POWER*	NOT *BENT*	→	*POWER*	*POWER*	*BENT*
T	T	F	T	T	T	?
T	F					
F	T	T	T	T	T	?
F	F					

However, when both statements are true, *BENT* can be true or false. So it is not correct to assert that *BENT* is *necessarily* true when the first two statements are true.

In a similar manner, consider the first example in this section.

A card is bent or the power is off	*BENT* OR *POWER*
The power is on	NOT *POWER*

Both of these statements are true only on line 2 in Table 4.13

TABLE 4.13

BENT	*POWER*	*BENT*.OR.*POWER*	.NOT.*POWER*	*BENT*
T	T		F	
T	F	T	T	?
F	T		F	
F	F	F		

Also on this line, the conclusion *BENT* is true. Thus, when *both* of the first two statements are true, the third statement is *necessarily* true.

This truth table method of determining if one statement follows from others may seem too complex to be of any value. However, note that each of the following sets of statements has the same form as the last example.

The flowchart is incorrect or the card is out of sequence.
The flowchart is correct.
Therefore, a card is out of sequence.

Prices will remain the same or inflation will continue.
Prices rose.
Therefore inflation continues.

Each of these is of the form:

P.OR.*Q*	is true
NOT *Q*	is true
Therefore *P*	is true

So a truth table check (similar to Table 4.13) would establish that the third statement is necessarily true when the first two statements are true. One need not treat each set of statements as a new problem but, by noting the structure of the combination of statements, they can be related to a general form.

EQUIVALENCE. A second form of implication is illustrated by the following statement:

You will get Medicare insurance *only if* you enroll for it.

This implication can be written in symbolic form as

$$Q \rightarrow P$$

where

Q: You will get Medicare insurance
P: You enroll for it.

Note the following implication, which, although similar to the previous one, is not the same.

If you enroll for it, you will get Medicare.

Some Federal employees are not eligible for Medicare; hence, even if they applied they would not receive it.

Whenever *only* precedes *if* in an implication, the statement following *only if* is the conclusion instead of the antecedent.

The more general form

$$P \text{ IF AND ONLY IF } Q$$

can be represented symbolically as

$$(P \rightarrow Q).\text{AND}.(Q \rightarrow P).$$

This is commonly abbreviated to

$$P \leftrightarrow Q \qquad (P \text{ equivalent to } Q).$$

The truth values of this form, called an **equivalence**, are easily determined from the truth values of an .AND. statement and an implication (Table 4.14).

TABLE 4.14

P	Q	$P \rightarrow Q$	$Q \rightarrow P$	$P \leftrightarrow Q$
T	T	T	T	T
T	F	F	T	F
F	T	T	F	F
F	F	T	T	T

The table can be summarized by noting that $P \leftrightarrow Q$ is true when the truth values of P and Q are the same; otherwise it is false.

Fortran IV, Cobol, Algol, Balgol, and Mad all include provisions for the use of the basic logical forms: NOT, OR, AND, *implication,* and *equivalence.* The symbolic representation used in each language is shown in Table 4.15.

TABLE 4.15

	NOT	.AND.	.OR.	*Implication*	*Equivalence*
Algol	$>$B1	B1 $\wedge$ B2	B1 $\vee$ B2	B1 $>$ B2	B1 $\equiv$ B2
Balgol	NOT B1	B1 AND B2	B1 OR B2	B1 IMPL B2	B1 EQUIV B2
Mad	.NOT.B	P.AND.R	P.OR.R	P.THEN.R	P.EQV.R
Fortran IV	.NOT.B	P.AND.R	P.OR.R		
Cobol	NOT B	P AND R	P OR R		

EXERCISE 4.5

Construct a truth table for each of the following conditionals. Use Table 4.11 as a guide.

1. $P \rightarrow$ .NOT.Q
2. .NOT.$P \rightarrow Q$
3. .NOT.$P \rightarrow$ .NOT.Q
4. IF .NOT.Q, THEN P
5. IF Q, THEN .NOT.P
6. IF .NOT.Q, THEN .NOT.P

Find an .OR. statement which has the same truth values as

7. $P \rightarrow Q$
8. $P \rightarrow$ .NOT.Q
9. IF .NOT.Q, THEN P
10. IF Q, THEN .NOT.P

11. Construct a truth table for
    (a) $P \rightarrow Q$
    (b) P
    (c) On which lines of the table are *both* $P \rightarrow Q$ and P true?
    (d) Is Q true on this line?
    (e) When $P \rightarrow Q$ and P are true, does it follow that Q is *necessarily* true?
12. Using a sequence of steps similar to those in problem 11, show that when .NOT.$P \rightarrow Q$ is true and also .NOT.P is true, then Q is also true.
13. Using problem 11 as a guide, show that when $P \rightarrow Q$ is true and .NOT.Q is true, then .NOT.P must be true.
14. Given
    (a) .NOT.P.OR.Q is true and P is true,
    (b) does it follow that Q is true?
15. Given
    (a) (IF .NOT.P THEN Q) is true, and .NOT.Q is true,
    (b) does it follow that P is true? (compare with problem 13)
16. Given
    (a) .NOT.$P \rightarrow Q$ is true and .NOT.Q is true,
    (b) does it follow that P is true?
17. Given
    (a) P.OR..NOT.Q is true and P is true,
    (b) does it follow that .NOT.Q is true?

Construct truth tables for each of the following equivalences. Use Table 4.14 as a guide.

18. $P \leftrightarrow$ .NOT.Q
19. .NOT.$P \leftrightarrow Q$
20. .NOT.$P \leftrightarrow$ NOT Q
21. .NOT.$(P \leftrightarrow Q)$
22. Construct truth tables for
    (a) P.EOR.Q (b) $P \leftrightarrow$ .NOT.Q (c) .NOT.$P \leftrightarrow Q$
    (d) Are these three forms logically equivalent?
23. Using Table 4.15 as a guide, construct truth tables for:
    (a) $B1 \wedge (B2 \vee B1)$ (b) $B1 > (B1 \wedge B2)$

4.6 SPECIAL USES OF AND, OR, AND EOR

In a preceding section, it was shown how .AND. and .OR. statements may occur in an instruction to a computer. These examples closely parallel the use of these statements in everyday language. We now consider a more specialized usage of these forms.

In the base 2 notation used by computers, 0 and 1 can also represent false and true for a particular statement. For example, consider the following classifications related to car insurance.

EXAMPLE 4.6

A Regular drivers are 25 or over.
B Unmarried male regular drivers are 30 or over.

C Normal use of car is for pleasure.
D Normal use of car includes not over 30 miles weekly to work.
E Annual mileage is over 7500 miles.
F Two or more cars are insured.

For each of our insured customers, we might set aside the rightmost six-bit positions of a computer word to store the required information. Using a 1 bit to represent a yes condition (true) and a 0 bit for a no condition (false), the record of a given insured would appear in storage as shown in Figure 4.14.

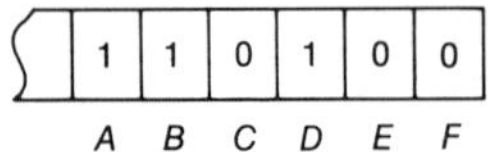

FIGURE 4.14

In this case, the binary number

1 1 0 1 0 0 representing T T F T F F

indicates statements *A, B* and *D* are true while statements *C, E* and *F* are false.

Any change in classification for a particular car owner would probably result in a different rate. With rising costs, the company and the car owner are interested in updating the binary number which acts as a record for essential information. We now indicate how the basic logical forms can be used for this purpose.

THE INCLUSIVE OR. Suppose the car owner with the above classification 110100 insures a second car with the company. If this is the only change,

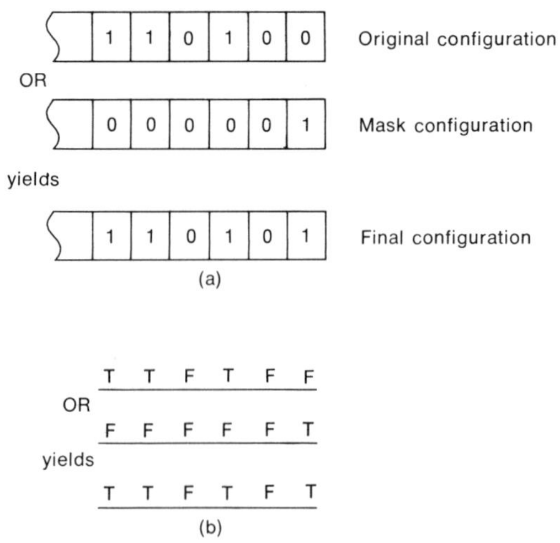

FIGURE 4.15 The logical OR

his record should be changed to 110101. This can be accomplished as illustrated in Figure 4.15. Recall that

1. An .OR. is true (has value of 1) if one component is true (has a value of 1);
2. An .OR. is false (has a value of 0) only when both components are false (have values of 0).

Thus we conclude that the first five zeros (false) produce no change.

1 .OR. 0 is 1 (T.OR.T is T)
0 .OR. 0 is 0 (F.OR.F is F)

This example demonstrates that an inclusive OR can be used to change a particular 0 to 1 and leave the remaining binary digits (bits) unchanged.

THE AND. Another often required operation is to change a particular 1 to 0. In the above example, suppose a 17-year old youth begins driving the family car but is not classified as a regular driver. Then the second 1 bit in 110100 should be changed to 0 (100100). This can be accomplished as illustrated in Figure 4.16. *Note*: An .AND. statement is true (has a value of 1) only when both components are true (have values of 1); an .AND. statement is false (has value of 0) if one component is false (has value of 0).

Thus we conclude that the first 1 and the last four 1s in the mask produce no change in the original configuration since

1 .AND. 1 is 1 or (T.AND.T is T)
0 .AND. 1 is 0 or (F.AND.T is F)

In this case, .AND. is used to change a 1 to 0 while leaving all other bits unchanged.

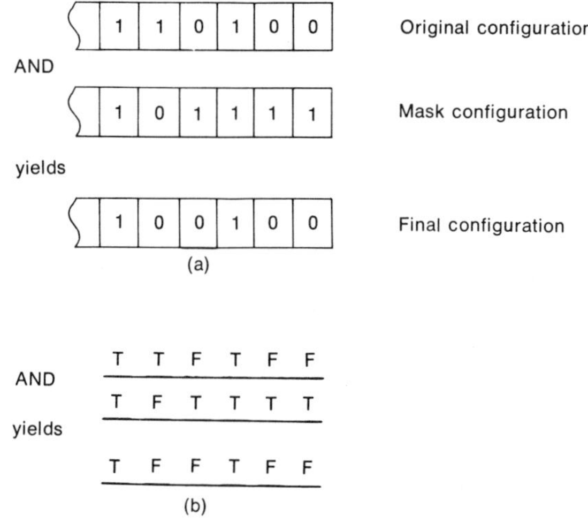

FIGURE 4.16 The logical AND

THE EXCLUSIVE OR. The exclusive OR also has a specialized application. The difference between the two logical OR forms is shown in Table 4.16.

TABLE 4.16

P	*Q*	*P*.OR.*Q*	*P*.EOR.*Q*	*P*	*Q*	*P*.OR.*Q*	*P*.EOR.*Q*
T	T	T	F	1	1	1	0
T	F	T	T	1	0	1	1
F	T	T	T	0	1	1	1
F	F	F	F	0	0	0	0

Note the one small difference when both bit components are 1. The EOR produces a 0 while OR produces a 1. This means EOR could have been used in place of OR in the above example since we were changing a 0 to 1. To illustrate another use of EOR, consider the examples in Figure 4.17.

No change in original   A complete reversal of original

As shown in 4.17(a), a 0 bit in the mask produces no change, but as shown in 4.17(b) a 1 bit changes 0 to 1 and 1 to 0. Thus in 4.17(b) the entire six-bit area is inverted meaning that the exclusive OR can be used to produce the 1s complement of a binary number.

TESTING BINARY DATA. Let us assume that we have the task of identifying all individuals who drive over 7500 miles annually and have more than one car. Referring to the storage register of Figure 4.14, we will then be

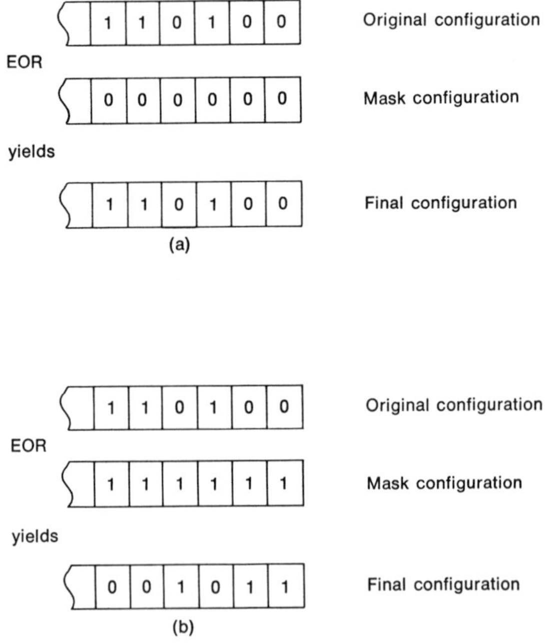

FIGURE 4.17 The logical exclusive OR

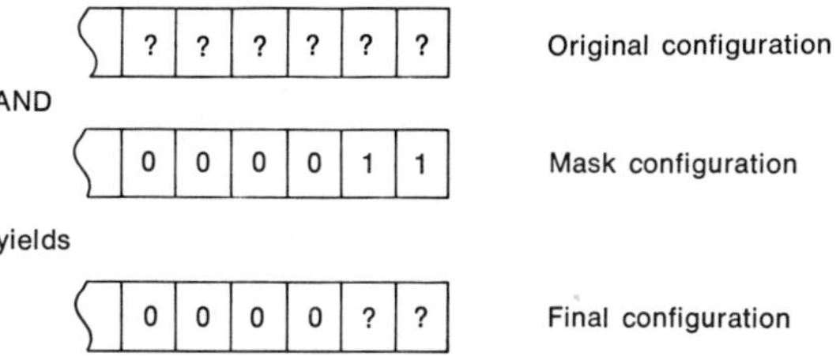

FIGURE 4.18

checking each record to find those with 1 bits in the *E* and *F* positions. This can be done using the logical instructions and the fact that in most computers special indicators are set after execution of a logical instruction to indicate whether or not the result contains all 0s. Since we are interested only in the rightmost two bits, we would first "eliminate" the other four using the **AND** and a mask of 000011 as shown in Figure 4.18. Now if we use an **EOR** with a mask of 000011, any record which satisfies the specified criteria (that is, has the entries 000011 after **AND**ing) will yield zero. Any other combination will yield a result other than 0. This is illustrated with the four possible configurations in Figure 4.19. Note that only in 4.19(d) does the resulting field consist of all 0 bits. Thus the programmer would include within his program at this point a test for 0. The overall sequence is illustrated by the flowchart of Figure 4.20.

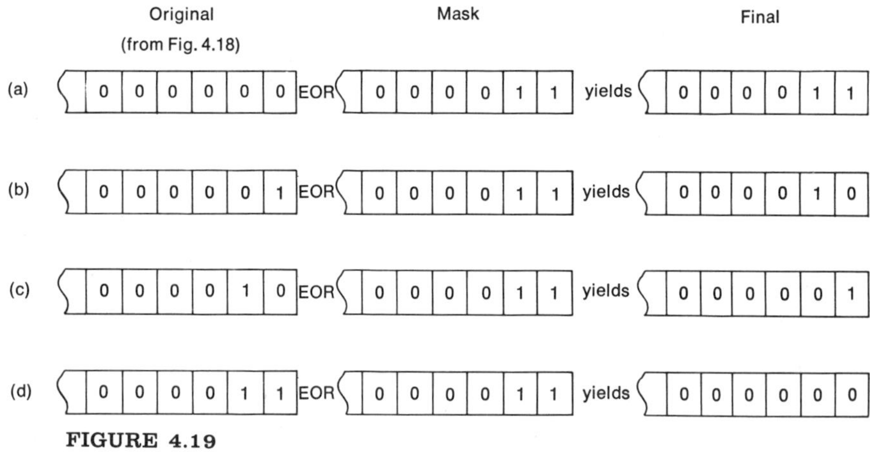

FIGURE 4.19

The IBM System/360 computer has a special instruction, **TEST UNDER MASK**, which provides the programmer with a convenient means for testing selected bits in a storage byte. In using the instruction, a test mask (which becomes part of the instruction itself) is defined. A mask bit of 1 indicates that the corresponding storage bit is to be tested; a mask bit of 0 indicates that the corresponding storage bit is to be ignored. A special *condition code* within the machine is then set as follows

All tested bits are 0	code set to 0
All tested bits are 1	code set to 1
Tested bits are mixed 0 and 1	code set to 3

The test mask to be used in this example would be 00000011. Records satisfying the specified conditions (as shown in the original field of 4.19(d)) would yield a condition code of 1; all others would produce a code of 0 or 3.

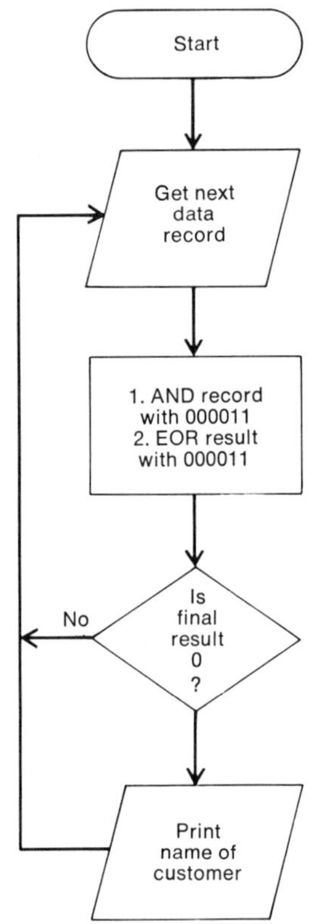

FIGURE 4.20

EXERCISE 4.6

1. Let *A* represent the four-bit word 1011.
   (a) List the results of combining *A* with 0000 using AND, OR, and EOR in that order. Use a format similar to that of Figure 4.16(a).
   (b) List the results of combining *A* with 1111 using AND, OR, and EOR in that order.
   (c) List the results of combining *A* with itself (1011) using AND, OR, and EOR in that order.
   (d) List the results of combining *A* with 0100 using AND, OR, and EOR in that order.
2. In problem 1, let *A* represent 1011, *Z* represent 0000, *N* represent 1111 and $\overline{A}$ represent 0100.

(a) Three of the results in problem 1 are all 0s or Z. One of these could be represented by A AND Z. Find representations for the other two.

(b) Which combinations produce the result N or all ones.

(c) Which combinations produce the result A again.

(d) Which combinations produce the result $\overline{A}$, the opposite of A.

3. Using the classifications, A through F, discussed in Example 6, let R represent the customer record 110010. Describe the changes which occur in each of the following:

   (a) R OR C where C represents 000100
   (b) R OR C where C represents 000001
   (c) R EOR C where C represents 010010
   (d) R AND C where C represents 101101
   (e) R EOR C where C represents 010001
   (f) R AND C where C represents 101110

Suppose the following classifications (used in problems 4-10) are used by a company for each employee.

		Age Group			Insurance Coverage			
Bit Value	Sex	Under 21	Over 65	Pay Category	Special	Extended	Family	Supple-mentary
0	M	No	No	Hourly	No	No	No	No
1	F	Yes	Yes	Salary	Yes	Yes	Yes	Yes

Thus, for example, an employee record of 11010100 indicates an employee who is female, under 21, on a salary, with extended insurance coverage.

4. Indicate how each record could be updated for the given change.
   (a) 11011010; changes—age 21, special injury insurance deleted, supplemental coverage added.
   (b) 00000110; changes—age 66, special injury and supplementary life insurances added
   (c) 10111001; changes—hourly, extended coverage added.

5. Using the classifications in problem 4, describe each change.

```
(a)      10000101          (b)      01011001
     EOR 00101100              AND  11100111
(c)      11000101          (d)      11000101
     OR  00010000              EOR  01010000
```

6. Indicate the result and condition code for each use of a TM instruction.

   Example:

```
        Use of TEST UNDER MASK (TM) instruction
                 11000101
        Mask     01010000
        Result   x1x0xxxx   Condition Code 1
```

```
(a)          10111001        (b)          10111001
      Mask   10010000              Mask   01000110
(c)          10000101        (d)          11000101
      Mask   01100000              Mask   01100000
```

7. For an unknown record, a TEST UNDER MASK instruction with mask (01100000) gives the condition code 0. What can be said of the employee's age?
8. A TEST UNDER MASK instruction with mask (01100000) is applied to the result of:

```
    xxxxxxxx
EOR 01000000
```

Explain the meaning of each condition code in terms of age. (Note second and third entries can only be 00, 01, or 10.)

9. Explain how to use an EOR and TEST UNDER MASK instruction with appropriate mask to determine if an employee is under 21 and female.
10. Explain how to use an EOR and TEST UNDER MASK instruction with appropriate mask to determine if an employee is male, over 65, and has special injury insurance.
11. Consider two storage bytes, call them *A* and *B*, which contain 1011 and 1101, respectively. Perform the following sequence of operations.
    (a) *A* EOR *B* with result replacing original value of *B*.
    (b) *B* EOR *A* with result replacing current value of *A*.
    (c) *A* EOR *B* with result replacing current value of *B*.

By comparing the new contents of *A* and *B* with the original values, describe what has occurred.

12. Let *A* represent 0110 and *B* represent 0101 and follow the sequence of steps in problem 11.
13. Let *A* represent 1100 and *B* represent 0011 and follow the sequence of steps in problem 11.
14. Let *A* and *B* represent simple statements with truth values. Referring to the intermediate results of (a), (b) and (c) in problem 11 as *C, D,* and *E*, respectively, construct a truth table for *C, D,* and *E*. How does the table verify the results of problems 11-13?

4.7 SIMPLIFICATION BY FLOWCHART

In earlier sections, we have noted the possibility of using one logical form in place of another if they had the same truth values (or were *logically equivalent*). Logically equivalent forms can be determined from truth tables or by using certain "rules" (Boolean properties) in simplifying complex forms. The first is frequently tedious and the second requires some understanding of Boolean algebra. As an alternative, we will use flowcharts to illustrate logical structure and to simplify more complex logical forms.

We consider a problem which can be represented by

$$(.\text{NOT}.P.\text{AND}.(P.\text{OR}.Q)) \rightarrow \text{PRINT}.$$

The flowchart begins with a check of the truth values of *P*. When *P* is true, .NOT.*P* is false and the overall .AND. statement is false; when *P* is false, *P*.OR.*Q* should be checked. But since *P* is false we need only check *Q*. Now looking

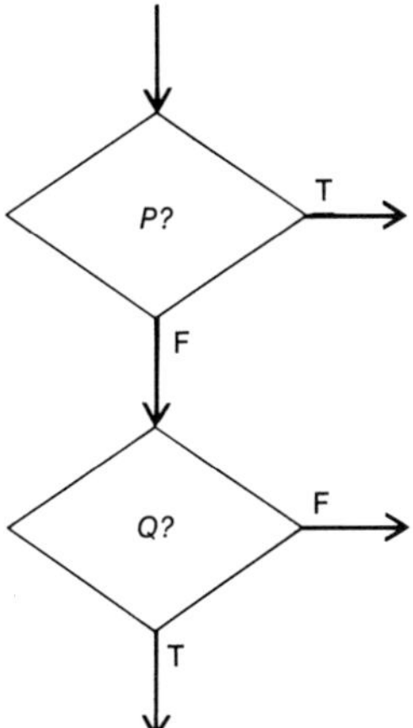

FIGURE 4.21 .NOT.*P*.AND.(*P*.OR.*Q*)

at the flowchart in Figure 4.21 the path from START to PRINT can be labelled

```
.NOT.P.AND.Q
```

This suggests that the last form is logically equivalent to

```
.NOT.P.AND.(P.OR.Q)
```

This can be verified by a truth table (Exercise 5.7, problem 1).

If a distributive property (from Boolean algebra) is used, we have

```
(.NOT.P.AND.P).OR.(.NOT.P.AND.Q)
```

Then we note that .NOT.*P*.AND.*P* is always false so the final truth value depends on .NOT.*P*.AND.*Q*; that is, the truth value of this AND statement will be the same as the original.

In a similar manner, the flowchart of Figure 4.22 which represents

(.NOT.*P*.OR.(*P*.AND.*Q*)) → PRINT

suggests that the following two forms are logically equivalent. Again

```
.NOT.P.OR.(P.AND.Q)
.NOT.P.OR.Q
```

this can be verified by a truth table or a distributive property (Exercise 5.7, problem 2).

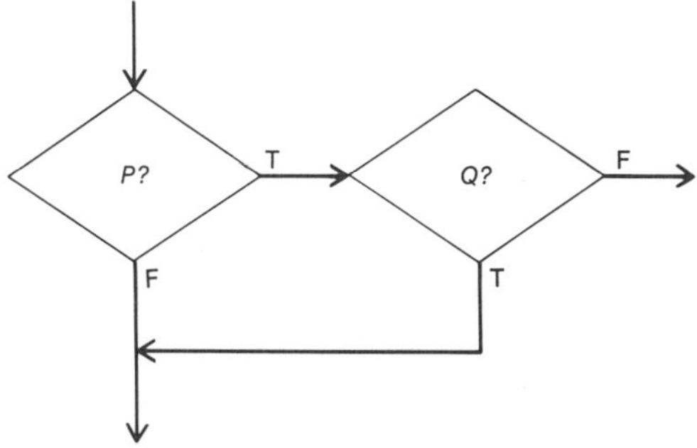

FIGURE 4.22 .NOT.*P*.OR.(*P*.AND.*Q*)

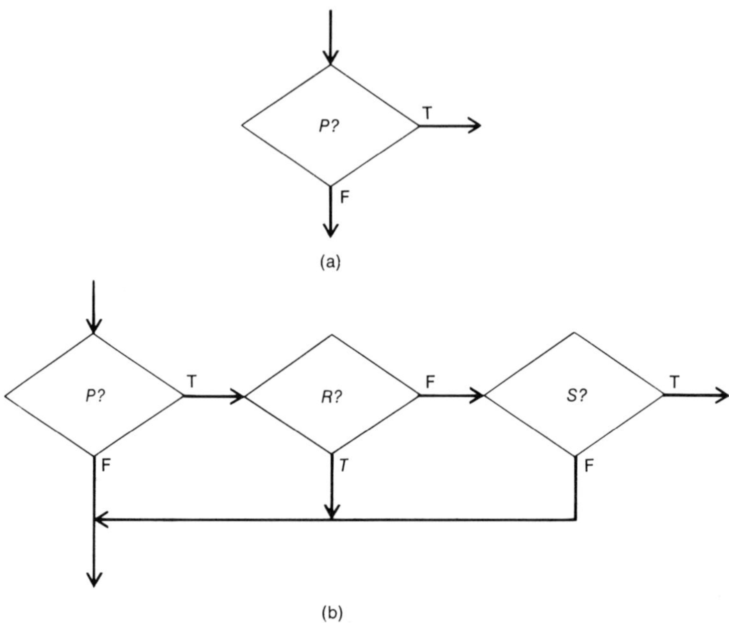

FIGURE 4.23 (a) P; (b) .NOT.P.OR.(R.OR..NOT.S)

Finally, we consider a more complex situation. Print a list if the following is true.

```
((P.OR.Q.AND.(R.OR..NOT.S)).OR..NOT.P
```

Should the flowchart begin with the .OR.S and .AND. in the parentheses or .NOT.*P* on the right? The overall form is an .OR. statement with the contents of the parentheses and .NOT.*P* as components. If .NOT.*P* is true, the overall .OR. statement is true. For this reason, *P* should be tested first as shown in Figure 4.23(a). Then, in the parentheses, note that *P*.OR.*Q* is true when *P* is true (no need to check *Q*, see Figure 4.23(a)). The truth value of the .AND. statement, then, depends on *R*.OR..NOT.*S*. The flowchart is completed by inserting the check of this .OR. statement as shown in Figure 4.23(b). Now the flowchart in Figure 4.23(b) reveals the simplified form

```
.NOT.P.OR.(R.OR..NOT.S)
```

When any one of the three components .NOT.*P*, *R*, or .NOT.*S* is true, a PRINT instruction should be performed.

EXERCISE 4.7

1. Show that .NOT.*P*.AND.(*P*.OR.*Q*) is logically equivalent to .NOT.*P*.AND.*Q* by constructing a truth table.
2. (a) Show that .NOT.*P*.OR.(*P*.AND.*Q*) is logically equivalent to .NOT.*P*.OR.*Q* by constructing a truth table.
   (b) Use a distributive property in part (a) and simplify.
   (c) Compare the results in parts (a) and (b) with Figure 4.22.

3. Draw a flowchart for each of the following. Then use the flowchart to simplify the condition or antecedent.

   (a) $(P.\text{OR}.(Q.\text{AND}.P)) \rightarrow$ PRINT
   (b) $(P.\text{AND}.(Q.\text{OR}.P)) \rightarrow$ PRINT
   (c) $(P.\text{OR}.Q).\text{OR}.Q \rightarrow$ PRINT
   (d) $(P.\text{AND}.Q).\text{AND}.Q \rightarrow$ PRINT

4. Complete:

   (a) $P.\text{OR}.P$ has the same truth values as __________.
   (b) $Q.\text{AND}.Q$ has the same truth values as __________.
   (c) $P.\text{OR}..\text{NOT}.P$ is always __________.
   (d) $P.\text{AND}..\text{NOT}.P$ is always __________.
   (e) $P.\text{OR}.(Q.\text{AND}.R)$ is logically equivalent to $(P.\text{OR}.Q).\text{AND}.$ __________.
   (f) $P.\text{AND}.(Q.\text{OR}.R)$ is logically equivalent to $(P.\text{AND}.Q).\text{OR}.$ __________.

5. Use problem 4 to simplify the conditions (or antecedents) in problem 3(a) and 3(b).

6. Complete: (compare with problem 4)

   (a) $A \cup A =$ __________
   (b) $A \cap A =$ __________
   (c) $A \cup A' =$ __________
   (d) $A \cap A' =$ __________
   (e) $A \cup (B \cap C) = (A \cup B) \cap$ __________
   (f) $A \cap (B \cup C) = (A \cap B) \cup$ __________

7. Simplify each condition by drawing a flowchart and noting the logical structure of the flowchart.

   (a) $(P.\text{AND}.Q).\text{OR}.(P.\text{AND}.R) \rightarrow$ PRINT
   (b) $(P.\text{OR}.Q).\text{AND}.(P.\text{OR}.R) \rightarrow$ PRINT
   (c) $(.\text{NOT}.P.\text{OR}.Q).\text{OR}.Q \rightarrow$ PRINT
   (d) $(P.\text{AND}..\text{NOT}.Q).\text{OR}.Q \rightarrow$ PRINT
   (e) $(P.\text{AND}..\text{NOT}.Q).\text{OR}.(R.\text{AND}..\text{NOT}.Q) \rightarrow$ PRINT
   (f) $(.\text{NOT}.P.\text{OR}.Q).\text{AND}.(.\text{NOT}.P.\text{OR}.R) \rightarrow$ PRINT

8. In four Venn diagrams where $A \cap B \neq \phi$, shade the following sets:

   (a) $A \cap B$
   (b) $(A \cap B)'$
   (c) $A' \cap B'$
   (d) $A \cup B$
   (e) $(A \cup B)'$
   (f) $A' \cup B'$

9. (a) Find an .OR. statement which is logically equivalent to $.\text{NOT}.(P.\text{AND}.Q)$.
   (b) Find an .AND. statement which is logically equivalent to $.\text{NOT}.(P.\text{OR}.Q)$.
   (c) What are the equations involving sets which correspond to (a) and (b)?

10. Verify the results of problems 9(a) and 9(b) by constructing flowcharts for each of the following:

    (a) $.\text{NOT}.(P.\text{AND}.Q) \rightarrow$ PRINT
    (b) $.\text{NOT}.(P.\text{OR}.Q) \rightarrow$ PRINT

11. Simplify by constructing flowcharts:

    (a) $(.\text{NOT}.(P.\text{AND}..\text{NOT}.Q).\text{OR}.Q) \rightarrow$ PRINT
    (b) $(.\text{NOT}.P.\text{AND}..\text{NOT}.(P.\text{OR}.Q)) \rightarrow$ PRINT
    (c) $(P.\text{OR}.(Q.\text{AND}.R).\text{OR}..\text{NOT}.P) \rightarrow$ PRINT
    (d) $(Q.\text{AND}.(P.\text{OR}.R).\text{AND}..\text{NOT}.R) \rightarrow$ PRINT

(e) (.NOT.*P*.OR.*R*).OR.(*Q*.AND.(*P*.OR.*R*)) → PRINT
(f) (*P*.AND.(*P*.OR.*Q*)) → PRINT
(g) (*P*.OR.(*P*.AND.*Q*)) → PRINT

REFERENCES

Bauer, F., Baumann, R., Feliciano, M., and Samelson, K., *Introduction to Algol.* Englewood Cliffs, N. J., Prentice-Hall, Inc., 1964.

Crowdis, D. G. and Wheeler, B. W., *Introduction to Mathematical Ideas.* New York, McGraw-Hill, Inc., 1970.

DeAngelo, S. and Jorgensen, P., *Mathematics for Data Processing.* New York, McGraw-Hill, Inc., 1970.

Galler, B., *The Language of Computers.* New York, McGraw-Hill, Inc., 1962.

Oakford, R. V., *Introduction to Electronic Data Processing Equipment.* New York, McGraw-Hill, Inc., 1962.

5 BASIC ALGEBRA

ALGEBRAIC EXPRESSIONS / 154

5.1 EVALUATING EXPRESSIONS / **156**
Hierarchy of Operations / **157**
Exercise 5.1 / **159**

5.2 ARITHMETIC OPERATIONS / **160**
Addition and Subtraction / **160**
Multiplication / **162**
Exercise 5.2 / **164**

5.3 FACTORING / **165**
Factoring Monomials from Polynomials / **165**
Factoring Quadratics / **166**
Special Quadratic Forms / **167**
Factoring for Computer Calculation / **168**
Exercise 5.3 / **169**

ALGEBRAIC FRACTIONS / 170

5.4 RELATIONSHIPS INVOLVING FRACTIONS / **171**
Equality / **171**
Fundamental Principle of Fractions / **172**
Signs / **172**
Multiplication / **173**
Addition / **174**
Representing Fractional Quantities in Fortran / **174**
Exercise 5.4 / **175**

5.5 REDUCING RATIONAL NUMBERS / **177**
Fractions Containing Binomials / **178**
Exercise 5.5 / **179**

5.6 MULTIPLICATION AND DIVISION OF RATIONAL NUMBERS / **179**
Exercise 5.6 / **181**

5.7 ADDITION AND SUBTRACTION OF RATIONAL NUMBERS / **181**
Least Common Denominator / **182**
Algebraic Forms / **182**
Exercise 5.7 / **183**

EXPONENTS AND RADICALS / 184

5.8 PROPERTIES OF EXPONENTS / **184**
Multiplication / **184**
Division / **185**
Zero and Negative Exponents / **185**
Other Properties of Exponents / **186**
Exponents in Fortran / **188**
Exercise 5.8 / **189**

5.9 FRACTIONAL EXPONENTS AND RADICALS / **191**
Exponents of Form 1/a / **191**
Exponents of Form a/b / **192**
Manipulating Radicals / **192**
Use in Fortran / **193**
Exercise 5.9 / **193**

ALGEBRAIC EXPRESSIONS

A good definition of the word *algebra* is *a generalization of arithmetic in which letters and other symbols representing numbers are combined according to the rules of arithmetic.* Much of the difficulty of algebra for beginners appears to stem from the fact that letters do not appear to combine in just quite the same way as do numbers. For example, in arithmetic, the sum of the number 3 and the number 5 is the number 8 (that is, $3+5=8$). In algebra, the letter x may be used to represent one number and the letter y to represent another. But their sum is written as $x+y$, which also represents a number.

In general, any collection consisting of numbers and symbols (representing numbers) which are related by arithmetic operations is said to be an **algebraic expression**. In the preceding discussion, $x+y$ is an algebraic expression consisting of x and y related by the operation of addition. In geometry, the area of a triangle is found from the expression $bh/2$, where b represents the base and h the height. In business, the total cost of several items of inventory might be expressed as $6a + 4b + 9c + d$. All of these algebraic quantities consist of numbers and of letters which we shall refer to as **variables** (later this word will be defined more concisely).

Algebraic expressions which involve no roots of the variable, in which all exponents on the variable are natural numbers, and in which no variable appears in the denominator, are called **polynomials**. Thus, of the following expressions, only $\sqrt{x} - (1/y)$ is not a polynomial.

$$ax + bx, \qquad 3y + 4xz - 6z^3 - 7, \qquad 2\pi r, \qquad \sqrt{x} - \frac{1}{y}\cdot$$

Each of these expressions consists of one or more **terms**, that is, quantities related by plus and minus signs. Thus, the expression $x+y$ consists of the two terms x and y; $bh/2$ consists of only one term, the expression itself; and $6a + 4b + 9c + d$ consists of four terms, $6a$, $4b$, $9c$, and d. Similarly, $ax + bx$ consists of two terms, $3y + 4xz - 6z^3 - 7$ of four terms, $2\pi r$ of one term, and $\sqrt{x} - (1/y)$ of two terms.

If we consider the individual terms of each expression, we will note that most of them consist of two or more quantities multiplied together. For example, the second term of the expression $3y + 4xz - 6z^3 - 7$ consists of the product of 4, x, and z. Thus we say that $4xz$ consists of the **factors** 4, x, and z. Similarly, the factors of $2\pi r$ are 2, π, and r.

Any single factor or group of factors is said to be a **coefficient** of the remaining factor or factors of the term. Thus it can be said, in referring to the term $4xz$, that 4 is a coefficient of xz, that $4x$ is the coefficient of z, that xz is a coefficient of 4, and so on. However, the general practice is to refer to the constant portion of a term as the coefficient, in which case the coefficient in $4xz$ would be 4, in $3y$ it would be 3, and so on. If no numerical coefficient appears on a term, it is assumed to be 1. In other words, we consider y as $(1)y$.

Whenever we wish to regard two or more terms as a group, we commonly use parentheses ( ). For example, the principal and interest accumulated in one year is the principal times 1 plus the interest rate, which may be represented by the expression

$$P(1 + r).$$

Other commonly used symbols to indicate grouping are brackets [ ] and braces { }, which appear as

$$P[1 + r] \qquad \text{and} \qquad P\{1 + r\}.$$

With reference to our previous definitions, we would normally consider the expression $P(1 + r)$ as consisting of one term. However, this term contains two factors, one of which consists of two terms. If the distributive law were applied, yielding $P + Pr$, the expression would be said to consist of two terms.

The beginner often wonders why it is necessary to study mathematics in general, and algebra in particular, with the wide availability of the computer. As pointed out in earlier chapters, the computer is a device which will perform operations precisely as it is told. A problem such as computing compound interest (see Example 1.1) can conveniently be represented in the form of an algebraic algorithm and then the steps programmed on a computer (Example 1.4). Just as the computer is a tool for performing calculations, algebra is a tool for clearly defining and solving problems. The Fortran language is a language designed to resemble algebra as closely as the card punching machine permits. The acronym Fortran is derived from *FOR*mula *TRAN*slation. Fortran, as does algebra, includes variables and constants. However, the notion of a variable in Fortran has different implications than a variable in algebra. That is, whenever a variable is used in a Fortran program, the variable is equated to a storage area into which one number can be stored (analogous to the pigeon-hole slots of Example 1.4). This might be a quantity which is read from an IBM card or one which is calculated within the program. Since Fortran variables are not restricted to one letter as is the common usage in algebra (most compilers allow at least five letters to be used), all arithmetic operations must be explicitly indicated with an arithmetic operator. For example, Figure 5.1 shows how the expression for simple interest $P(1 + rt)$ might appear on a Fortran coding sheet and as printed by a computer. Although the Fortran equivalent of the algebraic expression is slightly different, most users of algebra would probably recognize the Fortran form as an expression for

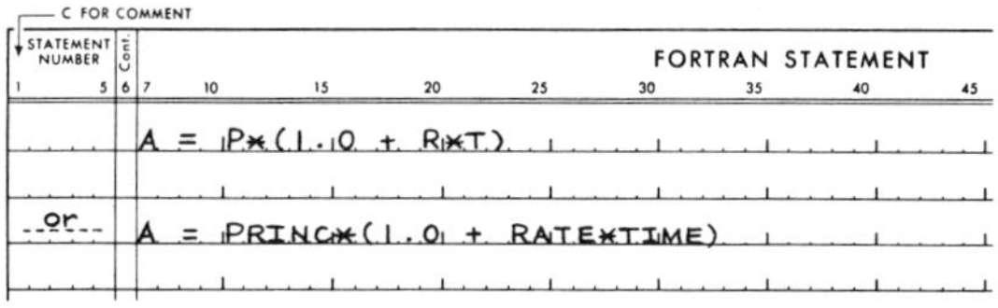

```
C FOR COMMENT
STATEMENT   Cont.
NUMBER             FORTRAN STATEMENT
1     5  6  7   10   15   20   25   30   35   40   45

            A = P*(1.0 + R*T)

  or        A = PRINC*(1.0 + RATE*TIME)

```

FIGURE 5.1 A Fortran expression

calculating interest. One important difference between algebra and Fortran is that multiplication in Fortran must always be indicated with an * symbol (asterisk) and is never implied. For instance, in algebra *lw* means *l* times *w*, whereas L*W must be used in Fortran. The operational symbols used in Fortran (which will be discussed in more detail as we proceed) are:

**	raise to a power
*	multiply
/	divide
+	add
−	subtract

Where the algebraic expression illustrates a basic form, the Fortran expression indicates a series of operations to be performed (according to the expressed form). Thus the machine will obtain the quantity from storage which we refer to as RATE, multiply it by TIME, add 1.0 to the product, then multiply the sum by the quantity it has stored as PRINC.

Although this discussion has centered on the comparison of Fortran and algebra, we must not get the idea that these forms are peculiar to the Fortran programming language. Indeed, the computer forms discussed in this chapter (and referred to as Fortran) are found in many other programming languages such as Algol (*ALGO*rithmic *L*anguage), Cobol (*CO*mmon *B*usiness *O*riented *L*anguage) and PL/1 (*P*rogramming *L*anguage/1). The computer language rules for operating with these forms are much the same as the corresponding rules in algebra, as we shall see in the next section.

5.1 EVALUATING EXPRESSIONS

Each of the preceding example expressions is a general representation and as such does not represent any specific number. Thus to refer to the **value** of the general expression $P(1 + i)$ without reference to given numbers for P and i is not especially meaningful. If numbers are substituted for each of the two symbols, the expression may be evaluated. For instance, if $P = 200$ and $i = 0.05$ (that is, 5%), the value of the expression is

$$\begin{aligned} P(1+i) &= 200(1+0.05) \\ &= 200(1.05) \\ &= 210.00. \end{aligned}$$

For other values of the variables P and i, the value of the expression will, in general, be different. When the algebraic expressions in which the variables may assume either positive or negative values are evaluated, the rules of signs discussed in Chapter 2 must be taken into account. For example, consider the algebraic expression $x - y$ evaluated for $x = +2$, $y = +4$. Upon substituting, we have, by the rules of Chapter 2,

$$\begin{aligned} x - y &= (+2) - (+4) \\ &= -2. \end{aligned}$$

On the other hand, the value of the expression is different if $x=+2$, $y=-4$:

$$\begin{aligned} x-y &= (+2)-(-4) \\ &= 2+4 \\ &= 6. \end{aligned}$$

Of a somewhat more confusing nature are expressions of the form

$$-y^2 \qquad \text{and} \qquad x-y^2.$$

In Chapter 2 it was emphasized that an exponential form such as $(-5)^2$ means $(-5)\,(-5)=+25$. The same situation is true with an algebraic expression. When $x=10$ and $y=2$, these expressions have the following values:

$$\begin{aligned} -y^2 &= -(2)^2 = -(2)(2) = -(4) = -4 \\ x-y^2 &= 10-(2)^2 = 10-4 = 6. \end{aligned}$$

On the other hand, if $x=10$ and $y=-2$,

$$\begin{aligned} -y^2 &= -(-2)^2 = -(-2)(-2) = -(4) = -4 \\ x-y^2 &= 10-(-2)^2 = 10-4 = 6. \end{aligned}$$

Obviously the important factor in evaluating algebraic expressions is to replace each variable by its appropriate value and carefully apply the rules for signed numbers when performing the arithmetic operations.

HIERARCHY OF OPERATIONS. When evaluating complex expressions, it is useful to perform arithmetic operations in a prescribed order. The general rule is to perform multiplications and divisions first, and then additions and subtractions. If this rule is followed, forms such as $2+3\times 4$ are not ambiguous:

$$2+3\times 4 = 2+(3\times 4) = 14,$$

but

$$2+3\times 4 \neq (2+3)\times 4 = 20.$$

Equivalent forms in Fortran are

```
2.0 + 3.0*4.0
WORK + FIELD*DATA
```

To dwell on this subject may appear trivial, but such simple rules are commonly overlooked in evaluating algebraic expressions and performing arithmetic operations on literal quantities. The relative order in which arithmetic operations are considered in common algebra and in Fortran programming is of utmost importance. The Fortran programming language is a language of algebra, and the basis for it is the arithmetic statement, which has been made as nearly identical to an actual algebraic equation as possible. The order in which the computer considers arithmetic operations is identical to that in ordinary algebra, that is,

1. groupings within parentheses,
2. exponentiations,
3. multiplications and divisions,
4. additions and subtractions.

Know these

For instance, the amount of money accumulated from P dollars invested at a compound rate of interest r for a period of t years may be represented by the algebraic expression $P(1+r)^t$ or the Fortran expression

```
P*(1.0+R)**T
```

The value of this expression when $P = \$200$, $r = 5\%$, and $t = 3$ years may be determined as shown below:

$$\begin{aligned} P(1+r)^t &= 200(1+0.05)^3 && (1)\\ &= 200(1.05)^3 && (2)\\ &= 200(1.157625) && (3)\\ &= 231.53 \quad \text{(rounded).} && (4) \end{aligned}$$

In evaluating this expression, the following sequence was followed:

1. Values for each algebraic quantity were substituted in the expression.
2. Operations were begun within the parentheses and the indicated sum obtained.
3. The exponentiation was performed.
4. The multiplication was performed.

The algebraic expression $3x - a(x - 3b^2)^2$, which has the following Fortran equivalent,

```
3.0*X-A*(X-3.0*B**2)**2
```

is more complex than the foregoing one. However, its value when $x=6$, $a=2$, and $b=-2$ may also be found by performing the operations in the required sequence:

$$\begin{aligned} 3x - a(x-3b^2)^2 &= 3(6) - (2)[6 - (3)(-2)^2]^2 && (1)\\ &= 3(6) - (2)[6 - (3)(4)]^2 && (2)\\ &= 3(6) - (2)[6 - 12]^2 && (3)\\ &= 3(6) - (2)[-6]^2 && (4)\\ &= 3(6) - (2)(36) && (5)\\ &= 18 - 72 && (6)\\ &= -54. && (7) \end{aligned}$$

Here the sequence of operations is extremely important; refer to each operation by the number to the right as follows:

1. Values for each algebraic quantity are substituted into the expression.
2. Operations are begun within the innermost grouping; the exponentiation is performed first.
3. The product of 3 and 4 is determined.
4. The subtraction is performed within the brackets, to give a value for this factor of -6.
5. The exponentiation is performed next.
6. The two products are obtained.
7. The subtraction is performed.

Note that evaluation of the expression within the brackets (steps 2, 3, and 4) was carried out by further application of the hierarchy rules.

Whenever parentheses within parentheses are encountered, the innermost grouping should be evaluated first. As an example, consider the following expression when $a = 3$, $b = 2$, $x = 4$:

```
2.0*(3.0*X-A*(X-3.0*(A-B**2)**2))
```

$$\begin{aligned} 2\{3x - a[x - 3(a - b^2)^2]\} &= 2\{3 \times 4 - 3[4 - 3(3 - 2^2)^2]\} && (1) \\ &= 2\{3 \times 4 - 3[4 - 3(3 - 4)^2]\} && (2) \\ &= 2\{3 \times 4 - 3[4 - 3(-1)^2]\} && (3) \\ &= 2\{3 \times 4 - 3[4 - 3(1)]\} && (4) \\ &= 2\{3 \times 4 - 3[4 - 3]\} && (5) \\ &= 2\{3 \times 4 - 3[1]\} && (6) \\ &= 2\{12 - 3\} && (7) \\ &= 2\{9\} && (8) \\ &= 18. && (9) \end{aligned}$$

The sequence of operations in this example is as follows:

1. Values for a, b, and x are substituted in the expression.
2. Since the innermost grouping is $(3 - 2^2)$, operations are begun here. Within the parentheses, the exponentiation is performed first.
3. The subtraction is performed to complete operations within the innermost grouping.
4. Operations within the brackets are begun by performing the exponentiation.
5. The multiplication within the brackets is performed.
6. The subtraction within the brackets is performed.
7. The two products within the braces are obtained.
8. The subtraction within the braces is carried out.
9. Multiplication gives the final result.

Large, complicated expressions may be evaluated with little more difficulty than the simple ones we have considered here.

The importance of the hierarchy of operations cannot be overstressed; it is fundamental to our algebraic system. The Fortran programming language for computers, being essentially an algebraic language, incorporates these rules for performing arithmetic operations. Thus, it is imperative that, in writing Fortran arithmetic statements, a programmer have a thorough understanding of this topic.

EXERCISE 5.1

Evaluate the following expressions when (*a*) $x = 2$, $y = 3$, $z = 5$; (*b*) $x = 4$, $y = -1$, $z = -2$; (*c*) $x = -3$, $y = -2$, $z = 1$; also write the Fortran equivalent of each.

1. $x + y$
2. $x + y + z$
3. $x - y + z$
4. $xy + z$
5. $x + 2xy$
6. xyz
7. $x - y$
8. $-x + y$
9. $-(x + y)$

10. $-(x - y + z)$
11. $-xy$
12. $x - 2xy$
13. $y - x^2$
14. $y^2 - x^2$
15. $y^2 - x^3$

Evaluate the following expressions when $a = 2$, $b = -1$, $c = 3$; also write the Fortran equivalent of each.

16. $a + 3(b - c)$
17. $a^2 + 5(b + 4c)$
18. $a - a(b + c^2)$
19. $c(b + 2) - (a + 1)$
20. $ab - (2bc - a)$
21. $ab(c - 1)^2$
22. $b(c - 1)^2 - c(b^2 + 1)$
23. $2(c^2 + 1)^2 - ab^2$
24. $2(-c^2 - 1)^2 - (ab)^2$
25. $2[(a + b)^2 - 3(ac)]$
26. $-3[(2a - b) - 4c^2]$
27. $2a[a - 3(c + 2)]$
28. $4a\{b - 2c[(2b - 1)^2 + 1]\}$
29. $-\{3a + 2[b^2 - (c + a)^2]\}$

5.2 ARITHMETIC OPERATIONS

ADDITION AND SUBTRACTION. From elementary algebra, we know that the sum of $3x$ and $2x$ is $5x$, and the sum of $4mn$ and $3mn$ is $7mn$. On the other hand, the sum of $4z$ and $5y$ can only be expressed as $4z + 5y$; without more information, there is no way to combine these expressions. It is the old story, "You cannot add apples and oranges." From a mathematical point of view the ability to add "apples and apples" though not "oranges and apples" is a direct consequence of the distributive law:

$$a(b + c) = ab + ac$$

which may be written as

$$ab + ac = a(b + c).$$

The expression $3x + 2x$ is of the same form as $ab + ac$; therefore,

$$\begin{aligned} 3x + 2x &= x(3 + 2) \\ &= x(5) \\ &= 5x. \end{aligned}$$

Similarly,

$$\begin{aligned} 4mn + 3mn &= mn(4 + 3) \\ &= mn(7) \\ &= 7mn. \end{aligned}$$

A convenient rule to remember is: *If two expressions have identical literal portions, that is, are like terms, their sum (or difference) is a like term whose coefficient is the sum (or difference) of the coefficients of the original terms.*

By application of the associative law, this rule can be extended to include the addition and/or subtraction of any number of like terms. For example,

$$\begin{aligned} 4x + 3x - 6x + 12x &= (4x + 3x) - 6x + 12x \\ &= 7x - 6x + 12x \\ &= (7x - 6x) + 12x \\ &= x + 12x \\ &= 13x. \end{aligned}$$

The inclusion of more than one unknown in the summation presents no problem, since only like terms are combined, as shown below:

$$\begin{aligned} 14ab + 5cd + 2cd - 7ab + ab &= 14ab - 7ab + ab + 5cd + 2cd \\ &= 8ab + 7cd. \\ (4z^2 - 3z + 2) + (3z - 4) &= 4z^2 - 3z + 3z + 2 - 4 \\ &= 4z^2 - 2. \end{aligned}$$

In the latter illustration, the parentheses were removed merely by rewriting the quantities without the parentheses. Had either of them been preceded by a minus sign, it would have been necessary to apply the rule for signs. For example, $-(3x - 4)$ may be considered in the following manner:

$$\begin{aligned} -(3x - 4) &= (-1)(3x - 4) \\ &= (-1)(3x) + (-1)(-4) \quad \text{by the distributive law} \\ &= -3x + 4. \end{aligned}$$

In other words, the removal of parentheses preceded by a minus sign requires changing the sign of each term within the parentheses. Another example is

$$\begin{aligned} (5x^3 - 2x^2 + 1) - (4x^2 - x + 1) &= 5x^3 - 2x^2 + 1 - 4x^2 + x - 1 \\ &= 5x^3 - 6x^2 + x. \end{aligned}$$

Whenever expressions such as $-[13x - (27x - y)]$ occur—that is, when quantities are nested—it is usually simplest to begin with the innermost group and work outward, in a manner very similar to that used in evaluating algebraic expressions. For instance,

$$\begin{aligned} (19y + 42x) - [13x - (27x - y)] &= (19y + 42x) - [13x - 27x + y] \\ &= 19y + 42x - 13x + 27x - y \\ &= 18y + 56x. \end{aligned}$$

Whenever a question arises regarding the manipulation of algebraic expressions, it is convenient to remember that they represent real numbers and involve the same operations as numbers.

These concepts may seem unrelated to the computer and to programming in general. A logical reaction is: "Why learn all of this? Let the computer do it." Unfortunately, the computer is not capable of performing algebraic operations. It must be directed in detail to solve each and every problem. Moreover, the good programmer attempts to write each set of instructions in the most efficient manner possible. For example, let us assume that the expression

$$(19y + 98x) - 13x - (29x - y)$$

represents the handling cost to a businessman of a particular inventory item, and he wants to know the cost for a wide range of values for x and y (for instance, 100 values for x and 100 values for y, which would give 100×100 or 10,000 evaluations). We can readily see that the equivalent form of the expression, $18y + 56x$, would be simpler to evaluate. In terms of arithmetic operations, the results are summarized below:

Expression	Multiplications	Additions or Subtractions
`(19.0*Y+98.0*X)-13.0*X-(29.0*X-Y)`	4/evaluation 40,000 total	4/evaluation 40,000 total
`18.0*Y+16.0*X`	2/evaluation 20,000 total	1/evaluation 10,000 total

The arithmetic operations are reduced by more than half if the equivalent form is used. This type of consideration can result in substantial savings of time and money in using computers.

MULTIPLICATION. In previous chapters we have, to a limited extent, studied multiplication of a polynomial by a **monomial** (a polynomial consisting of only one term). For instance, the distributive law

$$a(b+c) = ab + ac$$

represents just such an operation. Here a **binomial** (a polynomial consisting of two terms) is multiplied by a monomial. The scope of the distributive law can be further expanded by application to the product of a monomial and a polynomial. Following are three examples:

$$\begin{aligned} 3x(5y-2x) &= (3x)(5y) + (3x)(-2x) && (1)\\ &= 3 \times 5 \times x \times y + (-2) \times 3 \times x \times x \\ &= 15xy - 6x^2. \\ -2m^2(m^2-n^2) &= (-2m^2)(m^2) + (-2m^2)(-n^2) && (2)\\ &= (-2)(m^2)(m^2) + (-2)(m^2)(-n^2) \\ &= -2m^4 + 2m^2n^2. \\ -a(a^2-2a-1) &= -a[a^2 + (-2a-1)] && (3)\\ &= (-a)(a^2) + (-a)(-2a-1) \\ &= (-1)(a)(a^2) + (-a)(-2a) + (-a)(-1) \\ &= -a^3 + 2a^2 + a. \end{aligned}$$

More than normal detail is included in the third example to illustrate successive applications of the distributive rule. With practice, it is possible to write the required product of a polynomial and a monomial with no intermediate steps. However, great care must be taken to insure that each term of the polynomial is multiplied by the monomial, and that rules for signs are correctly applied.

Extension of the distributive law to multiplication of two polynomials is also possible, as shown in the following example:

$$\begin{aligned} (2m-3n)(5m+n) &= (2m)(5m+n) + (-3n)(5m+n) && (1)\\ &= (2m)(5m) + 2m(n) + (-3n)(5m) + (-3n)(n) && (2)\\ &= 10m^2 + 2mn - 15mn - 3n^2 && (3)\\ &= 10m^2 - 13mn - 3n^2. && (4) \end{aligned}$$

In step 1 the distributive law is applied, with $(5m+n)$ taken to be equivalent to a in $a(b+c)$. In step 2, it is again applied to each of the two terms resulting from step 1. The similar terms in step 3 are combined to give the final expression in step 4. This process of multiplying polynomials is commonly referred to as

expanding an expression, and the final form is frequently called the **expansion** of the original expression. Another example, this one illustrating the product of a binomial and a trinomial, is

$$\begin{aligned}(2x+3)(x^3-2x+1) &= (2x)(x^3-2x+1)+(3)(x^3-2x+1)\\ &= 2x^4-4x^2+2x+3x^3-6x+3\\ &= 2x^4+3x^3-4x^2-4x+3.\end{aligned}$$

The above expansion is a commonly encountered polynomial form. It is referred to as a **polynomial in one variable**. This specific example is called a fourth degree polynomial in the single variable x, because it contains no variable other than x and the largest exponent on x is 4.

The simplification of large nested expressions is a direct application of the foregoing principles. It is accomplished in evaluating an expression by beginning with the innermost grouping. For example,

$$\begin{aligned}-3a\{x+2[x-3x(x^2-1)]\} &= -3a\{x+2[x+(-3x)(x^2)+(-3x)(-1)]\}\\ &= -3a\{x+2[x-3x^3+3x]\}\\ &= -3a\{x+2[-3x^3+4x]\}\\ &= -3a\{x+(2)(-3x^3)+(2)(4x)\}\\ &= -3a\{x-6x^3+8x\}\\ &= -3a\{-6x^3+9x\}\\ &= (-3a)(-6x^3)+(-3a)(9x)\\ &= 18ax^3-27ax.\end{aligned}$$

Although some of the preceding examples illustrate how expressions can be rearranged to simplify the Fortran operations, we must not conclude that the most convenient form for algebraic use will be the best form for Fortran. For example, consider some of the preceding examples whose Fortran equivalents follow. (The arithmetic operations required in evaluating a quantity with an exponent is treated as successive multiplications. For example, X**4 is considered as equivalent to X*X*X*X requiring three multiplications.)

Expression	Multiplications	Addition or Subtraction
`3.0*X*(5.0*Y - 2.0*X)`	4	1
`15.0*X*Y - 6.0*X**2`	4	1
`(2.0*M - 3.0*N)*(5.0*M + N)`	4	2
`10.0*M**2 - 13.0*M*N - 3.0*N**2`	6	2
`(2.0*X + 3.0)*(X**3 - 2.0*X + 1.0)`	5	3
`2.0*X**4 + 3.0*X**3 - 4.0*X**2 - 4.0*X + 3.0`	10	4
`-3.0*A*(X+2.0*(X-3.0*X*(X**2-1.0)))`	6	3
`18.0*A*X**3 + 27.0*A*X`	6	1

In addition to considering numbers of operations to be performed, error propagation (Chapter 3) is sometimes a consideration in determining the best form to use. As is evident by the preceding, the programmer must exercise good judgment in considering using these forms.

EXERCISE 5.2

Simplify each of the following by combining similar terms:

1. $3x + x$
2. $5y - 3y$
3. $4x - x + 2x$
4. $3ab + ab - 2ab$
5. $2xy^2 + 2xy^2 - xy^2$
6. $-12z + 2z - 4z$
7. $-5xyz - 2xyz + 2xyz$
8. $7y - 7y$
9. $-2xz + 2xz$
10. $-3a^4 + 3a^4 - a^4$
11. $6m - 4m - 2m$
12. $12x - 11x$
13. $m + n - 2m$
14. $6x + (5x + 2)$
15. $(25x + 1) + (13x - 5)$
16. $(x^2 + 3x - 1) + (x^2 - 3x - 1)$
17. $(a + b) + (a - b)$
18. $(x + y) + (y + z) + (x + z)$
19. $(m^2 + n + 1) + (1 - n)$
20. $(x + y) + (2xy)$
21. $y - (3q + 2p)$
22. $-(-3a + 4b) + (a - b)$
23. $(8m + 4) - (8n + 4)$
24. $(x^2 + 3x + 2) - (x^3 + 2x - 2)$
25. $-(y + 3) + (y - 3)$
26. $(a + b + c) - (a - b) - (d - c)$
27. $(m - n) - (n - p) - (p - m)$
28. $-(b^2 + 4b - 2) + (b^2 - 2)$

Perform the indicated multiplications:

29. $2(x + y)$
30. $3(2a + b)$
31. $7(3m - 2n)$
32. $4(-2n + m)$
33. $-5(a + 3b)$
34. $-3(2x - y)$
35. $-9(-3x - 2y)$
36. $a(x - y)$
37. $2a(3x - 2z)$
38. $5xy(x - z - 2x)$
39. $-3m(x + y - z)$
40. $2a(a^2 - 2b)$
41. $3m(2m^3 - 4n^2)$
42. $-4x(3x^2 - y^2)$
43. $3x(2x^2 - x + 5)$
44. $-2y(y^2 + y - 6)$
45. $3a(a^3 - a^2 + 3a - 1)$
46. $3xy[xy - (xy)^2]$

Perform the indicated operations; simplify by collecting similar terms:

47. $(x + 1)(x + 2)$
48. $(y - 2)(y + 1)$
49. $(y + 3)(y + 4)$
50. $(z + 3)(z - 7)$
51. $(z + 3)(z + 3)$
52. $(z - 3)(z - 3)$
53. $(a + 5)(2a - 4)$
54. $(3m - 1)(4m - 2)$
55. $-(5a - 3)(2a - 1)$
56. $(3b - a)(2b + a)$
57. $(2x + y)(5x - 6y)$
58. $(3a + b)(3a - b)$
59. $2(3x + 1)^2$
60. $3(x - a)^3$
61. $-2(5x + y)(x - y)$
62. $-4(-2x - 1)(3x - 2)$
63. $a(a + 1)(a - 3)$
64. $(z^2 + 1)(z^2 + 2)$
65. $(a^3 - 3)(a^3 + 3)$
66. $-b(b^2 - 2)(2b^2 + 3)$
67. $-xy(x + a)(y - b)$
68. $3(x + 2)(x^2 - 3)$
69. $-a(z - 1)(z^3 + 3)$
70. $(x + 1)(x^2 + 2x - 3)$

71. $-(3a-b)(3a-b)$
72. $(x-1)^2$
73. $(y+a)^2$
74. $(z^2-3z+2)(z-1)$
75. $a(b-1)(n^3-n+1)$
76. $m(n+1)(n^3-n+1)$
77. $3[a+b(b^2-a)]$
78. $2m[x+2x(x^2-1)]$
79. $-[a-a(1-a)]$
80. $2x[3x+(2x-1)(x-2)]$
81. $ab[-a-(b-a)]$
82. $(x-y)[x-x(x-y)]$
83. $2x\{3a-2-[2ax-x(2a+2)]\}$
84. $-3a\{4b+a-[2(b-5a)-3a^2]\}$
85. $-4m\{y+2z-z[y+2(2-y)-y]\}$
86. $3z^2\{z[z^2-z(z-1)]\}$
87. (*a*) What is the difference between (a^2+b^2) and $(a+b)^2$?
    (*b*) When is $(a^2+b^2)<(a+b)^2$?
    (*c*) When is $(a^2+b^2)=(a+b)^2$?
    (*d*) When is $(a^2+b^2)>(a+b)^2$?
88. Write Fortran expressions for the original algebraic form and for the "simplified" form, and count the number of operations in each for problems 19, 22, 27, 40, 45, 47, 55, 60, 83.
89. Repeat problem 88 for problems 16, 17, 26, 41, 46, 50, 58, 73, 84.

5.3 FACTORING

In the process of solving algebraic problems it is frequently necessary to expand products and collect similar terms. **Factoring**, which is the inverse operation of expanding, is also a useful tool in algebra. The first examples we shall consider involve factoring monomials from polynomials, and represent a direct application of the distributive law.

FACTORING MONOMIALS FROM POLYNOMIALS. Just as $4 \times 7 = 28$ may be written as $28 = 4 \times 7$, the distributive law $a(b+c)=ab+ac$ may be written as

$$ab+ac=a(b+c).$$

The latter case represents a factoring of the common variable a from each of the two terms in which it appears. If the value for a were 2, the distributive rule would take the form

$$2b+2c=2(b+c).$$

The distributive rule can also be applied to longer and more complex expressions, such as

$$3bz^3+6baz^2-9b^2z.$$

Studying this expression, we can see that each of the numerical coefficients contains the common factor 3; further, each term contains b and z. Therefore, the expression may be factored as

$$\begin{aligned} 3bz^3+6baz^2-9b^2z &= (3bz)(z^2)+(3bz)(2az)+(3bz)(-3b) \\ &= 3bz(z^2+2az-3b). \end{aligned}$$

Another example is

$$4x^4 + 2ax^3 - 6x^2 = (2x^2)(2x^2) + 2x^2(ax) + (2x^2)(-3)$$
$$= 2x^2(2x^2 + ax - 3).$$

It is often useful to factor -1 when performing these operations, but it is important to properly apply the laws of sign. For example,

$$-2x^3 + 6x^2 - 4x = (-2x)(x^2) + (-2x)(-3x) + (2x)(-3)$$
$$= -2x(x^2 - 3x + 2).$$

In this example, we have factored $-2x$ from the expression, leaving each of the remaining terms with nothing in common. However, this trinomial factor can be further reduced to the product of two binomials, as we shall see in the next section.

FACTORING QUADRATICS. A polynomial containing only one variable with 2 as its largest exponent is commonly called a **quadratic**. Although quadratics such as $5x^2 + 3x - 6$ are commonly encountered, we shall restrict our discussion here to quadratics in which the coefficient of the squared term is 1, for example, $x^2 - 6x + 4$. The final results from exercises 47-52 were quadratics of this type. We can speak of this form in general by referring to

$$(x + a)(x + b) = x^2 + (a + b)x + ab. \tag{5.1}$$

Here x is being used to represent the variable and a and b are used to represent **arbitrary constants** (numbers which do not change in a given context).

Let us now return to the example of the previous section, where

$$-2x^3 + 6x^2 - 4x = -2x(x^2 - 3x + 2).$$

After factoring the monomial $-2x$ from this expression, the remaining factor $x^2 - 3x + 2$ appears to have the form of Eq. 5.1. Thus we may be able to factor it into the product of two binomials. By comparing the forms

$$x^2 + (a + b)x + ab$$
$$x^2 + (-3)x + 2,$$

we note that $a + b = -3$ and $ab = 2$ for our special case. Now the problem becomes one of finding two integers whose sum is -3 and whose product is 2.

Since we will be dealing with a limited number of integers, it is feasible to use simple trial and error. First of all we must recognize that there are only two possible sets of integer factors of 2; they are 1, 2, and $-1, -2$. That is, if $a = 1$, then $b = 2$, or if $a = -1$, then $b = -2$, since $ab = 2$. However, we know that the sum of these two factors is -3, since $a + b = -3$. This restricts us to the factors -1 and -2. The final result is

$$x^2 - 3x + 2 = (x - 1)(x - 2).$$

EXAMPLE 5.1

$$x^2 + 9x + 20$$

SOLUTION

Since both $a + b$ and ab are positive, both a and b must be positive. The positive factors of 20 are 1 and 20; 2 and 10; 4 and 5. By trial and error we find that

$$x^2 + 9x + 20 = (x + 4)(x + 5).$$

EXAMPLE 5.2

$$2x^2 - 10x - 72$$

SOLUTION

The first step is to factor 2 from each expression, giving

$$2x^2 - 10x - 72 = 2(x^2 - 5x - 36).$$

Since ab is negative, a and b must have opposite signs; further, because $a + b$ is also negative, the one with the negative sign must be the larger. Under these conditions the allowable factors for -36 are 1 and -36; 2 and -18; 3 and -12; 4 and -9. Again, by trial and error

$$2x^2 - 10x - 72 = 2(x + 4)(x - 9).$$

EXAMPLE 5.3

$$x^2 + 5x + 3$$

SOLUTION

Since both $a + b$ and ab are positive, a and b must be positive. The only possible factors of 3 are 3 and 1. However, $3 + 1 \neq 5$. Therefore, $x^2 + 5x + 3$ cannot be factored.

In Example 5.3, the expression $x^2 + 5x + 3$ is called a **prime** factor because it cannot be broken down into a product of other polynomials. (In studying the theory of equations, one of the important proofs involves showing that any polynomial of degree greater than two with integer coefficients may be factored into products of first and second degree polynomials.)

SPECIAL QUADRATIC FORMS. Although the techniques of the previous section will be adequate for most of the second degree polynomials we must factor, three other forms which are special cases of Eq. 5.1 will be useful to know. They are

$$(x - a)(x + a) = x^2 - a^2, \quad (5.2)$$
$$(x + a)^2 = x^2 + 2a + a^2, \quad (5.3)$$
$$(x - a)^2 = x^2 - 2a + a^2. \quad (5.4)$$

All three of these forms involve a perfect square for the constant term, illustrated by the following examples.

EXAMPLE 5.4

$$x^2 - 9$$

SOLUTION

Since both x^2 and 9 are perfect squares and are separated by a minus sign, this is the general form of Eq. 5.2. Thus

$$x^2 - 9 = (x + 3)(x - 3).$$

Note that the term in x does not occur in this type of quadratic, since it cancels when expanded, as shown below:

$$\begin{aligned}(x + 3)(x - 3) &= x(x - 3) + 3(x - 3) \\ &= x^2 - 3x + 3x - 9 \\ &= x^2 - 9.\end{aligned}$$

EXAMPLE 5.5

$$ax^2 + 6ax + 9a$$

SOLUTION

After factoring the a from each term, both the x^2 and 9 are perfect squares, which means this may be of the form of Eq. 5.3. If so, it may be factored into $(x + 3)^2$; expanding as a check gives

$$\begin{aligned}(x + 3)^2 &= (x + 3)(x + 3) \\ &= x^2 + 6x + 9.\end{aligned}$$

Thus

$$ax^2 + 6ax + 9a = a(x + 3)^2.$$

Note that the middle term coefficient is twice the square root of the last term.

EXAMPLE 5.6

$$ax^2 - 6ax + 9a$$

SOLUTION

This example is identical to Example 5.5, except for the negative middle term. Since the middle term is negative and the last term is positive, both constant terms in the factors must be negative. This suggests a polynomial of the type of Eq. 5.4, which is factored as

$$ax^2 - 6ax + 9a = a(x - 3)^2.$$

Although it is not necessary to memorize the special quadratic types of Eqs. 5.2-5.4, it is convenient to be able to recognize them immediately, since they are commonly occurring forms.

FACTORING FOR COMPUTER CALCULATION. In evaluating polynomials with a computer, considerable calculation time can be saved by careful factoring. For example, consider the following example.

EXAMPLE 5.7

$$5x^4 - 13x^3 + 11x^2 - 15x - 29$$

```
5.0*X**4 - 13.0*X**3 + 11.0*X**2 - 15.0*X - 29.0
```

SOLUTION

The most direct means for writing this as a Fortran statement would require four multiplications to obtain $5x^4$, three to obtain $13x^3$, and so on, for a total of 10 multiplications and four additions and subtractions. However, consider the following:

$$\begin{aligned} 5x^4 - 13x^3 + 11x^2 - 15x - 29 &= (5x^3 - 13x^2 + 11x - 15)x - 29 \\ &= [(5x^2 - 13x + 11)x - 15]x - 29 \\ &= \{[(5x - 13)x + 11]x - 15\}x - 29. \end{aligned}$$

```
(((5.0*X - 13.0)*X + 11.0)*X - 15.0)*X - 29.0
```

Evaluation of the final expression, which is equivalent to the original expression, requires only four multiplications and four additions and subtractions.

EXERCISE 5.3

Wherever possible, factor completely each of the following:

1. $6x + 4$
2. $6x - 2cx$
3. $x^2 - x$
4. $7y^4 + 14xy^3$
5. $6xy^3 - 3xy$
6. $x^2 + xy + y^2$
7. $ax - cx - dx^2$
8. $x^2yz - xy^2z + xyz^2$
9. $4xy - 9y^2 + 3y$
10. $3x^ny^n + 2x^{2n}$
11. $x^my - xy^n$
12. $z(a + b) - z(a + b)$
13. $x^2 + 6x + 5$
14. $x^2 - 5x + 6$
15. $x^2 - 1$
16. $x^2 - 4$
17. $8x^2 - 2$
18. $x^2 - 7x + 12$
19. $x^2 - x - 12$
20. $y^2 - 4y + 4$
21. $x^2 + x - 12$
22. $x^2 - 4x - 5$
23. $z^3 + 2z^2 + z$
24. $y^2 - 6$
25. $4x^2 - 16x + 16$
26. $x^2 + 4x - 5$
27. $z^4 - 1$
28. $x^2 + x + 1$
29. $3x^2 - 27x + 42$
30. $5x^2y^5 - 20y$
31. $z^5 - 6z^4 + 9z^3$
32. $2x^2 - 16x - 66$
33. $2ax^2 - 16ax + 32a$
34. $3x^4 + 3x^3 - 36x^2$
35. $(x - 1)^2 - 4(x - 1) + 4$
36. $2(x - 2)^2 + 12(2 - x) + 18$
37. $x^{n+2} - x^{n+1} - 6x^n$
38. $a(a^2 - 9) - 3(a^2 - 9)$
39. $(p^2 - 1) + p(p^2 - 1)$
40. $a^2(a + 3) + (6a + 9)(a + 3)$
41. $(x^2 - 4)(4x^2 - 9) - (x^2 + 4)(4x^2 - 9)$
42. $(x^2 + 2)(4x^2 - 4) - (6 - 3x^2)(4x^2 - 4)$

43. Write the equivalent Fortran expressions of the algebraic expressions of problems 4, 9, 23, 33, and 42.
44. The following expressions are to be factored as was Example 5.7 and written in Fortran to require a minimum number of arithmetic operations (specify how many multiplications and addition/subtractions are required in each): problems 4, 23, 31, and 34.
45. Factor problem 40 in two different ways and write the equivalent Fortran expressions. State how many arithmetic operations are required by each.

ALGEBRAIC FRACTIONS

The word *fraction*, like the word *numeral*, is frequently misused in everyday speech. Rigorously speaking, fraction is used to describe the symbol which represents a rational number when written in the form a/b. In other words, fraction is to rational number as numeral is to number. To be correct, the expression "rational number in fractional form" should be used rather than the word "fraction." However, for the sake of convenience, our usage will not be completely rigorous.

By now most of us are familiar with the operational rules involving fractional numbers, and have little difficulty with addition, changing form, and so on. For example, we know at a glance that

$$\frac{1}{3}=\frac{2}{6}$$

and

$$\frac{1}{2}+\frac{1}{4}=\frac{2}{4}+\frac{1}{4}=\frac{3}{4}.$$

However, the same operations and rules are frequently misapplied when dealing with algebraic fractions. This is unfortunate, since the algebraic symbols merely represent numbers. In performing arithmetic operations on algebraic fractions, certain rules are followed, some of which come about by definition and some of which are proved using basic definitions and previously derived relationships. With minor variations, they are identical to the laws and relationships used in working with numeric fractions.

A fraction, such as 2/5, actually involves three numbers: (1) the fraction itself represents a rational number which can be located on the real axis between 0 and 1, (2) the integer 2 is referred to as the **numerator**, and (3) the integer 5 is called the **denominator**. The numerator of a fraction can be any integer or zero, but the denominator can never be zero since division by zero is undefined. Thus, the algebraic fraction in the unknown x, must carry the

$$\frac{4}{3-x}$$

restriction that $x \neq 3$, since in this case the denominator vanishes (equals zero), and the quantity is not defined. Thus, whenever we speak of the general properties of a rational number a/b, we shall qualify our statement with $b \neq 0$.

Whereas in algebra we sometimes tend to think of a fraction as an entity in itself, it is important to recognize that in Fortran there is actually no such thing as a fraction. Remember that within the computer it is possible to store numbers which are often approximations to their desired values. These numbers can then be used for the arithmetic operations of addition, subtraction, multiplication, *and division*. For example, in arithmetic, we might be interested in the product of two quantities such as the fraction 2/3 and the whole number 3 which is, of course, equal to 2. However, considered in Fortran, this might be written as follows:

```
         (2/3)*3
OR       (2.0/3.0)*3.0
```

In both cases we are dealing with three distinct numbers, each of which will be assigned its unique storage position within the computer. The first example causes an integer division to take place yielding zero so the result of this integer evaluation would be zero. If we were using a decimal computer with eight-digit capacity, division in the second example would yield 0.66666666 and the final result would be 2.9999998. As we studied in Chapter 3, this can sometimes present serious problems to the programmer.

5.4 RELATIONSHIPS INVOLVING FRACTIONS

EQUALITY. We can see at a glance that the fractions 1/2 and 3/6 represent the same rational number. In general, two rational numbers are equal only when the following condition is satisfied:

Property 1. The rational numbers a/b and c/d (b and d $\neq$ 0) are equal if and only if ad = bc.

This is the means, frequently referred to as cross-multiplication, by which we identify fractions representing the same quantity. For example,

$$\frac{1}{4}=\frac{3}{12} \qquad \text{since } 1\times 12 = 4\times 3.$$

$$\frac{9}{15}=\frac{12}{20} \qquad \text{since } 9\times 20 = 15\times 12.$$

$$\frac{3}{5}\neq\frac{2}{3} \qquad \text{since } 3\times 3 \neq 5\times 2.$$

Similarly, for algebraic fractions, we have

$$\frac{1}{x}=\frac{4}{4x} \qquad \text{since } 1\times 4x = x\times 4.$$

$$\frac{x+1}{5}=\frac{x^2+x}{5x} \qquad \text{since } (x+1)5x = 5(x^2+x).$$

$$\frac{x+2}{x+3}\neq\frac{2}{3} \qquad \text{since } (x+2)(3) \neq (x+3)(2).$$

FUNDAMENTAL PRINCIPLE OF FRACTIONS. In manipulating equations containing fractions, the following property is often employed:

Property 2. Both the numerator and denominator of a rational number may be multiplied (or divided) by any nonzero number without changing its value. That is

$$\frac{a}{b} = \frac{ac}{bc} \qquad (b, c \neq 0).$$

The validity of this property can be verified by employing Property 1:

$$\frac{a}{b} = \frac{ac}{bc} \qquad \text{since } a(bc) = b(ac).$$

Each of the two examples illustrating Property 1 can be placed in the form displayed by Property 2 by factoring:

$$\frac{1}{4} = \frac{3}{12}, \qquad \frac{9}{15} = \frac{12}{20},$$

but

$$\frac{3}{12} = \frac{3 \times 1}{3 \times 4}, \qquad \frac{9}{15} = \frac{3 \times 3}{3 \times 5} \text{ and } \frac{12}{20} = \frac{4 \times 3}{4 \times 5}.$$

Therefore,

$$\frac{1}{4} = \frac{3 \times 1}{3 \times 4}, \qquad \frac{3 \times 3}{3 \times 5} = \frac{4 \times 3}{4 \times 5}.$$

Algebraic fractions may be handled in a like manner:

$$\frac{x+1}{5} = \frac{x^2 + x}{5x}$$

but

$$\frac{x^2 + x}{5x} = \frac{x(x+1)}{x(5)}.$$

Therefore,

$$\frac{x+1}{5} = \frac{x(x+1)}{x(5)}.$$

Here, of course, we must be careful to specify that $x \neq 0$, since the fraction $(x^2 + x)/(5x)$ represents a so-called **indeterminate** form when both the numerator and denominator become zero (that is, the fraction becomes 0/0).

SIGNS. In preceding paragraphs it was pointed out that three numbers are associated with a fraction. The rational number for which the fraction itself is the symbol, the integer which is the numerator, and the integer which is the denominator. As a result we would expect to find three signs associated with a fraction, that is, the sign of the rational number, the sign of the numerator, and the sign of the denominator. This leads to the following property.

Property 3. The negative of a rational number a/b, (b ≠ 0), may be expressed as

$$-\frac{a}{b}=\frac{-a}{b}=\frac{a}{-b}=-\frac{-a}{-b}.$$

If any two of these signs are changed, the value of the fraction is unchanged. We can justify changing the signs of both numerator and denominator by direct application of Property 2. For example,

$$\frac{-2}{3}=\frac{(-1)\times(-2)}{(-1)\times(3)}=\frac{2}{-3}.$$

By the rules for division of signed numbers developed in Chapter 2, either of these forms could be shown equal to $-(2/3)$. Although all of these forms are correct, we will generally manipulate our fractions to eliminate negative quantities from the denominators. In other words, $(-2)/3$ and $-2/3$ are preferred to $2/(-3)$.

Frequently, when the denominator consists of an algebraic expression containing more than one term, the desired form becomes a matter of convenience. For example, if the denominator of a fraction is $-2x+y$, it also could be written as $y-2x$ or $-(2x-y)$, and the fraction could appear as

$$\frac{3}{-2x+y}=\frac{3}{y-2x}$$

or

$$\frac{3}{-2x+y}=\frac{3}{-(2x-y)}.$$

The latter might be changed to

$$\frac{3}{-(2x-y)}=\frac{3(-1)}{-(2x-y)(-1)}=\frac{-3}{2x-y}.$$

In this particular case, the final form used is usually selected for convenience in performing other operations. However, it is important to recognize that the denominator consists of the complete expression, and if its sign is changed, the sign of *each* term must be changed. In other words, the fraction line acts as a grouping symbol and must be considered as such.

MULTIPLICATION. Multiplication of fractional forms is accomplished in accordance with the following property:

Property 4. The product of two rational numbers a/b and c/d (b, d ≠ 0) is

$$\frac{a}{b}\times\frac{c}{d}=\frac{ac}{bd}.$$

In other words, whenever it is necessary to multiply two rational numbers, either numeric or algebraic, the product is a rational number whose numerator is the product of the numerators and whose denominator is the product of the denominators. Some examples are

$$\frac{2}{3}\times\frac{7}{9}=\frac{14}{27},$$

$$\frac{9}{4}\times\frac{-3}{8}=\frac{9(-3)}{4\times 8}=\frac{-27}{32}=-\frac{27}{32},$$

$$\frac{x}{2y}\times\frac{3x^2}{y}=\frac{x(3x^2)}{2y\times y}=\frac{3x^3}{2y^2},$$

$$(x-y)\frac{4x}{2y}=\frac{(x-y)}{1}\times\frac{4x}{2y}=\frac{4x(x-y)}{2y}.$$

Note that in the second example a negative sign is encountered and is treated in the usual manner. In the last example, the first factor is $(x-y)$, which is handled as if it were the fraction $(x-y)/1$.

ADDITION. The addition of rational numbers with like denominators is a simple matter, as described by the following property:

Property 5. The sum of two rational numbers with like denominators is another rational number whose numerator is the sum of the numerators of the original fractions and whose denominator is the common denominator. That is,

$$\frac{a}{c}+\frac{b}{c}=\frac{a+b}{c}\qquad (c\neq 0).$$

From arithmetic, we undoubtedly remember additions of fractions, such as

$$\frac{1}{7}+\frac{3}{7}=\frac{1+3}{7}=\frac{4}{7},$$

$$\frac{6}{13}+\frac{-8}{13}=\frac{6-8}{13}=\frac{-2}{13}.$$

Addition of algebraic fractions is performed in the same way. For example,

$$\frac{1}{x}+\frac{3y-2}{x}=\frac{1+3y-2}{x}=\frac{3y-1}{x},$$

$$\frac{-6mn}{m-n}+\frac{-3m}{m-n}=\frac{-6mn-3m}{m-n}=-\frac{6mn+3m}{m-n}.$$

Note that this property says nothing of adding rational numbers when the denominators are not the same. Frequently the difficult part of adding rational numbers is to manipulate them, using Properties 2 and 3, so that the denominators are the same. This and other uses of these properties will be the subject of the following sections.

REPRESENTING FRACTIONAL QUANTITIES IN FORTRAN. One characteristic of the fraction line to which we are accustomed is that it serves to indicate grouping. For example, in the quantities

$$\frac{5}{8+3} \qquad \frac{6a}{a^2-3b} \qquad \frac{x^2-y^2+z^2}{z} \tag{1}$$

extension of the fraction line indicates that the complete expressions $8+3$ and a^2-3b are the denominators for the first two examples and that the expression $x^2-y^2+z^2$ is the numerator in the last example. We clearly see that the following forms are not equivalent to the forms of (1).

$$5/8+3 \qquad 6a/a^2-3b \qquad x^2-y^2+z^2/z \tag{2}$$

Because of hierarchy of operations (multiplication and division before addition and subtraction) the forms in (2) represent

$$\frac{5}{8}+3 \qquad \frac{6a}{a^2}-3b \qquad x^2-y^2+\frac{z^2}{z}. \tag{3}$$

Obviously the grouping characteristic of the fraction line is lost when the quantities are written on a "single line." On the other hand the desired groupings can be retained when parentheses are used as is illustrated by the following, which are equivalent to the examples in (1).

$$5/(8+3) \qquad 6a/(a^2-3b) \qquad (x^2-y^2+z^2)/z. \tag{4}$$

These notions are directly applicable to Fortran since the division operation symbol (/) is never used to imply a grouping of terms to form part of an expression; this must be done by the use of parentheses. Thus the quantities in (1) must be put in the form of (4) in order to produce the following equivalent Fortran expressions.

```
5.0/(8.0+3.0)   OR   5/(8+3)
6.0*A/(A**2-3.0*B)
(X**2-Y**2+Z**2)/Z
```

In Fortran, the forms of (2)—which are *not* equivalent to (1)—appear as follows.

```
5.0/8.0+3.0
6.0*A/A**2-3.0*B
X**2-Y**2+Z**2/Z
```

EXERCISE 5.4

Using Property 1, identify which of the pairs of rational numbers in 1-16 are equal.

1. $\frac{3}{4}, \frac{27}{34}$
2. $\frac{7}{6}, \frac{42}{36}$
3. $\frac{5}{8}, \frac{15}{25}$
4. $\frac{7}{13}, \frac{28}{65}$
5. $\frac{1}{x}, \frac{xy}{x^2y}$
6. $\frac{3x}{5y}, \frac{9x^3}{15x^2y}$
7. $\frac{a-b}{a}, \frac{a^2-b^2}{a^2+ab}$
8. $\frac{z+1}{z^2+3z}, \frac{z^2+2z+1}{z^3+3z^2}$

9. $\frac{b-a}{b^2-a^2}, \frac{a}{b}$

10. $\frac{a+b}{a-b}, -\frac{b+a}{b-a}$

11. $\frac{m+2}{m^2-m-6}, \frac{-1}{3-m}$

12. $\frac{2+3}{2+5}, \frac{3}{5}$

13. $\frac{1-7}{1-8}, \frac{7}{8}$

14. $\frac{1+x}{1+y}, \frac{x}{y}$

15. $\frac{x+y}{x-y}, \frac{y}{-y}$

16. $\frac{a}{a-b}, \frac{1}{-b}$

17. Change each of the following fractions to equivalent fractions with denominators of 120:

(a) $\frac{1}{2}$ (b) $\frac{1}{3}$ (c) $\frac{5}{12}$ (d) $\frac{9}{8}$ (e) $\frac{13}{15}$ (f) $\frac{5}{24}$ (g) $\frac{1}{15}$

18. Change each of the following fractions to equivalent fractions with denominators of $x^3 + 4x^2 + 3x$:

(a) $\frac{1}{x}$ (b) $\frac{x}{x+1}$ (c) $\frac{x^2+1}{x^2+x}$ (d) $\frac{x^2}{x+3}$ (e) $\frac{x-1}{x^2+3x}$

19. Change each of the following fractions to equivalent fractions with denominators of $a^2 - ab$:

(a) $\frac{1}{a-b}$ (b) $\frac{1}{b-a}$ (c) $\frac{a}{a-b}$ (d) $\frac{a^3(a-b)}{b-a}$

Using Property 3 regarding the signs of a fraction, $1/(a-b)$ may be written as

$$-\frac{-1}{a-b}, -\frac{1}{-(a-b)}, -\frac{1}{b-a}, \frac{-1}{-(a-b)}, \text{ or } \frac{-1}{b-a}$$

In 20-27, express the fractions in alternate forms by manipulating signs.

20. $\frac{x}{y}$

21. $\frac{1}{b}$

22. $\frac{a}{1}$

23. b

24. $\frac{1}{x-y}$

25. $\frac{a}{b-a}$

26. $\frac{m-n}{n}$

27. $\frac{a+1}{a}$

Find the product of each pair of rational numbers:

28. $\frac{x}{1} \times \frac{1}{y}$

29. $\frac{a}{b} \times \frac{ac}{d}$

30. $\frac{1}{x} \times \frac{-y}{x}$

31. $a \times \left(-\frac{1}{b}\right)$

32. $-x \times \frac{x}{x+1}$

33. $\frac{x+1}{-y} \times \frac{x-1}{y^2+2}$

34. $\frac{xy}{x+y} \times \frac{3x^2}{4(x-y)}$

35. $\frac{3x^3}{2y} \times \frac{-z^2}{3(x+y)}$

36. $(x-1)\times\frac{x-1}{x+1}$

37. $\frac{a^3-b^3}{a^2}\times(-5)$

Find the sum of each pair of rational numbers whenever possible without using Property 2:

38. $\frac{1}{x}+\frac{x+1}{x}$

39. $\frac{2+a}{a}+\frac{-2}{a}$

40. $\frac{3x^2}{x+1}+\frac{1}{1+x}$

41. $\frac{2x}{x+1}+\frac{3x+1}{x}$

42. $\frac{x^2+3x-1}{x+1}+\frac{x-1}{1+x}$

43. $\frac{1}{a-b}+\frac{b}{b-a}$

44. $\frac{x}{x-a}+\frac{x}{a-x}$

45. $\frac{-1+x}{x^2-1}+\frac{x}{(1-x)(x+1)}$

46. Write Fortran expressions for the original algebraic forms and for the results in problems 32, 34, 37, 38, 42, and 45. Count the number of arithmetic operations in each case.

47. Repeat problem 46 for problems 31, 33, 35, 39, 41, and 43.

5.5 REDUCING RATIONAL NUMBERS

In working with rational numbers, we would automatically convert the sum of 1/8 and 3/8 (which is 4/8) to its simplified form of 1/2. In arithmetic, we say the fraction has been reduced to its lowest terms, and we may justify the operation in the following way:

$$\frac{1}{8}+\frac{3}{8}=\frac{4}{8} \qquad \text{(Property 5)}$$

$$\frac{4}{8}=\frac{1\times 4}{2\times 4} \qquad \text{(factoring)}$$

$$\frac{1\times 4}{2\times 4}=\frac{1}{2}\times\frac{4}{4} \qquad \text{(Property 4)}$$

$$\frac{1}{2}\times\frac{4}{4}=\frac{1}{2}\times 1=\frac{1}{2}.$$

In algebra we do the same thing. First, each numerator and denominator is factored completely. Then similar factors in the numerator and denominator may be eliminated by the application of Property 4. For example,

$$\frac{2a^2}{4a^3b}=\frac{2a\times a}{2\times 2a\times a\times a\times b}$$

$$=\frac{1}{2ab}\times\frac{2aa}{2aa}$$

$$=\frac{1}{2ab}.$$

FRACTIONS CONTAINING BINOMIALS. When dealing with algebraic fractions containing binomials, the procedure is similar to the foregoing one. Generally speaking, the probability of finding similar factors when dealing with polynomials is smaller, but it is usually necessary to factor both the numerator and denominator to be certain:

$$\frac{3x(x^2-y^2)}{(2x-2y)x^2y} = \frac{3x(x-y)(x+y)}{2x^2y(x-y)}$$

$$= \frac{x(x-y) \times 3(x+y)}{x(x-y) \times 2xy}$$

$$= \frac{3(x+y)}{2xy};$$

$$\frac{2x^3+8x^2-10x}{(2x+10)xy} = \frac{2x(x^2+4x-5)}{2xy(x+5)}$$

$$= \frac{2x(x+5) \times (x-1)}{2x(x+5) \times y}$$

$$= \frac{x-1}{y}.$$

Whenever two rational numbers are multiplied together (according to Property 4), the product should be checked to see that it is in its simplest form. Frequently, a resulting product can be reduced considerably. For example,

$$\frac{3x^2y}{x-y} \times \frac{x^2-y^2}{2xy} = \frac{3x^2y(x^2-y^2)}{2xy(x-y)}$$

$$= \frac{xy(x-y) \times 3x(x+y)}{xy(x-y) \times 2}$$

$$= \frac{3x(x+y)}{2}.$$

Generally speaking, we will avoid discussing incorrect forms in this text. However, incorrect handling of algebraic fractions is sufficiently common to warrant emphasizing the nature of fractions themselves. In dealing with numeric fractions, few algebra students fail to realize that

$$\frac{2+3}{2+5} \neq \frac{\not{2}+3}{\not{2}+5} \neq \frac{3}{5},$$

or that

$$\frac{2+3}{2} \neq \frac{\not{2}+3}{\not{2}} \neq 3.$$

Again it is emphasized that the fraction line serves as a grouping symbol and that the numerator is the *quantity* $(2+3)$. Similarly, in the first example, the denominator also consists of a quantity, which is the sum of the two terms 2 and 5. Many students who would never consider such an operation on numbers will commit this error with algebraic expressions:

$$\frac{x+y}{x+z} \neq \frac{\cancel{x}+y}{\cancel{x}+z} \neq \frac{y}{z}$$

$$\frac{x+y}{x} \neq \frac{\cancel{x}+y}{\cancel{x}} \neq y$$

The above fractions are already in their simplest form.

EXERCISE 5.5

Reduce to lowest terms:

1. $\dfrac{5x^3}{2x}$
2. $\dfrac{6x^5}{3x^4y}$
3. $\dfrac{24x^3y^2}{-6xyz}$
4. $-\dfrac{-2abc^2}{a^2bc}$
5. $\dfrac{(-a)(-b)c}{a^2b^2c^2}$
6. $\dfrac{3x(-y)z}{12(-x)z^2}$
7. $\dfrac{2x-4}{2}$
8. $\dfrac{3x^2-5xy}{2x}$
9. $\dfrac{2x^2z^3y-4xz^2y}{xzy}$
10. $\dfrac{a-b}{b-a}$
11. $\dfrac{a^2-b^2}{b^2-ab}$
12. $\dfrac{1-x^2}{x^2+2x+1}$
13. $\dfrac{x^2-3x+2}{2x-x^2}$
14. $\dfrac{2(y-x)^3}{4x-4y}$
15. $\dfrac{x^2-x-20}{5x-x^2}$

Multiply and reduce to lowest terms:

16. $\dfrac{x}{y} \times \dfrac{y}{z}$
17. $\dfrac{xy}{4z} \times \dfrac{2xz}{y}$
18. $\dfrac{a^2}{bb} \times \dfrac{3ba}{2}$
19. $\dfrac{-32ab^2}{33c^2} \times \dfrac{11c^2b}{16a^3}$
20. $\dfrac{x-1}{3} \times \dfrac{5x}{x^2-1}$
21. $(y-4) \times \dfrac{5y}{y^2-16}$
22. $\dfrac{b^2}{3b^2+9b+6} \times 6(b+1)$
23. $\dfrac{-3x^3yz}{x^2+2x+1} \times \dfrac{x^2+x}{-6x^2yz}$
24. Write Fortran expressions for the original algebraic forms and for the results in problems 5, 9, 15, and 19. Count the number of arithmetic operations in each case.
25. Repeat problem 24 for problems 7,11, 17, and 21.

5.6 MULTIPLICATION AND DIVISION OF RATIONAL NUMBERS

Multiplication of rational numbers is a relatively simple operation. In fact, it is usually easier than the operations of addition and subtraction, as we shall discover in Section 5.7.

The process of dividing one number by another is normally defined by multiplication. That is, if x and y are any two numbers ($y \neq 0$), we can divide x by y, giving

$$\frac{x}{y} = p$$

where p satisfies the relationship

$$x = yp.$$

For example,

$$\frac{18}{6} = 3$$

implies that $18 = 6 \times 3$. If x and y are the rational numbers a/b and c/d, respectively, the problem becomes one of division:

$$\frac{x}{y} = \frac{a}{b} \div \frac{c}{d} = p.$$

But the nature of p must be such that

$$\frac{a}{b} = \frac{c}{d} \times p.$$

Multiplying by d/c gives

$$\frac{d}{c} \times \frac{c}{d} \times p = \frac{a}{b} \times \frac{d}{c}.$$

Solving this for p gives

$$p = \frac{a}{b} \times \frac{d}{c}.$$

Comparing this to the original form,

$$p = \frac{a}{b} \div \frac{c}{d},$$

we see that to divide one rational number by another it is necessary to invert the divisor and multiply. For example,

$$\begin{aligned}
\frac{4xy^2}{3} \div \frac{2x}{3y(x+y)} &= \frac{4xy^2}{3} \times \frac{3y(x+y)}{2x} \\
&= \frac{4xy^2 \times 3y(x+y)}{3 \times 2x} \\
&= \frac{2x \times 2y^2 \times 3 \times y(x+y)}{3 \times 2x} \\
&= \frac{2x \times 3 \times 2y^2 \times y(x+y)}{2x \times 3} \\
&= \frac{2y^3(x+y)}{1} \\
&= 2y^3(x+y).
\end{aligned}$$

Occasionally it is necessary to perform mixed operations of multiplication and division on several fractions. This is merely more of the same, as shown by the following example:

$$x^2 \div \frac{x^2+2x+1}{x} \times \frac{3x+3}{x^2} = \frac{x^2}{1} \times \frac{x}{x^2+2x+1} \times \frac{3x+3}{x^2}$$

$$= \frac{x^2}{1} \times \frac{x}{(x+1)(x+1)} \times \frac{3(x+1)}{x^2}$$

$$= \frac{x^2(x+1) \times 3x}{x^2(x+1) \times (x+1)}$$

$$= \frac{3x}{x+1}.$$

EXERCISE 5.6

Perform the indicated operations and simplify:

1. $\frac{5}{7} \div \frac{7}{9}$
2. $\frac{3}{8} \div 2$
3. $\frac{4xy}{z} \div \frac{2z}{x}$
4. $\frac{5a^2b^3c}{9} \div \frac{10a^2b}{3c}$
5. $\frac{4rs}{6t^3} \div \frac{6t^3}{-rs^2}$
6. $\frac{1}{5x^3} \div \frac{1}{4xy^2}$
7. $\frac{-a}{a-b} \div \frac{a}{b-a}$
8. $\frac{x-1}{x+1} \div \frac{x^2-1}{x}$
9. $\frac{a^2+b^2}{a-b} \div \frac{a+b}{a-b}$
10. $\frac{x^2-4}{x^2-2x+1} \div (1-x)$
11. $(x+3) \div \frac{x+4x+3}{x}$
12. $\frac{x-1}{x} \times \frac{x^3}{x^2+2x-3} \div \frac{1}{x+3}$
13. $\frac{xy}{5z^2} \div \frac{1}{x+1} \times \frac{10z^3}{x^2-1}$
14. $\frac{a-b}{a} \times \frac{b+a}{3a^2} \times \frac{6a^2b}{b^2-a^2}$
15. Write Fortran expressions for the original algebraic forms and for the results in problems 5 and 13. Count the number of arithmetic operations in each case.
16. Repeat problem 15 for problems 9 and 11.

5.7 ADDITION AND SUBTRACTION OF RATIONAL NUMBERS

Property 5 describes how to add rational numbers when the denominators are the same; however, in practice this usually is not the case. Thus it is necessary to change the form of each fraction, using Property 2, so that they have identical denominators. In adding and subtracting numeric fractions, the process of obtaining equivalent forms with a common denominator is practically automatic. For example,

$$\frac{1}{6} + \frac{1}{10} = \frac{10}{60} + \frac{6}{60}$$

$$= \frac{16}{60}$$

$$= \frac{4 \times 4}{4 \times 15}$$

$$= \frac{4}{15}.$$

In this case the common denominator used was simply the product of both denominators. Although this is valid and gives the correct result, it is not the simplest or the most efficient quantity to use.

LEAST COMMON DENOMINATOR. The smallest number which is common to both or all denominators of two or more fractions is the **least common denominator** (*LCD*). In the foregoing example, we can see by inspection that 30 is the smallest number into which both 6 and 10 will divide; it is thus the least common denominator. That the final sum is unaffected by the choice of the common denominator is evident from the following:

$$\begin{aligned}\frac{1}{6}+\frac{1}{10}&=\frac{5}{30}+\frac{3}{30}\\&=\frac{8}{30}\\&=\frac{4}{15}.\end{aligned}$$

To obtain the smallest possible common denominator, each denominator is first factored, giving

$$\begin{aligned}6&=3\times 2,\\10&=5\times 2.\end{aligned}$$

We can see that any common denominator must consist of at least the factors 2, 3, and 5; thus the LCD is $2\times 3\times 5=30$. Each of the original fractions is changed to the required form:

$$\frac{1}{6}=\frac{5\times 1}{5\times 6}=\frac{5}{30},$$

$$\frac{1}{10}=\frac{3\times 1}{3\times 10}=\frac{3}{30}.$$

ALGEBRAIC FORMS. Any algebra student should be completely familiar with the common operations of arithmetic. However, when these same operations must be applied to algebraic fractions, confusion often results. The LCD of the fractions $1/a$ and $1/b$ is obviously the product ab. The addition is performed as follows:

$$\begin{aligned}\frac{1}{a}+\frac{1}{b}&=\frac{b\times 1}{b\times a}+\frac{a\times 1}{a\times b}\\&=\frac{b}{ab}+\frac{a}{ab}\\&=\frac{a+b}{ab}.\end{aligned}$$

The operation is similar for the following fractions, which have an LCD of ab^2:

$$\frac{1}{a}+\frac{1}{b}-\frac{1}{ab^2}=\frac{b^2\times 1}{b^2\times a}+\frac{ab\times 1}{ab\times b}-\frac{1}{ab^2}$$

$$=\frac{b^2+ba-1}{b^2a}.$$

Other examples are

$$\frac{x}{y}+\frac{1}{x}=\frac{x}{y}\times\frac{x}{x}+\frac{1}{x}\times\frac{y}{y}$$

$$=\frac{x^2}{xy}+\frac{y}{xy}\qquad\text{(since } xy \text{ is the LCD)}$$

$$=\frac{x^2+y}{xy},$$

$$\frac{2}{x-1}+\frac{3x-1}{x^2-3x+2}-\frac{5}{x}=\frac{2}{x-1}+\frac{3x-1}{(x-1)(x-2)}+\frac{-5}{x}.$$

In the latter case the LCD is $x(x-1)(x-2)$, which gives

$$=\frac{2}{x-1}\times\frac{x(x-2)}{x(x-2)}+\frac{3x-1}{(x-1)(x-2)}\times\frac{x}{x}+\frac{-5}{x}\times\frac{(x-1)(x-2)}{(x-1)(x-2)}$$

$$=\frac{2x(x-2)+(3x-1)x-5(x-1)(x-2)}{x(x-1)(x-2)}$$

$$=\frac{2x^2-4x+3x^2-x-5x^2+15x-10}{x(x-1)(x-2)}$$

$$=\frac{10(x-1)}{x(x-1)(x-2)}$$

$$=\frac{10}{x(x-2)}.$$

EXERCISE 5.7

Combine the following fractions as indicated, and simplify:

1. $\frac{4}{3}+\frac{5}{6}$
2. $\frac{5}{7}+\frac{4}{21}$
3. $\frac{7}{10}-\frac{2}{5}$
4. $\frac{7}{12}-\frac{5}{8}$
5. $\frac{1}{4}-\frac{-1}{11}$
6. $\frac{1}{x}-\frac{2}{y}$
7. $\frac{3}{ab}-\frac{b}{x}$
8. $\frac{x}{ay}-\frac{y}{bx}$
9. $\frac{a}{b}-\frac{b}{a}$
10. $\frac{1}{a^2x}-\frac{-b}{ax}$

11. $\frac{x-1}{4}+\frac{2x+4}{3}$

12. $\frac{x-1}{6}+\frac{1-2x}{3}$

13. $\frac{3x^2-2x+3}{3}-(x^2+1)$

14. $(x-1)-\frac{x^2+2x+4}{2}$

15. $\frac{1}{x-a}+\frac{1}{x+b}$

16. $\frac{1}{x+b}-\frac{1}{x}$

17. $\frac{1}{z-1}-\frac{1}{1-z}$

18. $\frac{4}{ax-b}+\frac{3}{b-ax}$

19. $\frac{x+2}{x-1}-\frac{x-1}{x+2}$

20. $\frac{1}{1-n}+\frac{n-1}{n^2-2n+1}$

21. $\frac{x}{xy-yz}+\frac{x}{xz-x^2}$

22. $\frac{2xy}{x^2y^2}+\frac{1}{y-x}+\frac{1}{x+y}$

23. $\frac{a+b}{a-b}+\frac{a^2+3b^2}{b^2-a^2}+\frac{a-2b}{b+a}$

24. $\frac{2}{2x+1}+\frac{3}{1-2x}+\frac{1-2x}{4x^2-1}$

EXPONENTS AND RADICALS

5.8 PROPERTIES OF EXPONENTS

In Chapter 3, the basic principles of exponents were introduced for use in discussing number systems in Chapters 3 and 4. Remember that quantities such as 10^4 and a^n actually represent

$$10^4 = 10 \times 10 \times 10 \times 10,$$
$$a^n = \underbrace{a \times a \times a \times \ldots \times a}_{n \text{ times}}. \tag{5.5}$$

Four basic properties of exponents were used in the earlier chapters; these will be reviewed and justified here.

MULTIPLICATION. If we have two quantities, such as x^3 and x^4, and wish to know their product, what is the correct procedure? In this type of situation, it is usually best to refer back to the basic definition (Eq. 5.5), which gives

$$(x \times x \times x) \times (x \times x \times x \times x) = x \times x \times x \times x \times x \times x \times x$$

or

$$x^3 \times x^4 = x^{3+4} = x^7.$$

This can be generalized as the first property of exponents, as follows:

Property 1. For any real number $x \neq 0$ (x not equal to zero) and natural numbers a and b.

$$x^a \times x^b = x^{a+b}.$$

In general, when multiplying numbers having the same base, we add the exponents and use the common base. Other examples are shown below:

$$10^2 \times 10^5 = 10^7,$$
$$2^4 \times 2^9 = 2^{13},$$
$$(xy)^5 \times (xy) = (xy)^6.$$

Note that $x^a \times x^b = x^b \times x^a$, since x^a and x^b represent real numbers and real numbers are commutative. In other words, the exponential forms are commutative under multiplication.

DIVISION. If it is necessary to divide one exponential quantity by another, this can be accomplished by again referring to Eq. 5.5. For instance,

$$\frac{x^6}{x^4} = \frac{\not{x} \times \not{x} \times \not{x} \times \not{x} \times x \times x}{\not{x} \times \not{x} \times \not{x} \times \not{x}} = \frac{x \times x}{1} = x^2.$$

Here the final exponent on x is equal to the number of x's "left over," or the difference between the exponent on the numerator and that on the denominator. This can be generalized as the second property of exponents:

Property 2. For any real number $x \neq 0$ and natural numbers a and b

$$\frac{x^b}{x^a} = x^{b-a}.$$

That is, when dividing numbers having the *same* base, we subtract the exponents and use the common base. Additional examples are illustrated below:

$$\frac{10^7}{10} = 10^{7-1} = 10^6, \qquad \frac{2^4}{2^4} = 2^{4-4} = 2^0,$$
$$\frac{2^3}{2^2} = 2^{3-2} = 2.$$
$$\frac{10^5}{10^4} = 10^{5-4} = 10, \qquad \frac{10^2}{10^5} = 10^{2-5} = 10^{-3},$$

As is evident, whenever $a = b$ or $a > b$, situations arise which have not yet been discussed. This is the topic of the next section.

ZERO AND NEGATIVE EXPONENTS. We will first consider the case of equal exponents. Assume that $a = b = 3$. The quotient of x^b/x^a may be determined by applying Eq. 5.5 to both the numerator and denominator:

$$\frac{x^3}{x^3} = \frac{\not{x} \times \not{x} \times \not{x}}{\not{x} \times \not{x} \times \not{x}} = 1.$$

However, we know by Property 2 that

$$\frac{x^3}{x^3} = x^{3-3} = x^0.$$

In general it is apparent that any nonzero quantity divided by itself is equal to 1, so we have

$$\frac{x^a}{x^a} = 1.$$

But by application of Property 2 we have

$$\frac{x^a}{x^a} = x^{a-a} = x^0.$$

Therefore, in order that the structure of this system of exponents be consistent, it is necessary to define any nonzero real number raised to the zero power as 1:

Property 3. For any real number $x \neq 0$

$$x^0 = 1.$$

If the exponents, rather than being equal, have the relationship $a > b$, a similar problem arises, which can also be disposed of by reference to Eq. 5.5. Suppose $b = 3$ and $a = 5$. We then have

$$\frac{x^b}{x^a} = \frac{x^3}{x^5} = \frac{\not{x} \times \not{x} \times \not{x}}{\not{x} \times \not{x} \times \not{x} \times x \times x} = \frac{1}{x^2}$$

But again by Property 2

$$\frac{x^3}{x^5} = x^{3-5} = x^{-2}.$$

Similarly,

$$\frac{10^4}{10^9} = 10^{4-9} = 10^{-5} = \frac{1}{10^5}.$$

Thus, as before, the need for a consistent structure makes it necessary to define negative exponents as follows:

Property 4. For any real number $x \neq 0$ and any integer a,

$$x^{-a} = \frac{1}{x^a}.$$

OTHER PROPERTIES OF EXPONENTS. As a direct consequence of the definition of the form x^a, we have

Property 5. For any real number $x \neq 0$,

$$(x^a)^b = x^{ab}.$$

For example,

$$\begin{aligned}(10^3)^4 &= 10^{3 \times 4} \\ &= 10^{12}.\end{aligned}$$

The logic of this becomes apparent when the original form is considered relative to the definition:

$$\begin{aligned}(10^3)^4 &= (10^3)(10^3)(10^3)(10^3) \\ &= 10^{3+3+3+3} \\ &= 10^{3 \times 4} \\ &= 10^{12}.\end{aligned}$$

The general case appears as

$$(x^a)^b = \underbrace{x^a \times x^a \times x^a \cdots \times x^a}_{b \text{ factors}}$$

$$= x^{\overbrace{a + a + a + \cdots + a}^{b \text{ terms}}}$$

$$= x^{ab}.$$

Two other useful properties are as follows:

Property 6. For any real numbers x and $y \neq 0$,

$$(xy)^a = x^a y^a.$$

This follows from the definition of exponent, since

$$(xy)^a = \underbrace{(xy)\,(xy)\,(xy) \cdots (xy)}_{a \text{ factors}}$$

$$= \underbrace{x \times x \times x \cdots \times x}_{a \text{ factors}} \times \underbrace{y \times y \times y \cdots \times y}_{a \text{ factors}}$$

$$= x^a y^a.$$

Property 7. For any real numbers x and $y \neq 0$,

$$\left(\frac{x}{y}\right)^a = \frac{x^a}{y^a}.$$

This property also follows from the definition of the exponent, since

$$\left(\frac{x}{y}\right)^a = \underbrace{\frac{x}{y} \times \frac{x}{y} \times \frac{x}{y} \cdots \times \frac{x}{y}}_{a \text{ factors}}$$

$$= \frac{\overbrace{x \times x \times x \cdots \times x}^{a \text{ factors}}}{\underbrace{y \times y \times y \cdots \times y}_{a \text{ factors}}}$$

$$= \frac{x^a}{y^a}.$$

Examples illustrating the application of these properties are as follows:

$$(4x^2)^3 = 64x^6,$$

$$\left(\frac{ab}{c^2}\right)^4 = \frac{a^4b^4}{c^8},$$

$$\left(\frac{x^n}{y^m}\right)^2 = \frac{x^{2n}}{y^{2m}},$$

$$\left(\frac{-a^2}{b}\right)^3 = \frac{(-a^2)^3}{b^3} = \frac{-a^6}{b^3},$$

$$\left(\frac{y^{2n-1}}{y^{m+2n-1}}\right)^2 = [y^{(2n-1)-(m+2n-1)}]^2$$

$$= [y^{-m}]^2$$

$$= \frac{1}{y^{2m}}.$$

Successive application of these properties of exponents often results in a considerably simplified expression, as shown in the following example:

$$\left(\frac{2x^3y^a}{z^2}\right)^3 \times \left(\frac{z^2y}{2x^4}\right)^2 = \frac{(2x^3y^a)^3}{z^6} \times \frac{(z^2y)^2}{(2x^4)^2}$$

$$= \frac{2^3x^9y^{3a}}{z^6} \times \frac{z^4y^2}{2^2x^8}$$

$$= 2^{3-2} \times x^{9-8} \times z^{4-6} \times y^{3a+2}$$

$$= \frac{2xy^{3a+2}}{z^2}.$$

EXPONENTS IN FORTRAN. As we have already seen, an exponent is indicated in Fortran by use of the double asterisk. For example, the quantities ax^2 and $(b^2 - c^3)/2$ can be represented, respectively, as

```
A*X**2
(B**2 - C**3)/2.0
```

Perhaps the reader has noted that all examples to this point have used integer exponents whereas the constants and variables have all been real (floating point). Technically speaking, a Fortran expression consisting of both integer and real variables and/or constants is termed a *mixed mode* expression, which although not incorrect, is inefficient. However, any quantities used as exponents in a real expression may be integer or real without causing the expression to be interpreted as mixed mode. If, on the other hand, the exponents will always be whole numbers, then it is best to use integer exponents due to the manner in which the Fortran system performs the computations. For example, consider the following:

```
A**5          A**5.0
```

In most Fortran systems, the first expression will be evaluated by performing successive multiplications, that is, it will be treated as $A*A*A*A*A$. However, the real exponent (5.0) forces the Fortran system to convert to logarithms to perform the operation. Since special Fortran subroutines must be used, this is relatively slow, less accurate, and requires more storage in some instances. Consequently, integer quantities should be used whenever possible.

Although preceding Fortran examples and problems have all used simple constants as exponents, the Fortran language is not so restricted. But then again, we must give particular emphasis to the principle of hierarchy of operations. For example, the following pair of expressions might, at first glance be considered as representing a^{2n}.

```
A**2*N          A**(2*N)
```

Applying the hierarchy rules (remember exponentiation, then multiplication and division), we see that the first expression actually represents a^2n. Through the use of parentheses, the exponent can be virtually any expression within itself. Negative exponents must also be enclosed with parentheses since it is illegal to write two operational symbols adjacent to one another (except that ** is recognized as indicating exponentiation). Thus b^{-n} would be written as the first form following, not the second.

```
B**(-N)          B**-N    (INVALID)
```

EXERCISE 5.8

Express in exponential form:

1. 2×2
2. $2 \times 2 \times 2 \times 2$
3. 10
4. $10 \times 10 \times 10 \times 10 \times 10 \times 10 \times 10$
5. $a \times a \times a$
6. $x \times x \times x \times x \times x$
7. $z \times z \times z \times z$
8. y

Express the following in exponential form. Write all answers with positive exponents.

9. $\dfrac{1}{2 \times 2 \times 2}$
10. $\dfrac{1}{8 \times 8 \times 8 \times 8}$
11. $\dfrac{1}{10 \times 10}$
12. $\dfrac{1}{a \times a \times a}$
13. 2^{-3}
14. 2^{-1}
15. 2^{-5}
16. 2^{-a}
17. x^{-3}
18. x^{-b}

Perform the indicated operations and express in exponential form:

19. $2^3 \times 2$
20. $2^5 \times 2^1$
21. $2^0 \times 2^5$
22. $8^2 \times 8^3$
23. $10^5 \times 10^4$
24. $10^0 \times 10^n$
25. $10^0 \times 10^0$
26. $x^2 \times x$
27. $x^3 \times x^2$
28. $a^2 \times a^2$
29. $a^{12} \times a$
30. $b^0 \times b$
31. $b^0 \times b^2$
32. $b^0 \times b^{14}$
33. $x^0 \times x^0$

Perform the indicated operations and express in exponential form:

34. $10^{-2} \times 10^2$
35. $10^{-5} \times 10^{-3}$
36. $10^{-1} \times 10$
37. $10^{-1} \times 10^{-2}$
38. $y^3 \times y^{-3}$
39. $x^4 \times x^{-5}$
40. $x^{-1} \times x^2$
41. $x^{-2} \times x^{-3}$
42. $x^{-1} \times x^{-1}$

Perform the indicated operations and express in exponential form:

43. $\dfrac{10^3}{10^2}$
44. $\dfrac{10^5}{10^5}$
45. $\dfrac{2^4}{2^5}$
46. $\dfrac{2^3}{2^2}$
47. $\dfrac{2^{-1}}{2^{-1}}$
48. $\dfrac{8^4}{8^{-4}}$

49. $\dfrac{x^4}{x^3}$

50. $\dfrac{x^9}{x^{10}}$

51. $\dfrac{x^0}{x^0}$

52. $\dfrac{y^a}{y}$

53. $\dfrac{z^{-1}}{z}$

54. $\dfrac{z^2}{z^{-2}}$

55. Show that exponential forms are associative under the operation of multiplication; that is, show that $(x^a \times x^b) \times x^c = x^a \times (x^b \times x^c)$.

56. Demonstrate the inconsistency which would arise if addition of exponential quantities were defined as $x^a + x^b = x^{a+b}$.

57. Demonstrate the inconsistency which would result if multiplication of exponential quantities were defined as $x^a \times x^b = x^{ab}$.

58. Use the basic definition of exponents and the commutative law to show that $(xy)^4 = x^4y^4$. Generalize this to $(xy)^a$.

Perform the indicated operations and simplify. All results are to be written without zero or negative exponents.

59. $(2ax^2)^2$

60. $(2b^2x^3)^2$

61. $(-5a)^2$

62. $(-2ab^2)^3$

63. $-(a^2b^3c)^4$

64. $-(2ax^5)^3$

65. $(bx^n)^{2m}$

66. $(6a^{2m}y^m)^n$

67. $(2^xx^2y)^z$

68. $(-3ab^2c^3)^a$

69. $[(-1)ax^2]^m$

70. $[(-3)by^m]^n$

71. $(-xy^2)^a$

72. $(-ab^{-1})^x$

73. $(x^2y^{-2})^{-2}$

74. $(3ab^{-2})^{-n}$

75. $(x^ay^{-2}z^b)^{-5}$

76. $[(a^2)^2]^3$

77. $[(2ax^{-1})^2]^n$

78. $[(-3by^2)^2]^3$

79. $\left[\dfrac{a^4}{b^2}\right]^3$

80. $\left[\dfrac{2xy}{z^3}\right]^4$

81. $\left[\dfrac{x^{-2}}{y^{-2}}\right]^a$

82. $\left[\dfrac{-2x}{y^{-2}}\right]^2$

83. $\left[\dfrac{-1}{2x^3}\right]^m$

84. $\left[\dfrac{-3}{2x^{-1}}\right]^a$

85. $\left[\dfrac{ax^{2m}}{3y^{-3}}\right]^n$

86. $\left[\dfrac{x^y}{y^y}\right]^x$

87. $\dfrac{(3x^2)^3(6xy)^2}{9(x^4y^3)^2}$

88. $\dfrac{(xy)^2(x)^3}{(x^3y)^2}$

89. $\dfrac{(a^2b^{-1})^3(b^2a^{-1})^2}{(ba)^2}$

90. $\dfrac{(-9xz^3)^2}{(-xz)^2(27xz^2)}$

91. $\left[\dfrac{3x^2}{2y^3}\right]^2 \times \left[\dfrac{y^3}{x^4}\right]$

92. $\left[\dfrac{2x^3}{3y^2}\right]^2 \times \left[\dfrac{-2x^2}{3y}\right]^{-3}$

93. $\left[\dfrac{3ab^2}{c^3}\right]^2 \times \left[\dfrac{1}{2c^2}\right]^{-2}$

94. $\left[\dfrac{1}{xy^2}\right]^{-3} \times \left[\dfrac{2y}{x^3z}\right]^2$

95. Write Fortran expressions for the original algebraic forms and for the

results in problems 65, 69, and 93. Use integer exponents; prefix any variable name which is not integer with the letter *I*.

96. Repeat problem 95 for problems 67, 77, 83, and 87.

5.9 FRACTIONAL EXPONENTS AND RADICALS

EXPONENTS OF FORM 1/a. Up to this point, exponents have been designated as being integers. The very nature of the definition $x^a = x \times x \cdots \times x$ (a factors) makes it difficult to imagine fractional exponents. However, all of these rules apply whenever the exponents are any real numbers.

We shall now evolve an interpretation for rational exponents by starting with reciprocals of the integers, exponents of the form $1/a$. For example, consider the form $x^{1/2}$; if it is squared, the result is

$$(x^{1/2})^2 = x^{1/2 \times 2} = x \qquad \text{for } x \geq 0 \qquad \text{(by Property 5).}$$

On the other hand, by the definition of square root, we know that the square of $\sqrt{x}$ is x or $(\sqrt{x})^2 = x$. Thus, in order for the notion of fractional exponents to be consistent with other algebraic concepts, we must define $x^{1/2}$ as

$$x^{1/2} = \sqrt{x}.$$

Similarly, we find for $x^{1/3}$ that

$$(x^{1/3})^3 = x^{1/3 \times 3} = x.$$

Thus $x^{1/3}$ is referred to as the third root of x and is represented by $\sqrt[3]{x}$. For example, a square root (second root) of 9 is 3, since $3^2 = 9$; a cube root (third root) of 8 is 2, since $2^3 = 8$. In general we have

$$(x^{1/a})^a = x^{1/a \times a} = x,$$

and we thus define $x^{1/a}$ as

$$y = x^{1/a} \qquad \text{if} \qquad y^a = x.$$

For example,

$$2 = 8^{1/3} \qquad \text{since} \qquad 2^3 = 8.$$

Another commonly used method for defining $x^{1/a}$ involves a special radical symbolism as follows

$$x^{1/a} = \sqrt[a]{x}.$$

The symbol $\sqrt{}$ is normally referred to as the **radical,** a as the **index** or **order** of the radical, and x as the **radicand**; the expression is called the "a^{th} root of x." Thus $\sqrt[4]{81} = 3$ is the fourth root of 81, $\sqrt[6]{64} = 2$ is the sixth root of 64, and $\sqrt{36} = 6$ is the second root of 36. The absence of an index is used to indicate the second root (square root).

In advanced mathematics, it is shown that roots of the form $\sqrt[a]{x}$ exist for x as any real number. However, we will work only with radicands which are positive rational numbers. We will also adopt the convention that $\sqrt{x}$ refers to the positive real root, which is called the **principal root**. For example, both 3 and -3 give 9 when squared; thus, both are second roots (or square roots) of 9.

However, by convention, 3 is referred to as the principal second root of 9 and is written $\sqrt{9} = 3$. If it is necessary to indicate both roots, the form $\pm\sqrt{9} = \pm 3$ is normally used. The symbol $\pm$ is read "plus or minus," and implies that first the plus sign and then the minus sign are applied.

EXPONENTS OF FORM a/b. The general form for rational exponents follows immediately from the preceding definition of $x^{1/a}$ and Property 5:

$$\begin{aligned} x^{a/b} &= x^{a \times 1/b} \\ &= (x^a)^{1/b} \\ &= \sqrt[b]{x^a}. \end{aligned}$$

This is summarized as

Property 8. For any real number $x \neq 0$,

$$x^{a/b} = \sqrt[b]{x^a} \qquad (a \text{ and } b \text{ integers, } b > 0).$$

Examples are

$$\begin{aligned} x^{3/4} &= (x^3)^{1/4} = \sqrt[4]{x^3}, \\ 10^{3/5} &= (10^3)^{1/5} = \sqrt[5]{10^3} = \sqrt[5]{1000}. \end{aligned}$$

All of the operations studied in previous sections may be used with fractional exponents, as illustrated by the following example:

$$\begin{aligned} \left(\frac{x^2y^4}{4z^2}\right)^{1/2} \times \left(\frac{z^3}{x^6}\right)^{1/3} &= \frac{xy^2}{2z} \times \frac{z}{x^2} \\ &= \frac{y^2}{2x}. \end{aligned}$$

The restriction of b to the positive integers in no way limits Property 8, since fractions such as $2/(-3)$ and $(-4)/(-5)$ may be converted to $(-2)/3$ and $4/5$, respectively. For instance,

$$x^{2/(-3)} = x^{(-2)/3} = (x^{-2})^{1/3} = \sqrt[3]{x^{-2}} = \sqrt[3]{\frac{1}{x^2}} = \frac{1}{\sqrt[3]{x^2}}$$

$$2^{-3/8} = (2^{-3})^{1/8} = \sqrt[8]{2^{-3}} = \sqrt[8]{\frac{1}{2^3}} = \frac{1}{\sqrt[8]{2^3}}$$

MANIPULATING RADICALS. All of the relationships and techniques which we have studied concerning fractional exponents can be applied to radical forms. Indeed we should expect this, since the radical form is merely another means for expressing exponential forms. Actually, in Fortran programming the exponential form is used almost exclusively. However, a brief exposure to operations with radical forms will provide us with an insight to those operations.

Basic to manipulating radicals are the following three relationships, which are a direct consequence of properties of exponents:

$$\sqrt[a]{x^a} = x \qquad (4.5)$$

$$\sqrt[a]{xy} = \sqrt[a]{x}\ \sqrt[a]{y} \tag{4.6}$$

$$\sqrt[a]{\frac{x}{y}} = \frac{\sqrt[a]{x}}{\sqrt[a]{y}} \tag{4.7}$$

The reader should express each of these relationships in exponential form, and then justify them on the basis of properties of exponents.

Application of these relationships are illustrated by the following simple examples:

$$\sqrt[3]{64} = \sqrt[3]{4^3} = 4,$$

$$\sqrt{x^2y^4} = \sqrt{(xy^2)^2} = xy^2,$$

$$\sqrt{21} = \sqrt{7} \times \sqrt{3},$$

$$\sqrt{x} \times \sqrt{xy} = \sqrt{x^2y} = \sqrt{x^2} \times \sqrt{y} = x\sqrt{y},$$

$$\sqrt{8} = \sqrt{2^2 \times 2} = \sqrt{2^2}\ \sqrt{2} = 2\sqrt{2},$$

$$\sqrt[3]{8xy^3} = \sqrt[3]{(2y)^3 \times x} = \sqrt[3]{(2y)^3}\ \sqrt[3]{x} = 2y\sqrt[3]{x},$$

$$\sqrt{\frac{10}{4}} = \frac{\sqrt{10}}{\sqrt{4}} = \frac{\sqrt{10}}{2},$$

$$\sqrt[3]{\frac{x}{y^3}} = \frac{\sqrt[3]{x}}{\sqrt[3]{y^3}} = \frac{\sqrt[3]{x}}{y}.$$

USE IN FORTRAN. Since the radical symbol is nonexistent in Fortran, the programmer must convert all radicals to the exponential form for the program. For example, following are some algebraic forms and the Fortran equivalents.

$\sqrt[3]{a^2} = a^{2/3}$ `A**(2.0/3.0)`

or

$= a^{0.6667}$ `A**0.6667`

$\sqrt[5]{b^n} = b^{n/5}$ `B**(AN/5.0)`

$\sqrt[4]{15ax^3} = (15ax^3)^{1/4}$ `(15.0*A*X**3)**0.25`

Note that in all cases the exponents are expressed as real (floating point) quantities; if integer values were used then the integer arithmetic would yield invalid results (for example, 2/3 is 0).

EXERCISE 5.9

Express each of the following as a rational number:

1. $16^{1/2}$ 2. $64^{1/2}$ 3. $(-8)^{1/3}$

4. $81^{1/4}$
5. $32^{2/5}$
6. $(-128)^{3/7}$
7. $125^{2/3}$
8. $216^{2/3}$
9. $\left(\frac{1}{4}\right)^{1/2}$
10. $\left(\frac{1}{81}\right)^{3/4}$
11. $\left(\frac{4}{9}\right)^{1/2}$
12. $\left(\frac{27}{125}\right)^{2/3}$
13. $\left(\frac{1}{25}\right)^{-(1/2)}$
14. $\left(\frac{1}{8}\right)^{-(1/3)}$
15. $\left(\frac{1}{16}\right)^{-(3/4)}$
16. $\left(\frac{4}{9}\right)^{-(1/2)}$

Remove all possible factors from the radicand by applying Eqs. 4.5 and 4.6:

17. $\sqrt{18}$
18. $\sqrt{72}$
19. $\sqrt[3]{24}$
20. $\sqrt[4]{80}$
21. $\sqrt{5ax^3}$
22. $\sqrt{18b^2y^5}$
23. $\sqrt{\frac{3}{4}}$
24. $\sqrt[3]{\frac{9}{8}}$
25. $\sqrt[3]{\frac{3}{8}}$
26. $\sqrt{\frac{7}{18}}$
27. $\sqrt{\frac{1}{a^2}}$
28. $\sqrt[3]{\frac{2}{x^3}}$
29. $\sqrt[3]{\frac{4x^2}{8y^3z^3}}$
30. $\sqrt{\frac{a+b}{a^2b^2}}$

Perform the indicated operations and simplify the resulting quantity:

31. $\sqrt{6} \times \sqrt{10}$
32. $\sqrt[3]{18} \times \sqrt[3]{3}$
33. $\sqrt[3]{25} \times \sqrt[3]{50}$
34. $\frac{\sqrt[3]{18}}{\sqrt[3]{9}}$
35. $\sqrt{2xy} \times \sqrt{xy^3}$
36. $\sqrt[3]{3xy^2} \times \sqrt[3]{x^2y^5}$
37. $\frac{\sqrt{xy}}{\sqrt{xz}}$
38. $\frac{\sqrt[3]{uv^2}}{\sqrt[3]{uv^5}}$
39. Write Fortran expressions for the original algebraic forms and for the results in problems 29, 30 and 36.

REFERENCES

Apostle, H. G., *Survey of Basic Mathematics.* Boston, Little, Brown & Company, 1960.

Barnett, R. A., *Elementary Algebra: Structure and Use.* New York, McGraw-Hill, Inc., 1968.

Dubisch, R., Howes, V. E., and Bryant, S. J., *Intermediate Algebra,* 2d ed. New York, John Wiley & Sons, Inc., 1969.

Wooten, W. and Drooyan, I., *Intermediate Algebra,* 2d alt. ed. Belmont, Calif., Wadsworth Publishing Company, Inc., 1968.

6 EQUATIONS

SIMPLE EQUATIONS / 196

6.1 BASIC NOTIONS OF EQUATIONS / **196**
Conditional and Identical Equations / **196**
The Importance of the Conditional Equation / **197**
The Equal Sign in Fortran / **199**
Equivalent Equations / **200**
Basic Properties of Equalities / **200**
The Additive Inverse / **202**
The Multiplicative Inverse / **203**
Exercise 6.1 / **204**

6.2 MORE ON EQUATIONS / **205**
Combining Operations / **205**
Equations Containing Fractions / **206**
Word Problems / **208**
Exercise 6.2 / **210**

6.3 EQUATIONS IN SEVERAL VARIABLES / **211**
The Mixture Problem / **212**
Simple Interest and Simple Discount / **213**
Computer Applications / **214**
Exercise 6.3 / **214**

6.4 SIMPLE INEQUALITIES / **216**
Addition and Inequalities / **217**
Multiplication and Inequalities / **218**
Solving Inequalities / **218**
Exercise 6.4 / **220**

POLYNOMIALS / 220

6.5 SOLVING QUADRATICS BY FACTORING / **221**
Exercise 6.5 / **222**

6.6 COMPLETING THE SQUARE AND THE QUADRATIC FORMULA / **223**
Completing the Square / **223**
The Quadratic Formula / **225**
The Discriminant / **227**
The Quadratic Formula and Fortran / **228**
Exercise 6.6 / **229**

SIMPLE EQUATIONS

One of the most important components of an algebraic system is the concept of an equation. We have found many occasions to indicate the equality of two algebraic expressions through the use of the equal sign. Arithmetic equations are useful to indicate that the sum of 2 and 3 is 5 or the product of 4 and 7 is 28:

$$2 + 3 = 5 \qquad 4 \times 7 = 28.$$

In algebra, we define the exponential form x^3 as

$$x^3 = x \times x \times x.$$

Furthermore, we find that the product of the binomials $(x + 3)$ and $(x - 4)$ is

$$(x + 3)\,(x - 4) = x^2 - x - 12.$$

In this chapter we shall consider equations in more detail, and study various means for manipulating them.

6.1 BASIC NOTIONS OF EQUATIONS

CONDITIONAL AND IDENTICAL EQUATIONS. The equation

$$x(x + 1) = x^2 + x$$

is called an **identical** equation, or **identity**, because for any value of x the left side equals the right side. Checking the equation for $x = 3$, 4, and 5, we have

$$\begin{aligned} 3(3 + 1) &= 3^2 + 3 \\ 12 &= 12, \\ 4(4 + 1) &= 4^2 + 4 \\ 20 &= 20, \\ 5(5 + 1) &= 5^2 + 5 \\ 30 &= 30. \end{aligned}$$

Even when limited to integers, this equation represents an infinite number of arithmetic equations. The commutative, associative, and distributive laws discussed in Chapter 2 are also identities, since they define relationships that are true for all real numbers. For example, $a + b = b + a$ is a true statement for any pair of real numbers.

On the other hand, the left side of equation $x + 2 = 7$ will equal the right side for only one value of x. In other words, this algebraic equation represents the single arithmetic equation, $5 + 2 = 7$. An equation that is true only on the condition that x represents a particular number is called a **conditional** equation. The value or values that make the equation a true statement are called **solutions** or **roots** of the equation. The equation

$$(x - 3)\,(x - 2) = 0$$

is a conditional equation, since the left side equals the right side only when $x = 2$ or 3. When $x = 2$,

$$\begin{aligned}(x-2)(x-3) &= (2-2)(2-3) \\ &= 0\,(-1) \\ &= 0.\end{aligned}$$

When $x = 3$,

$$\begin{aligned}(x-2)(x-3) &= (3-2)(3-3) \\ &= 1\,(0) \\ &= 0.\end{aligned}$$

When a real number is substituted for x and makes the left side of an equation equal to the right side, we say this real number **satisfies** the equation. In the last example, the real numbers 2 and 3 satisfied the given equation.

THE IMPORTANCE OF THE CONDITIONAL EQUATION. Most of our attention up to this point has centered on identical equations or algebraic expressions. However, when algebra is used as a tool in solving practical problems, the conditional equation plays a prominent role.

The first and sometimes most difficult task in problem solving is to translate the problem into a mathematical form, the end result being a conditional equation. Note that an algebraic expression such as $3x + 4$ carries with it no requirement that x be any particular value. If $x = 1$, $3x + 4 = 7$; if $x = 2$, $3x + 4 = 10$—but there is no reason to conclude that x must be equal to 1, 2, or any other number.

An identical equation, such as $x(x - 1) = x^2 - x$, is also deficient in this respect because it is true for all values of x. To determine the solution to a given problem we need a conditional equation and the particular values which satisfy the equation.

In order to thoroughly understand the concept of the conditional equation and to be able to manipulate it for solving problems, it is important to note that the equal sign always asserts that two numerical quantities, whether in symbolic or numeric form, are the same. This may seem trivial, but the beginning student often loses sight of it in dealing with algebraic equations. For example, the equal sign in the equation.

$$2x + 5 = x + 17$$

should be considered as asking the question, "For what values of x does $2x + 5$ equal $x + 17$?" If we were given the additional information that the value of x satisfying this condition is a natural number, then we could easily define the trial and error algorithm of Figure 6.1 to obtain the solution. Proceeding through this algorithm, we would find that when $x = 12$, both $2x + 5$ and $x + 17$ are equal to 29; therefore, 12 is the solution of this equation. Although this is a simple, intuitive approach for solving equations (it is using a general approach commonly employed in computers as is described in Chapter 13), classical algebra techniques yield the solution much more quickly in this case.

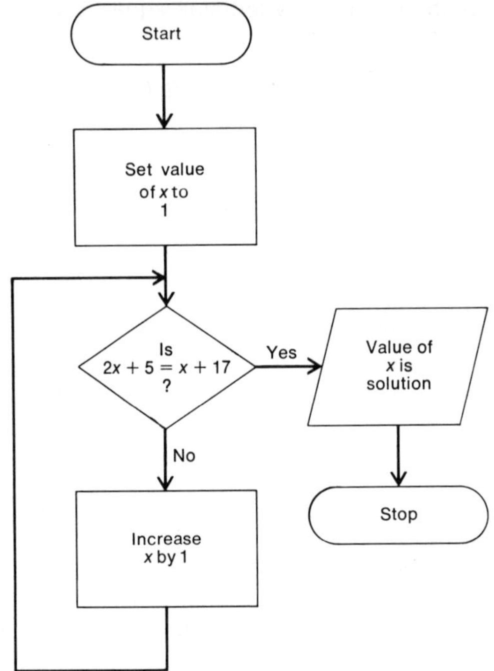

FIGURE 6.1 A trial and error algorithm

In Chapter 1, Example 1.1 involved a word statement for calculating interest; an alternate form used the language of algebra to show the same relationship. Actually, as was pointed out in that example, the language of algebra and English are similar. Using a simplified version of the example, the relationship between quantities is described by the sentence "the interest is the product of the principle and the sum of 1.0 plus the interest rate." The algebra equivalent is $I = P(1 + r)$. We can see that the equation in algebra is analogous to the sentence in English; the variable I serves as the noun (interest), the symbol $=$ serves as the verb (is) and the algebraic expression $P(1 + r)$ serves as the clause (the product of···). Both the clause in English and the expression in algebra are not normally considered complete within themselves; they form components of a larger entity. This parallel carries through to both Fortran (an algebraic language) and Cobol (which is designed to appear like English). For instance, complete statements in both Fortran and Cobol showing this relationship follow.

```
A = P*(1.0+R)
```

```
ADD ONE, R GIVING PART.
MULTIPLY PART BY P GIVING A.
```

As we shall see in a later paragraph, there are important and significant differences between the algebraic equation and its Fortran equivalent.

The goal of this chapter is to relate concepts of the algebraic equation to the so-called Fortran arithmetic statement and to develop efficient procedures for solving conditional equations.

THE EQUAL SIGN IN FORTRAN. The beginning programmer is commonly confused by the use of the equal sign in programming languages. On one hand he learns that Fortran is a language patterned after algebra but on the other he learns that care must be taken when interpreting the meaning of Fortran quantities. Remember that a Fortran expression represents a set of instructions defining arithmetic operations to be carried out on quantities in storage. The Fortran *arithmetic statement* (also referred to as *arithmetic assignment* statement and *assignment* statement) directs the computer to carry out designated operations and *then store the result in a particular part of its storage unit.* For example, statement 4 in Figure 6.2 will cause the computer to:

1. Use the values currently stored for PRINC, TIME, and RATE to evaluate the expression on the right,
2. Store the result of this evaluation in the area of storage reserved for AMNT.

Prior to execution of statement 4, the value stored in AMNT will have been the remains of some previous calculation and, in all probability, equality in the algebraic sense would *not* exist. A similar action takes place in statement 5 where the current value of X would be used to evaluate this expression with the result being placed in the storage area Y.

When the arithmetic statement is interpreted in this fashion, the function of the following statements becomes apparent.

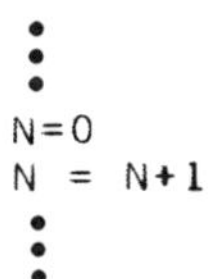

Assuming that the first statement is carried out only once (assigning a value of zero to N) and that the other is repetitively executed, then each execution of this statement will see the value of N increased by 1. That is, N serves as a *counter.*

The beginning programmer is often confused by this form and concludes that either: (1) these forms are inconsistent since N equals zero in one case and something else in the other, or (2) this is impossible since N cannot equal

C FOR COMMENT

STATEMENT NUMBER	Cont.	FORTRAN STATEMENT
4		AMNT = PRINC*(1.0 + RATE*TIME)
5		Y = 4.0*X**3 - 5.0*X + 6.0

FIGURE 6.2 Fortran arithmetic statements

N + 1. The latter conclusion shows a good understanding of algebra since, by algebra, we can solve for n obtaining

$$n = n + 1$$
$$n - n = n - n + 1$$
$$0 = 1$$

which is obviously false, therefore the original equation is not valid. Both of these incorrect conclusions are resolved by recognizing the nature of the Fortran arithmetic assignment statement.

EQUIVALENT EQUATIONS. We have noted that the number 5 is a solution or root of the equation

$$2x - 7 = 3 \tag{6.1}$$

if, when x is replaced by 5, the resulting number on the left $(2 \times 5 - 7)$ is the same as the number on the right. Since $2 \times 5 - 7$ does equal 3, we can write

$$x = 5. \tag{6.2}$$

Equations 6.1 and 6.2 are frequently called **equivalent** equations, because they have the same root. Basically, the procedure of solving any equation is to produce an equivalent equation of the form displayed by Eq. 6.2. The unknown x is isolated on one side of the equation, with the solution appearing on the other. The equations

$$5x + 3 = 7x - 5, \tag{6.3}$$
$$x = 4 \tag{6.4}$$

are equivalent equations because the number 4 is a root of each equation; that is, $5 \times 4 + 3 = 7 \times 4 - 5$ and $4 = 4$. The importance of equivalent equations lies in the fact that the common root is obvious in Eq. 6.4, while it is not in Eq. 6.3.

BASIC PROPERTIES OF EQUALITIES. We shall now consider two properties of equations which, when used judiciously, will simplify equations of great complexity.

1. *The Addition Property of Equality. For any real numbers a, b, and c, if $a = b$,*

$$a + c = b + c.$$

EXAMPLE 6.1

$$4 = 4,$$
$$4 + 2 = 4 + 2,$$
$$6 = 6.$$

EXAMPLE 6.2

$$x = 4,$$
$$x + 2 = 4 + 2,$$
$$x + 2 = 6.$$

EXAMPLE 6.3

$$x + 2 = 3,$$
$$x + 2 + (-2) = 3 + (-2),$$
$$x + 0 = 1,$$
$$x = 1.$$

EXAMPLE 6.4

$$x - a = b,$$
$$x - a + a = b + a,$$
$$x + 0 = b + a,$$
$$x = b + a.$$

The addition property of equality asserts that, if two numbers are equal, after adding the same number to each, the resulting sums will be equal. The arithmetic Example 6.1 is trivial. Examples 6.3 and 6.4 demonstrate the value of this property if one notes that the resulting equivalent forms represent solutions. Application of this property need not be limited to numbers or constants, as is evident by the following example:

EXAMPLE 6.5

$$3x - 2 = 2x - 5$$
$$3x - 2 + 2 = 2x + 2 - 5 \qquad (\textit{Adding } 2)$$
$$3x = 2x + 2 - 5$$
$$3x + (-2x) = 2x + (-2x) + 2 - 5 \qquad [\textit{Adding } (-2x)]$$
$$x = -3.$$

The second property of equality is

2. *The Multiplication Property of Equality. For any real numbers a, b, and c, if* $a = b$,

$$ac = bc.$$

EXAMPLE 6.6

$$4 = 4,$$
$$2 \times 4 = 2 \times 4,$$
$$8 = 8.$$

EXAMPLE 6.7

$$x = 4,$$
$$2 \times x = 2 \times 4,$$
$$2x = 8.$$

EXAMPLE 6.8

$$\frac{1}{2}x = 5$$

$$2\left(\frac{1}{2}x\right) = 2 \times 5$$

$$x = 10.$$

EXAMPLE 6.9

$$3x = 12$$

$$\frac{1}{3}(3x) = \frac{1}{3} \times 12$$
$$x = 4.$$

The multiplication property of equality asserts that, if two numbers are equal, after multiplying both by the same number, the resulting products will be equal. The arithmetic Example 6.6 may not appear to be impressive, but Examples 6.8 and 6.9 demonstrate the usefulness of this property as an aid in obtaining equivalent equations which are in the form of the solution. In general, care should be taken when multiplying or dividing by the unknown itself in an equation. Normally multiplication will introduce an extra root which never existed in the original equation and division will eliminate one of them as illustrated by the following examples.

EXAMPLE 6.10

$$a = a + 1 \qquad \text{(a contradiction—no solution)}$$
$$a \times a = a(a + 1) \qquad \text{(multiplying by } a\text{)}$$
$$a^2 = a^2 + a$$
$$a = 0. \qquad \text{(solving for } a\text{)}$$

EXAMPLE 6.11

$$x^2 = x \qquad \text{(which has two solutions, } x = 0 \text{ and } x = 1\text{)}$$
$$x = 1. \qquad \text{(dividing by } x\text{)}$$

THE ADDITIVE INVERSE. In working with the equation $x + 2 = 3$, we obtained an equivalent form by eliminating 2 from the left side of the equation. This was accomplished by adding the additive inverse of 2 to both sides of the equation:

$$x + 2 = 3$$
$$x + 2 + (-2) = 3 + (-2)$$
$$x + 0 = 1$$
$$x = 1.$$

Recall that (-2) is called the additive inverse of 2 because $2 + (-2) = 0$. In general, the additive inverse of any number a is $(-a)$, because

$$a + (-a) = 0.$$

The significance of adding a number to its inverse lies in the fact that their sum, which is zero, is the identity element for addition; that is, for any real number c,

$$c + 0 = c.$$

The addition of other than the inverse to both sides of an equation will produce an equivalent equation, but one which may be of little more use than the original. For instance,

$$\begin{aligned} x + 2 &= 3 \\ x + 2 + (-4) &= 3 + (-4) \\ x - 2 &= -1. \end{aligned}$$

As we can see, this operation does not isolate x on the left side of the equation. The expression $x + 0$ simplifies to x, while $x - 2$ does not.

Note that the concept of inverse applies whether we find it necessary to add a quantity to both sides of an equation or subtract a quantity from both sides of an equation, since the inverse of a is $-a$ and the inverse of $-b$ is b. The following two examples illustrate this:

$$\begin{aligned} x + a &= c \\ x + a + (-a) &= c + (-a) \\ x + 0 &= c - a \\ x &= c - a. \end{aligned} \qquad \begin{aligned} x - b &= d \\ x - b + b &= d + b \\ x + 0 &= d + b \\ x &= d + b. \end{aligned}$$

THE MULTIPLICATIVE INVERSE. In demonstrating the multiplication property of equality, we considered the equation $x/2 = 5$ and found the equivalent equation $x = 10$. This was accomplished by multiplying both sides of the original equation by the multiplicative inverse of 1/2, which is 2.

$$\begin{aligned} \frac{1}{2}x &= 5, \\ 2 \times \frac{1}{2}x &= 2 \times 5, \\ 1 \times x &= 10, \\ x &= 10. \end{aligned}$$

The above solution indicates that the numbers 2 and 1/2 are inverses under multiplication because their product is the identity for multiplication, 1. Recall that for multiplication $1/a$ is the inverse of a, since

$$\frac{1}{a}a = 1.$$

The usefulness of inverses and identities is demonstrated again in the following examples:

$$\frac{x}{b} = d \qquad\qquad ax = c$$

$$\frac{1}{b} \times x = d \qquad \frac{1}{a} \times ax = \frac{1}{a}c$$

$$b \times \frac{1}{b}x = b \times d \qquad 1 \times x = \frac{c}{a}$$

$$1 \times x = b \times d \qquad x = \frac{c}{a}.$$

$$x = bd.$$

In these examples, intermediate steps are included to demonstrate the use of some of the basic properties. As the student gains more experience, it is natural to omit writing many of these steps. However, this is only after the basic concepts are thoroughly understood.

EXERCISE 6.1

Identify each of the following as being either an identity or a conditional equation.

1. $x = 4$
2. $x + 9 = (x + 3) + 6$
3. $x^2 + 4x + 4 = (x + 2)(x + 2)$
4. $3y - 5 = 2y + 9$
5. $a(b + c) = ab + ac$
6. $a + b + c = a + b + 2$
7. $a(x^2 + 2x + c) = ax^2 + 2ax + ac$
8. $\dfrac{x}{x - 4} = 4$
9. $(p - 2)(p + 2) = p^2 - 4$
10. $x^2 - 7x + 10 = 0$

Find the additive inverse for each value of *a* and use the inverse to solve the given equation.

11. $a = 7 \qquad x + 7 = 8$
12. $a = -2 \qquad x - 2 = -5$
13. $a = \dfrac{3}{4} \qquad x + \dfrac{3}{4} = \dfrac{7}{4}$
14. $a = -\dfrac{1}{3} \qquad x - \dfrac{1}{3} = \dfrac{2}{3}$
15. $a = 10 \qquad x + 10 = 14$
16. $a = -17 \qquad x - 17 = 52$

Find the multiplicative inverse for each value of *b* and use the inverse to solve the given equation.

17. $b = 4 \qquad 4x = 8$
18. $b = 7 \qquad 7x = 28$
19. $b = \dfrac{1}{3} \qquad \dfrac{1}{3}x = 5$
20. $b = \dfrac{2}{5} \qquad \dfrac{2}{5}x = 4$
21. $b = 15 \qquad 15x = 210$
22. $b = \dfrac{7}{4} \qquad \dfrac{7x}{4} = 28$

Which of the following pairs of equations are equivalent?

23. $x = 3,\ 2x - 7 = 2$
24. $x = 5,\ 2x - 7 = 3$
25. $5x + 2 = 7x - 3,\ x = 2$
26. $\dfrac{x}{x - 4} = 4,\ x = 4$
27. $x^2 - 4 = 0,\ x = -2$
28. $(x - 2)(x - 5) = 0,\ x = 3$
29. $x - 4 = 7,\ x + 2 = 13$

Write Fortran arithmetic statements for each of the following equations. (Use floating point quantities except for exponents.)

30. $C = \frac{5}{9}(F - 32)$

31. $A = P\left(1 + \frac{r}{n}\right)^{nt}$

32. $y = x^3 - 2x^2 + 5x + 7$

33. $Z = 2v^5 - 5v^2 + 8v - 9$

34. $S = \frac{P}{1 - dt}$

35. $d = \frac{S - P}{St}$

36. The equation $2x + 5 = x + 17$ was solved by trial and error using the algorithm of Figure 6.1. Would this method always be successful with similar such equations (assuming that they do have solutions)? What would occur if the solution were negative? What if the solution were positive but not an integer?

6.2 MORE ON EQUATIONS

COMBINING OPERATIONS. Equations containing several operations can be solved by the procedures of the last section if the appropriate inverses are applied in sequence. The equation $2x + 5 = 7$ can be solved by first applying the additive inverse of 5 and then the multiplicative inverse of 2, that is,

$$2x + 5 + (-5) = 7 + (-5)$$

$$2x + 0 = 2$$

$$\frac{1}{2}(2x) = \frac{1}{2}(2)$$

$$x = 1.$$

The order of application of the inverses can be reversed only if the distributive property is used on the left side.

$$\frac{1}{2}[(2x) + 5] = \frac{1}{2}(7)$$

$$\frac{1}{2}(2x) + \frac{1}{2}(5) = \frac{7}{2} \qquad \textit{(Distributive property)}$$

$$x + \frac{5}{2} = \frac{7}{2}$$

$$x = \frac{7}{2} - \frac{5}{2} \qquad \textit{(Apply additive inverse of 5/2)}$$

$$x = 1$$

In general, all additive inverses should be applied first and similar terms collected before applying the necessary multiplicative inverse.

The following examples further illustrate the use of inverses in solving equations. Since the sum, $a + (-a)$, is always 0 and the addition of 0 produces no change, several steps can be eliminated by indicating on only one side of the equations that the inverse has been added. A similar remark applies to the multiplicative inverse and the product $b(1/b)$.

EXAMPLE 6.12

$$2x - 7 = 5x + 2$$

$$2x = 5x + 2 + 7 \qquad \textit{(Apply additive inverse of } -7\textit{)}$$

$$2x + (-5x) = 9 \qquad \textit{(Apply additive inverse of } 5x\textit{)}$$

$$-\frac{1}{3}(-3x) = -\frac{1}{3}(9) \qquad \textit{(Apply multiplicative inverse of } -3\textit{)}$$

$$x = -3.$$

The first and last equations are equivalent, since

$$2(-3) - 7 = -13 = 5(-3) + 2.$$

EXAMPLE 6.13

$$7x + 5 = 9 - 4x$$

$$7x + 4x = 9 - 5 \qquad \textit{(Apply additive inverses of 5 and } -4x\textit{)}$$

$$11x = 4$$

$$x = \frac{1}{11}(4) \qquad \textit{(Apply multiplicative inverse of 11)}$$

$$x = \frac{4}{11}.$$

EXAMPLE 6.14

$$3(2x - 4) = 7 - (x - 2)$$

$$6x - 12 = 7 - x + 2 \qquad \textit{(Distributive property)}$$

$$6x - 12 = 9 - x \qquad \textit{(Combine like terms)}$$

$$6x + x = 9 + 12 \qquad \textit{(Apply additive inverses of } -12, -x\textit{)}$$

$$7x = 21$$

$$x = 3. \qquad \textit{(Apply multiplicative inverse of 7)}$$

EQUATIONS CONTAINING FRACTIONS. In Chapter 5, the addition of two algebraic fractions, such as

$$\frac{x}{x+3} + \frac{2}{x-1}$$

was considered in some detail. The important first step was to determine the least common denominator (or LCD), and then to convert them to equivalent fractions, each containing the common denominator. In solving equations containing fractions, the LCD can be used to eliminate fractions, thus simplifying the equations. For example, consider the following equation:

$$\frac{x}{x+3} + \frac{2}{x-1} = 12.$$

The LCD is $(x+3)(x-1)$, so multiplying each term of the equation of this expression yields

$$(x+3)(x-1)\frac{x}{x+3}+(x+3)(x-1)\frac{2}{x-1}=12(x+3)(x-1)$$

or

$$(x-1)x+(x+3)(2)=12(x+3)(x-1).$$

The latter equation contains no fractions and can be solved by removing parentheses and applying the appropriate inverses. That this form contains no fractions is an automatic and direct consequence of the fact that the LCD contains all factors of each individual denominator.

The following additional examples demonstrate the use of the LCD in solving equations containing fractions:

EXAMPLE 6.15

$$\frac{x}{4}+\frac{3x}{5}=\frac{1}{2},$$

$$20\left(\frac{x}{4}+\frac{3x}{5}\right)=20\left(\frac{1}{2}\right), \qquad \textit{(Multiply by LCD)}$$

$$20\left(\frac{x}{4}\right)+20\left(\frac{3x}{5}\right)=10, \qquad \textit{(Distributive property)}$$

$$5x+12x=10$$

$$17x=10$$

$$x=\frac{1}{17}(10) \qquad \textit{(Apply multiplicative inverse of 17)}$$

$$x=\frac{10}{17}.$$

EXAMPLE 6.16

$$\frac{5}{2x+4}-\frac{x}{x^2-4}=\frac{3}{x-2}$$

$$\frac{5}{2(x+2)}-\frac{x}{(x-2)(x+2)}=\frac{3}{x-2} \qquad \textit{(Factoring)}$$

The LCD is $2(x-2)(x+2)$.

$$2(x-2)(x+2)\frac{5}{2(x+2)}-2(x-2)(x+2)\frac{x}{(x-2)(x+2)}$$

$$=2(x-2)(x+2)\frac{3}{x-2}$$

$$5(x-2)-2x=6(x+2)$$

$$5x-10-2x=6x+12 \qquad \textit{(Distributive property)}$$

$$3x - 6x = 12 + 10$$ *(Applying the additive inverses of 6x and −10)*

$$-3x = 22$$

$$x = -\frac{22}{3}$$ *(Applying the multiplicative inverse of −3)*

WORD PROBLEMS. Thus far the primary emphasis has been on techniques for solving equations. In practice, equations to be solved are often evolved from a problem which is stated in verbal form, a *word* problem. On one hand, a conditional equation can be solved through use of a somewhat mechanical set of steps. On the other, converting a problem from its English statement form to its mathematical representation (sometimes referred to as the *mathematical model*) is not nearly so mechanical and is usually much more difficult. This process often requires a great deal of intuition and a degree of creativity and is not easily learned. A careful, systematic approach is usually necessary to achieve any degree of success in solving word problems. General guidelines for this type of problem solving are:

1. Carefully define all quantities which are involved and assign variable names to them;
2. Express all relationships between variables and other quantities in a mathematical form;
3. Check to insure that all features of the problem statement are represented by the mathematical forms.

As the beginner studies mathematics, whether it be geometry, algebra, or some of the advanced topics discussed later in this book, he is commonly discouraged that the logical, step-by-step derivations simply do not appear very logical to him. It is important to realize that most mathematical relationships are achieved through a great deal of trial and error and often with considerable frustration on the part of the mathematician. In a sense, mathematics is much like a jigsaw puzzle; much trial and error combined with available evidence and experience. The general processes of formulating the mathematical representation and solving the resulting equations are directly analogous to the task confronting the programmer in preparing a problem for computer solution. From the overall needs, he must formulate the problem, then convert it to an algorithm in the form of a flowchart. Once the problem logic is clearly defined by a flowchart, the task of performing the coding is considerably easier.

Let us now consider some algebra examples to give an insight into this technique of problem solving.

EXAMPLE 6.17

The sum of two numbers is 47. If one number is 5 more than the other, what are the numbers?

SOLUTION

If we let x be the smaller number, $x + 5$ is the larger number, but their sum is 47; therefore,

$$x + (x + 5) = 47.$$

The equation can be solved by the procedures of preceding sections. Note that the equation becomes apparent after a representation for each variable is clearly defined.

EXAMPLE 6.18

A druggist has 1000 cc (cubic centimeters) of a 10% acid solution (that is, 10% acid and 90% water). How much water must be added to make it a 2% solution?

SOLUTION

In this problem, we are to determine how much water should be added. Thus the water to be added is the unknown. We also know that the amount of acid will remain unchanged, although its concentration will diminish. Further, the amount of acid before addition of the water will be 10% of the total and the amount of acid afterwards will be 2% of the new total. Thus we can set the problem up as follows: Let x be the number of cubic centimeters of water to be added. Then $(1000 + x)$ cc will represent the total amount of the final solution, and $0.02(1000 + x)$ cc will represent the amount of acid in final solution. Since 0.10(1000) cc is the amount of acid in the original solution, and the amount of acid remains unchanged,

$$\begin{aligned} 0.10(1000) &= 0.02(1000 + x) \\ 100 &= 20 + 0.02x \\ 0.02x &= 80 \\ x &= 4000 \text{ cc.} \end{aligned}$$

EXAMPLE 6.19

Two computers are used to perform a sales analysis. Machine A, which rents for \$250 per hour, logged 0.3 hour more than machine B, which rents for \$100 per hour. The total charge for the analysis was \$880. Determine logged time for each machine and cost for using each machine.

SOLUTION

$$\begin{aligned} \text{Let } x &= \text{logged time for machine } B, \\ x + 0.3 &= \text{logged time for machine } A, \\ 100x &= \text{charge for machine } B, \\ 250(x + 0.3) &= \text{charge for machine } A, \\ 100x + 250(x + 0.3) &= \text{total charge}, \\ 100x + 250x + 75 &= 880, \\ x &= 2.3; \end{aligned}$$

$$\begin{aligned}\text{logged time for } B &= 2.3 \text{ hr},\\ \text{logged time for } A &= 2.6 \text{ hr},\\ \text{charge for } B &= 2.3(100) = \$230,\\ \text{charge for } A &= 2.6(250) = \$650.\end{aligned}$$

EXAMPLE 6.20

A sum of \$3000 is invested, part at 3% per year and the rest at 4% per year. Find the amount invested at each rate, if the income is \$102 for one year.

SOLUTION

$$\begin{aligned}\text{Let } x &= \text{the amount invested at 3\%},\\ 3000 - x &= \text{the amount invested at 4\%},\\ 0.03x &= \text{income from the 3\% investment},\\ 0.04(3000 - x) &= \text{income from the 4\% investment},\\ 0.03x + 0.04(3000 - x) &= 102,\\ 0.03x + 120 - 0.04x &= 102,\\ x &= 1800,\\ \text{amount invested at 3\%} &= \$1800,\\ \text{amount invested at 4\%} &= \$1200.\end{aligned}$$

EXERCISE 6.2

Solve for x, y, or z:

1. $2x - 17 = 13 - 5x$
2. $5z = 6 - 2z$
3. $y + 9 = 7y - 3$
4. $2(x - 5) = 3x - 14$
5. $x - 3x + 5 = 17 - 2x + 8$
6. $8(2x + 4) = 18x$
7. $4 - 2(3x - 5) = 6(x + 2) - 12$
8. $x(x + 5) - 2 = x(6 + x)$
9. $\frac{1}{3}x = 14$
10. $\frac{1}{3}(3x - 9) = 5 - (x - 2)$
11. $3z = 7(2z + 5) - 13(z - 2)$
12. $\frac{5x}{7} = 45$
13. $2(x - 5) = 2x - 10$
14. $4 = 5y - 3(y - 7)$
15. $\frac{3x}{5} - \frac{2x}{3} = \frac{7}{10}$
16. $\frac{31}{3y} - \frac{3}{2} = \frac{2(y - 4)}{2y}$
17. $\frac{2}{y - 5} = \frac{4}{3y + 7}$
18. $\frac{3z}{6z - 9} = \frac{1}{2}$
19. $\frac{5}{2y + 4} - \frac{2}{3} = \frac{7}{2}$
20. $\frac{1}{2x - 3} + \frac{1}{x + 1} = \frac{2}{x + 1}$
21. $\frac{5}{z} - \frac{1}{z - 7} = \frac{1}{2z}$
22. $\frac{3z}{z^2 - 9} - \frac{2}{z - 3} = \frac{4}{z + 3}$
23. $\frac{1}{2x - 6} = \frac{1}{2} - 5$
24. $\frac{y}{3 - y} - \frac{2}{3} = \frac{4y}{3y}$
25. How much money must be invested at 5% simple interest in order to have a total amount of \$5000 after 5 years?
26. A sum of \$17,000 is invested, part at 3% simple interest and the remainder

at 4% simple interest. If the annual interest is $600, how much is invested at each rate?

27. Three sorters, *A*, *B*, and *C*, are used in a data processing installation; sorter *B* operates twice as fast as sorter *A*, and sorter *C* operates 2.5 times as fast as *A*. Their combined speed is 2200 cards per minute. What is the speed of each sorter?
28. Consider that the speeds of the sorters in problem 27 are such that *B* is 2.5 times faster than *A*, and the speed rate of *C* is only 80% that of *B*. If their combined rate is 2200 cards per minute, what is the speed of each sorter?
29. Two sorters are to be used to sort a deck of 32,000 cards. Sorter 1 operates at a rate of 250 cpm (cards per minute) and sorter 2 at a rate of 150 cpm (assume that these speeds include card handling and so on). How much time will be required to sort the entire deck?
30. Two identical decks are sorted on different sorters. The first sorter, which operates at a rate of 400 cards per minute, completes the operation 5 minutes before the second sorter, which operates at a rate of 350 cards per minute. Determine (a) the time required for each sorter, (b) the number of cards in the decks.
31. If problem 30 had stated that the slower sorter had finished 5 minutes sooner (all other conditions unchanged), the entire statement would be incorrect. However, it would still be possible to obtain a solution based on this assumption. Determine the solution. What is the significance of the results?
32. A card-punching job consisting of 1000 cards can be completed by one operator in 4 hours; another operator would require 6 hours to complete the same job. How much time would be required to complete the job if both of them were working on it?
33. Find a number, such that 10 times the number when deducted from 880 is 25 times 3 more than the number.
34. A businessman making a trip travels part of the way on a high-speed train going 100 miles per hour. He then boards an aircraft to complete his trip, flying at 250 miles per hour. If the portion of the trip traveled by air required 0.3 hour more than the portion traveled by rail, and the total distance covered was 880 miles, how many hours did each portion of the trip require?
35. A nationally known consultant in data processing charges $25 an hour for his own services and $10 an hour for those of his assistant. On a given job, for which the consultant's total bill is $880, he works 3 hours more than his assistant does. How many hours did each man put in on the job?
36. Compare problems 33, 34, and 35.

6.3 EQUATIONS IN SEVERAL VARIABLES

Problems as simple as those in Section 6.2, and requiring only one calculation, are seldom programmed for solution on a computer. This would be like using a steam shovel to pick up a pebble; the computer is too powerful, sophisticated, and expensive a device to be used for simple, nonrepetitive calculations. However, there is often only a slight difference between a problem that is economical for the computer and one that is not. For example, let us consider a minor variation of the problem given in Example 6.18 of Section 6.2.

THE MIXTURE PROBLEM. A druggist has 1000 cc of a 10% acid solution (10% acid and 90% water), and he wants to know how much water to add to obtain a series of varying acid values. Compile a table for him showing the amounts of water that must be added to obtain acid solutions of 9.9% to 2.0%, in increments of 0.1%.

The problem may be analyzed as follows: Let $x =$ the cubic centimeters of water to be added for each solution, $1000 + x =$ the cubic centimeters of solution after each dilution, and $c =$ the final required concentration of acid for each solution. Since the amount of acid remains unchanged,

$$c(1000 + x) = 0.1(1000). \tag{6.5}$$

This formula is an equation in two variables, c and x. We must solve it for x in terms of c before we can proceed with computation:

$$1000 + x = \frac{0.1(1000)}{c} \tag{6.6}$$

$$x = \frac{100}{c} - 1000 \tag{6.7}$$

Since the objective is to solve for x, the standard techniques of Section 6.1 are employed in the following order:

1. The original equation, Eq. 6.5.
2. Both sides of the equation multiplied by the multiplicative inverse of c to eliminate c from the left side, giving Eq. 6.6.
3. The additive inverse of 1000 added to both sides to isolate x, giving Eq. 6.7.

It is obvious that a great many nearly identical computations must be performed to obtain a complete solution to this problem. It is also obvious that performing these computations by hand would be very time consuming.

Equation 6.7 was programmed and run by a beginning student on an IBM 1620. The total time required to write, prepare, and run the program was 14 minutes. The completion of many more sets of calculations in the computer would have required very little additional time.

What is the significance of this illustration? It is that the computer is capable of performing simple arithmetic operations at very high rates of speed. However, if the programmer does not know how to solve a problem, the machine is of no help. At the present state of the computer art, the algebra must be done by the programmer before the computer can calculate. To better illustrate the algebraic techniques required, let us consider a business mathematics problem.

SIMPLE INTEREST AND SIMPLE DISCOUNT. In elementary business math, simple interest and simple discount are basic concepts. The formula

$$S = \frac{P}{1 - dt} \tag{6.8}$$

gives the amount due S in terms of the present value P, the discount d, and the time t. If we are to calculate tables of d for various sets of values for t and given values of S and P, we must solve Eq. 6.8 for the variable d, to yield

$$d = \frac{S - P}{St}. \tag{6.9}$$

Now let us consider those operations of addition and multiplication by which Eq. 6.9 was obtained from Eq. 6.8.

$$S = \frac{P}{1 - dt} \tag{6.8}$$

$$(1 - dt)S = P \qquad [\textit{Multiplying by } (1 - dt)] \tag{6.10}$$

$$S - Sdt = P \qquad [\textit{Distributive law}] \tag{6.11}$$

$$-Sdt = P - S \qquad [\textit{Adding } -S] \tag{6.12}$$

$$Sdt = -P + S \qquad [\textit{Multiplying by } -1] \tag{6.13}$$

$$Sdt = S - P \qquad [\textit{Commutative law}] \tag{6.14}$$

$$d = \frac{S - P}{St} \qquad \left[\textit{Multiplying by } \frac{1}{St}\right] \tag{6.15}$$

Although the operations performed here are no different than those in the preceding section, the apparent accumulation of terms which do not readily combine (as do numbers) is frequently confusing to the student; however, this need not be the case. As before, this solution can be broken down into the following logical steps:

1. Clear all fractions, as in Eq. 6.10,
2. Expand all terms, as in Eq. 6.11,
3. Collect all terms containing the unknown on the left side of the equation and all other terms on the right,
4. Factor the unknown from each term on the left,
5. Multiply by the appropriate inverse to obtain the desired solution.

The following examples further illustrate this procedure.

EXAMPLE 6.21

Solve for d in $L = a + (n - 1)d$.

SOLUTION

$$L - a = (n - 1)d \qquad [\textit{Add } (-a)]$$

$$d = \frac{L - a}{n - 1} \qquad \left[\textit{Multiply by } \frac{1}{n - 1}\right]$$

EXAMPLE 6.22

Solve for r in $S = \dfrac{a - rd}{1 - r}$.

SOLUTION

$$
\begin{aligned}
S(1-r) &= a - rd && [\textit{Multiply by } (1-r)] \\
S - Sr &= a - rd && [\textit{Distributive property}] \\
rd - Sr &= a - S && [\textit{Isolate terms containing } r] \\
r(d - S) &= a - S && [\textit{Distributive property}] \\
r &= \frac{a-S}{d-S} && \left[\textit{Multiply by } \frac{1}{d-S}\right]
\end{aligned}
$$

COMPUTER APPLICATIONS. Equation 6.6 is typical of the algebraic equations which *cannot* be programmed in Fortran without modification. Remember that the Fortran arithmetic statement involves an expression on the right and a single variable into which the result is placed on the left. Thus it is meaningless to consider an expression on the left; the equation must be modified to the form of 6.7 before writing the Fortran statement. Furthermore, if we desired to calculate values of d for various values of S, P and t in Eq. 6.8, it would be necessary to solve for d (Eq. 6.9) by conventional algebra before we could write the necessary Fortran statement. Again we see that the computer does not eliminate the need for conventional algebraic methods.

To illustrate a practical type of operation, let us refer to the original Eq. 6.8 and assume that we must calculate a table of values of S for values of P, d and t as follows:

$$
\begin{aligned}
P &= 100, \\
d &= 0.05 \text{ to } 0.20 \text{ in increments of } 0.01, \\
t &= 1 \text{ to } 40 \text{ in increments of } 1.
\end{aligned}
$$

Now we have a problem which is compatible with the computer because of the repetitive nature of the calculations (the evaluation must be made for 16 values of d and 40 of t—total is $16 \times 400 = 6400$. A flowchart to perform this operation is shown in Figure 6.3.

EXERCISE 6.3

Solve for the letter indicated:

1. $S = \dfrac{P}{1 - nd}$; d
2. $y = \dfrac{x}{x - 2}$; x
3. $A(B - C) = D$; C
4. $r = \dfrac{d}{1 - dt}$; d
5. $z = 5 - \dfrac{3}{y}$; y
6. $\dfrac{1}{a} - \dfrac{1}{s} = i$; s
7. $a = \dfrac{b - c}{k - c}$; c
8. $S = \dfrac{n}{2}[2a + (n - 1)d]$; a
9. The relationship between the Fahrenheit and centigrade temperature scale is expressed by the formula $C = 5(F - 32)/9$. Solve for F.

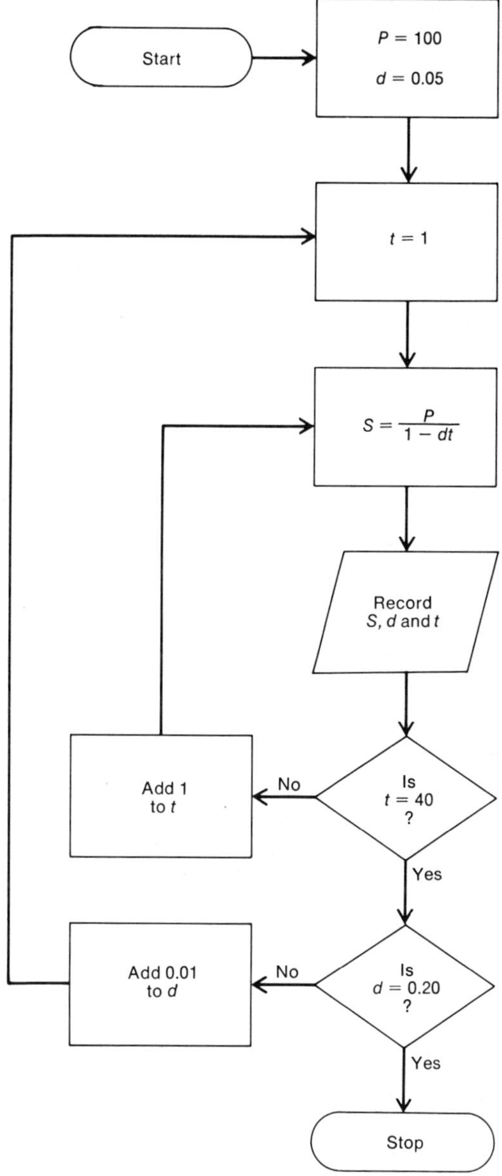

FIGURE 6.3 Compiling a table

10. If three resistors with values of R_1, R_2, and R_3 are connected in parallel, their overall resistance is R, where

$$\frac{1}{R}=\frac{1}{R_1}+\frac{1}{R_2}+\frac{1}{R_3}.$$

Solve for R.

11. Using the equation of problem 10, solve for R_2.

12. In finding a relationship between simple interest and simple discount, the starting point is the equation

$$P(1 + it) = \frac{P}{1 - dt}.$$

Solve for *i*.

13. Using the equation of problem 12, solve for *d*.

14. Write Fortran statements for the original equations and the new equations in problems 1, 7, 9 and 13. For example, in problem 2, the Fortran statement would have the variable Y on the left whereas in the new equation X would be on the left.

15. Draw a flowchart to illustrate calculating a table of values for R (problem 10) incrementing each of R_1, R_2 and R_3 from 100 to 1000 by units of 100.

6.4 SIMPLE INEQUALITIES

In studying real numbers (Chapter 2) we spoke of one number as being greater than another or less than a third and represented this symbolically as $a > b$ or $a < c$. In some fields of mathematics there is often a need to work with algebraic quantities that are related by inequalities. For example, consider a small cabinet shop whose owner manufactures a certain type of table. In order to make a profit on the table, he must determine a minimum price below which he cannot sell the table without breaking even or taking a loss. Let us assume that the price must be greater than $30. This condition may be represented algebraically by the inequality $x > 30$. On the other hand, the owner would recognize that a practical upper price limit exists, beyond which he is unlikely to sell many tables. This limit is somewhat arbitrary, but if we assume it to be $50, he has a defined range within which he can determine his selling price.

The relationship between the allowable selling price and its limits may be represented by the following pair of inequalities:

$$x > 30 \qquad x < 50.$$

We speak of these inequalities as being opposite in sense, because one is *greater than* and the other is *less than*. Opposite inequalities are commonly combined:

$$30 < x < 50.$$

This can be shown on the real axis as illustrated in Figure 6.4. Note that the end points of 30 and 50 are excluded, since x must be greater than 30 but less than 50. If, on the other hand, they are included we would say x may be *equal to or greater than* 30, satisfying the minimum condition, and *equal to or less than* 50, satisfying the maximum condition. These may be expressed symbolically as

$$x \geq 30 \qquad x \leq 50 \qquad \text{or} \qquad 30 \leq x \leq 50.$$

The solution *set* of these inequalities includes all numbers that satisfy the stated conditions. That is, 30, 50, and all real numbers in between are solutions to these inequalities.

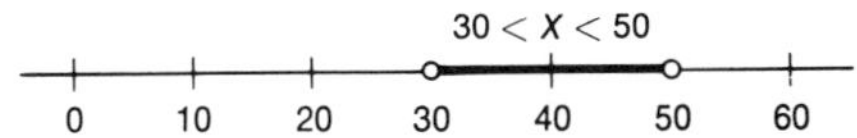

FIGURE 6.4

ADDITION AND INEQUALITIES. Often the solution set for a pair of inequalities is not obvious, making it necessary to perform algebraic operations to determine it. Just as the equations

$$x - 3 = 2 \qquad \text{and} \qquad x = 5$$

are equivalent and related by basic operations, so also are equivalent inequalities related by similar operations. For example, considering the inequality $4 < 6$ intuitively, we can see that it is possible to add any number to both sides without affecting the validity of the inequality. That is,

$$\begin{aligned} 4 &< 6, \\ 3 + 4 &< 3 + 6, \\ 7 &< 9, \end{aligned} \qquad \text{or} \qquad \begin{aligned} 6 &> 4, \\ 21 + 6 &> 21 + 4, \\ 27 &> 25. \end{aligned}$$

In effect, all this does is shift both points to the right on the real axis without changing their relative positions, as shown in Figure 6.5. In a similar manner we can see that the addition of a negative quantity to both sides of an inequality also results in an equivalent inequality. For example,

$$\begin{aligned} 3 &< 6, \\ -5 + 3 &< -5 + 6, \\ -2 &< 1. \end{aligned}$$

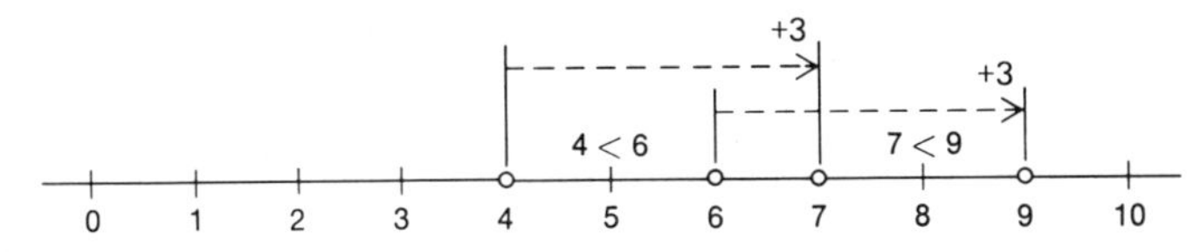

FIGURE 6.5

The effect in this case is to shift both points to the left an equal amount, which is evident by inspection of Figure 6.6. The following addition property of inequalities summarizes the preceding development:

Property 1. The addition of the same quantity to both sides of an inequality results in an equivalent inequality of the same sense.

EXAMPLE 6.23

$$\begin{aligned} x + 9 &> -27 \\ x + 9 - 9 &> -27 - 9 \qquad \textit{(Adding } -9 \textit{ to both sides)} \\ x &> -36 \end{aligned}$$

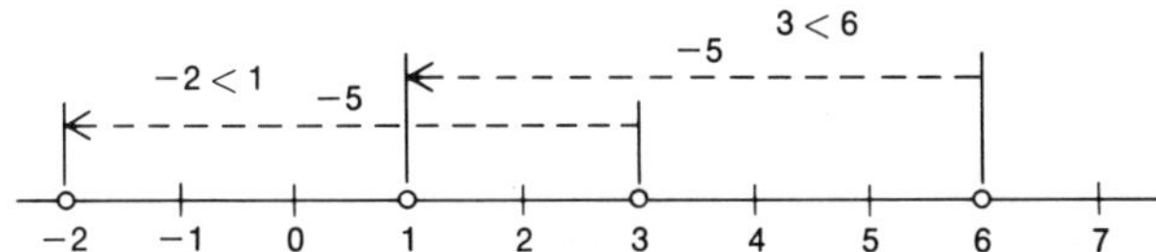

FIGURE 6.6

EXAMPLE 6.24

$$4x - 5 < 3x + 8$$
$$4x - 5 + (-3x + 5) < 3x + 8 + (-3x + 5) \quad \text{(Adding } -3x \text{ and 5 to both sides)}$$
$$x < 13$$

MULTIPLICATION AND INEQUALITIES. If we multiply both sides of an inequality by the same positive number, the overall effect will be to spread them out without changing the nature of the inequality. For instance,

$$5 < 7$$
$$3 \times 5 < 3 \times 7$$
$$15 < 21.$$

Intuitively, this operation should be obvious. On the other hand, if our multiplier is negative, a different situation will exist. Let us consider multiplication by -1:

$$5 < 7$$
$$(-1) \times 5 \ ? \ (-1) \times 7$$
$$-5 > -7.$$

In other words, multiplication by -1 changes the sense of the original inequality; it can readily be shown that any negative number will have a similar effect. Thus the multiplication property of inequalities depends upon the sign of the multiplier:

> *Property 2a. Multiplication of both sides of an inequality by the same positive number results in an equivalent inequality of the same sense.*
>
> *Property 2b. Multiplication of both sides of an inequality by the same negative number results in an inequality of the opposite sense.*

EXAMPLE 6.25

$$5x > 10$$
$$\frac{1}{5} \times 5x > \frac{1}{5} \times 10 \quad \text{(multiplying both sides by 1/5)}$$
$$x > 2.$$

EXAMPLE 6.26

$$-\frac{x}{6} > 3$$
$$(-6)\left(-\frac{x}{6}\right) < (-6) \times 3 \quad \text{(multiplying by } -6 \text{ and changing sense)}$$
$$x < -18$$

SOLVING INEQUALITIES. We may now operate on inequalities in much the same manner as we operate on equalities to obtain more desirable forms and solutions, as illustrated by the following examples.

EXAMPLE 6.27

Find the solution set and sketch the result on the real axis of the following inequality: $3x - 5 > x + 2$.

SOLUTION

$$3x - 5 + (-x + 5) > x + 2 + (-x + 5) \qquad \textit{(Adding } -x \textit{ and 5)}$$

$$2x > 7 \qquad \textit{(Combine terms)}$$

$$x > \frac{7}{2}. \qquad \textit{(Multiply by 1/2)}$$

The solution to this inequality is shown graphically in Figure 6.7.

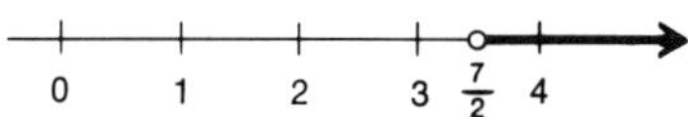

FIGURE 6.7

EXAMPLE 6.28

Find the solution set and sketch the result on the real axis of the following inequality: $(x/2) + 3 \leq x - 2$.

SOLUTION

$$-\frac{x}{2} \leq -5 \qquad [\textit{Add } -x - 3]$$

$$(-2)\left(-\frac{x}{2}\right) \geq (-2)(-5) \qquad [\textit{Multiply by } -2]$$

$$x \geq 10.$$

This solution is shown graphically in Figure 6.8.

FIGURE 6.8

EXAMPLE 6.29

Find the solution set and sketch the result on the real axis of the following pair of inequalities:

$$-3x + 2 > -4x + 5 \qquad \text{and} \qquad 3x - 5 < x + 9.$$

SOLUTION

In this example we have two inequalities, imposing restrictions upon the solution set. This is referred to as a **simultaneous system** of inequalities. Any solution set must satisfy both of the inequalities. In solving the problem, they are considered separately:

$$\begin{aligned} -3x + 2 &> -4x + 5 \\ -3x + 4x &> 5 - 2 \\ x &> 3 \end{aligned} \qquad \text{and} \qquad \begin{aligned} 3x - 5 &< x + 9 \\ 3x - x &< 9 + 5 \\ 2x &< 14 \\ x &< 7. \end{aligned}$$

The solution set is $3 < x < 7$, as is shown in Figure 6.9.

FIGURE 6.9

These principles of inequalities will form the basis for more extensive studies of linear inequalities and simultaneous systems in Chapter 9. Furthermore, they will be put to practical use in studying linear programming concepts (Chapter 11).

EXERCISE 6.4

Find the solution to each of the following inequalities and sketch it on the real axis.

1. $2x - 3 > 0$
2. $3x + 1 \leq 0$
3. $y + 5 > 2$
4. $3w - 4 \geq w - 2$
5. $4x + 4 \leq 6x + 2$
6. $3 - 9z < z + 8$
7. $3x + 4 < 6x + 4$
8. $\frac{x}{2} - 4 > 2x - 1$
9. $\frac{x}{3} + \frac{1}{6} \leq \frac{x}{6} - 1$
10. $\frac{x+2}{3} < \frac{1}{2}$
11. $\frac{3x-1}{4} \geq \frac{3x+1}{3}$
12. $\frac{x-2}{2} < x + \frac{3x-1}{2}$

Find the solution set to each of the following pairs of inequalities and sketch it on the real axis:

13. $3x - 5 > 2x - 3 \qquad 2x - 2 > 3x - 5$
14. $\frac{8z}{3} + 4 > \frac{2z-18}{3} \qquad \frac{5z-6}{3} < \frac{z}{7} - 2$
15. $\frac{3w+8}{2} < \frac{w+20}{5} \qquad \frac{5w-4}{4} < \frac{10w-1}{3}$

Recall from Chapter 2 the notion of absolute value. For example, $|-3| = 3$, $|3| = 3$, and so on. In a similar manner, the equation $|x| = 9$ can be considered as the two equations $x = 9$ and $x = -9$, without the absolute value signs (try them). With this in mind and by experimenting, consider the following exercises:

16. What is the solution set for $|x| < 3$? Show it on the real axis.
17. What is the solution set for $|x - 1| > 2$? Show it on the real axis.
18. Show that the inequality $|x| < a$ may be written as $-a < x < a$ whenever a is a positive real number.
19. Explain why there is no value for x which satisfies the inequality $|x| < a$ whenever a is a negative real number.

POLYNOMIALS

Up to this point we have concentrated on finding the roots of simple equations. In each problem there has been a single value for the unknown, which, when substituted in the equation, satisfied the equality. On the other hand, at the beginning of the chapter, reference was made to the equation

$$(x - 2)(x - 3) = 0. \tag{6.16}$$

This equation has not one, but two roots, or values, for which the equality is satisfied (that is, $x = 2$ and $x = 3$). Behind this interesting occurrence lies a very important and basic principle.

Let us consider two real numbers a and b whose product is 10, that is, $ab = 10$. If we solve the equation for a in terms of b, or vice versa, we see that if a is 10, b is 1; if a is 2, b is 5; and so on. On the other hand, if their product had been 0 (that is, $ab = 0$), then we know that either $a = 0$ or $b = 0$ or that both a and $b = 0$. In other words, **if the product of two quantities is zero then at least one of the quantities must be zero.** Note that this conclusion can be drawn only if the product is zero.

How does this relate to Eq. 6.16? As can be seen, Eq. 6.16 consists of the product of two factors, which equals zero. Thus, we can conclude that $(x - 2)(x - 3) = 0$ only when one of the factors is zero. In other words, it is true if

$$(x - 2) = 0 \tag{6.17}$$

or if

$$(x - 3) = 0. \tag{6.18}$$

Now we have two simple equations, each of which may be solved by the methods described earlier in this chapter, giving the roots

$$\begin{aligned} x &= 2 \qquad \text{for Eq. 6.17} \\ x &= 3 \qquad \text{for Eq. 6.18.} \end{aligned}$$

In general, an equation of the form

$$(x + a)(x + b) = 0 \tag{6.19}$$

has the two roots $x = -a$ and $x = -b$, arising from the fact that the equation is satisfied when $(x + a) = 0$ and when $(x + b) = 0$.

Note that the left sides of Eqs. 6.16 and 6.19 are the factored forms of second degree polynomials, since

$$\begin{aligned} (x - 2)(x - 3) &= x^2 - 5x + 6 \\ (x + a)(x + b) &= x^2 + (a + b)x + ab. \end{aligned}$$

6.5 SOLVING QUADRATICS BY FACTORING

The technique of finding roots of a polynomial by factoring and setting each factor equal to zero is a useful one. Whenever a quadratic equation is to be solved by factoring, it is necessary to put it in the standard polynomial form, with all nonzero terms on one side of the equal sign and zero on the other. The result of solving a quadratic will be two quantities.

EXAMPLE 6.30

$$2x^2 + 5x - 3 = 0.$$

SOLUTION

$$(2x - 1)(x + 3) = 0 \qquad \textit{(Factor)}$$

$$(2x - 1) = 0 \qquad (x + 3) = 0 \qquad \textit{(Set each factor equal to zero)}$$

$$x = \frac{1}{2} \qquad x = -3$$

or

$$x = \frac{1}{2}, -3.$$

EXAMPLE 6.31

$$4x^2 + x = 4x.$$

SOLUTION

$$4x^2 - 3x = 0 \qquad \textit{(Write in standard form)}$$

$$x(4x - 3) = 0 \qquad \textit{(Factor)}$$

$$x = 0 \qquad 4x - 3 = 0 \qquad \textit{(Set each factor equal to zero)}$$

$$x = \frac{3}{4}$$

$$x = 0, \frac{3}{4}.$$

EXAMPLE 6.32

$$-x^2 + 1 = -3x^2 + 4x - 1.$$

SOLUTION

$$2x^2 - 4x + 2 = 0 \qquad \textit{(Write in standard form)}$$

$$x^2 - 2x + 1 = 0 \qquad \textit{(Multiply both sides by 1/2)}$$

$$(x - 1)(x - 1) = 0 \qquad \textit{(Factor)}$$

$$x - 1 = 0 \qquad x - 1 = 0 \qquad \textit{(Set each side equal to zero)}$$

$$x = 1 \qquad x = 1$$

$$x = 1, 1.$$

In the last example both of the factors provide the same root, that is, $x = 1$. For the sake of consistency, this equation is still considered to have two roots; in this case they are identical.

EXERCISE 6.5

Find the roots of the following equations by factoring:

1. $x^2 + 3x + 2 = 0$
2. $x^2 + 6x + 8 = 0$
3. $2x^2 + 3x + 1 = 0$
4. $3x^2 + 2x - 5 = 0$
5. $3x^2 + 5x = 0$
6. $2x^2 - 9x = 0$
7. $-3x^2 + 3 = 0$
8. $8 = 2x^2$
9. $x^2 = -4x - 4$
10. $2x^2 - 5x = -2$

11. $x^2 - x - 10 = -4$
12. $2x^2 = 6x + 8$
13. $5 = x^2 + 4x$
14. $x^2 + 4x + 4 = 2x^2 - 4x + 16$
15. $x^2 - 4 = 9x - 24$
16. $4x^2 = 4x + 80$
17. $3x^2 + 3x = +60$
18. $2x^2 = 2x$
19. $16 = x^2 + 6x$
20. $2x^2 - 24x + 72 = 0$
21. $4x^2 + 16x + 16 = 0$
22. $8x = x^2 + 16$
23. $x^2 + 8x - 16 = 8x$
24. $x^2 + 3x - 4 = 3x$
25. $2x^2 - 36 = x^2$
26. $x^2 - 9 = 12x - 36$
27. A positive number has the following characteristics: If the sum of the number and 1 are multiplied by the sum of the number and 2, the product is 42. Find the number. (Let x be the number; then $x + 1$ will be the first factor.)
28. A positive number has the following characteristics: If 3 is subtracted from twice the number and this resulting difference is multiplied by the sum of the number and 2, the product will be equal to the square of the number. Find the number.

6.6 COMPLETING THE SQUARE AND THE QUADRATIC FORMULA

COMPLETING THE SQUARE. It would be most convenient if all the quadratic equations we encountered could be solved by factoring. Unfortunately, factorable quadratics are the exception rather than the rule. For example, the equation

$$x^2 + 4x + 2 = 0$$

might appear to be almost factorable but, alas, it is not. Thus, we will devise another technique, using a form of the quadratic that results from squaring a binomial (Chapter 5):

$$x^2 + 2ax + a^2 = (x + a)^2.$$

Note that the quadratic which we desire to solve is almost like the above special case. In fact, if 2 were added to the polynomial, the left side would indeed be a perfect square. Let us consider this approach to a possible solution:

$$\begin{aligned} x^2 + 4x + 2 &= 0 \\ x^2 + 4x + 4 &= 2 && \textit{(Add 2 to each side)} \\ (x + 2)^2 &= 2. && \textit{(Factor the left side)} \end{aligned}$$

We now wish to consider how this equation could be solved for the unknown x.

If, in our thinking, we were dealing with

$$z^2 = 2,$$

we could take the square root of both sides and find that

$$z = \sqrt{2}, -\sqrt{2} \qquad \text{or} \qquad z = \pm\sqrt{2}.$$

Similarly, for the polynomial equation we can take the square root of both sides, that is,

$$(x+2)^2 = 2$$

$$(x+2) = \pm\sqrt{2} \qquad \textit{(Take the square root)}$$

$$x+2 = -\sqrt{2} \qquad x+2 = +\sqrt{2} \qquad \textit{(Write as two equations)}$$

$$x = -2-\sqrt{2} \qquad x = -2+\sqrt{2} \qquad \textit{(Solve for each } x\textit{)}$$

$$x = -2-\sqrt{2}, -2+\sqrt{2}$$

or

$$x = -2 \pm \sqrt{2}.$$

Substituting both of these values into the original equation will verify that they are proper solutions.

Another example is

$$2x^2 + 9x + 3 = 0 \tag{6.20}$$

$$x^2 + \frac{9}{2}x + \frac{3}{2} = 0. \qquad \textit{(Multiply by 1/2)} \tag{6.21}$$

In order to simplify the operation of converting this polynomial to a perfect square, the coefficient of x^2 has been reduced to 1. It is necessary to find the value of a constant which, when added to 3/2, will make the polynomial a perfect square. Probably the easiest approach is to move 3/2 to the right hand side and start from scratch:

$$x^2 + \frac{9}{2}x = \frac{-3}{2}. \tag{6.22}$$

Comparing $x^2 + 9x/2$ to $x^2 + 2ax + a^2$, we note that 9/2 in the first expression must be equivalent to $2a$ in the second expression. This means that the required value of a for this case is 9/4. Since the constant term in the general expression is a^2, our required constant term must be $(9/4)^2$. Adding this to both sides of Eq. 6.22 gives

$$x^2 + \frac{9}{2}x + \left(\frac{9}{4}\right)^2 = -\frac{3}{2} + \left(\frac{9}{4}\right)^2$$

$$\left(x + \frac{9}{4}\right)^2 = -\frac{24}{16} + \frac{81}{16}$$

$$x + \frac{9}{4} = \pm\sqrt{\frac{57}{16}}$$

$$= \pm\frac{\sqrt{57}}{4}$$

$$x = \frac{-9 \pm \sqrt{57}}{4}.$$

Basically, the technique of finding roots of a quadratic equation in x by *completing the square* involves the following steps:

1. Place the terms involving the unknown on the left of the equal sign and the constant term on the right;
2. If the coefficient of the x^2 term is other than 1, multiply by the appropriate inverse, thus changing it to 1;
3. Determine one half the coefficient of the x term;
4. Add the square of this result to both sides of the equation;
5. Factor the left side of the equation;
6. Take the square root of both sides;
7. Solve for x.

THE QUADRATIC FORMULA. The real value of the method of completing the square lies not in the fact that it is commonly used to solve quadratic equations which cannot be factored, but that a general formula may be developed from it. In order that the formula be general in nature we will consider a general representation for a quadratic: $ax^2 + bx + c = 0$, where a, b, and c represent any real numbers. Used in this sense they are usually called **arbitrary** constants.

$$ax^2 + bx + c = 0 \qquad (a \neq 0)$$

$$x^2 + \frac{b}{a}x + \frac{c}{a} = 0 \qquad \text{(Multiply by } 1/a\text{)}$$

$$x^2 + \frac{b}{a}x = -\frac{c}{a} \qquad [\text{Add } -(c/a)]$$

$$x^2 + \frac{b}{a}x + \left(\frac{b}{2a}\right)^2 = -\frac{c}{a} + \left(\frac{b}{2a}\right)^2 \qquad \text{(Complete the square)}$$

$$\left(x + \frac{b}{2a}\right)^2 = -\frac{c}{a} + \frac{b^2}{4a^2} \qquad \text{(Factor)}$$

$$= \frac{b^2 - 4ac}{4a^2} \qquad \text{(Combine fractions on the right)}$$

$$x + \frac{b}{2a} = \pm\sqrt{\frac{b^2 - 4ac}{4a^2}} \qquad \text{(Take the square root)}$$

$$x + \frac{b}{2a} = \frac{\pm\sqrt{b^2 - 4ac}}{2a} \qquad \text{(Simplify the fraction)}$$

$$x = \frac{-b \pm \sqrt{b^2 - 4ac}}{2a}. \qquad \text{(Solve for } x\text{)}$$

This equation for the solution of a second degree polynomial is commonly called the **quadratic formula**. Although this formula appears complex, its use in finding the roots of quadratic equations simply involves substituting the proper coefficients and performing arithmetic simplifications. For instance, consider the following examples.

EXAMPLE 6.33

$$x^2 - 12x + 36 = 0.$$

SOLUTION

This corresponds to the general form of $ax^2 + bx + c = 0$, where $a = 1$, $b = -12$, and $c = 36$.

$$x = \frac{-b \pm \sqrt{b^2 - 4ac}}{2a} \qquad \textit{(Quadratic formula)}$$

$$= \frac{-(-12) \pm \sqrt{(-12)^2 - 4(1)(36)}}{2(1)} \qquad \textit{(Substitute for a, b, and c)}$$

$$= \frac{12 \pm \sqrt{144 - 144}}{2}$$

$$= \frac{12 \pm 0}{2}$$

$$x = 6.$$

In this example the roots are identical, which results from the fact that the quantity under the square root symbol is zero. The solution can be verified by recognizing that the original equation can be factored to

$$x^2 - 12x + 36 = (x - 6)(x - 6).$$

EXAMPLE 6.34

$$12x^2 - 13x - 14 = 0.$$

SOLUTION

Here we have $a = 12$, $b = -13$, and $c = -14$.

$$x = \frac{-b \pm \sqrt{b^2 - 4ac}}{2a}$$

$$= \frac{-(-13) \pm \sqrt{(-13)^2 - 4(12)(-14)}}{2(12)}$$

$$= \frac{13 \pm \sqrt{841}}{24}$$

$$= \frac{13 \pm 29}{24}$$

$$x = \frac{7}{4}, -\frac{2}{3}.$$

Both of these values are verified by substitution into the original equation:

$$12\left(\frac{7}{4}\right)^2 - 13\left(\frac{7}{4}\right) - 14 = 0 \qquad 12\left(-\frac{2}{3}\right)^2 - 13\left(-\frac{2}{3}\right) - 14 = 0$$

$$12\left(\frac{49}{16}\right) - 13\left(\frac{7}{4}\right) - 14 = 0 \qquad 12\left(\frac{4}{9}\right) - 13\left(-\frac{2}{3}\right) - 14 = 0$$

$$\frac{147}{4} - \frac{91}{4} - \frac{56}{4} = 0 \qquad \frac{16}{3} + \frac{26}{3} - \frac{42}{3} = 0$$

$$0 = 0. \qquad 0 = 0.$$

EXAMPLE 6.35

$$5x^2 + 4x + 1 = 0.$$

SOLUTION

In this example, $a = 5$, $b = 4$, and $c = 1$.

$$x = \frac{-b \pm \sqrt{b^2 - 4ac}}{2a}$$

$$= \frac{-4 \pm \sqrt{(4)^2 - 4(5)(1)}}{2(5)}$$

$$x = \frac{-4 \pm \sqrt{-4}}{10}.$$

The portion $\sqrt{-4}$ of this solution is unlike any quantity appearing in Chapter 3. No real number exists which, when multiplied by itself, gives -4. Thus -4 has no real square root (nor does any other negative number). Numbers such as these are commonly called **imaginary numbers**. Roots containing imaginaries are usually called **complex roots**. The study and use of complex numbers constitutes an entire field of mathematics with wide applications in engineering and science.

THE DISCRIMINANT. It is useful to recognize that the general character of the roots of a quadratic equation can be determined by reference to the radicand, $b^2 - 4ac$, of the quadratic formula. Because of its importance, it is commonly referred to as the **discriminant** of the equation. In Example 6.33, for instance, we saw that when the discriminant is zero, the roots are equal. The relationships between the discriminant and the roots of a quadratic equation may be summarized as follows:

1. The roots are **equal** if $b^2 - 4ac = 0$ (that is, if $b^2 = 4ac$).
2. The roots are **real and unequal** if $b^2 - 4ac > 0$ (that is, if $b^2 > 4ac$).
   (a) If $b^2 - 4ac$ is a perfect square, the roots are **rational**.
   (b) If $b^2 - 4ac$ is not a perfect square, the roots are **irrational**.
3. The roots are complex if $b^2 - 4ac < 0$ (that is, if $b^2 < 4ac$).

We have already seen, from Example 6.33, that equal roots result if the discriminant is zero. In Example 6.34 the discriminant is

$$\begin{aligned} b^2 - 4ac &= (-13)^2 - 4(12)(-14) \\ &= 169 + 672 \\ &= 841. \end{aligned}$$

In this case the roots must be real and unequal, since the discriminant is greater than zero. Further, inspection shows that the discriminant is a perfect square (that is, $29^2 = 841$). Thus, we also conclude that the roots are rational, which is consistent with the results.

In Example 6.35 we had $b^2 - 4ac = -4$. Since the discriminant is less than zero, the roots must be complex.

THE QUADRATIC FORMULA AND FORTRAN. The quadratic formula used to find roots of second degree polynomials is a common illustration of various techniques in Fortran. The formula might take the following form in Fortran, where X1 represents one root and X2 the other:

```
X1 = (-B+SQRT(B**2 - 4.0*A*C))/(2.0*A)
X2 = (-B-SQRT(B**2 - 4.0*A*C))/(2.0*A)
```

Although this pair of statements for calculating the two roots is valid, it represents a very inefficient approach from a Fortran point of view. The computer will first calculate the discriminant in computing X1, determine its square root,

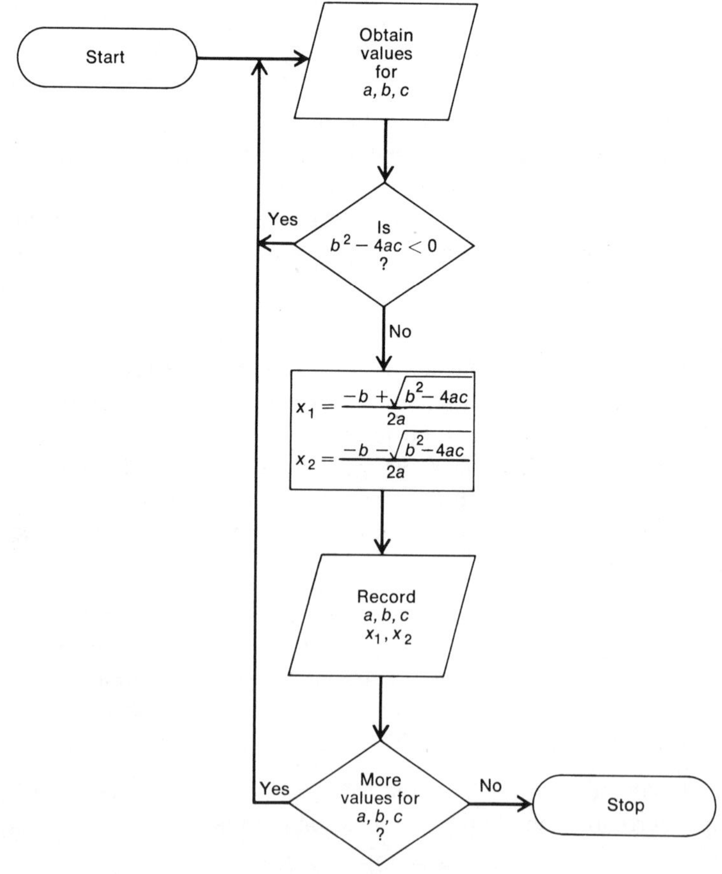

FIGURE 6.10 Roots of a quadratic

then complete the calculation storing the result in X1. In executing the second statement the same discriminant will again be calculated, since it will not have been "saved" from the preceding evaluation. The efficiency of this operation can be improved by first calculating the square root of the discriminant for use in calculating both roots, as is done in the following sequence.

```
ROOT = SQRT(B**2 - 4.0*A*C)
X1 = (-B+ROOT)/(2.0*A)
X2 = (-B-ROOT)/(2.0*A)
```

If, when evaluating roots on a computer, the possibility of a negative discriminant exists, then a test should be made to protect against attempting to calculate the square root. If we assume that no determination is necessary for a negative discriminant, then the overall process can be represented in the flowchart of Figure 6.10.

EXERCISE 6.6

What constant must be added to each of the following expressions to make the resulting trinominal a perfect square? Assume *a* and *b* represent constants.

1. $x^2 - 4x$
2. $y^2 - 5y$
3. $z^2 + \frac{1}{2}z$
4. $y^2 - ay$
5. $y^2 - 2by$
6. $z^2 - b^2z$

Solve the following equations by completing the square.

7. $x^2 = 6x$
8. $2z^2 = 5z$
9. $3x^2 + 2x - 1 = 0$
10. $2a^2 + 3a - b = 0$
11. $x^2 = -ax - b$
12. $\frac{x^2}{a} + \frac{x}{b} = 0$
13. $mx^2 - nx + q = 0$

Referring to the general form $ax^2 + bx + c$, what are the values for the coefficients *a*, *b*, and *c* in the following quadratics?

14. $2x^2 - 2x + 3$
15. $5x^2 - 3x + 1$
16. $x^2 - x - 2$
17. $3x^2 + 9rx - 5s$
18. $-x^2 + px - 4q$
19. $mx^2 - nx - 1$

In the following exercises, first examine the discriminant and state the nature of the roots, and then, using the quadratic formula, determine the roots.

20. $x^2 + x + 1 = 0$
21. $3x^2 + 5x + 1 = 0$
22. $x^2 - 4x + 2 = 0$
23. $x^2 - x - 7 = 0$
24. $x^2 - 3x + 8 = 0$
25. $4x^2 + 20x + 5 = 0$
26. $-81x^2 + 54x - 9 = 0$
27. $15x^2 - 14x - 8 = 0$
28. $-x^2 + 9x - 1 = 0$
29. If a natural number is multiplied by the sum of itself and 8, the product is 128. What is the number?
30. An integer has the following characteristics: If 3 more than twice the number is multiplied by the number plus 1, the product is 28. Find the integer.

31. Find three consecutive positive integers the sum of whose squares is 509.

32. The quantity 3/2 times the square of a positive number is equal to the product of the number plus 15 and the number plus 12. Find the number.

33. The distance d that an object falls when dropped can be described by the equation

$$d = \frac{16t^2}{1 + (t/20)}$$

where t is the time in seconds. How long would it take an object to fall from 3200 ft?

34. The solution of the equation in problem 33 consists of two roots. Both of them have meaning relative to the equation, but only one of them has meaning relative to the physical problem. The other is frequently called an *extraneous* root, which may be determined by context. Which is the extraneous root and why?

35. A computer program is written to evaluate an expression for various values of A and B. For example, 20 values for A and 20 values for B would require $20 \times 20 = 400$ evaluations. Running time is directly related to the number of evaluations performed. A programmer makes a run with the number of values for A and B the same. He then increases the number of values for A by 15 and increases the number of values for B by 12 and finds that the computer requires 50 percent more computing time (that is, 50 percent more calculations were made). How many values were run for A and B in each of the two runs?

36. Compare problems 32 and 35.

37. If the roots obtained by using the quadratic equation are rational, argue that the original quadratic equation could have been factored.

38. The quadratic equation $21x^2 - 2x - 8 = 0$ has the roots 2/3 and $-4/7$ as determined by the quadratic formula. Since these roots are rational, the implication is that the equation can be factored. Show how the factored form may be obtained from the roots.

39. If a quadratic equation has the roots of -1 and n/m, show that its factored form would be $(x + 1)(mx - n) = 0$.

40. If a quadratic equation has roots of p and q, determine the original equation.

REFERENCES

Apostle, H. G., *Survey of Basic Mathematics.* Boston, Little, Brown & Company, 1960.

Barnett, R. A., *Intermediate Algebra: Structure and Use.* New York, McGraw-Hill, Inc., 1971.

Dubisch, R., Howes, V. E., and Bryant, S. J., *Intermediate Algebra,* 2d ed. New York, John Wiley & Sons, Inc., 1969.

Groza, V. S. and Shelley, S., *Modern Intermediate Algebra for College Students.* New York, Holt, Rinehart and Winston, Inc., 1969.

Richardson, M., *Fundamentals of Mathematics,* 3d ed. New York, The MacMillan Company, 1966.

Wooten, W. and Drooyan, I., *Intermediate Algebra,* 2d alt. ed. Belmont, Calif., Wadsworth Publishing Company, Inc., 1968.

7 FUNCTIONS

7.1 TABLES AND GRAPHS / **233**
A Table of School Days / **233**
A Functional Relationship / **233**
The Table in Computer Storage / **235**
Subscripted Variables in Fortran / **236**
Graphing / **237**
Continuity / **238**
Exercise 7.1 / **239**

7.2 EQUATIONS AND FORMULAS / **242**
Restricting the Domain / **243**
Functions Represented by Equations / **243**
Functional Notation / **244**
Functions in Computer Languages / **245**
Exercise 7.2 / **246**

7.3 THE LINEAR FUNCTION / **248**
Rectangular Coordinates / **248**
A Special Case of the Linear Function / **249**
Slope / **251**
The Y Intercept / **251**
Negative Slope / **253**
General Form of the Linear Function / **254**
Exercise 7.3 / **255**

7.4 LINEAR INEQUALITIES / **256**
Furniture and Inequalities / **256**
Additional Examples / **258**
Functions and Inequalities / **259**
Exercise 7.4 / **259**

7.1 TABLES AND GRAPHS

A TABLE OF SCHOOL DAYS. In programming computers, it is often necessary to provide information in tabular form. For example, assume that we have written an extensive program to calculate the projected monthly operating cost for each school building of a public school district. Since this monthly cost would depend, in part, on the number of school days in each month, it would be necessary to provide the computer with information similar to that of Table 7.1.

TABLE 7.1 SCHOOL DAYS BY MONTH

Month	School Days
January	20
February	18
March	23
April	17
May	20
June	9
July	0
August	0
September	18
October	21
November	19
December	13

During processing for the month of January, the computer would be programmed to search the table for January, and then use the corresponding 20 school days. Upon completing processing for January, the machine would continue to February and use 18 days, then to March, and so on for each of the twelve months. In using this table, we should note that it consists of two parts, the collection or set of months and the collection or set of days. Also, we choose a particular month, and then find the corresponding number of days. In other words, the number of days to be used in a given calculation depends upon the particular month.

A FUNCTIONAL RELATIONSHIP. In order to provide insight into the mathematical nature of this table, it is useful to consider it as representing two sets. The first set consists of elements defined as months of the year; the second set consists of elements defined as a number of school days. These may be represented abstractly as

$$M_1, M_2, M_3, \cdots M_{12} \qquad \text{and} \qquad D_1, D_2, D_3, \cdots D_{12}.$$

The important point here is that these two sets are not independent and unrelated. They are, in fact, related in the sense that for each M of the first set there is a corresponding D of the second set. This can be represented schematically, as shown in Figure 7.1. Such a relationship between elements of

$$\begin{array}{cccc} M_1, & M_2, & M_3, \cdots & M_{12} \\ \downarrow & \downarrow & \downarrow & \downarrow \\ D_1, & D_2, & D_3, \cdots & D_{12} \end{array}$$

FIGURE 7.1

two sets is commonly called a **function** and is one of the most fundamental notions in mathematics. The term "function" is frequently defined as *a correspondence between the elements of two sets such that each element of the first set is associated with exactly one element of the second set.* Normally we consider sets whose elements are numbers, but, as we have illustrated, the basic notion of function is not limited to numbers.

Although the inclusion of the month name in Table 7.1 is convenient, a month, when used in a computer, is normally coded to save machine storage. If January is coded 1, February is coded 2, and so on, Table 7.1 would appear as shown in Table 7.2.

TABLE 7.2 SCHOOL DAYS BY MONTH NUMBER

Month	School Days
1	20
2	18
3	23
4	17
5	20
6	9
7	0
8	0
9	18
10	21
11	19
12	13

The first set or column is called the **domain** of the function, and the second set or column is called the **range** of the function. In this illustration the elements that constitute the domain are the integers 1, 2, 3, 4, 5, 6, 7, 8, 9, 10, 11, 12, and the elements that constitute the range are the integers 0, 9, 13, 17, 18, 19, 20, 21, 23. In using the table we choose a month from the domain set either arbitrarily or as required, and then determine the corresponding number of days from the range set; this is the operation which the computer would perform in this hypothetical problem. Since the range element depends upon the choice of the domain element, we can refer to the second set (or column in the illustration) as the **dependent** set and the first set (or column) as the **independent** set. Thus for each element of the independent set there is always one and only one corresponding element in the dependent set. This is simply a restatement of the definition of a function.

THE TABLE IN COMPUTER STORAGE. Although the information in the format of Table 7.2 is convenient for us to use, it would normally be stored in consecutive storage locations within the computer. For instance, if we consider a computer with storage registers each capable of storing one decimal number, then the table would require 24 positions of storage. Assuming that for a given application the table has been assigned storage registers 750 through 773, the storage contents might appear as shown in Figure 7.2. We can see that the numbers are stored in pairs consisting of the month number and number of days. (Alternate pairs are shaded for convenience.) Common terminology used by programmers is to refer to the independent variable (month number) as the *argument* and the corresponding dependent variable (number of days) as the *function.* If, in performing school accounting operations, the number of school days in a given month were required, the computer would be programmed to search this table. First the required month number would be compared to the first argument (at location 750); if they are identical then the desired function is in 751. If they are different, the search would continue to the next argument and so on until a match was found. This table searching procedure is illustrated by the flowchart segment of Figure 7.3. In writing a program for table searching, it is always wise to include some type of check to insure that searching does not continue beyond the bounds of the table in the event incorrect data is fed to the machine (for example, a month number of 13). Otherwise, the programmer will likely be in for an unpredictable type of surprise.

Since values of the argument in this example are conveniently ordered and vary in increments of 1, it is actually unnecessary to store values of the argument since this can be determined by position. For example, in Figure 7.4 consecutive values of the function (number of school days) have been placed in storage beginning in location 825. Knowing that the function values are ordered and in consecutive positions, the function corresponding to June (month 6) can be found by the simple calculation

$$\begin{aligned} \text{address} &= 825 + (6 - 1) \\ &= 830. \end{aligned}$$

This so-called *direct table look-up* is much more efficient since it involves no searching. However, if values of the argument do not form a convenient sequence, this technique is usually impractical.

	1	20	2	18	3	23	• • •	11	19	12	13
Storage addresses	750	751	752	753	754	755	• • •	770	771	772	773

↑ Beginning of table

FIGURE 7.2 Arguments and functions in storage

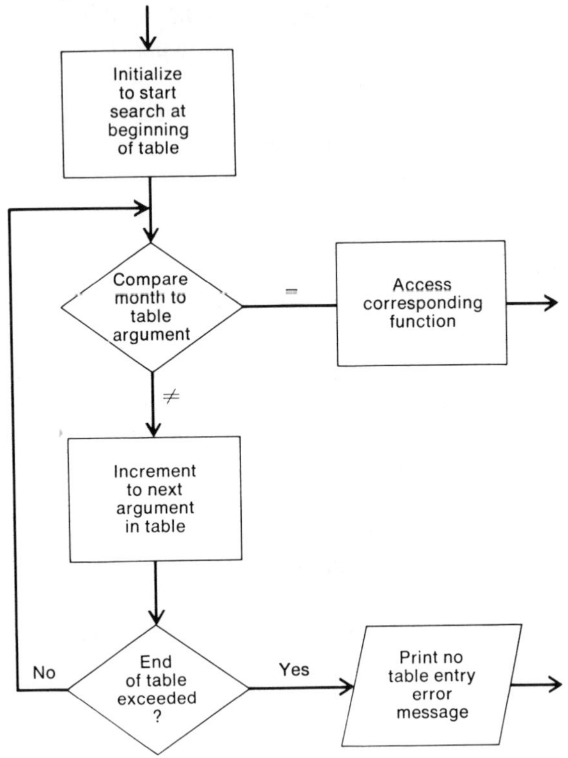

FIGURE 7.3 Flowchart to search a table

SUBSCRIPTED VARIABLES IN FORTRAN. This method of direct look-up is conveniently usable in Fortran through *subscripted variables*. Using subscripts in a manner similar to that used in Figure 7.1, a single array name can be used to define storage for the table. For example, the Fortran statement

```
DIMENSION D(12),TABLE(38)
```

will cause the Fortran system to set aside storage for two arrays D consisting of 12 elements and TABLE consisting of 38. The array D relates to the school days example and will consist of the elements D(1),D(2), ··· ,D(12). If the desired school month number is read into storage as NUM, then the programmer would access the number of days by specifying D(NUM). The table itself will be stored in consecutive locations as in Figure 7.4 (except that Fortran arrays are actually stored within the machine in reverse).

20	18	23	17	20	9	0	0	18	21	19	13
825	826	827	828	829	830	831	832	833	834	835	836
					↑ Month 6 (June)						

FIGURE 7.4 A direct look-up table

GRAPHING. The construction of tables to illustrate a function is frequently useful for displaying the information, and often necessary whenever the information is to be used in computer processing. However, to obtain insight into the nature of a function it is usually necessary to study a table very carefully. On the other hand, considerable insight may be obtained at a glance by referring to a **graph** of the function. For instance, Figure 7.5 is a **bar** graph (also commonly called a *histogram*) of the information contained in Table 7.1.

Note that the independent quantity is associated with the horizontal line or **axis** and the dependent quantity with the vertical axis. In order to use the bar graph, we locate the appropriate month (element of the domain) on the horizontal axis, then read the corresponding number of days (element of the range) from the vertical axis. Graphs of this type are very common; we see them in advertisements, news articles, and reports as well as in textbooks of various disciplines. They are usually quite useful in conveying information at a glance.

There are other types of graphs useful for conveying information, such as that shown in Figure 7.6. In keeping with the general forms of functions as we shall study them, both the domain and range are represented in Figure 7.6 as numbers (from Table 7.2). The location on the graph of each point is determined by two numbers, one a domain element and the other a range element. The totality of these points is, of course, the function.

The graphical representation of a function in Figure 7.6 is consistent with the definition of a function as "a correspondence between the elements of two sets such that each element of the first set is associated with exactly one element of the second set." If a domain element had no corresponding range element, the point could not be represented on the graph. On the other hand, for a domain element to correspond to two or more range elements would

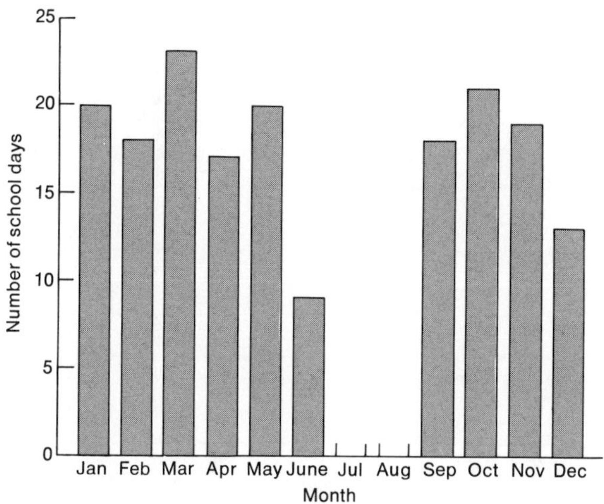

FIGURE 7.5 Bar graph of school days by month

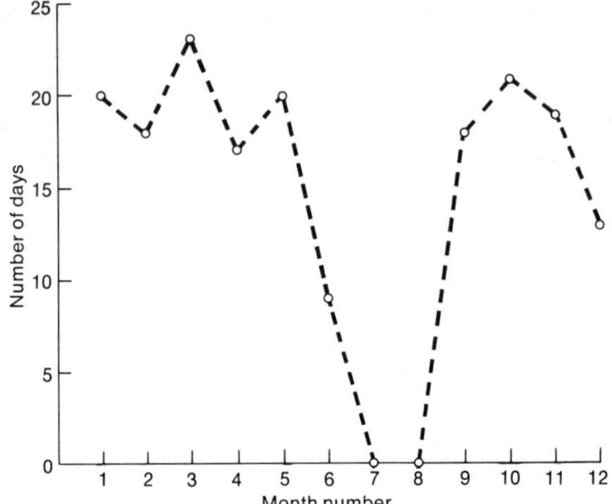

FIGURE 7.6

be meaningless. That is, month number 1 has 20 school days, not 20, 21, and 23.

CONTINUITY. The points on the graph of Figure 7.6 are connected by broken lines. In this case, the broken line is used to indicate that it is not meaningful to attempt to use the graph for domain points between those indicated. For example, we could estimate the point on the horizontal axis which corresponds to $3\frac{1}{2}$ and extend a line upward to a point of intersection with the graph, thus obtaining a corresponding range point (in this case, approximately 20). However, to do so with this example is meaningless since the domain elements represent months, 3 being March and 4 being April. In other words, we must temper our use of the graph by the original limitation placed on the domain of this function, that is, integers 1 through 12.

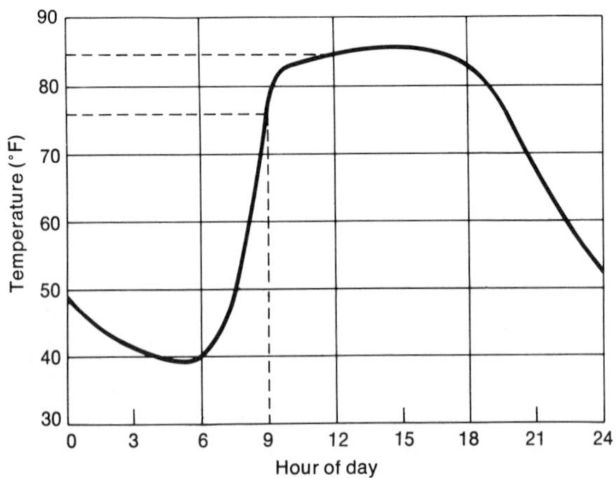

FIGURE 7.7 Temperature record

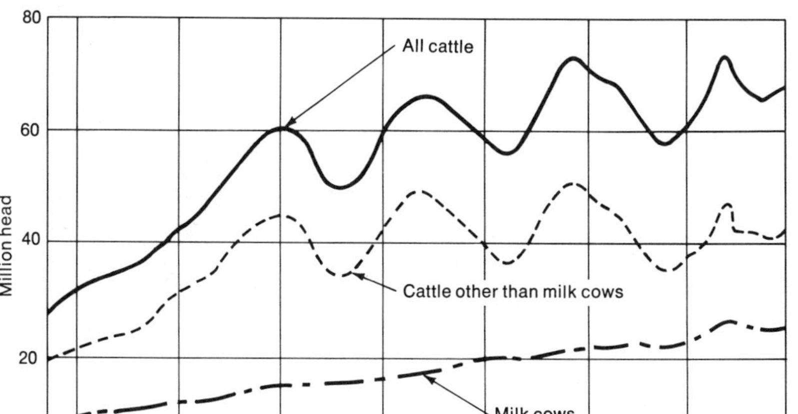

FIGURE 7.8

Because this function consists of isolated points, it is said to be **discontinuous**. Most of the functions that we will study in this book are **continuous** which, loosely speaking, means that they form a continuous line or curve on a graph.

A typical continuous function graph is shown in Figure 7.7, representing a record of outside air temperature over a 24-hour period. Such graphs are commonly drawn by automatic recording machines, thus giving a true record of the temperature at all times. Here it is just as valid to read the temperature 76° at 9.5 (corresponding to 9:30 AM) as the temperature 84° at 12 (corresponding to 12:00 noon).

An example of a different nature is shown in Figure 7.8. Here three functions are shown on one graph. Note that the curve labeled *all cattle* represents the sum of the other two curves. We would probably think of and use this as a continuous function; however, it is actually discontinuous. For example, at 6:00 PM on March 1, 1879, the cattle population may have been 40,000,362. An instant later, with the birth of a new calf, the population would have been 40,000,363. Thus at the instant of birth, a slight discontinuity would exist in the function. Of course in this example, populations are estimated, and are in no way intended to be used with such accuracy. Thus, for all practical purposes, this graph is continuous.

EXERCISE 7.1

1. For the years 1951 through 1964, what was the average amount collected by the Bureau of Internal Revenue? Which year was nearest this average? (Use Figure 7.9.)
2. Redraw Figure 7.9 using a single point for each yearly collection, and connect the points by a broken line.

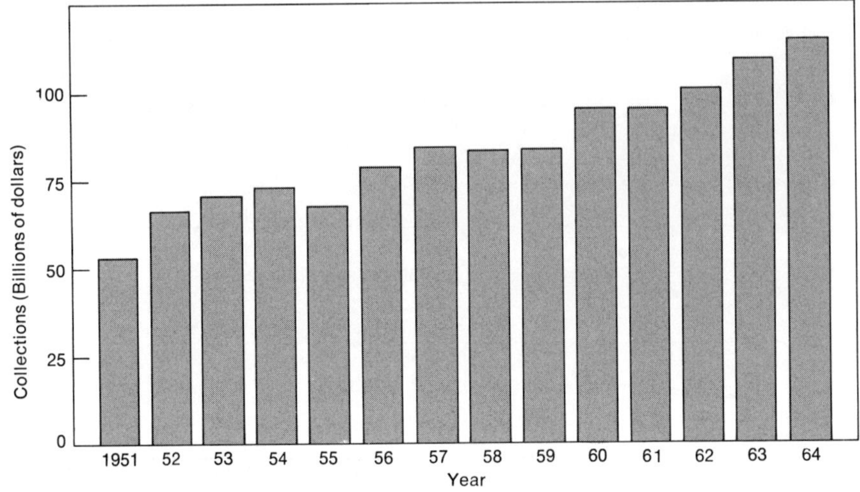

FIGURE 7.9 Yearly internal revenue collections

3. What is the significance of the negative point at 1961 in Figure 7.10?
4. Make a histogram from the graph of Figure 7.10.
5. From Figure 7.10 make a table of the year-to-year change in net profit. Which two consecutive years showed the greatest increase in net profit? How is it possible to tell at a glance, by reference to the graph, which year showed the greatest increase?
6. If a medical study included 250,000 typical men in the 60-year age bracket, approximately how many cases of cancer would be found? How many cases among 75,000 women in the 50-year age bracket? (Use Figure 7.11.)
7. Make a graph of the number of military personnel not stationed overseas by determining the difference between the two curves of Figure 7.12.
8. From Figure 7.13 make a table of rainfall, using mileage distances of 0, 100, 200, and 300. Connect each of these points by a broken line, and then superimpose on this the original graph, and compare the two. Repeat this procedure, using increments of 50 miles beginning at 0. In

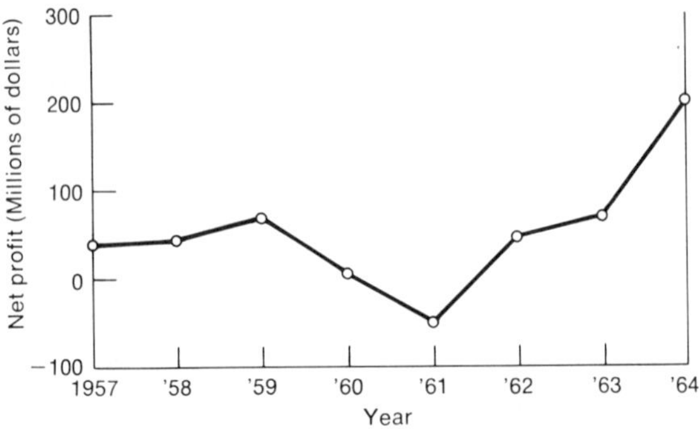

FIGURE 7.10 Net profit and loss—scheduled U. S. airlines

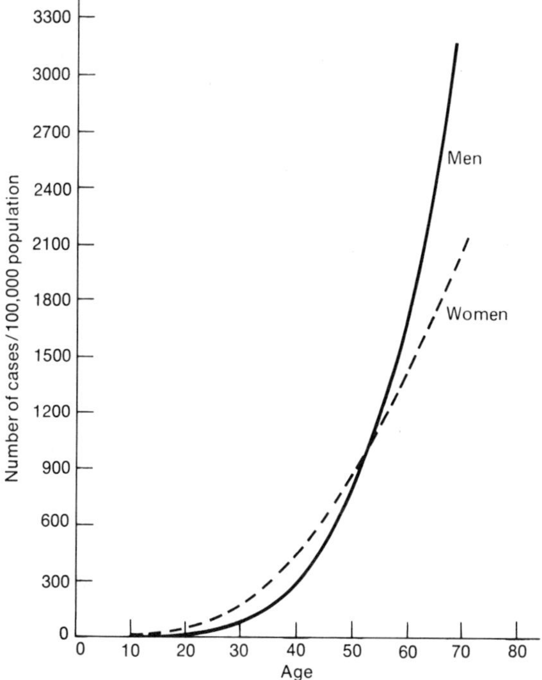

FIGURE 7.11 Incidence of cancer

interpreting graphs, note the importance of including a sufficient number of points and using caution in estimating between points.

9. Referring to Figure 7.14, what percentage of the earth's surface is above 3000 meters? −5000 meters? What percentage is below 2000 meters? −5,000 meters? What percentage is between 4000 meters and −5000 meters?

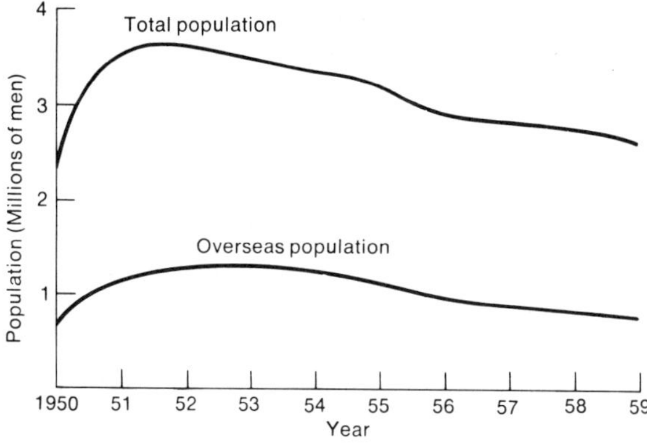

FIGURE 7.12 Population of armed forces

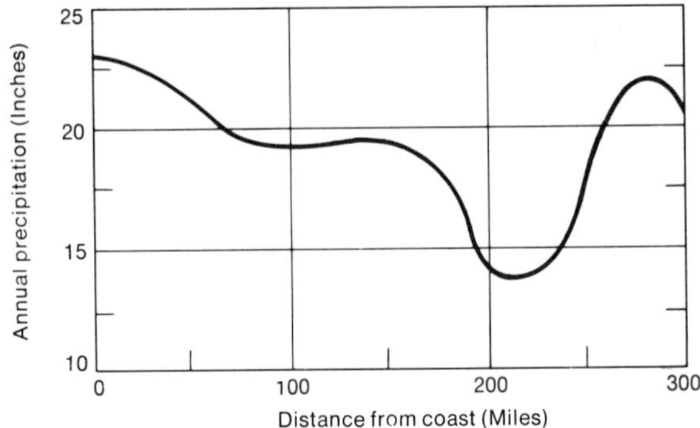

FIGURE 7.13 Annual precipitation

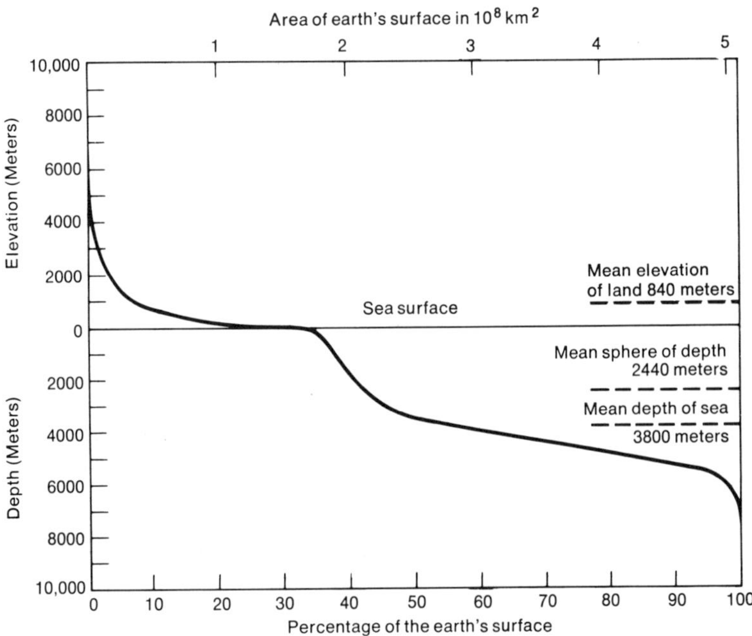

FIGURE 7.14 Distribution of earth surface by elevation from the sea

7.2 EQUATIONS AND FORMULAS

In developing the idea of a function, we used two sets defined by a table, and also showed how the function can be interpreted graphically. Functional relationships defined by tables and by graphs are frequently encountered in everyday life, and are often used in computer programming. However, there are many other ways in which a function may be defined, the most common being *an equation or formula*, or *a verbal statement or description of a relationship*. For the present, we shall concentrate on functions defined by an equation or a formula.

The equation $P = 4s$ relates the variable P to the variable s. For any chosen value of s we can determine one and only one value of P. If s may be any real number, the domain of this function consists of the real numbers; furthermore, it can be shown that the range (values which P may assume) also consists of the real numbers. When the equation is expressed in this form, the variable s is called the **independent variable** and the variable P the **dependent variable**. This is consistent with our notion of sets, since the equation defines the two sets and their relationship. Table 7.3 consists of several corresponding elements of the two sets related by $P = 4s$. Of course, Table 7.3 contains only a few members of the domain and range sets. The inclusion of the full sets is obviously impossible because there are infinitely many real numbers.

TABLE 7.3 A FUNCTION IN TABULAR FORM

s	P
-4	-16
-1	-4
0	0
$\frac{1}{2}$	2
1	4
$\frac{4}{3}$	$\frac{16}{3}$
$\sqrt{10}$	$4\sqrt{10}$
6	24

RESTRICTING THE DOMAIN. If we consider the equation $P = 4s$ as the formula for the perimeter of a square, where P represents the perimeter and s the length of a side, it is meaningless to speak of a square with a negative or zero side, so we would define the domain of the formula as the positive real numbers; the corresponding range would also be the positive real numbers.

In defining a function we can restrict the domain in any desired manner. For instance, a manufacturing company might produce square enclosures, varying from 6 ft to 15 ft in 1/4-ft increments. The domain of the function would be the 37-element set:

$$24/4,\ 25/4,\ 26/4,\ \cdots\ 60/4,$$

and the corresponding range would be the set

$$24,\ 25,\ 26,\ \cdots\ 60.$$

Note that the domain and range would not necessarily consist of the same types of numbers, since in this case the range consists only of integers, whereas the domain does not.

FUNCTIONS REPRESENTED BY EQUATIONS. Generally speaking, the domain of a function expressed by an equation is considered to be the real numbers unless specified otherwise. The corresponding range then depends upon the nature of the function itself. For example, consider

$$y = x^2 \qquad \text{and} \qquad m = \frac{1}{n-1}.$$

In the first example, x is the independent variable and y the dependent variable; the domain consists of all the real numbers, but the range consists only of zero and the positive real numbers. In the second example, n is the independent variable and m the dependent variable; the domain consists of all the real numbers except 1, since m is not defined if $n = 1$ (that is, division by zero is undefined). However, the corresponding range is the entire system of real numbers.

FUNCTIONAL NOTATION. In preparing a problem for computer solution it is often necessary to consider a physical situation, interpret it in the form of a verbal statement or series of statements, and translate the statements into mathematical equations. During the early stages of examining a problem, it is necessary to define the variables and determine dependence. For instance, the owner of a small manufacturing firm would be interested in all the variables that affect the selling price of his product. One variable might be the shipping cost to a distant sales outlet. The shipping cost would probably vary with weight. In this case we could say that shipping cost is a *function* of weight. Using C for shipping cost and w for weight, this is represented mathematically as

$$C = f(w).$$

This is read as "C is a function of w," or "C equals f of w," and does not imply that a variable f is multiplied by a variable w. The form $C = f(w)$ does not tell us *what* the equation is; it only serves to indicate that the value of C *depends* upon the value of w. For example, the exact relationship might be any of the following:

$$C = 2w \qquad C = \sqrt{w} + 3 \qquad C = 2w^2 + 1.$$

In each of the above, w represents the independent variable and the values it can assume comprise the domain; similarly, C represents the dependent variable and the corresponding values it may assume comprise the range.

In the second example, $\sqrt{w}$ implies by convention the positive root only. If both roots are intended, the equation would be written

$$C = \pm\sqrt{w} + 3.$$

It is important to note that the latter equation does *not* satisfy our definition of a function. For instance, if $w = 4$,

$$C = \pm\sqrt{4} + 3 = \pm 2 + 3 = 5,\ 1.$$

In other words, each element of the domain would have two corresponding elements of the range, whereas the definition states "each element of the first set is associated with exactly one element of the second set." A relationship of this type is sometimes referred to as a **multivalued** function.

The choice of the letter f in functional notation is logical. However, any convenient symbol may be used, and the letters $g(w)$, $F(w)$, $G(w)$, $Q(w)$, and so on, are often encountered; in fact, in this example the letter C itself would be useful. In this case we could say

$$C = C(w).$$

That is, the cost C depends upon the weight w. Each of the preceding equations could be written as

$$C(w) = 2w \qquad C(w) = \sqrt{w} + 3 \qquad C(w) = 2w^2 + 1.$$

This form is especially useful because we may replace w by a particular number in the domain, say a, and $C(a)$ will represent the corresponding element in the range. For example, if we use the function

$$C(w) = 2w^2 + 1,$$

then

$$C(3) = 2(3)^2 + 1 = 19$$

This may be interpreted as "the value of the function C is 19 when w is 3." For consistency with our previous definitions, we think of w as the independent variable, and the values which it may assume as the domain; similarly, $C(w)$ is the dependent variable and its corresponding values make up the range. Other terminology commonly used in computer programming is to call the independent variable the **argument**. Thus for the function $F(x) = 2x + 3$, x is the argument and $F(x)$ is the function. In evaluating the function for $x = 5$, we have

$$F(5) = 2(5) + 3 = 13.$$

Thus the argument is 5 and the corresponding function is 13.

FUNCTIONS IN COMPUTER LANGUAGES. The versatility of most programming languages is greatly increased through the use of *subprograms* (commonly called *subroutines*) which are prewritten sets of instructions for performing particular operations. Many standard types of operations are programmed by the computer manufacturer to form the *computer library.* Most languages also have provisions to allow the computer user to store his own subprograms in the library. In particular, the Fortran language has extensive provisions allowing the computer user to prepare several types of subprograms. One such type is the Fortran *function*, which was used to illustrate computing roots with quadratic formula (Chapter 6).

Several of the commonly used functions in Fortran and their Fortran names are summarized in Table 7.4.

TABLE 7.4 FORTRAN FUNCTIONS

Function	Fortran Name	Mathematical Operation	
Square root	SQRT	$\sqrt{x}$	
Absolute value (real)	ABS	$\lvert x \rvert$	
Absolute value (integer)	IABS	$\lvert i \rvert$	
trignometric sine	SIN	$\sin x$	(x in radians)
trignometric cosine	COS	$\cos x$	(x in radians)
exponential	EXP	e^x	
natural logarithm	ALOG	$\ln x$	

In algebra the variable y is said to be a function of x when some relationship or correspondence is given whereby for each value of x there is a corresponding value of y. The term function as used in Fortran is virtually identical to this usage in algebra. For example, the SQRT function (which is a function since it only yields the principal root) might be used in a program as illustrated by the first of the following examples.

```
ROOT = SQRT(X)
AMNT = 3.0 + 5.0*SQRT(DATA)
FIELD = ABS(X*Y-6.0)
```

In the first, an argument X is enclosed in parentheses following the name SQRT. The calculated result, that is the required function of X, will be returned to the program in place of the name ROOT. Thus we can think of the variable X as representing the independent variable (constituting the domain) and the expression SQRT(X) as representing the dependent variable (constituting the range). In this example, the result or corresponding function is assigned to the storage area defined as ROOT. In the second example, the argument is DATA; the resulting function is first multiplied by 5, then this product is increased by 3 before storing it in AMNT. The argument supplied to a Fortran function need not be a single variable as is evident by examining the third example in which the expression XY-6 is used. Indeed, the argument of a Fortran function can be any valid Fortran expression including those which themselves involve Fortran functions.

EXERCISE 7.2

Which of the following relationships are not functions (y a function of x)? Why not?

1. $y = x^2$
2. $y = -x^3 - 2$
3. $y = \frac{1}{x}$
4. $y = \frac{1}{\pm\sqrt{x-1}}$
5. $y = \begin{cases} 5 \text{ if } x < 0 \\ 0 \text{ if } x = 0 \\ x \text{ if } x > 0 \end{cases}$
6. $y = \begin{cases} x^2 \text{ if } x \leqq 0 \\ -x^2 \text{ if } x > 0 \end{cases}$

7. $y = \begin{cases} -(x+1) \text{ if } x \leqq 0 \\ (x+1) \text{ if } x \geqq 0 \end{cases}$

8.

x	y
0	1
1	0
2	1
3	0
4	1

9.

x	y
1	0
2	1
4	2
8	3
16	4

10.

x	y
0	0
−1	0
1	1
0	1

11. Which of the relationships in problems 1-10 are continuous?

If the domains of the following functions are to be the positive real numbers (excluding 0 or 1 where appropriate), what will be the corresponding ranges?

12. $y = 3x - 2$
13. $y = -2x + 1$
14. $y = x^2 - 5$
15. $y = -x^2 + 1$
16. $y = -x^2$
17. $y = |x|$
18. $y = \dfrac{1}{(1-x)}$
19. $y = \dfrac{1}{x}$
20. $y = x - 3$
21. $y = \dfrac{1}{1 - x^3}$
22. $y = \dfrac{1}{-x^3}$

In problems 23-29 solve each of the following equations for y and indicate which is the independent and which is the dependent variable.

23. $y + 3 = x - y$
24. $3x - 2y = 2y$
25. $y + 4 = z - 3y$
26. $y - z^2 = 3$
27. $z^2 - y^2 = y - y^2 + 2$
28. $xy = 1$
29. $(1 - z)y = 1$
30. In problems 23-29, solve for the other variable (x or z as the case may be) in terms of y. In this form which is the independent variable? Which are not functions in this form? Note that a function $y = f(x)$ may often be manipulated to solve for x in terms of y. Sometimes this so-called **inverse** function is not a function at all. Whenever it is, we can say that $y = f(x)$ and $x = F(y)$, implying that the inverse is also a function.
31. If $f(x) = x^2 - 3x + 2$, find (a) $f(0)$, (b) $f(1)$, (c) $f(a)$, (d) $f(b)$, (e) $f(b) - f(a)$.
32. If

$$f(y) = \frac{y-3}{y-5},$$

find (a) $f(0)$, (b) $f(3)$, (c) $f(5)$, (d) $f(a)$.

33. A computer installation has the following rate structure for one of its machines: \$120 per hour for the first 5 hours each month, \$110 per hour for the next 15 hours each month, and \$100 per hour for all time over

20 hours per month. Let y represent the total charges and x the total running time. Express y in terms of x (refer to the form of problems 5, 6, and 7). Is y a function of x? Is this relationship continuous? Using this relationship, compute the charge for 28 hours per month.

34. Write a sequence of Fortran statements to define the function of problem 5.

35. Write a sequence of Fortran statements to define the function of problem 6.

36. Write Fortran statements for each of the following algebraic equations.
    (a) $c = \sqrt{a^2 + b^2}$
    (b) $P = EI \cos \theta$
    (c) $y = |a + b - c^2|$
    (d) $d = \sqrt{a^2 + |y + z|}$
    (e) $a = \sqrt{b^2 + c^2 - 2bc \cos \theta}$
    (f) $F = a + b \ 1\mathrm{n} \dfrac{1 + e^2}{1 - e^2}$

7.3 THE LINEAR FUNCTION

RECTANGULAR COORDINATES. At the beginning of this chapter we discussed the graphical or geometric interpretation of a function as defined by means of a table of values. The graph of Figure 7.6 is a limited one, since values of the domain and range are restricted to a few positive integers. On the other hand, an equation such as $y = 4x$, where x is any real number, allows for both positive and negative values of the independent and dependent variables. Thus we extend the axes of the previously studied graphs in both the positive and negative directions. (As before, the horizontal axis represents the independent variable and the vertical axis the dependent variable.)

Figure 7.15 is such a coordinate system. Note that each of the two axes is identical to the real axis described in Chapter 2. Since they intersect at right angles, they are often called a **rectangular cartesian coordinate** system. Their intersection, which represents the zero point on each axis, is commonly called the **origin**. Just as a single number corresponded to a point on the real axis in Chapter 2, here a pair of numbers (an element of the domain and one of the range) corresponds to a point in the plane. For example, a point represented by ($x = 2$, $y = 4$) is designated by its **coordinates** (2,4), where the first quantity in the parenthesis is always the independent variable and the second is the dependent variable. The horizontal distance 2 to the point (2,4) is referred to as the **abscissa** of the point, and the vertical distance 4 as the **ordinate** of the point. Each of the four regions into which the plane is divided is called a **quadrant**, and they are referred to as the *first quadrant* (I), the second quadrant (II), and so on, as shown in Figure 7.15. (The graph of Figure 7.6 used only the first quadrant.)

Although coordinate systems were used thousands of years ago by the

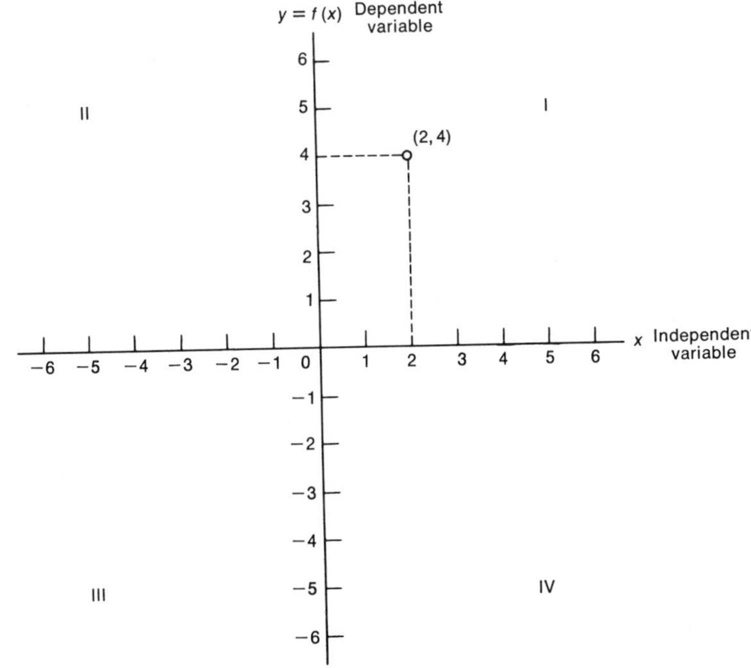

FIGURE 7.15 The rectangular cartesian coordinate system

ancient Egyptians for surveying, the idea of using a graphical coordinate system to represent algebraic entities (which we shall study next) is relatively new. It was first proposed by the mathematician and philosopher Rene Descartes in 1637, and represented the beginning of a new era in mathematics. The cartesian coordinate system is so named in honor of Descartes (the Latin equivalent of Descartes is Cartesius).

A SPECIAL CASE OF THE LINEAR FUNCTION. The first degree polynomial whose general form is

$$y = ax + b$$

is extremely important in most applications of mathematics. It is commonly referred to as the **linear** function, because its graph is a straight line. To gain insight into the linear function, we will consider the special case

$$y = 2x.$$

In order to graphically represent a function defined by an equation, it is convenient to make an abbreviated table, as shown below, and then plot the points from the table(A table for $y = 2x^2$ has also been included for contrast.) In compiling both of these tables, domain points (values for x) are obtained by increasing the previous values for x by 1 Note the regularity in the range points of the function $y = 2x$; here each value is 2 greater than the preceding value.

$y = 2x$		$y = 2x^2$	
x	y	x	y
−1	−2	−1	2
0	0	0	0
1	2	1	2
2	4	2	8
3	6	3	18

This is in contrast to the table for $y = 2x^2$, where even though values for x each increase by 1, no obvious pattern exists for successive values of y. The graphs of these two functions are plotted by determining the position of each point, as shown in Figure 7.16, and connecting them by a smooth curve. The graph for $y = 2x$ is apparently a straight line, whereas that for $y = 2x^2$ obviously is not. In contrast to the graph of Figure 7.6, every point along both of these curves is meaningful and constitutes a point of that function. That is, by locating the point $x = 2.5$ on the x axis, a line may be extended upward to the graph of $y = 2x$ and across to the intersection of the y axis. The point of intersection with the y axis is the corresponding value for y, which is 5. That this is consistent with the original equation can be shown by substitution:

$$y = 2x = 2(2\tfrac{1}{2}) = 5.$$

The following linear equations all have the same general form as $y = 2x$:

$$y = 4x, \qquad y = \frac{1}{2}x, \qquad y = x, \qquad y = \frac{1}{4}x.$$

and are shown on the graph of Figure 7.17.

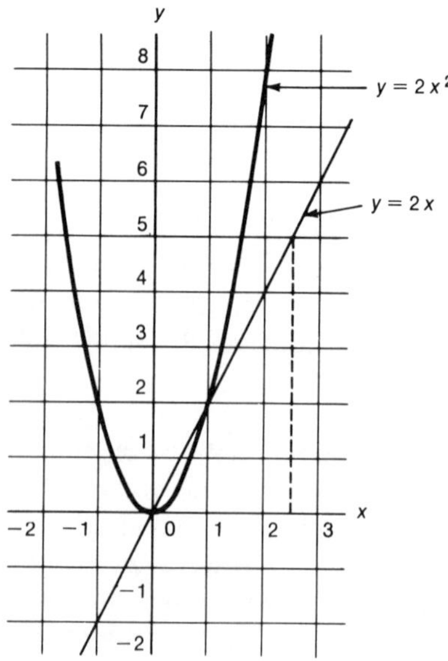

FIGURE 7.16

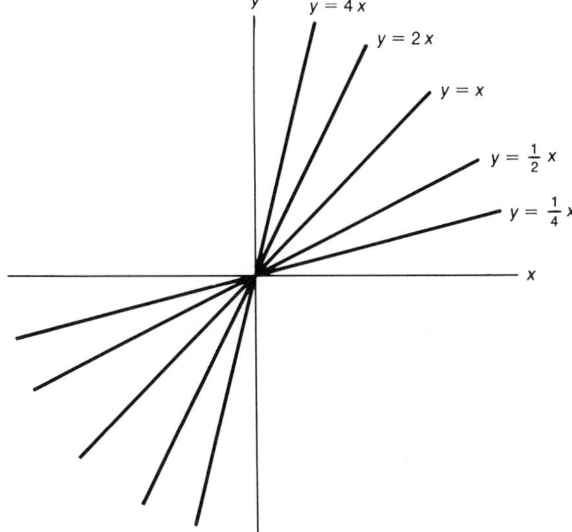

FIGURE 7.17

SLOPE. All of the functions of Figure 7.17 have one characteristic in common; that is, they all pass through the origin. Algebraically, they have the form $y = mx$ and differ only in the value of m, the coefficient on x; the larger this coefficient, the steeper the graph. The coefficient m is commonly called the **slope**, as suggested by the graph. It is, in a manner of speaking, the distance up divided by the distance out. In designing roads, a civil engineer thinks of the slope or grade of a roadbed as the rise divided by the run, as shown in Figure 7.18.

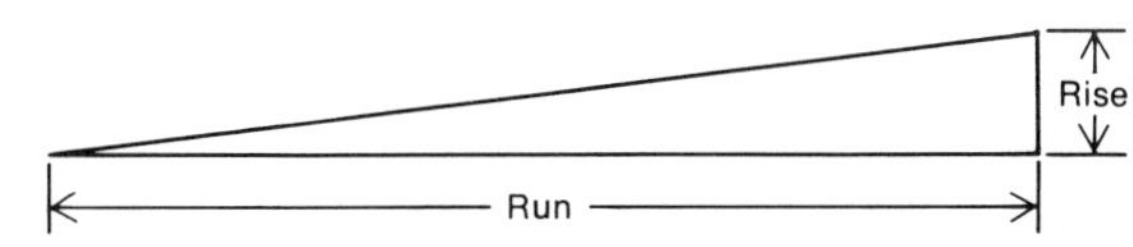

FIGURE 7.18

THE Y INTERCEPT. Although an infinite number of linear functions of the form $y = mx$ exist, a more general form of the linear equation is $y = mx + b$. The distinction between these forms is best illustrated by referring to the following additional examples:

$$
\begin{array}{ll}
y = x + 0 = x & m = 1,\ b = 0 \\
y = x + 2 & m = 1,\ b = 2 \\
y = x - 1 & m = 1,\ b = -1.
\end{array}
$$

Several sets of values for each of these functions are shown in Tables 7.5-7.7.

In Table 7.5 the values for y each differ from the corresponding values in Table 7.6 by -1. Similarly, values for y in Table 7.7 differ from corresponding

TABLE 7.5
$y = x - 1$

x	y
−2	−3
−1	−2
0	−1
1	0
2	1

TABLE 7.6
$y = x$

x	y
−2	−2
−1	−1
0	0
1	1
2	2

TABLE 7.7
$y = x + 2$

x	y
−2	0
−1	1
0	2
1	3
2	4

values in Table 7.6 by the amount +2. Since each of these functions differs from the other by a constant amount, we might guess that they represent parallel straight lines, as illustrated by Figure 7.19.

The point at which each of these functions crosses the y axis corresponds to the value of b in the general form $y = mx + b$. This is no coincidence, since whenever $x = 0$ in the general form $y = mx + b$, the equation reduces to $y = b$. In each of the tables, note that when $x = 0$, the corresponding value for y is b. From the graphical interpretation, b is called the **y intercept**. Thus we can quickly deduce the nature of any linear equation by putting it in the form $y = mx + b$. For example, the linear equation

$$2y - 4x = 2$$

can be reduced, using the methods of algebra, to its algebraic equivalent form of

$$y = 2x + 1.$$

With the equation in this form, we can see that this function has a slope of 2, which means its rise is 2 units for each unit of run; we can also see that its y

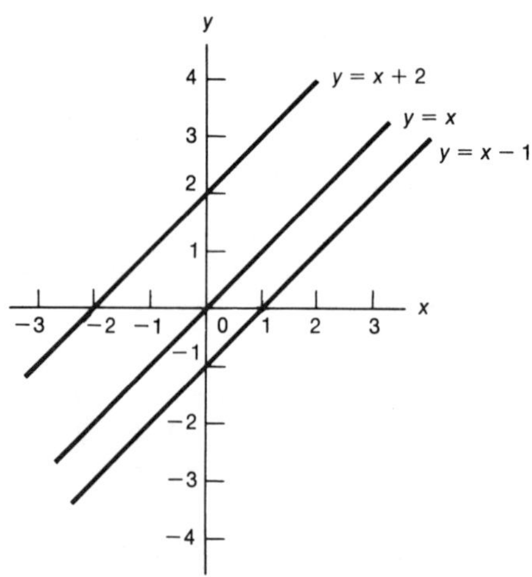

FIGURE 7.19

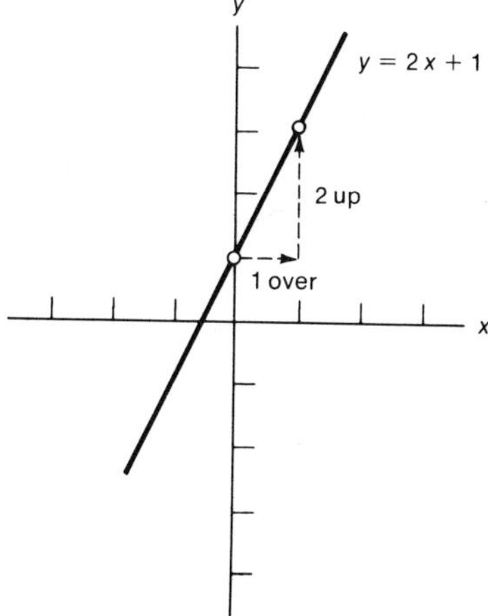

FIGURE 7.20

intercept is 1, which means that it crosses the y axis at a value $y = 1$. We can readily sketch this function without making a table, since it is determined by the slope and a point through which it passes. We merely locate the intercept point, and then move over one unit and up two, corresponding to the slope (see Figure 7.20).

NEGATIVE SLOPE. In the equation

$$y = \left(-\frac{1}{2}\right)x + 1$$

x has a coefficient of $-\frac{1}{2}$; in other words, the slope is $m = -\frac{1}{2}$. In the following table of this function, note that as values for x increase, the corresponding values for y decrease:

x	y
−2	2
0	1
2	0
4	−1
6	−2

The implication of the table, and of negative slope in general, is apparent when the function is plotted as shown in Figure 7.21.

As a general rule we can say that, for the linear function $y = mx + b$, positive slope ($m > 0$) means that the function increases for increasing values of x, and negative slope ($m < 0$) means that it decreases for increasing values of x.

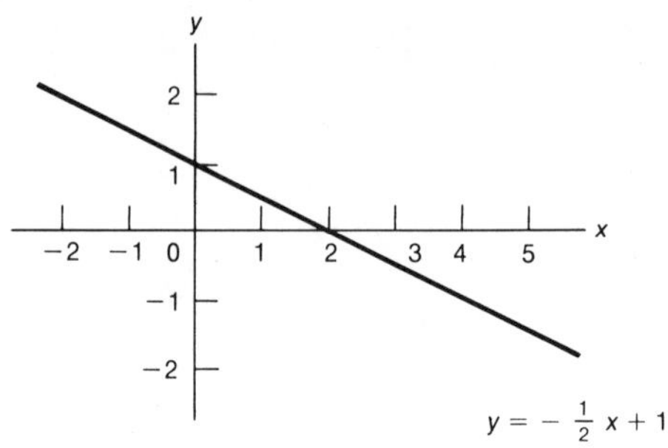

FIGURE 7.21

GENERAL FORM OF THE LINEAR FUNCTION. Although the form $y = mx + b$ is convenient for obtaining quick insight into the nature of a linear function, it is not the most general form, nor is it always convenient, as we shall discover in Chapter 9. The general linear equation in two variables is commonly studied in the form

$$Ax + By + C = 0.$$

Whenever we encounter an equation in this form, we can change it to the slope-intercept form by simple algebraic manipulations.

EXAMPLE 7.1

$$-3x + 4y - 5 = 0.$$

SOLUTION

$$4y = 3x + 5,$$

$$y = \frac{3}{4}x + \frac{5}{4}.$$

EXAMPLE 7.2

$$-2y - x = -5.$$

SOLUTION

$$2y + x = 5,$$

$$2y = -x + 5,$$

$$y = \frac{1}{2}x + \frac{5}{2}.$$

For the general case, we have

$$Ax + By + C = 0, \qquad B \neq 0,$$

$$By = -Ax - C,$$

$$y = -\frac{A}{B}x - \frac{C}{B}.$$

Whenever $B=0$, the general form cannot be changed to the slope-intercept form since division by zero is not defined. Further study of this phenomenon is left as an exercise.

EXERCISE 7.3

Express each of the following equations in the form $y=f(x)$. Determine the y intercept and the slope of each, and graph them.

1. $y=2x+1$
2. $y=3x-5$
3. $x+y=1$
4. $3x-3y=9$
5. $x=y+3x-1$
6. $2y-1+x=1$
7. $y+5=3$
8. $x+y=0$
9. $y-4=0$
10. $2y=7$
11. $y=-3$
12. $2(x-y)=x+3$
13. $y^2+x=(2-y)^2$
14. $(x+1)^2+y^2=(y-2)^2+x^2$

Convert the following linear forms to the general linear form of $Ax+By+C=0$ with A, B, and C as integers.

15. $y=-3x+2$
16. $y=2x-2$
17. $y=4x+3$
18. $y=8x-1$
19. $y=3x+2$
20. $y=x$
21. $y=2x$
22. $y=-4$
23. $y=5$
24. $y=-3$
25. $y=\frac{1}{2}$
26. $y=-\frac{1}{3}$
27. $y=\frac{1}{2}x+2$
28. $y=\frac{1}{3}x+4$
29. $y=\frac{4}{3}x+\frac{1}{2}$
30. $y=\frac{5}{2}x-\frac{2}{3}$
31. Make a graph of the equation $x=2$. Is this a linear equation? If so, what are the values of A, B, and C relative to the general form $Ax+By+C=0$?
32. Can the equation of problem 31 be represented in the slope-intercept form? What is the value of the slope? What is the value of the y intercept?
33. Make a table of values of x and y from the graph of problem 31. Is this equation a function? Justify your answer.
34. A small business leases a photocopying machine for a flat fee of \$60 a month plus \$3 per one hundred copies made. Express this as a linear equation with the total number of copies run (in hundreds) as the independent variable, and total monthly cost as the dependent variable. What is the significance of the slope? Of the y intercept?
35. A small computer leases for \$2000 per month. However, a charge of \$20 per hour is made for all operations over 100 hours per month. Express total monthly charge in terms of operating time. Is this a linear function?

36. A programmer finds that his program requires 20 minutes running time if he uses 10 variables, and 11 minutes running time if he uses 4 variables. Assuming a linear relationship between the number of variables and running time, express this as a linear function with the number of variables as the independent variable. (It may help to sketch the function.) What is the significance of the y intercept? Of the slope?

7.4 LINEAR INEQUALITIES

FURNITURE AND INEQUALITIES. In Chapter 6 we studied simple inequalities, such as $x > 5$, and related them to the real axis. The linear inequality of the general form

$$Ax + By + C > 0$$

has a similar geometric interpretation. Furthermore, it forms the basis for linear programming, which has become an extremely important tool in the use of high-speed digital computers. As an illustration of the linear inequality, let us consider the following example:

EXAMPLE 7.3

A businessman decides to manufacture two similar types of tables, table X and table Y. He has \$1200 to invest in this project; his cost for table X is \$30 and for table Y is \$20. What is the greatest possible number of tables he can manufacture without exceeding \$1200 in total cost?

SOLUTION

Making only table X, he can turn out $1200/30 = 40$ without exceeding the \$1200. Of course, he can also make 39, 38, . . . and still not exceed \$1200. Manufacturing only table Y, his output can be $1200/20 = 60$ (or fewer). Calculation will also show that he can make 15 of Y and 30 of X and not exceed the allowable limit.

This restriction can be represented with a linear inequality by allowing x to represent the number of X tables and y to represent the

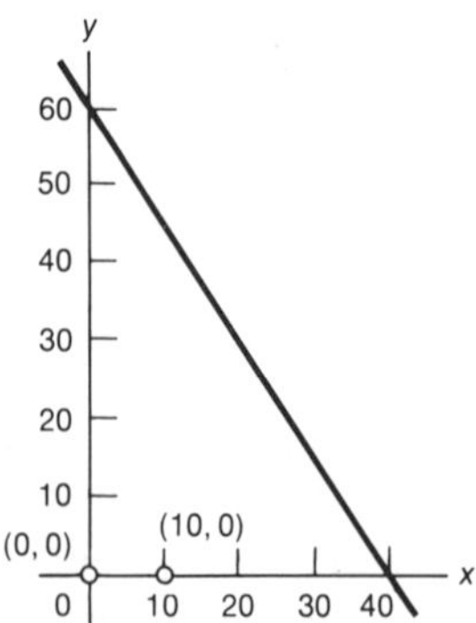

FIGURE 7.22

number of Y tables. The cost will be

$$\text{Cost} = 30x + 20y. \tag{7.1}$$

But the cost cannot exceed \$1200, so we have

$$30x + 20y \leqq 1200. \qquad (a)$$

Using the rules of Chapter 6 for operating on inequalities, this can be rewritten in the form

$$y \leqq -\frac{3}{2}x + 60. \qquad (b)$$

If, for the moment, we consider the relationship as an equality

$$y = -\frac{3}{2}x + 60, \qquad (c)$$

we have a simple linear equation with the graph shown in Figure 7.22.

To graph the inequality, it is only necessary to observe that any point below the line of Figure 7.22 satisfies Eq. (a). For example, consider the following points, both of which are below and to the left of the linear function:

$$(x = 0, y = 0) \qquad (x = 10, y = 0).$$

Substituting these values into Eq. (a) gives

$$30(0) + 20(0) \overset{?}{\leqq} 1200, \qquad 30(10) + 20(0) \overset{?}{\leqq} 1200,$$
$$0 < 1200. \qquad 300 < 1200.$$

Thus the graph of this linear inequality is represented by the entire region to the left of and below the line, including the line itself, since the symbol $\leqq$ is used. This is shown graphically in Figure 7.23.

The inequality is only meaningful for zero or positive values of x and y, so we are only interested in the shaded area of the first quadrant.

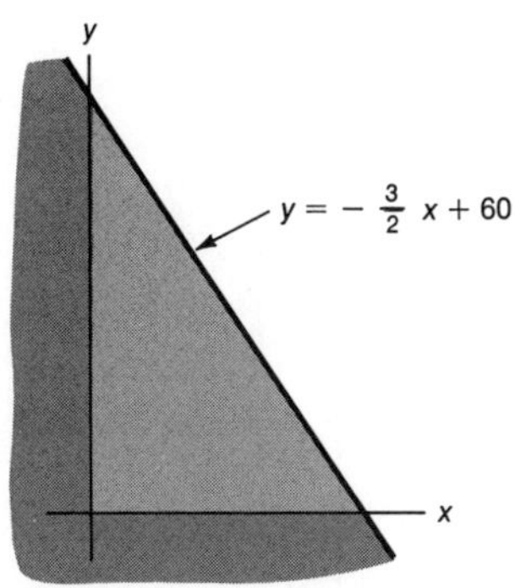

FIGURE 7.23

ADDITIONAL EXAMPLES. The following examples of linear inequalities each illustrates a different principle.

EXAMPLE 7.4

$$y > \frac{1}{2}x.$$

SOLUTION

In this example the boundary $y = \frac{1}{2}x$ of the region is not included in the inequality, since $y > \frac{1}{2}x$. Consequently, the boundary is shown as a broken line in Figure 7.24.

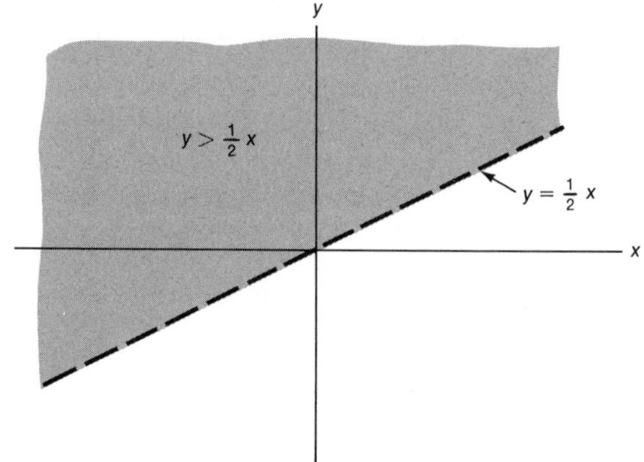

FIGURE 7.24

EXAMPLE 7.5

$$x \geqq 1.$$

SOLUTION

The boundary, which is included in the inequality, is the vertical straight line $x = 1$. In Figure 7.25, the shaded region to the right of the boundary, as well as the boundary, represents the graphical interpretation of this inequality.

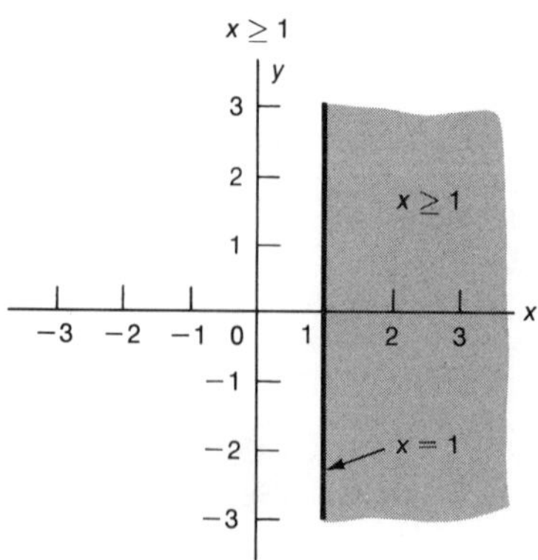

FIGURE 7.25

FUNCTIONS AND INEQUALITIES. Early in this chapter a function was defined as "a correspondence between the elements of two sets such that each element of the first set is associated with exactly one element of the second set." In other words, if we choose a value of the independent variable, we should find one and only one value of the dependent variable. That an inequality is not a function may be illustrated by considering the inequality

$$y \leqq x - 1,$$

which is graphed in Figure 7.26. If we consider some value for x, say $x = 2$, the value of y can be any number less than or equal to 1, and similarly for any other value of x. Since this violates the foregoing definition, the linear inequality is not a function.

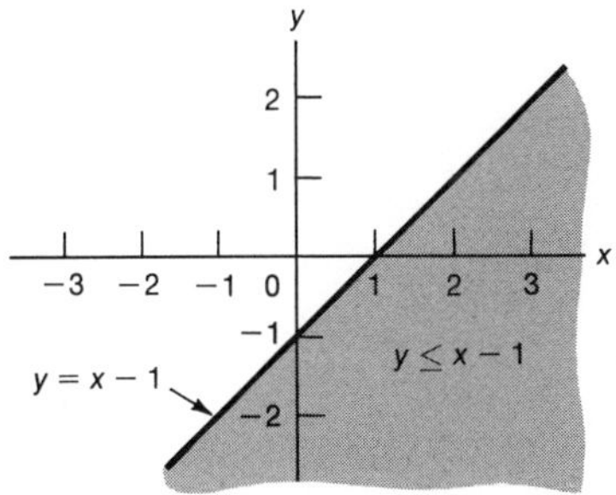

FIGURE 7.26

EXERCISE 7.4

Graph the following linear inequalities:

1. $x > 2$
2. $x \geqq 3$
3. $y \leqq -1$
4. $y - 3 > 0$
5. $y + 1 < 0$
6. $y > x$
7. $x > y$
8. $x + y > 0$
9. $x - y - 1 \leqq 0$
10. $2x + 3y > 3$
11. $4y - 2 \leq 6x$
12. $3y - 2 \leqq 4y + x$
13. Graph the inequality $y > |x|$.
14. A progressive farmer decides to use a special food supplement in the feed for his cattle. The recommended amount is 3000 units, which may be obtained using additive X (100 units per ounce) or additive Y (150 units per ounce). Express the relationship between the amounts of X and Y and the recommended dosage as a linear inequality. Solve this inequality in terms of the amount of Y and graph it.
15. A manufacturer has 1000 metal blanks from which he makes two items, brackets and levers. Each bracket requires 4 blanks and each lever requires 5 blanks. Represent the combinations of items (brackets and levers) which he may manufacture as a linear inequality. Solve for the number of brackets and graph as a linear inequality.

REFERENCES

Barnett, R. A., *Intermediate Algebra: Structure and Use.* New York, McGraw-Hill, Inc., 1971.

DeAngelo, S. and Jorgenson, P., *Mathematics for Data Processing.* New York, McGraw-Hill, Inc., 1970.

Dubisch, R., Howes, V. E. and Bryant, S. J., *Intermediate Algebra,* 2d ed. New York, John Wiley & Sons, 1969.

Richardson, M., *Fundamentals of Mathematics,* 3d ed. New York, The MacMillan Company, 1966.

Wooten, W. and Drooyan, I., *Intermediate Algebra;* 2d alt. ed. Belmont, Calif., Wadsworth Publishing Company, Inc., 1968.

8 NONLINEAR FUNCTIONS

8.1 THE QUADRATIC FUNCTION / **262**
Profit and Output Relationships / **262**
Characteristics of the Function / **263**
Cost Relationships / **264**
Equal Roots / **265**
The General Form of the Quadratic Function / **266**
Higher Degree Polynomials / **266**
Exercise 8.1 / **266**

8.2 TRIGONOMETRIC FUNCTIONS / **267**
Angles and the Coordinate System / **267**
Definition of Trigonometric Functions / **269**
Characteristics of Trigonometric Functions / **270**
Radian Measurement / **272**
Exercise 8.2 / **272**

8.3 THE EXPONENTIAL AND LOGARITHMIC FUNCTIONS / **273**
The Exponential Function / **273**
The Logarithmic Function / **274**
Properties of Logarithms / **275**
Exercise 8.3 / **277**

8.4 COMMON LOGARITHMS / **278**
Arithmetic Operations Using Logarithms / **279**
Characteristic and Mantissa / **280**
Exercise 8.4 / **281**

8.5 LINEAR INTERPOLATION / **282**
Graphical Interpretation / **283**
The Nature of Linear Approximation / **284**
A Convenient Format / **285**
Exercise 8.5 / **286**

8.6 LOGARITHM TABLES / **287**
Using the Table / **287**
Negative Characteristics / **288**
Applications of Logarithms / **289**
Logarithms and Computers / **290**
Exercise 8.6 / **290**

Emphasis in the latter portions of Chapter 7 was placed on the geometric (graphic) interpretation of linear functions. However, the adage that a picture is worth a thousand words is not limited to linear functions. In fact, the study of nonlinear functions, those whose graphs are not straight lines, is considerably enhanced by reference to their graphs. The first nonlinear function we shall consider is the quadratic function, which is a familiar form from Chapter 6.

8.1 THE QUADRATIC FUNCTION

PROFIT AND OUTPUT RELATIONSHIPS. In economics it is frequently possible to determine functional relationships between supply, demand, unit price, profit, and so on, for a given item to be sold. Often an attempt is made to linearize the relationships over small intervals in order to obtain a solution. However, in many cases nonlinear functions result. For example, let us assume operation of a small company which processes maple syrup and, being in business, we are anxious to maximize our profit. Although such a determination depends upon many factors and mathematically would be very complex, we may desire to determine the relationship between unit profit and quantity manufactured. Intuitively we might expect that a minimum output would exist, below which it would be impossible to make a profit because of high unit operating costs. If our plant had a rated capacity, it would also be likely that a maximum output would exist beyond which it would be impossible to make a profit because of overloaded facilities, overtime for the work force, and so on. Suppose that the functional relationship between unit profit P and the weekly output unit w is

$$P = -100w^2 + 800w - 700, \tag{8.1}$$

where w is measured in thousands of pounds (that is, one unit is 1000 lb) and P is dollars per unit. Thus, if 3000 lb (3 units) are produced, the profit would be

$$\begin{aligned} P &= -100w^2 + 800w - 700 \\ &= -100(9) + 800(3) - 700 \\ &= \$800. \end{aligned}$$

Table 8.1 is a typical set of values for P and w obtained from Eq. 8.1. This table may be plotted in order to obtain a better insight into the general nature of the function. The graph of Figure 8.1 clearly shows that to be operating at a profit, the weekly output of the plant must be greater than one unit (1000 pounds) but less than 7 units (7000 pounds).

Further inspection shows that this particular function reaches a maximum value at (or about) the point $w = 4$ [this point (4, 9) is commonly referred to as the **vertex**]. Thus, if we desired to maximize unit profit we should operate with an output of 4000 pounds per week. However, in practice there would be many other factors to be considered and total net profit would probably occur at some other point.

TABLE 8.1

w	P
0	−700
1	0
2	500
3	800
4	900
5	800
6	500
7	0
8	−700

CHARACTERISTICS OF THE FUNCTION. Pursuing the nature of this function further, we note that the values of P preceding $w = 4$ exactly correspond to the values of P following $w = 4$. In other words, the function is symmetric about the line $w = 4$, which appears to be consistent with the graph of Figure 8.1. That this relationship is indeed a function is evident by examining the equation and noting that substitution of a value for w will always result in a single, unambiguous value for P. Intuitively we can justify connecting the points by a smooth curve (this implies that the function is continuous) by expanding Table 8.1 to include additional intermediate points.

The right-hand side of Eq. 8.1 has the form of the quadratic expression studied in Chapter 6; thus, it is called the **quadratic function**. Its graph shown in Figure 8.1 is called a **parabola**. Substitution of larger and larger values for w causes P to continually increase in the negative sense; similarly, smaller values of w also cause P to increase in the negative sense. In other words, the slopes of both legs of the parabola become increasingly steeper.

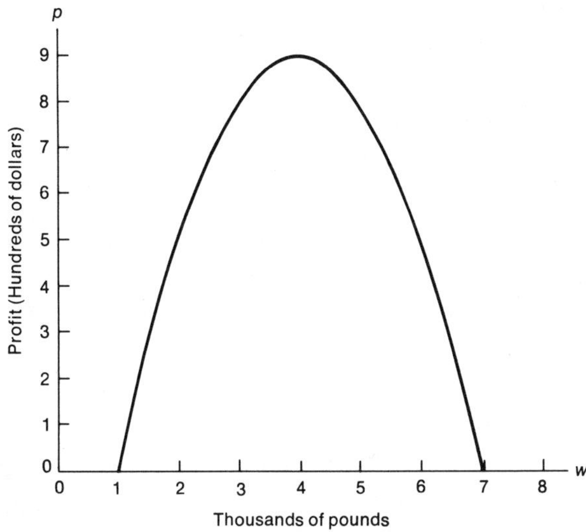

FIGURE 8.1

Of particular interest in this example are values of w, which result in the function being zero ($P = 0$). These are commonly called the **zeros** of the function, and are found by setting the function equal to zero and solving the resulting quadratic equation, using the methods of Chapter 6. For example,

$$-100w^2 + 800w - 700 = 0$$
$$w^2 - 8w + 7 = 0$$
$$(w - 1)(w - 7) = 0$$
$$w = 1, 7.$$

That these are the zero profit points can be verified by referring to Table 8.1.

COST RELATIONSHIPS. Another relationship important in economics is that between unit cost and output. Here, we are interested in lowering unit cost to a minimum. Assume that we are producing the well-known Brand *X* and have established that the following quadratic function relates unit cost u (in dollars) to output quantity q (in hundreds of units):

$$u = q^2 - 6q + 11.$$

Figure 8.2 includes the graph of this function and the table from which the graph was prepared. The graph of this function is very similar to the graph of Figure 8.1, except that it is turned over. The vertex in this case is a minimum value corresponding to the point of minimum unit cost. Both legs continue upward with slopes of increasingly greater magnitude. In contrast to the previous example, there are no zero cost points; that is, the curve never crosses the x-axis; mathematically speaking, this implies that the function has no zeros. Let us check this by setting the function equal to zero and solving for roots, using the quadratic formula of Chapter 6:

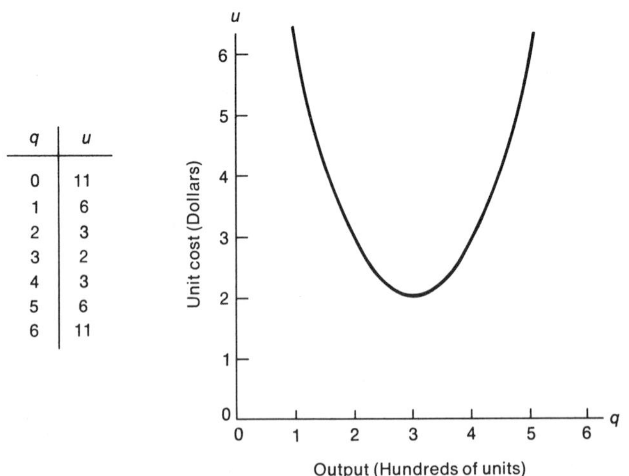

q	u
0	11
1	6
2	3
3	2
4	3
5	6
6	11

FIGURE 8.2

$$q^2 - 6q + 11 = 0$$

$$q = \frac{-b \pm \sqrt{b^2 - 4ac}}{2a}$$

$$= \frac{-(-6) \pm \sqrt{(-6)^2 - 4(1)(11)}}{2(1)}$$

$$= \frac{6 \pm \sqrt{-8}}{2}$$

$$= 3 \pm \sqrt{-2}.$$

Thus, in an attempt to determine the roots of this equation, we obtain complex numbers that cannot be interpreted as points on the q-axis. This is consistent with the graphic results of Figure 8.2.

EQUAL ROOTS. A special case which we have not yet encountered occurs when the roots are identical. For example, consider the equation

$$y = x^2 - 2x + 1,$$

which has the double root 1, as calculated below:

$$x^2 - 2x + 1 = 0$$
$$(x - 1)(x - 1) = 0$$
$$x = 1.$$

The graph of this function is shown in Figure 8.3. In other words, the graphical interpretation of a quadratic function with equal roots is a parabola, which is tangent to the x-axis at its vertex.

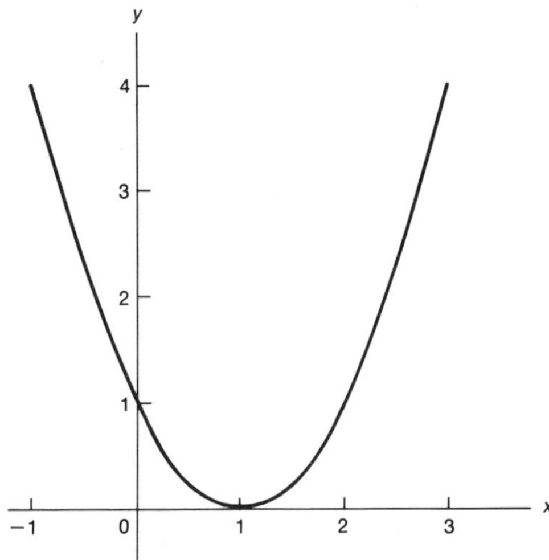

FIGURE 8.3

THE GENERAL FORM OF THE QUADRATIC FUNCTION. All of the foregoing examples display the form

$$f(x) = ax^2 + bx + c \qquad (a \neq 0),$$

which is the general form of the quadratic function. Note that whenever the coefficient a is positive (for instance, see the second example), the function exhibits a local minimum and the legs extend upward. If the coefficient a is negative, the function has a local maximum and the legs extend downward. If the quadratic equation obtained by setting the function equal to zero has real roots, the function crosses the x-axis. If it has complex roots, the function does not cross the x-axis. If the quadratic has identical roots, the function is tangent to the x-axis.

HIGHER DEGREE POLYNOMIALS. The graphical interpretation of second-degree polynomials can be extended to polynomials of any degree. A third-degree polynomial has three zeros, a fourth-degree polynomial four zeros, and so on (not all of the zeros need be real, however). Similarly, a third-degree polynomial has two vertices, a fourth-degree polynomial three vertices, and so on. Figures 8.4 and 8.5 illustrate the general shape of the third- and fourth-degree polynomials.

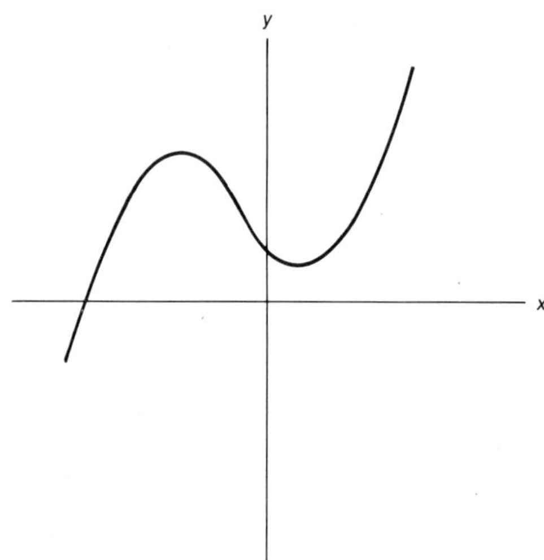

FIGURE 8.4 Third degree polynomial

EXERCISE 8.1

In each of the following exercises perform the following: (a) Compile a table for the given quadratic function, using values of the independent variable to illustrate the maximum or minimum point. (b) Graph the function from the table.

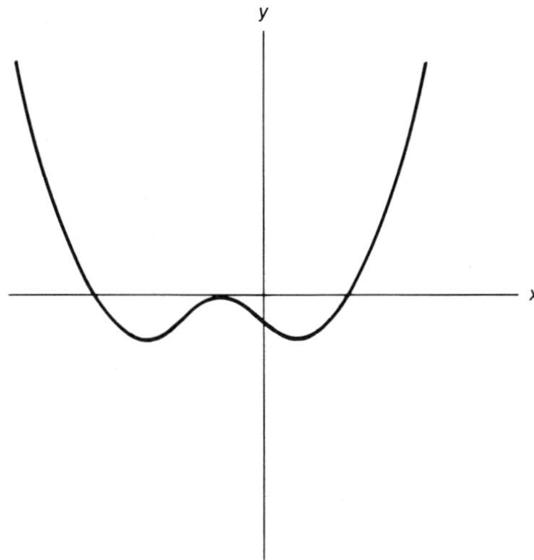

FIGURE 8.5 Fourth degree polynomial

(c) Estimate the coordinates of the minimum (or maximum) point of the function and the points at which the function crosses the x axis.

1. $y = x^2 + 3x + 2$
2. $y = x^2 - 4x + 4$
3. $f(x) = x^2$
4. $f(x) = x^2 + x + 1$
5. $f(x) = -x^2 - x + 1$
6. $y = -x^2 + x$
7. $f(z) = 2z^2 + 3z + 1$
8. $f(q) = 4q^2 - 4q + 1$
9. $y = -x^2 + 1$
10. $f(x) = -x^2 - 5x - 4$
11. $y = -x^2 + x - 1$
12. $y = x^2 - 4$
13. $f(x) = x^2 - x - 20$
14. $f(x) = x^2 - 7x - 8$
15. $y = x^3 - 5x^2 + 2x + 8$
16. $y = x^3 + 5x^2 - 22x - 56$
17. $y = x^3 + 4x^2 - 4x - 16$
18. $f(x) = -x^3 - 3x^2 + 9x + 27$

8.2 TRIGONOMETRIC FUNCTIONS

ANGLES AND THE COORDINATE SYSTEM. In Chapter 7 we studied the rectangular cartesian coordinate system and recognized that any point can be located on the plane by its x and y coordinates, as shown in Figure 8.6(a). Another means for specifying the location of this point is by the distance r, which is measured from the origin to the point, and the angle θ which is measured from the positive x-axis to the line segment r. The line segment r is commonly referred to as the **radius vector** and is always considered to be positive regardless of which of the four quadrants the point falls. Although measurement of the angle could be from either the positive or the negative portion of either the X- or Y-axis, standard convention is to use the positive X-axis as the initial side and measure in a counterclockwise direction. Although the point in Figure 8.6(b) lies in the first quadrant, this method of specifying a

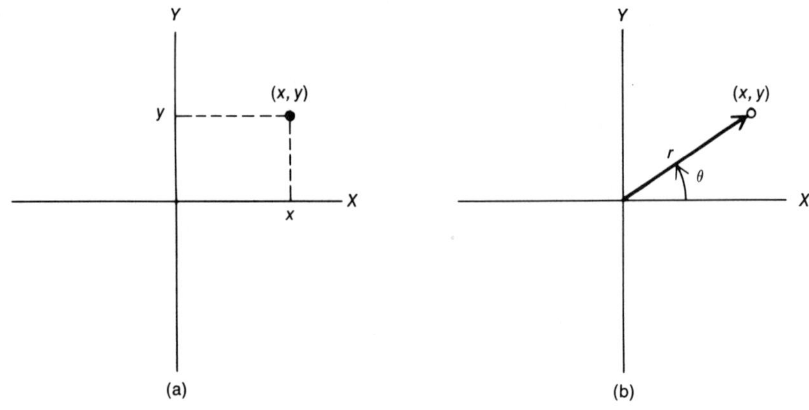

FIGURE 8.6

point is not so limited as is evident in Figure 8.7. Each of the four points P_1, P_2, P_3, and P_4 can be located by their respective radius vectors and angles from the positive X-axis (angles are commonly named with lower case Greek letters as shown in 8.7). The reader will note that both the angle β and the angle α, which makes an additional rotation, can be used to describe the location of the radius vector r_1. The common unit measure of angles is in *degrees* where a complete rotation represents 360°. Thus the cartesian coordinate system is divided into four quadrants each consisting of 90°. We can see by referring to Figure 8.7 that the general range of values for the angles is as follows:

β between 0° and 90°—approximately 25°
γ between 90° and 180°—approximately 120°

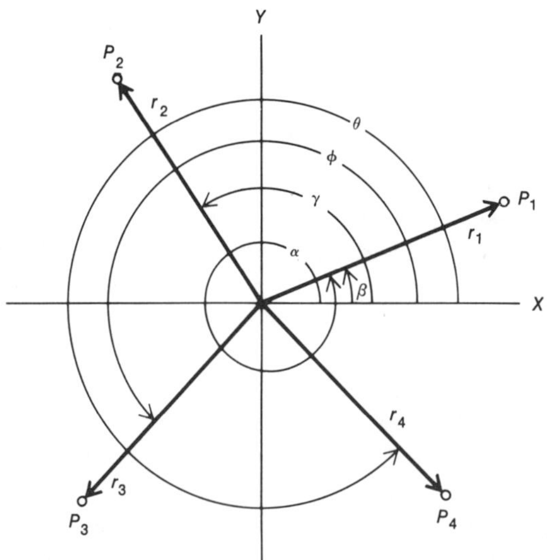

FIGURE 8.7

ϕ between 180° and 270°—approximately 225°
θ between 270° and 360°—approximately 315°
α between 360° and 450°—approximately 385°

This method of graphing using a radius vector and an angle is commonly used for certain applications and forms the basis for the **polar** coordinate system. However, our attention here will be directed to the fact that these principles form the basis for the subject of ***trigonometry.***

DEFINITION OF TRIGONOMETRIC FUNCTIONS. With these notions of angles and their relationship to the coordinate system, we can now consider three trigonometric functions basic to the subject of trigonometry. In Figure 8.8, the radius vector to the point P (at an angle θ) is shown. Additionally, a vertical line has been dropped to X-axis from the point P, which has the co-ordinates x and y. This forms the right triangle OPN, which includes the angle θ. By constructing the triangle in this manner, we see that $ON = x$, $NP = y$ and $OP = r$. These three lengths are used to define the three basic trigonometric functions **sine, cosine,** and **tangent** (normally abbreviated **sin, cos,** and **tan**) as follows:

$$\sin \theta = \frac{y}{r}$$

$$\cos \theta = \frac{x}{r}$$

$$\tan \theta = \frac{y}{x}.$$

It is important to recognize that these ratios are dependent *only* on the angle and are independent of the length r chosen for the measurement (that is, the distance of point P from the origin). The choice of some other point, say P_1 shown in Figure 8.9, will yield the same result since, by geometry, the indicated ratios are the same. Although other functions are commonly defined in trigonometry, they can all be related to these three.

Although the illustrations of Figure 8.8 and 8.9 are limited to angles in

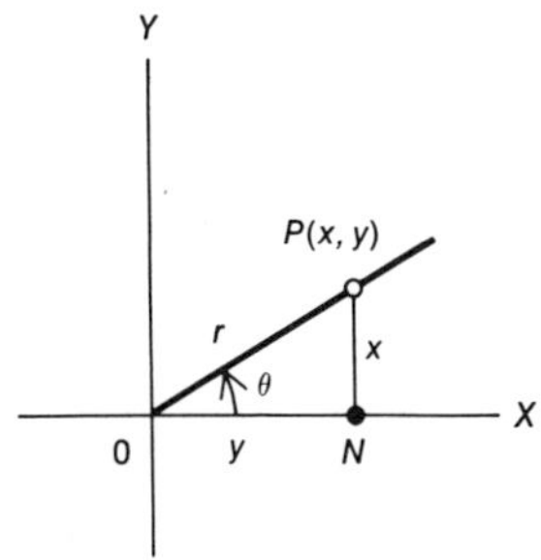

FIGURE 8.8

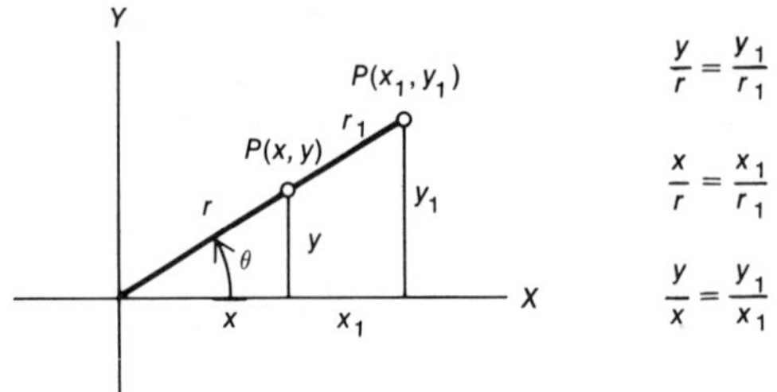

FIGURE 8.9

the first quadrant, trigonometry is not so restricted, as illustrated in Figure 8.10. However, it is important that we recognize (1) the values for x in 8.10(a) and 8.10(b) and for y in 8.10(b) and 8.10(c) will be negative, and (2) the radius vector is always considered positive. Thus, for example, cos θ is positive in the first and fourth quadrants and negative in the second and third.

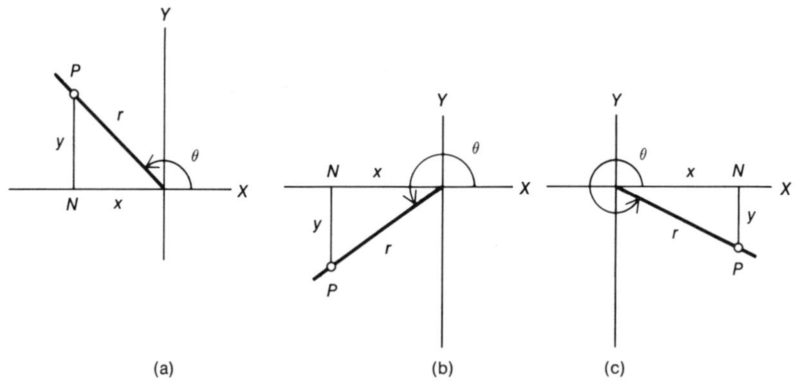

FIGURE 8.10

CHARACTERISTICS OF TRIGONOMETRIC FUNCTIONS. By careful inspection of these illustrations, we can see that the sine of 0°, 30° and 90° is 0.0, 0.5 and 1.0, respectively; that is sin 0° = 0.0, sin 30 ° = 0.5, sin 90° = 1.0. Although these and a few other values are obvious from geometry, trigonometric tables are commonly used for hand computations and special numerical techniques are used in conjunction with the Fortran functions SIN and COS. Table 8.2, which includes values at 10° increments, is an example of a trigonometric function table. We can see, for example, that sin 40° = .6428 and tan 60° = 1.7321; however, note that tan 90° is not defined since, in the ratio y/x, the value of x becomes zero. This table is easily extended beyond 90° simply by recognizing that the values themselves carry over but appropriate adjustments must be made to the sign. For example, the values of sine for the angles 0° through 90° correspond exactly to those for angles 180° through 270° except for opposite sign.

TABLE 8.2 VALUES OF TRIGONOMETRIC FUNCTIONS

Angle	Sine	Cosine	Tangent
0	0.0000	1.0000	0.0000
10	.1737	0.9848	.1763
20	.3420	.9397	.3640
30	.5000	.8660	.5774
40	.6428	.7660	.8391
50	.7660	.6428	1.1918
60	.8660	.5000	1.7321
70	.9397	.3420	2.7475
80	.9848	.1737	5.6713
90	1.0000	.0000	———

Some important characteristics of the trigonometric functions become apparent by graphing the functions from the forms

$$s = \sin\theta$$

and

$$c = \cos\theta.$$

Values from Table 8.2 have been expanded to plot the graphs of Figure 8.11. A very important feature of the trigonometric functions is that they are **periodic** in nature; roughly speaking, they repeat themselves at 360° intervals.

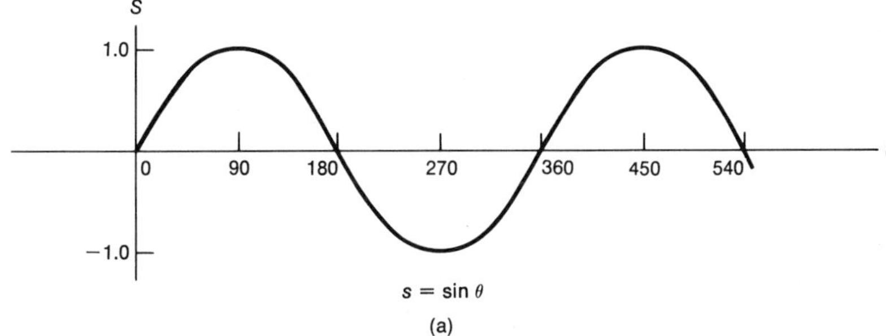

$s = \sin\theta$

(a)

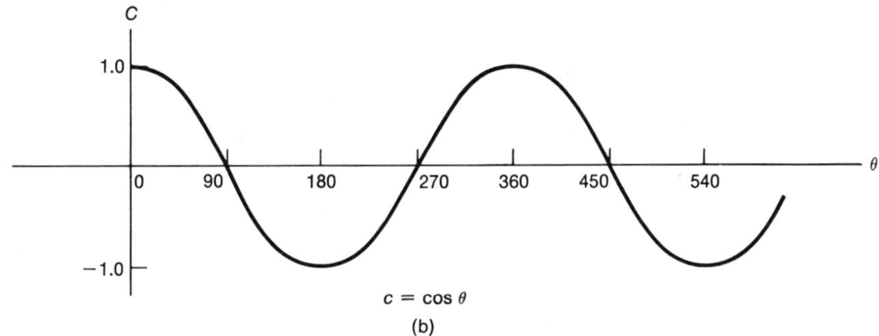

$c = \cos\theta$

(b)

FIGURE 8.11 (a) $s = \sin\theta$; (b) $c = \cos\theta$

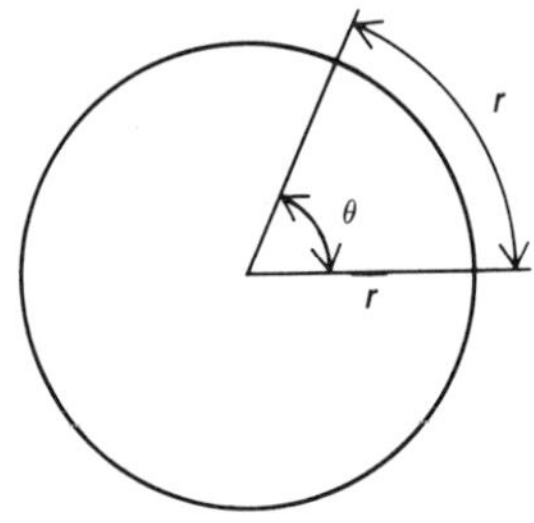

FIGURE 8.12

RADIAN MEASUREMENT. The common unit of measurement for angles is the degree, where a complete rotation includes 360°. A convenient unit for use in mathematics is based on the length of the radius of a circle as shown in Figure 8.12. The angle θ is determined by an arc on the circle whose length is exactly equal to the radius of the circle. Since the circumference of a circle (of radius r) is $2\pi r$, it follows that there are 2π radians in 360° or that one radian is $360/2\pi$ or approximately 57°. In using the Fortran functions SIN and COS, the arguments must be in radians. Thus in program calculations using angles measured in degrees, the programmer must take care to make appropriate adjustments when using trigonometric functions. For example, $y = 5\sin\theta$ (θ degrees) would appear in Fortran as

```
Y = 5.0*SIN(THETA/57.296)
```

EXERCISE 8.2

Sketch the following angles on a rectangular cartesian coordinate system and label them as in Figure 8.7; state in which quadrant the radius vector lies.

1. 60°, 120°, 180°, 210°, 315°, 420°
2. 45°, 150°, 210°, 270°, 300°, 405°

Each of the following points $P(x,y)$ terminates a radius vector. Find the three functions sine, cosine, and tangent for each. (Remember the theorem of Pythagoras concerning right triangles—referring to Figure 8.8, $r^2 = x^2 + y^2$.)

3. (4,3)
4. (3,4)
5. (−4,3)
6. (−4,−3)
7. (2,−2)
8. (−2,2)
9. (0,4)
10. (−4,0)
11. $(\sqrt{3}, 1)$
12. $(-1, \sqrt{3})$
13. $(1, -\sqrt{15})$
14. $(-\sqrt{15}, -1)$

Determine in which quadrants the radius vector can be positioned and find the other possible function values for each angle θ as follows.

15. $\tan\theta = 1.0$
16. $\tan\theta = -1.0$
17. $\sin\theta = \frac{3}{5}$
18. $\cos\theta = -\frac{4}{5}$
19. $\sin\theta = -\frac{\sqrt{3}}{3}$
20. $\cos\theta = \frac{\sqrt{3}}{3}$
21. $\sin\theta = \frac{5}{13}$
22. $\tan\theta = 2.0$

23. Expand Table 8.2 in 30° increments from 0°–450°.

Using Table 8.2, find the indicated functions; if the table does not include the exact point, then make an estimate using the table.

24. sin 20°, cos 70°, tan 50°, sin 140°, cos 130°, sin 45°, cos 25°

25. sin 50°, cos 40°, tan 160°, sin 190°, sin 350°, cos 75°, sin 105°

Using Table 8.2 in reverse, find the appropriate values for θ. If the table does not include the exact point, then make an estimate using the table.

26. $\tan\theta = 1.1918$; $\sin\theta = .8660$; $\cos\theta = .9397$; $\sin\theta = .4226$; $\tan\theta = .7002$

27. $\cos\theta = .1737$; $\tan\theta = .3640$; $\sin\theta = .6428$; $\cos\theta = .2588$; $\tan\theta = 1.428$

28. Using the basic representation of Figure 8.8, show that

$$(\sin\theta)^2 + (\cos\theta)^2 = 1.$$

Note that $(\sin\theta)^2$ and $(\cos\theta)^2$ are commonly written as $\sin^2\theta$ and $\cos^2\theta$, respectively.

29. Show that $\tan\theta = \dfrac{\sin\theta}{\cos\theta}$.

30. What is the relationship between $\cos\theta$ and $\sin(\theta + 90°)$?

31. The law of cosines states that for any triangle with sides *a, b,* and *c* and an angle *C* formed by sides a and b,

$$c = \sqrt{a^2 + b^2 + 2ab\cos C}.$$

Assuming that values for each of the two sides *a* and *b* and the angle *C* (in degrees) are stored in the computer as A, B, and AC, respectively, write a Fortran expression to compute *c*.

32. Within the computer, trigonometric functions are calculated by use of special numerical methods which yield approximate values. A classical mathematical representation for the sine is

$$\sin x = x - \frac{x^3}{3!} + \frac{x^5}{5!} - \frac{x^7}{7!} + \cdots \qquad x \text{ in radians.}$$

Note: $3! = 3 \cdot 2 \cdot 1$; $5! = 5 \cdot 4 \cdot 3 \cdot 2 \cdot 1$, and so on.

Write a Fortran program to calculate and print sin *x* for a range of angles from 0° to 90° in increments of 10°. Perform the calculations using the first two terms in the equation and also using the first four terms as shown. Print results for both calculations and compare them to Table 8.2.

8.3 THE EXPONENTIAL AND LOGARITHMIC FUNCTIONS

THE EXPONENTIAL FUNCTION. In earlier chapters we developed the principles of exponents and defined the expression x^a for *a* as any rational number. Although we did not carry the development beyond rational numbers, all of the rules for exponents do apply for any real number exponent. This means, then, that we can speak of the **exponential function**

$$y = b^x,$$

where *b* is the base and *x* may be any real number. Let us explore the domain and corresponding range of this function (that is, values which *x* and *y* may

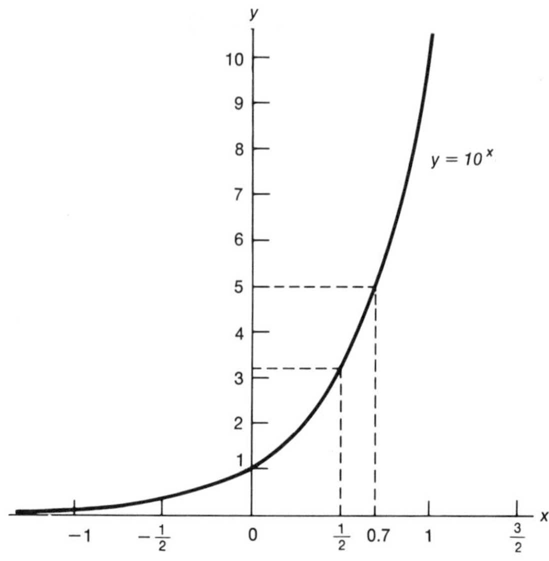

FIGURE 8.13

assume) for $b > 1$. If $x = 0$, $y = 1$; for values of $x > 0$, we see that $y > 1$. On the other hand, if $x < 0$, $y < 1$, but no matter how small the exponent becomes, y can never be zero (or negative). In other words, the domain of this function consists of all the real numbers, but the corresponding range is limited to the positive real numbers. The graph of the function $y = 10^x$ (which is continuous) is shown in Figure 8.13.

In using the graph of Figure 8.13 to obtain a value y for any given value of x, we locate a point on the x-axis, project a line upward to an intersection with the function, and then project a line across to the y-axis. For example, if $x = 1/2$, $y \simeq 3.2$, as shown in Figure 8.13. On the other hand, if we chose a value for y, say 5, we could project across to the function and then down to the x-axis, and obtain $x \simeq 0.7$. In other words, the inverse of the exponential function is also a function, since a given value for y results in one and only one value for x.

THE LOGARITHMIC FUNCTION. In speaking of the inverse function of

$$y = b^x,$$

it would be convenient to express x in terms of y. To manipulate this equation to obtain the desired form is impossible using the algebraic tools studied in earlier chapters. Thus we define the inverse of the exponential function as the **logarithmic function** and write it as

$$x = \log_b y.$$

We say that "x equals the log to base b of y"; in other words, x is the exponent to which the base b must be raised in order to obtain y. For example, the exponents to which 10 must be raised to obtain 100 and 3.162 are 2 and 1/2, respectively. In equation form we have

$$2 = \log_{10} 100 \qquad \text{since} \qquad 100 = 10^2$$

and

$$\frac{1}{2} = \log_{10} 3.162 \qquad \text{since} \qquad 3.162 = 10^{1/2}$$

Similarly,

$$3 = \log_2 8 \qquad \text{since} \qquad 8 = 2^3$$

$$-2 = \log_5 \frac{1}{25} \qquad \text{since} \qquad \frac{1}{25} = 5^{-2}$$

$$3 = \log_{1/2} \frac{1}{8} \qquad \text{since} \qquad \frac{1}{8} = \left(\frac{1}{2}\right)^3$$

In view of the relationship between the exponential function and the logarithmic function, the logarithmic function is defined in order to produce a representation for the inverse of the exponential function. When we speak of the logarithm, we are using the term almost as a synonym for exponent; the logarithm **is** the desired exponent.

PROPERTIES OF LOGARITHMS. The real value of logarithms lies in the fact that they short cut time-consuming multiplication or division and long tedious exponentiations. Even in computers which perform arithmetic calculations at extremely high speeds, it is sometimes desirable to use logarithms rather than perform the exponentiations by repeated multiplications. In fact, subroutines for performing calculations with logarithms are an important part of such programming languages as Fortran and Algol.

The fact that such operations are possible is derived directly from the basic rules of exponents and the fact that logarithms are actually exponents. In Chapter 5, we studied the nature of exponents and found that

1. $$b^x b^y = b^{x+y}.$$

2. $$\frac{b^x}{b^y} = b^{x-y}.$$

3. $$(b^x)^y = b^{xy}.$$

From these we can justify the following properties of logarithms:

Property 1. $\log_b xy = \log_b x + \log_b y.$

To show this, let

$$\log_b x = u \qquad \text{and} \qquad \log_b y = v.$$

But

$$\log_b x = u \qquad \text{implies that} \qquad x = b^u$$

and

$$\log_b y = v \qquad \text{implies that} \qquad y = b^v.$$

By the laws of exponents we have

$$xy = b^u \times b^v = b^{u+v}.$$

Applying the definition of the logarithmic function gives

$$\log_b xy = u + v.$$

But

$$u = \log_b x \qquad \text{and} \qquad v = \log_b y.$$

Therefore,

$$\log_b xy = \log_b x + \log_b y.$$

For example, using the familiar base 10, we have

$$\log_{10} (100 \times 1000) = \log_{10} 100 + \log_{10} 1000.$$

That this relationship is valid can easily be shown, since

$$\log_{10} (100 \times 1000) = \log_{10} (100000) = 5,$$

and

$$\log_{10} 100 + \log_{10} 1000 = 2 + 3 = 5.$$

Property 2. $\log_b \dfrac{x}{y} = \log_b x - \log_b y.$

Once again, using base 10, this property is illustrated by the following example:

$$\log_{10} \frac{1000000}{100} = \log_{10} 1000000 - \log_{10} 100 = 4.$$

It can readily be seen that the left side of the equation is

$$\log_{10} \frac{1000000}{100} = \log_{10} 10000 = 4,$$

whereas the right side is

$$\log_{10} 1000000 - \log_{10} 100 = 6 - 2 = 4.$$

Property 3. $\log_b x^a = a \log_b x.$

Using base 2 as an illustration, we have

$$\log_2 4^3 = 3 \log_2 4.$$

Here the left side of the equation is

$$\log_2 4^3 = \log_2 64 = 6.$$

On the right side of the equation we have

$$3 \log_2 4 = 3(2) = 6.$$

EXERCISE 8.3

Compile a table; graph each of the following exponential functions:

1. $y = 2^x$
2. $y = 3^x$
3. $y = 1^x$
4. $y = \left(\frac{1}{2}\right)^x$
5. $y = \left(\frac{1}{4}\right)^x$

6. Referring to the table compiled in Problem 1, graph the logarithmic function $y = \log_2 x$. Note that this is the inverse function of $y = 2^x$.

Express the following in logarithmic form:

7. $2^3 = 8$
8. $2^2 = 4$
9. $2^0 = 1$
10. $10^3 = 1000$
11. $10^{-2} = 0.01$
12. $2^{-4} = \frac{1}{16}$
13. $16^{3/4} = 8$
14. $\sqrt{9} = 3$
15. $\left(\frac{1}{4}\right)^{-2} = 16$
16. $\left(\frac{1}{2}\right)^4 = \frac{1}{16}$
17. $a^3 = b$
18. $e^x = z$

Express each of the following in exponential form:

19. $\log_2 16 = 4$
20. $\log_2 1 = 0$
21. $\log_8 \frac{1}{8} = -1$
22. $\log_{10} 1000 = 3$
23. $\log_{10} 10 = 1$
24. $\log_{10} 0.01 = -2$
25. $\log_5 \frac{1}{125} = -3$
26. $\log_{1/2} 8 = -3$
27. $\log_{1/8} 8 = -1$
28. $\log_a b = 5$
29. $\log_e 5 = x$
30. $\log_1 1 = 5$

In each of the following determine the value of the unknown:

31. $\log_{10} 100 = y$
32. $\log_{10} 10 = y$
33. $\log_2 8 = y$
34. $\log_2 1 = y$
35. $\log_5 0 = y$
36. $\log_{10} 0.001 = y$
37. $\log_{10} x = 4$
38. $\log_{10} x = -2$
39. $\log_{10} x = 0$
40. $\log_2 x = 0$
41. $\log_8 x = -3$
42. $\log_5 x = 1$
43. $\log_b 8 = 3$
44. $\log_b 1000 = 3$
45. $\log_b \frac{1}{64} = -2$

46. Show that $\log_b \frac{x}{y} = \log_b x - \log_b y$.

47. Show that $\log_b x^a = a \log_b x$.

8.4 COMMON LOGARITHMS

In defining the logarithmic function, no restriction is placed on the base b other than that it be positive; for example, Exercise 8.3 used several bases. However, for purposes of computation, the base of our number system (10) and the irrational number e (approximately 2.718) are the most commonly used. Normally, whenever a logarithm is designated with no base indicated, the base is assumed to be 10. That is,

$$\log 1000 = \log_{10} 1000 = 3$$
$$\log 0.01 = \log_{10} 0.01 = -2$$

Logarithms to the base 10 are frequently referred to as **common** logarithms (in contrast to the **natural** logarithms which have the base e). Table 8.3 consists of common logarithms for integral values of the argument. If these points were plotted and the portion of the function defined by the limits $1 < x < 10$ were shown on a fine scale, the result would appear as shown in Figure 8.14.

TABLE 8.3 COMMON LOGARITHMS

x	$\log x$
$0.001 = 10^{-3}$	-3
$0.01 = 10^{-2}$	-2
$0.1 = 10^{-1}$	-1
$1 = 10^{0}$	0
$10 = 10^{1}$	1
$100 = 10^{2}$	2
$1000 = 10^{3}$	3

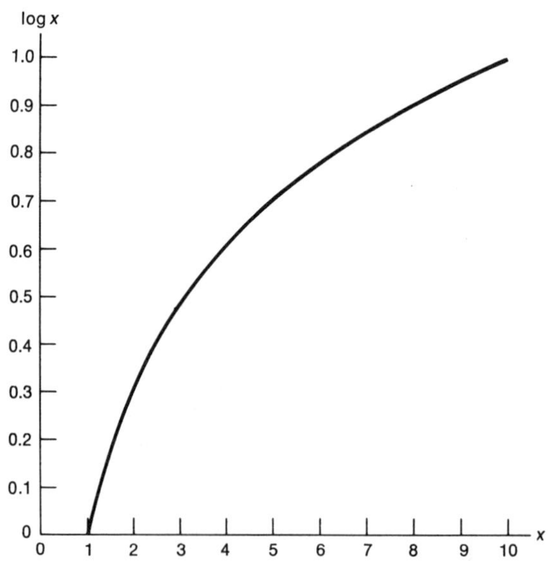

FIGURE 8.14

From this graph we could estimate the more detailed set of values for the function as shown in Table 8.4. In other words, $10^0 = 1$, $10^{0.3} \simeq 2$, $10^{0.48} \simeq 3$, and so on. Note that *these are approximate values which have been estimated from a graph and should not be used as exact.*

TABLE 8.4

x	log x (approximate values)
1	0.00
2	0.30
3	0.48
4	0.60
5	0.70
6	0.78
7	0.85
8	0.90
9	0.95
10	1.00

ARITHMETIC OPERATIONS USING LOGARITHMS. Using the abbreviated table of logarithms (Table 8.4), we can perform simple arithmetic operations. To illustrate the process of multiplication using Table 8.4, we shall employ the simple factors of 2 and 3 to demonstrate the principles involved:

$$\begin{aligned} 2 \times 3 &= 10^{0.30} \times 10^{0.48} \\ &= 10^{0.30+0.48} \\ &= 10^{0.78}. \end{aligned}$$

But

$$10^{0.78} = 6.$$

The final result for $10^{0.78}$ was obtained by using the table in reverse. It is not necessary to change to exponential form for each such multiplication; the logarithmic form can be used throughout, as shown by the following:

$$\begin{aligned} \log (2 \times 3) &= \log 2 + \log 3 \\ &= 0.30 + 0.48 \\ &= 0.78. \end{aligned}$$

In Table 8.4, the number whose logarithm is 0.78 is 6; or $\log (2 \times 3) = 0.78$. Thus, $2 \times 3 = 6$. The above development uses Property 1 for logarithms, but note that the forms

$$\log (2 \times 3) = \log 2 + \log 3,$$

and

$$10^{0.30} \times 10^{0.48} = 10^{0.30+0.48}.$$

are essentially equivalent because of the relationship between the exponential and logarithmic functions. In both cases, the multiplication of 2 by 3 has been performed by: (1) obtaining the logarithm of each factor from the table, (2) by adding the logarithms, and (3) by using the table in reverse and obtaining the number corresponding to that logarithm. The latter operation is usually referred to as obtaining the **antilogarithm**, where x is often called the antilogarithm (for example, 0.78 is the log of 6, whereas 6 is the antilog of 0.78, or antilog $0.78 = 6$).

The quotient of two numbers can also be obtained by using logarithms in conjunction with Property 2. For example,

$$\frac{8}{2} = 4.$$

Using logs, we have

$$\begin{aligned} \log \frac{8}{2} &= \log 8 - \log 2 \\ &= 0.90 - 0.30 \\ &= 0.60. \end{aligned}$$

But

$$\text{antilog } 0.60 = 4.$$

Therefore,

$$\frac{8}{2} = 4$$

Finally, a number may be raised to a power using Property 3, as shown in the following example:

$$2^3 = 2 \times 2 \times 2 = 8.$$

Using logs,

$$\begin{aligned} \log 2^3 &= 3 \log 2 \\ &= 3(0.30) \\ &= 0.90, \end{aligned}$$

and

$$\text{antilog } 0.90 = 8.$$

Therefore,

$$2^3 = 8.$$

CHARACTERISTIC AND MANTISSA. At first it might appear that the table of Figure 8.4 can be used only for values of x between 1 and 10. However, the logarithm of 30 or 300 (and so on) can be obtained by applying the basic properties of logarithms. For example,

$$\begin{aligned} \log 300 = \log(3 \times 100) &= \log 3 + \log 100 \\ &= 0.48 + 2 \\ &= 2.48; \\ \log 50000 = \log(5 \times 10000) &= \log 5 + \log 10000 \\ &= 0.70 + 4 \\ &= 4.70. \end{aligned}$$

In exponential form, we write

$$10^{2.48} = 300$$

and

$$10^{4.70} = 50000.$$

In these examples the logarithm (exponent) consists of a whole number part and a decimal part. The whole number portion is usually referred to as the **characteristic** and the decimal portion as the **mantissa**. Thus the characteristic of 2.48 is 2 and the mantissa is 0.48. Whenever it is necessary to determine the logarithm of a number, say 60,000, it should first be converted to scientific form, that is, $60{,}000 = 6 \times 10^4$; then the characteristic is the exponent and the mantissa may be obtained from the table. Thus, $\log 60{,}000 = \log 10000 + \log 6 = 4.78$.

Obtaining an antilog of numbers greater than 1.0 involves a similar process. For example, antilog 5.95 can be obtained by recognizing that the whole number portion will govern the placement of the decimal (that is, it represents the appropriate power of 10), and the mantissa will determine the number. In this case the mantissa is 0.95, whose antilog is 9; the characteristic is 5, whose antilog is 10^5, so the required number is

$$\begin{aligned} \text{antilog } 5.95 &= 9 \times 10^5 \\ &= 900{,}000. \end{aligned}$$

Most of the principles studied thus far are illustrated by the following example:

$$\begin{aligned} \log \frac{300 \times 4000}{20} &= \log 300 + \log 4000 - \log 20 \\ &= 2.48 + 3.60 - 1.30 \\ &= 4.78 \\ \text{antilog } 4.78 &= 6 \times 10^4. \end{aligned}$$

Therefore,

$$\frac{300 \times 4000}{20} = 60{,}000.$$

EXERCISE 8.4

From Table 8.4, find the common logarithm of the following numbers:

1. 2	3. 30	5. 9000	7. 8000
2. 5	4. 700	6. 10,000	8. 80

From Table 8.4, find the antilogarithm of the following numbers:

9. 0.30	11. 1.78	13. 5.00	15. 2.48
10. 0.95	12. 4.30	14. 8.48	16. 1.90

Using Properties 1, 2, and 3 and Table 8.4, perform the indicated arithmetic below. In some cases, due to the approximate nature of the table, the logarithm may not correspond exactly to that in the table. Use the closest values.

17. 2×4
18. 3×3
19. 3^2
20. 4×5
21. 5×8
22. 200×40
23. 50×600
24. $2 \times 3 \times 50$
25. $9 \times 20 \times 500$
26. $\dfrac{900}{30}$
27. $\dfrac{2 \times 6}{3}$
28. $\dfrac{1000}{50}$
29. $\dfrac{5 \times 400}{20}$
30. $\dfrac{400 \times 90}{20 \times 3}$
31. 5×2^3
32. $4^2 \times 5$
33. $\dfrac{5 \times 2^6}{8}$
34. $\dfrac{600}{2 \times 20}$

8.5 LINEAR INTERPOLATION

The simple table of logarithms constructed for Table 8.4 has been convenient for illustrating basic properties, but is limited in scope. If we desire the logarithm of 2 we locate $x = 2$ and read across to find that $\log 2 = 0.30$, but if we need the logarithm of 2.5, we see that no entry exists. However, we can obtain an approximate value by using the table plus a little ingenuity (for convenience, Table 8.4 is repeated here). Since 2.5 is halfway between 2 and 3, log 2.5 should be approximately halfway between log 2 and log 3 (we will study implications of this assumption shortly). Setting this up in a convenient form gives

TABLE 8.4

x	$\log x$
1	0.00
2	0.30
3	0.48
4	0.60
5	0.70
6	0.78
7	0.85
8	0.90
9	0.95
10	1.00

$$\begin{array}{ll} x = 2 & \log 2 = 0.30 \\ x = 2.5 & \log 2.5 = ? \\ x = 3 & \log 3 = 0.48. \end{array}$$

Since the difference between log 3 and log 2 is $0.48 - 0.30 = 0.18$, and half of that is 0.09, we have

$$\log 2.5 \simeq 0.30 + 0.09 = 0.39.$$

As a test of this estimate let us determine 5/2.5 and 4×2.5, using logarithms:

$$\begin{aligned} \log \frac{5}{2.5} &= \log 5 - \log 2.5 \\ &= 0.70 - 0.39 \\ &= 0.31 \end{aligned}$$

but

$$\log 2 = 0.30;$$

$$\begin{aligned}\log (4 \times 2.5) &= \log 4 + \log 2.5\\ &= 0.60 + 0.39\\ &= 0.99\end{aligned}$$

but

$$\log 10 = 1.0.$$

In both cases the results are reasonably close to the correct values and appear to be consistent with the overall approximation we have made.

Finally, we will carry this process one step further in considering the quotient of two numbers, such as 10/3:

$$\begin{aligned}\log \frac{10}{3} &= \log 10 - \log 3\\ &= 1.00 - 0.48\\ &= 0.52\\ \text{antilog } 0.52 &= ?\end{aligned}$$

Referring to the table, we find no entry under log x for 0.52; however, we see that the following entries bracket our logarithm.

$$\begin{array}{ll} x = 3 & \left.\begin{array}{l}\log 3 = 0.48\\ \log x = 0.52\end{array}\right\} 0.04 \\ x = ? & \\ x = 4 & \log 4 = 0.60 \end{array} \quad \left.\vphantom{\begin{array}{l}a\\b\\c\end{array}}\right\} 0.12.$$

In fact, 0.52 appears to be 1/3 of the way between 0.48 and 0.60 (since 0.04/0.12 = 1/3). Thus we estimate that the quantity we desire is 1/3 of the way between 3 and 4, or

$$\text{antilog } 0.52 = 3\frac{1}{3}.$$

Referring back to the original problem, we see that the approximate method gives the exact answer this time, that is,

$$\frac{10}{3} = 3\frac{1}{3}.$$

Intuitively this method of interpolation appears to be a good one, but let us delve into the nature of what is being done and attempt to justify the practice.

GRAPHICAL INTERPRETATION. To illustrate the nature of the foregoing approximation, the logarithmic function in the interval $x = 3$ to $x = 4$ is shown in Figure 8.15. In keeping with our earlier development, we will read x to the nearest tenth and log x to the nearest one hundredth (that is, two digits in both cases). Note that over this small increment the function may be approximated by the straight line segment shown as a broken line connecting the points (3.0, 0.48) and (4.0, 0.60). By inspection of the difference between

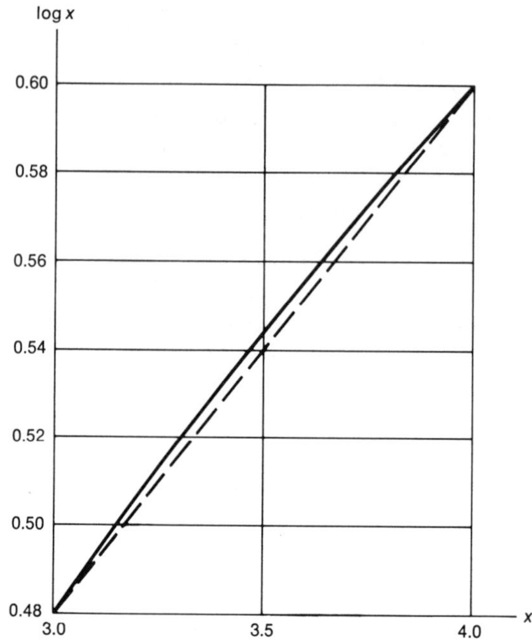

FIGURE 8.15

the function and the straight line approximation, we see that they differ by less than 0.01 in log x. For example, when

$$x = 3.5$$

$$\log x \simeq 0.54 \qquad \textit{(From the straight-line segment)}$$

$$\log x \simeq 0.545. \qquad \textit{(From the true function)}.$$

Of course it is imperative to recognize that *all* of the values used in this study are crude approximations, which were estimated from a graph. We should continually bear this factor in mind when using these numbers. Thus, for our rough computations we may assume that the function is linear in the interval $3.0 < x < 4.0$, and we may now use this linear approximation to find log x when $3.0 < x < 4.0$ without drawing a graph of the function.

THE NATURE OF LINEAR APPROXIMATION. To understand the process of linear approximation, consider the triangle ABC formed by the linear segment of the function and the dotted lines in Figure 8.16. The points A and B represent the end points of the linear approximation and correspond to (3.0, 0.48) and (4.0, 0.60), respectively. The determination of log x for some value of x, say $x = 3.6$, is considerably simplified if we note that the two triangles formed by ABC are similar and thus have proportioned sides. That is, we have

$$\frac{AE}{AC} = \frac{ED}{CB},$$

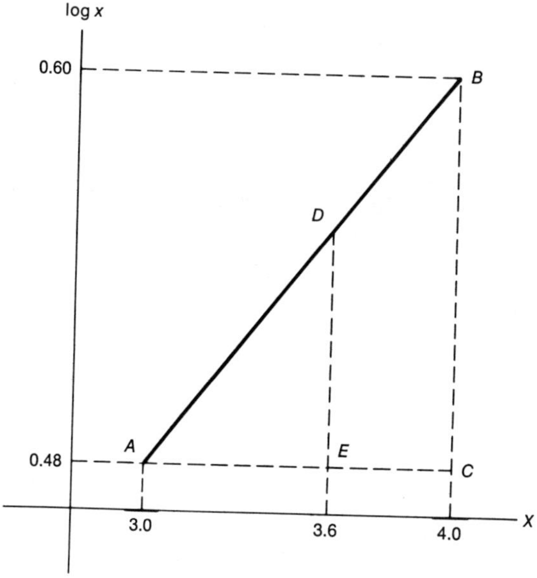

FIGURE 8.16

which is useful, since ED corresponds to the quantity that must be added to 0.48 to obtain log 3.6. Thus we have

$$AE = 3.6 - 3.0 = 0.6$$
$$AC = 4.0 - 3.0 = 1.0$$
$$CB = 0.60 - 0.48 = 0.12,$$

and

$$\frac{0.6}{1.0} = \frac{ED}{0.12}$$

$$ED = 0.072 \simeq 0.07.$$

Thus

$$\log 3.6 = 0.48 + 0.07 = 0.55.$$

Referring back to our previous intuitive approach, we see that this development justifies the method. In other words, since 3.6 is 6/10 the distance between 3.0 and 4.0, log 3.6 is 6/10 the distance between log 3.0 and log 4.0.

A CONVENIENT FORMAT. It is often possible to simplify procedures and minimize errors by using an orderly format, such as the one shown below:

$$\begin{array}{l} x \\ \left.\begin{array}{l} \left.\begin{array}{l} 3.0 \\ \\ 3.6 \end{array}\right\} 0.6 \\ \\ 4.0 \end{array}\right\} 1.0 \end{array} \qquad \begin{array}{l} \log x \\ \left.\begin{array}{l} \left.\begin{array}{l} 0.48 \\ \\ ? \end{array}\right\} b \\ \\ 0.60 \end{array}\right\} 0.12 \end{array}$$

$$\frac{0.6}{1.0} = \frac{b}{0.12}$$

$$b = 0.07$$

$$\log 3.6 = 0.48 + 0.07 = 0.55.$$

The problem of finding antilogs by interpolation is basically the same. For example, antilog 0.59 may be obtained as follows:

x			$\log x$		
3.0	a	1.0	0.48	0.11	0.12
?			0.59		
4.0			0.60		

$$\frac{a}{1.0} = \frac{0.11}{0.12}$$

$$a \simeq 0.9$$

$$\text{antilog } 0.59 = 3.0 + 0.9$$

$$= 3.9.$$

The general technique just described is referred to as **linear interpolation**, based on the assumption that the logarithmic function may be approximated as a series of straight line segments providing the intervals are sufficiently short. In advanced mathematics, the computer programmer studies many different interpolation techniques for finding intermediate points from tabular functions. However, linear interpolation is the most fundamental of these and is commonly used for functions which do not exhibit rapidly changing values.

EXERCISE 8.5

By interpolation in Table 8.4, find the logarithms of the following numbers (to two decimal places):

1. 3.2
2. 5.8
3. 36
4. 920
5. 2700
6. 170
7. 46000
8. 1100

By interpolation in Table 8.4, find the antilogarithms of the following numbers (to two significant digits):

9. 0.97
10. 1.56
11. 2.56
12. 5.06
13. 2.33
14. 0.03
15. 4.79
16. 1.54

Perform the following arithmetic operations using Table 8.4. Use the method of linear interpolation whenever necessary to obtain a logarithm or antilogarithm.

17. 2.5×63

18. 22×790

19. 24×21

20. $\frac{7800}{130}$

21. $\frac{2 \times 5 \times 300}{15}$

22. $\frac{4500}{900}$

23. $\frac{24 \times 1600}{80}$

24. 2^6

25. $\frac{90^2}{30}$

26. 1.2^5

27. $5500 \times (1.1)^{25}$

28. $\frac{23000}{(1 \times 2)^8}$

8.6 LOGARITHM TABLES

The abbreviated table of logarithms used in the preceding section was a satisfactory learning device, but would prove to be insufficient for accurate calculations. A much more accurate table is given in the appendix, a section of which is reproduced here as Table 8.5.

TABLE 8.5 PARTIAL TABLE OF LOGARITHMS

x	0	1	2	3	4	5	6	7	8	9
30	4771	4786	4800	4814	4829	4843	4857	4871	4886	4900
31	4914	4928	4942	4955	4969	4983	4997	5011	5024	5038
32	5051	5065	5079	5092	5105	5119	5132	5145	5159	5172
33	5185	5198	5211	5224	5237	5250	5263	5276	5289	5302
34	5315	5328	5340	5353	5366	5378	5391	5403	5416	5428

USING THE TABLE. This is a four-place logarithm table because it consists of four-digit mantissas for three-digit numbers. Under the column labeled x are the first two digits, and across the top row is the third. The corresponding mantissa of log x is in the body of the table. To find the logarithm of a number, say 326, locate 32 in the left column and 6 in the top row. The desired mantissa is at the intersection of that row and column, as indicated by the shaded sections. Then

$$\begin{aligned} 326 &= 3.26 \times 10^2 \\ \log 326 &= 2 + 0.5132 \\ &= 2.5132. \end{aligned}$$

Similarly, to obtain antilog 3.5024, locate the mantissa 0.5024 in the body of the table; its row is 31 and column is 8. Since the log has a characteristic of 3,

$$\text{antilog } 3.5024 = 3180.$$

Linear interpolation may also be accomplished by using this table, as illustrated by the following examples:

EXAMPLE 8.1

Determine log 34.14.

SOLUTION

x			$\log x$		
34.1	0.04 (34.1 to 34.14)	0.1 (34.1 to 34.2)	2.5328	b (2.5328 to ?)	0.0012 (2.5328 to 2.5340)
34.14			?		
34.2			2.5340		

$$\frac{0.04}{0.1}=\frac{b}{0.0012}$$

$$\begin{aligned} b &= 0.00048 \\ &\simeq 0.0005 \\ \log 34.14 &= 1.5328 + 0.0005 \\ &= 1.5333. \end{aligned}$$

EXAMPLE 8.2

Determine antilog 1.4841.

SOLUTION

x			$\log x$		
30.4	a (30.4 to ?)	0.1 (30.4 to 30.5)	1.4829	0.0012 (1.4829 to 1.4841)	0.0014 (1.4829 to 1.4843)
?			1.4841		
30.5			1.4843		

$$\frac{a}{0.1}=\frac{0.0012}{0.0014}$$

$$\begin{aligned} a &\simeq 0.09 \\ \text{antilog } 1.4841 &= 30.4 + 0.09 \\ &= 30.49. \end{aligned}$$

NEGATIVE CHARACTERISTICS. In using logarithms for calculations, negative characteristics occur whenever the number whose logarithm is sought is less than 1. For example, the number 0.00322 should be treated as 3.22×10^{-3}, which has a logarithm with a mantissa of 0.5079 (from Table 8.5) and a characteristic of −3. In other words, we have

$$\begin{aligned} \log 0.00322 &= \log (3.22 \times 10^{-3}) \\ &= \log 3.22 + \log 10^{-3} \\ &= 0.5079 - 3. \end{aligned}$$

If these two numbers were combined, the result would be a negative mantissa;

Table 8.5 could not be used. In order to avoid this inconvenience and yet maintain proper "bookkeeping," the following arithmetic device is used:

$$\begin{aligned}\log 0.00322 &= 0.5079 - 3\\ &= 0.5079 + (7 - 10)\\ &= 7.5079 - 10.\end{aligned}$$

APPLICATIONS OF LOGARITHMS. One of the most common applications of logarithms involves computing compound interest, using the formula

$$A = P(1 + i)^t,$$

where P is the principal (initial deposit), i is the interest rate compounded yearly, t is the number of years deposited, and A is the accumulated amount after t years. Consider

$$\begin{aligned}P &= 2750\\ i &= 4\% = 0.04\\ t &= 25.\end{aligned}$$

Then

$$A = 2750(1.04)^{25}.$$

The use of logarithms gives

$$\begin{aligned}\log A &= \log 2750 + 25 \log 1.04\\ &= 3.4393 + 25(0.0170)\\ &= 3.4393 + 0.4250\\ &= 3.8643.\end{aligned}$$

By linear interpolation, we have

$$\begin{aligned}\log 3.8643 &= 7317\\ A &= 7317.\end{aligned}$$

Another simple application is in finding roots; for example, the fifth root of 25 may be found as follows:

$$R = \sqrt[5]{25} = 25^{1/5}$$

$$\log R = \frac{1}{5} \log 25$$

$$= \frac{1}{5}(1.3979)$$

$$= 0.2796.$$

By interpolation,

$$R = 1.904.$$

Just as was the simple table we studied earlier, the four-place table is also an approximation. If more accurate results are desired, five-place logarithms are available. However, they are less convenient to use since they are several pages long.

LOGARITHMS AND COMPUTERS. In Chapter 7, the notion of the function was developed and mention was made that many programming languages provide subroutines for automatically calculating commonly used mathematical functions. The Fortran programming language is provided with a special function to determine the natural logarithm (base e) of any number. Thus, if it is desired to find the natural logarithm of a quantity called Z, it is only necessary to write LOG(Z) and the appropriate logarithm will be determined. It is more economical to calculate the desired logarithm (using techniques beyond the scope of this text) rather than require the computer to store a large table of logarithms.

When a program written in Fortran involves exponents, the computer normally goes about the calculations by brute force. For example, if $y = x^{19}$, the computer will likely multiply the value of x the appropriate number of times to obtain y. However, if the calculation is $y = x^{873}$, the Fortran system causes the computer automatically to switch to logarithms in performing the calculations for faster results. The determination of the break-even exponent is a matter evaluated by computer systems engineers, who write the programming systems for each computer. For more on this, refer to problem 50 in Exercise 8.6.

EXERCISE 8.6

1. log 2.47
2. log 247
3. log 24700
4. log 0.247
5. log 0.000247
6. log 98.7
7. log 98.73
8. log 0.01552
9. log 521.6
10. log 997500
11. log 0.3987
12. log 0.05921
13. log 87.15
14. log 1.732
15. log 2.718

Using the log table in the appendix, find the value of x:

16. $\log x = 0.8779$
17. $\log x = 1.9978$
18. $\log x = 3.6964$
19. $\log x = 9.5416 - 10$
20. $\log x = 7.4900 - 10$
21. $\log x = 1.9936$
22. $\log x = 2.6024$
23. $\log x = 5.4722$
24. $\log x = 7.6600 - 10$
25. $\log x = 3.6598 - 10$
26. $\log x = 0.8397$
27. $\log x = 6.9891$
28. $\log x = 7.1799 - 10$
29. $\log x = 2.5810$
30. $\log x = 0.0083$

Perform the following calculations using logarithms:

31. 6.56×32.1
32. 79.5×0.0632
33. 23.11×861.4
34. $(4.714)^3$

35. $0.0615(.9877)^5$

36. $4834(1.013)^{20}$

37. $\sqrt[3]{685}$

38. $\sqrt[3]{9713}$

39. $273\sqrt[4]{1001}$

40. $(\sqrt[3]{599}) \times (2.177)^3$

41. $\dfrac{4.976}{2.811}$

42. $\dfrac{0.000651}{0.1419}$

43. $\dfrac{0.01522 \times (6.831)^3}{2.444}$

44. $\dfrac{(3.123)^2(4.123)^2}{(5.123)^3}$

45. If $1000 is invested at 5% compounded annually, how much would this amount to after 10 years? 100 years?

46. Annual compounding of interest by savings institutions appears to have followed the same path as the ancient dinosaur (that is, the path to extinction). The general practice now appears to be at least quarterly compounding. In fact, daily compounding is a fairly common practice now that the computer has taken over the chore of calculating. In any event, interest rate is still quoted on an annual basis; the interest formula then becomes

$$A = P\left(1 + \frac{i}{n}\right)^{nt},$$

where i is the annual interest rate, n is the number of annual compounding periods, and t is the number of years. Using this formula, calculate to how much $10 will amount after 10 years at an interest rate of 6%: (a) compounded once a year, (b) compounded twice a year, (c) compounded four times a year (quarterly), (d) compounded twelve times a year (monthly), (e) compounded 60 times a year (approximately every week).

47. Find the seventh root of 200.

48. An approximation of atmospheric pressure in pounds per square inch may be calculated from the formula $p = 14.7(10)^{-0.09a}$, where a is the altitude above sea level in miles. What is the pressure at 1 mile above sea level? At 2 miles? At 10 miles? What is the pressure at 1 mile below sea level (for example, at the bottom of a mine shaft 1 mile deep)?

49. Show that common logarithms and natural logarithms (logarithms to the base e) are related by

$$\log_e x = 2.303 \log_{10} x.$$

Use the approximate value for e of 2.718, and assume that the following relationship holds:

$$\log_{10} x = \log_{10} e^{\log_e x}.$$

50. When using exponents in Fortran, the programmer should use integer quantities rather than real whenever possible. The reason is that Fortran always uses logarithms when the exponent is real whereas successive multiplication usually results from integer exponents. Thus, for example, X**3 is more efficient than X**3.0. Write a program to compute and print powers of two (from the first to the twentieth power) using both

integer and floating point exponents. To provide a good basis for comparison of the two methods print approximately four digits to the right of the decimal point.

REFERENCES

Barnett, R. A., *Intermediate Algebra: Structure and Use.* New York, McGraw-Hill, Inc., 1971.

Dubisch, R., Howes, V. E. and Bryant, S. J., *Intermediate Algebra,* 2d ed. New York, John Wiley & Sons, 1969.

Richardson, M., *Fundamentals of Mathematics,* 3d ed. New York, The MacMillan Company, 1966.

Wooten, W. and Drooyan, I., *Intermediate Algebra,* 2d alt. ed. Belmont, Calif., Wadsworth Publishing Company, Inc., 1968.

9 SIMULTANEOUS SYSTEMS OF EQUATIONS

9.1 SYSTEMS IN TWO VARIABLES / **294**
Graphical Solution / **294**
Exercise 9.1-1 / **296**
Solution by Substitution / **296**
Exercise 9.1-2 / **298**
Solution by Linear Combination / **298**
Exercise 9.1-3 / **301**

9.2 DEPENDENT AND INCONSISTENT SYSTEMS / **302**
Dependent Systems / **302**
Inconsistent Systems / **302**
Using the Computer / **303**
Exercise 9.2 / **305**

9.3 SYSTEMS IN THREE VARIABLES / **305**
Solution by Linear Combination / **306**
Solution by The Method of Gauss / **306**
A Tabular Form of Solution / **307**
Exercise 9.3 / **309**

9.4 APPLICATIONS OF LINEAR SYSTEMS / **310**
Exercise 9.4 / **312**

9.5 SOLUTION BY DETERMINANTS / **312**
Systems in Two Variables / **312**
Exercise 9.5-1 / **315**
Systems in Three Variables / **316**
Solving Linear Equations / **317**
Exercise 9.5-2 / **320**

In Chapter 6 we studied simple equations in one unknown and were able to find a value or values for the unknown which, when substituted in the equation, satisfied that equation. For example, the equation $2x + 3 = x - 1$ has the solution $x = -4$. However, the linear equations consisting of two unknowns, in Chapter 7, have no such unique solutions. For example, the equation

$$x - y = 2$$

has an infinite number of solutions, which is readily apparent when the equation is changed to the following form:

$$y = x - 2.$$

It is easily seen that whenever a value is chosen for x, a corresponding value may be calculated for y. Any pair of these numbers satisfies the equality, and is therefore a solution.

On the other hand, it is frequently necessary to consider pairs of equations and to determine whether or not they contain any *common* solutions.

9.1 SYSTEMS IN TWO VARIABLES

GRAPHICAL SOLUTION. Insight into the nature of the pair of equations

$$x - y = 1 \qquad x + y = 5$$

may be gained by reference to their graphical forms shown in Figure 9.1. The points $(0, -1)$, $(1, 0)$, $(2, 1)$, and $(3, 2)$ all lie on the line $y = x - 1$, and therefore represent four of the possible solutions to this equation. Similarly, $(0, 5)$, $(1, 4)$, $(2, 3)$, and $(3, 2)$ lie on the line $y = -x + 5$, and represent four of the possible solutions to this equation. However, it is apparent that the only point

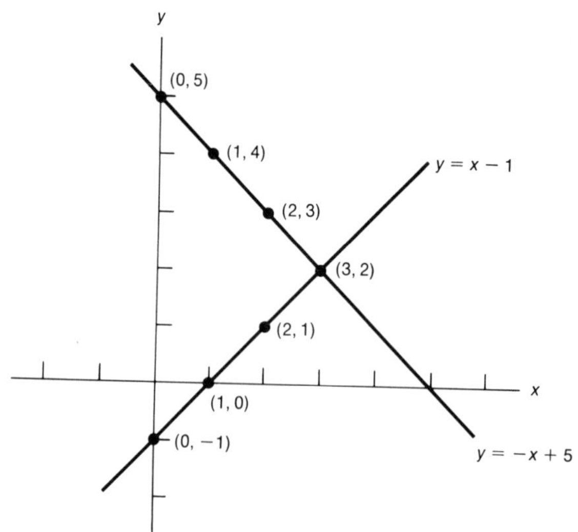

FIGURE 9.1

which these two lines have in common (3, 2) is their intersection. Two equations, when considered in this manner, are referred to as a **simultaneous system**; when the equations are both linear the system is called a **simultaneous system of linear equations**. It is apparent, from the graphical interpretation, that two linear equations representing different straight lines can have no more than one solution. In other words, two lines can intersect at no more than one point.

The use of a graph to illustrate a solution of simultaneous linear equations is valuable as an intuitive aid, but has limited value in practice because of the inconvenience and inaccuracies which usually result. The following example, which has been graphed in Figure 9.2, illustrates this:

$$5x - 3y = -6 \qquad \text{or} \qquad y = \frac{5}{3}x + 2$$

$$3x + 4y = 4 \qquad\qquad y = -\frac{3}{4}x + 1.$$

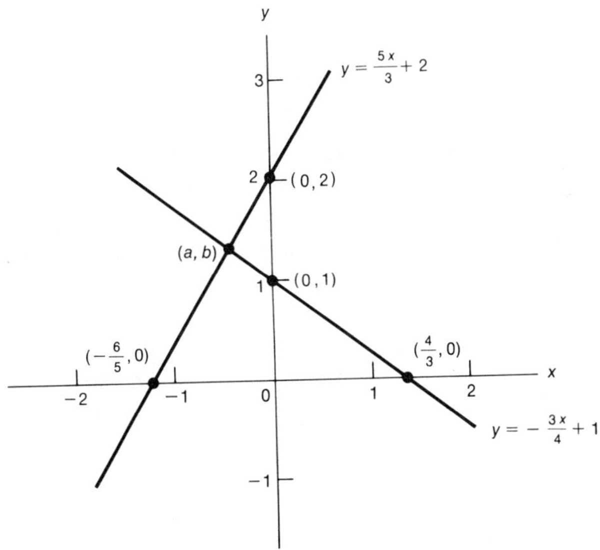

FIGURE 9.2

The intersection (a, b) has an abscissa of approximately $-1/3$ and ordinate of approximately 4/3. It is difficult to determine the exact values from the graph, although we could use accurate drawing and measuring instruments to obtain a fairly close approximation to the true values, which are $(-12/29, 38/29)$. This would be a needless refinement, since we will normally solve the equations using analytic methods. However, we should not overlook the insight to be gained by studying the graphs of equations or systems of equations.

EXERCISE 9.1-1

Solve the following pairs of linear equations graphically:

1. $-x + 2y = 3$
   $2x + 3y = 8$
2. $y - 2x + 3 = 0$
   $2x - 4y = -6$
3. $2x = 3y - 13$
   $5x + 2y + 4 = 0$
4. $6x + y + 1 = 0$
   $y + 7 = 0$
5. $5x = 2y$
   $y - 4x = 3$
6. $x = -2y - 7$
   $1 = x$
7. $x = 9y - 4$
   $2x + 7y = -8$
8. $2x - 3y - 1 = 0$
   $5x + 7 = -2y$

SOLUTION BY SUBSTITUTION. In Chapter 6 we studied simple problems, such as: The length of a rectangular field is three times the width. If 400 feet of fencing are required to enclose it, what are the dimensions?

This type of problem is solved by procedures such as the following: Let x represent the width. The length is then $3x$. But the perimeter is twice the width plus twice the length, or

$$2x + 2(3x) = 400$$
$$8x = 400$$
$$x = 50.$$

Thus the width = 50 feet and the length = 150 feet.

Actually, the problem involves two unknowns even though only one variable is used in the equation. Also, two relationships are involved, one of which is expressed by the equation and the other by the word statement relating the width and length to each other. Thus, in solving problems of this type, we have unknowingly been solving simultaneous systems in two unknowns.

To show this, let us reconsider the problem in the following way: Let x represent the width. Let y represent the length. As before, the perimeter is

$$2x + 2y = 400.$$

Also, the length is three times the width, so

$$y = 3x.$$

These two equations thus represent a simultaneous system. The solution was obtained, in essence, by the method of substitution; that is,

$$2x + 2y = 400 \qquad y = 3x.$$

Substituting the expression for y in the first equation for y in the second equation gives

$$2x + 2(3x) = 400,$$

which is exactly the form we obtained using the techniques of Chapter 6.

The following example further illustrates solving linear systems by the method of substitution:

$$x - 2y = -8 \qquad (1)$$
$$11x - 3y = 7 \qquad (2)$$
$$x = 2y - 8 \qquad \text{[Solve (1) for } x\text{]} \qquad (3)$$
$$11(2y - 8) - 3y = 7 \qquad \text{[Substitute (3) in (2)]} \qquad (4)$$
$$22y - 88 - 3y = 7$$
$$19y = 95$$
$$y = 5.$$

But

$$x = 2y - 8$$
$$= 2(5) - 8$$
$$= 2.$$

Therefore, the solution is (2, 5). Substituting in (1) and (2) to check validity gives:

$$2 - 2(5) = -8 \qquad (1a)$$
$$-8 = -8$$
$$11(2) - 3(5) = 7 \qquad (2a)$$
$$7 = 7.$$

The choice of solving for x in Eq. 1 is purely arbitrary, and in this case it was a matter of convenience. Equation 1 could as easily have been solved for y or Eq. 2 for x or y, all yielding the same results. For instance, let us solve Eq. 2 for y:

$$x - 2y = -8 \qquad (1)$$
$$11x - 3y = 7 \qquad (2)$$
$$y = \frac{11}{3}x - \frac{7}{3} \qquad \text{[Solve (2) for } y\text{]} \qquad (5)$$
$$x - 2\left(\frac{11}{3}x - \frac{7}{3}\right) = -8 \qquad \text{[Substitute (5) in (1)]} \qquad (6)$$
$$x - \frac{22}{3}x + \frac{14}{3} = -8$$
$$3x - 22x + 14 = -24$$
$$-19x = -38$$
$$x = 2.$$

But

$$y = \frac{11}{3}x - \frac{7}{3}$$
$$= \frac{11}{3}(2) - \frac{7}{3}$$
$$= 5.$$

Therefore, the solution is (2, 5). Although the form of Eq. 5 is not as convenient as the previous expression for x of Eq. 3, the end results are identical.

EXERCISE 9.1-2

Solve the following pairs of linear equations by substitution:

1. $x+2y+7=0$
   $1=x$
2. $6x+y=-1$
   $7+y=0$
3. $x-2y+3=0$
   $2x+3y=8$
4. $x-4=9y$
   $2x+7y-8=0$
5. $y-2x+3=0$
   $2x+6=4y$
6. $y-3=4x$
   $2y=5x$
7. $13-3y=2x$
   $5x=-2y-4$
8. $\frac{x}{3}-\frac{y}{2}=-4$
   $\frac{x}{2}+y=11$
9. $\frac{3x}{5}=\frac{2y}{3}+\frac{5}{3}$
   $\frac{x}{3}-\frac{3y}{2}=-\frac{4}{3}$
10. $\frac{x}{3}+3y=-3$
    $3x=y-5$
11. $2x=5y-2$
    $2y+2=x$
12. $4x+3=2y$
    $3y-1=4x$

SOLUTION BY LINEAR COMBINATION. In Chapter 6 we studied the process of solving equations by adding the same quantity to both sides of an equation to obtain an equivalent equation. In dealing with simultaneous systems such as

$$x-y=5 \qquad y=3,$$

we have used the method of substitution to obtain the solution. Recall that the solution is that pair of values for x and y [in this case (8, 3)] which satisfy both equations, that is, a single value for x and a single value for y. In the first equation above, $x-y$ and 5 can be considered as different means for representing the same thing; also the same can be said of y and 3 in the second equation. Intuitively, by combining these thoughts with the methods of Chapter 6, we see a logical basis for adding the second equation to the first, with the following effect:

$$x-y+y=5+3$$
$$x=8.$$

Note that the end result, $x=8$, obtained by collecting similar terms, is part of the solution. This result is not a coincidence but is based on an important principle, which we shall examine after considering another example:

$$x-2y=-5 \qquad (1)$$
$$7x+2y=13. \qquad (2)$$

By adding these two equations, we eliminate y from the resulting equation, giving

$$7x + x = -5 + 13 \quad (3)$$
$$8x = 8$$
$$x = 1. \quad (4)$$

We can now substitute for x in either (1) or (2) and obtain y; that is,

$$1 - 2y = -5 \qquad [\textit{Substitute in } (1)]$$
$$-2y = -6$$
$$y = 3.$$

Therefore the solution is (1, 3). Substitution of these values into both (1) and (2) confirms the solution.

In obtaining this solution, we combined (1) and (2) to obtain (3) and then (4). By combining the pair of equations (1) and (4) we obtained the remainder of the solution. Its graph, as we can see, is a vertical straight line. In this case we actually treated (1) and (4) as a linear system to obtain y.

Equation 4 is said to be a **linear combination** of (1) and (2), and the pair of equations (1) and (4) **equivalent** to the original system. Although this intuitive approach to solving systems of linear equations by using linear combinations is helpful, the principles used are sufficiently important that we shall justify the general case.

The general form for a pair of linear equations in two variables is

$$a_1x + b_1y + c_1 = 0 \quad (9.1)$$
$$a_2x + b_2y + c_2 = 0. \quad (9.2)$$

In the preceding sections, we have been concerned with the solution of these two equations, that is, the single point (m, n) on the graph at which the two lines intersect. By definition, whenever the solution is substituted for x and y in (9.1) and (9.2), the equality is satisfied. Thus we should be able to multiply either of these equations by any nonzero constant without affecting the equality. In other words,

$$A(a_1x + b_1y + c_1) = 0 \quad (9.3)$$
$$B(a_1x + b_1y + c_1) = 0, \quad (9.4)$$

where A and B are nonzero constants.

The complexity of the general form represented in (9.3) and (9.4) need not be confusing. Reference to the following example should be reassuring:

$$3x - 2y + 4 = 0 \quad (1)$$
$$11x - 3y - 7 = 0. \quad (2)$$

The solution of this system can readily be determined as (2, 5) by the foregoing methods. To illustrate the effect of multiplying by a constant, we will arbitrarily use $A = 4$ and $B = -3$, giving

$$4(3x - 2y + 4) = 4(0) \quad (1a)$$
$$-3(11x - 3y - 7) = (-3)(0). \quad (2a)$$

Substituting ($x = 2$, $y = 5$) gives

$$4[3(2) - 2(5) + 4] = 0 \quad (1b)$$
$$-3[11(2) - 3(5) - 7] = 0 \quad (2b)$$
$$4 \times (0) = 0 \quad (1c)$$
$$-3 \times (0) = 0. \quad (2c)$$

That the solution to (1) and (2) is independent of the constant multipliers A and B is suggested by (1c) and (2c). No matter what multipliers we use on (1) and (2), the left sides of the equations remain zero when $x = 2$ and $y = 5$, the only point satisfying both of these equations.

The same argument follows for the general system (9.1) and (9.2) since, for its solution (m, n), (9.3) and (9.4) give

$$A \times 0 = 0 \qquad B \times 0 = 0.$$

If we accept (9.3) and (9.4) as true, we can go one step further and combine these two equations to obtain

$$A(a_1x + b_1y + c_1) + B(a_2x + b_2y + c_2) = 0, \quad (9.5)$$

which is also satisfied for the solution (m, n).

Equation 9.5 is called a **linear combination** of Equations 9.1 and 9.2. This principle forms part of the foundation of the subject of linear algebra and provides a convenient means for solving systems of linear equations.

Let us consider the pair of equations

$$3x - 2y + 4 = 0 \quad (1)$$
$$11x - 3y - 7 = 0. \quad (2)$$

Although either or both of these equations may be multiplied by any nonzero constant without changing the solution, a proper choice of these multipliers will readily lead to a solution. For example, if Eq. (1) is multiplied by 3 and Eq. (2) by -2, the resulting linear combination will not contain y. That is,

$$9x - 6y + 12 = 0 \quad (1d)$$
$$-22x + 6y + 14 = 0 \quad (2d)$$
$$-13x + 26 = 0 \qquad [\textit{Add } (1d) \textit{ and } (2d)]$$
$$x = 2.$$

Substituting for x in (1) gives

$$6 - 2y + 4 = 0 \qquad y = 5.$$

Therefore the solution is (2, 5).

From the standpoint of convenience, systems of linear equations are usually arranged with the constant terms on the right, as shown in (1a) and (1b) below:

$$3x = 2y + 16 \quad (1)$$
$$5x - 3y = 26 \quad (2)$$

$$3x - 2y = 16 \qquad (1a)$$
$$5x - 3y = 26. \qquad (2a)$$

Here we will combine these two equations to eliminate y (an arbitrary choice). Since the coefficient on y in (1a) is -2 and in (2a) is -3, it will be necessary to multiply (1a) by 3 and (2a) by -2; that is,

$$9x - 6y = 48 \qquad [\textit{Multiply by } 3] \qquad (1b)$$
$$-10x + 6y = -52 \qquad [\textit{Multiply by } -2] \qquad (2b)$$
$$-x = -4 \qquad [\textit{Add } (1b) \textit{ and } (2b)] \qquad (3)$$
$$x = 4.$$

Substituting 4 for x in (1) gives

$$3(4) = 2y + 16 \qquad y = -2.$$

Therefore the solution to this set of equations is $(4, -2)$, which can be verified by substituting in (1) and (2).

The choice of the multipliers 3 and -2 for this system is not the only choice which would yield a solution. For instance, Eq. (1a) could have been multiplied by -5 and Eq. (2a) by 3; the addition would have caused a variable x to be eliminated. The choice is entirely arbitrary and is normally based on convenience.

EXERCISE 9.1-3

Solve the following systems by linear combinations:

1. $x - 3y = -1$
   $-x + 5y = 5$
2. $y = x + 1$
   $2y - 8x = 10$
3. $2x + 5 = 7y$
   $3x + 3y = 0$
4. $13 - 2y = -2x$
   $5x + 2y = -4$
5. $2y - 3 = 4x$
   $3y = 4x + 1$
6. $7x + 5y - 3 = 0$
   $6y - 4x = -46$
7. $8x - 3y = 9$
   $11y - 3 = 6x$
8. $\frac{x}{2} + 3 = 3y$
   $3x = 6y - 5$
9. $\frac{3x}{5} - \frac{2y}{3} = \frac{5}{3}$
   $\frac{x}{3} = \frac{3y}{2} - \frac{4}{3}$
10. $\frac{2x}{3} - \frac{5y}{2} = -\frac{31}{2}$
    $\frac{x}{4} = \frac{2y}{7} - \frac{3}{2}$
11. $5x - 7 = 8y$
    $4y = 6 - 2x$
12. $\frac{3x}{2} - 5 = \frac{y}{4}$
    $\frac{2y}{5} = \frac{2x}{5} - \frac{1}{2}$
13. $x + y = b$
    $x - y = c$
14. $ax + y = b$
    $x + y = 1$
15. $a_1x + b_1y = c_1$
    $a_2x + b_2y = c_2$

9.2 DEPENDENT AND INCONSISTENT SYSTEMS

Occasionally, when an attempt is made to solve a linear system, useless identities or inconsistent results are obtained. In both cases the implication is that the system has no unique solution. Equations of this type fall in two categories, **dependent** systems and **inconsistent** systems.

DEPENDENT SYSTEMS. To illustrate dependent equations, consider the system

$$6x - 3y = 12 \quad (1)$$
$$-2x + y = -4. \quad (2)$$

Using the method of substitution, we may solve Eq. 2 for y and continue with the solution:

$$y = 2x - 4 \quad (3)$$
$$6x - 3(2x - 4) = 12 \qquad [\textit{Substitute } (3) \textit{ in } (1)] \quad (4)$$
$$6x - 6x + 12 = 12$$
$$0 = 0.$$

If we multiply Eq. (2) by 3 and use the method of linear combination in attempting to solve this system, the result will always be $0 = 0$, or some equivalent identity. Thus we are thwarted in finding the single value for each x and y that satisfies (1) and (2) simultaneously. In order to understand why this occurs, we may multiply (2) by the constant -3 to obtain

$$6x - 3y = 12. \quad (2a)$$

But (2*a*) is identical to (1), which suggests that the graphs of (1) and (2) are the same straight line. If this is not immediately clear, rearrange both (1) and (2) to the slope-intercept form. The essential element here is that these two equations differ only by a constant multiplier (in this case, -3).

Equations of this type are called **dependent** equations and can be made identical by rearranging terms and multiplying one of them with the appropriate constant. Since their graphs are the same line, they have *all* of their points in common. The fact that the analytic solution resulted in an identity implies that the system is satisfied for all values of x (and corresponding values of y).

INCONSISTENT SYSTEMS. A problem of a different type arises with equations similar to the following:

$$2x - y = 3 \quad (1)$$
$$-6x + 3y = 4. \quad (2)$$

Solving this system by linear combination, we have

$$6x - 3y = 9 \qquad [\textit{Multiply } (1) \textit{ by } 3] \quad (3)$$
$$-6x + 3y = 4 \quad (2)$$
$$0 = 13 \qquad [\textit{Add } (3) \textit{ and } (2)]$$

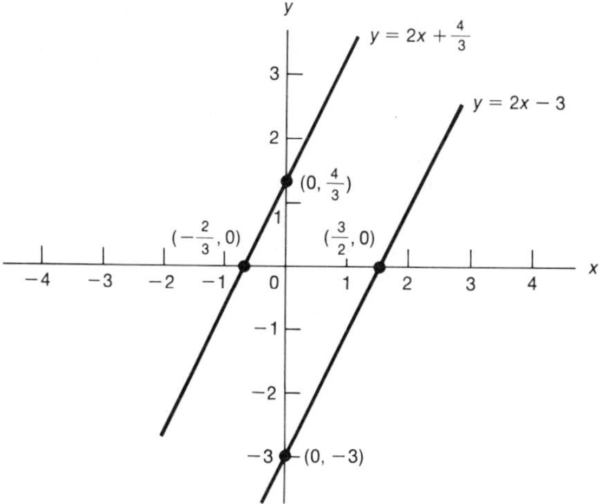

FIGURE 9.3

Here we have a contradiction which is never true, regardless of the values assigned x and y. An attempted solution by substitution will yield a similar result. At this point, we should back off and consider the premise that leads to this contradiction. The basis for solving linear systems is that a given pair of values for x and y exist which will satisfy *both* equations. Since the procedure used in solving the system is valid but leads to a contradiction, we must conclude that our basic assumption (that the equations have a point in common) must be false. In other words, these equations have no solution and are thus said to be *inconsistent*. To gain an insight into their nature, we rearrange (1) and (2) to the slope-intercept form, and consider their graphical interpretation.

$$y = 2x - 3 \tag{4}$$

$$y = 2x + \frac{4}{3}. \tag{5}$$

Equations (4) and (5) correspond to (1) and (2), respectively. Note that the slopes are identical (that is, $m = 2$) but the y intercepts are different. These forms suggest that the graphs for the equations be parallel, as shown in Figure 9.3.

Although constructing and measuring the distances of a graph cannot be used to prove that the lines are parallel, the proof is a basic exercise in geometry. Certainly we know from geometry that two parallel lines do not intersect, which is consistent with our previous conclusion that these equations have no solution.

USING THE COMPUTER. It would seem that with the computer, we could avoid the cumbersome techniques of the preceding sections. However, as we have already learned, the algebra portion of problem solving is up to the programmer. Before deciding to use the computer, we must recognize that if we have only one or two sets of equations to solve (such as problems 1 and 2 in Exercise 9.1-3) then it would be a waste of time to use the computer. Considerably more time would be required to write the program, prepare it for input to the machine, and so on. However, if we had several hundred such sets of equations then it would be to our advantage to write the program. Assuming the latter, it would be necessary to solve the general form illustrated in problem 15 of Exercise 9.1-3 before writing the program.

$$a_1x + b_1y = c_1$$
$$a_2x + b_2y = c_2.$$

Using either the method of linear combination or of substitution, these may be solved yielding

$$x = \frac{c_1b_2 - c_2b_1}{a_1b_2 - a_2b_1}$$

$$y = \frac{a_1c_2 - a_2c_1}{a_1b_2 - a_2b_1}.$$

The existence of a dependent or inconsistent system will result in a value for the denominator $(a_1b_2 - a_2b_1)$ of zero, thus yielding no solution. The flow-

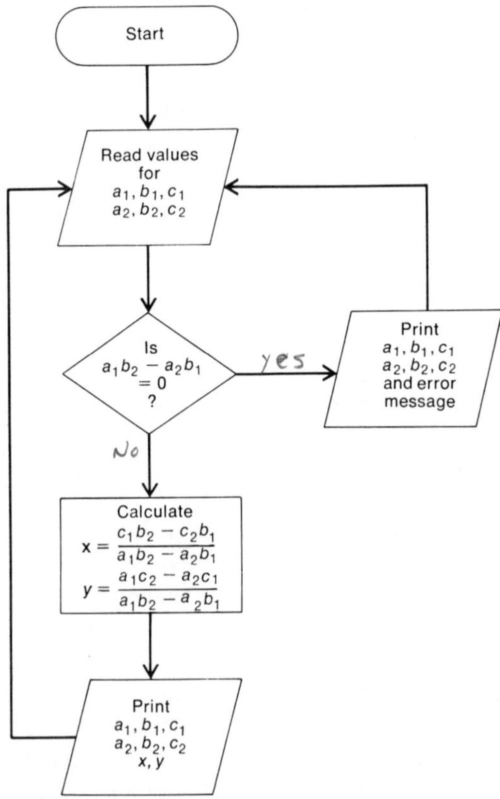

FIGURE 9.4 Calculating a solution

chart of Figure 9.4 illustrates the basic logic involved in calculating the solutions for various sets of equations.

EXERCISE 9.2

Identify which of the following systems are dependent and which are inconsistent:

1. $3x+2y=-4$
   $x-3y=6$
2. $x-4y=7$
   $-2x+8y=13$
3. $5x-y+3=0$
   $y=5x$
4. $x+2y=-3$
   $-6y=3x+9$
5. $x-8y=9$
   $2x-17y=9$
6. $x+y=1$
   $x-y=1$
7. $x=4y+3$
   $4y-x=1$
8. $2x+y=2-x$
   $x+2=2(x-2)-y$
9. $ax+by=1$
   $bx+ay=ab$
10. $\frac{x}{a}+\frac{y}{b}=1$
    $\frac{x}{b}+\frac{y}{a}=1$

9.3 SYSTEMS IN THREE VARIABLES

The linear systems studied thus far have involved two variables. However, physical problems commonly occur which are best described by three or more unknowns. For example, the equation

$$2x - 3y - 2z = 6$$

relates the three variables x, y, and z in much the same manner that $5x-2y=4$ relates the two variables x and y. It is apparent that any number of sets of values may be found which satisfy the equality. For example, when $x = 4$ and $y = 2$, the equality is satisfied if $z = -2$, and we represent the solution as $(4, 2, -2)$. On the other hand, if $x = 1$ and $y = -4$, $z = 4$, this gives $(1, -4, 4)$ as another solution. In other words, no limit exists to the number of such solutions.

In studying the linear function in two variables, we found that it can be graphed as a line in a two-dimensional coordinate system. Similarly, a graphical interpretation may be given to a linear equation in three variables. However, since one coordinate is required for each variable, it is necessary to use a three-dimensional coordinate system, with the result that the equation represents a plane rather than a line. In other words, each of the points $(4, 2, -2)$ and $(1, -4, 4)$ lie in a plane determined by the equation $2x-3y-2z=6$.

Two such equations represent two planes. If the planes are not parallel, their intersection is a line and the planes have an infinite number of points in common. On the other hand, three equations represent three planes with only a single point in common (except for special cases, which we shall not consider here). Intuitively this is easy to picture, by considering the front wall, side wall, and ceiling of a room as three planes. The only point which they have in common is the corner. Although this geometric interpretation of three equations in three variables provides a good insight into the nature of such systems, it is not useful in obtaining a solution.

SOLUTION BY LINEAR COMBINATION. Systems in three or more variables may be solved by using any of the techniques discussed for solving systems in two variables. For example, we can solve the following system by linear combination:

$$2x + 6y + 2z = 4 \tag{1}$$
$$3x + 7y + z = 4 \tag{2}$$
$$x - 2y - 2z = 1. \tag{3}$$

By inspection we can see that the sum of (1) and (3) results in another equation containing only the variables x and y. Similarly, $2 \times (2) + (3)$ also yields an equation in x and y.

$$3x + 4y = 5 \qquad \text{Add (1) and (3)} \tag{4}$$
$$7x + 12y = 9 \qquad \text{Add } 2 \times (2) \text{ to (3).} \tag{5}$$

The result is a system in two variables, easily solved by the methods of the preceding section to yield

$$x = 3 \qquad y = -1.$$

Substitution of these values for x and y in any of the original equations yields $z = 2$. The solution $(3, -1, 2)$ may be checked by substitution into each of the original equations:

$$2(3) + 6(-1) + 2(2) = 6 - 6 + 4 = 4$$
$$3(3) + 7(-1) + (2) = 9 - 7 + 2 = 4$$
$$(3) - 2(-1) - 2(2) = 3 + 2 - 4 = 1.$$

The foregoing sequence of steps in solving the system is by no means the only one which could be used. For example, we might combine (1) and (3) and (2) with (3) to eliminate x.

SOLUTION BY THE METHOD OF GAUSS. Although the techniques which we have used in solving linear equations are convenient, it is necessary to develop a more systematic approach before such methods can be programmed on a computer. A fundamental process was developed by the German mathematician Karl Friedrich Gauss. It involves the following sequence of steps (as applied to three equations in three variables, x, y and z):

1. Divide first equation by coefficient on x to reduce this coefficient to 1.
2. Combine appropriate multiples of first equation with second and third equations to eliminate x from them.
3. Divide second equation by coefficient on y to reduce this coefficient to 1.

4. Combine the appropriate multiple of second equation with third equation to eliminate y from third equation (which now contains only z).
5. Divide third equation by coefficient on z to reduce this coefficient to 1. This now represents the z part of the solution.
6. Substitute the result for z in the second equation to solve for y.
7. Substitute the results for y and z in the first equation to solve for x.

As an illustration, let us use this method to solve the system of the preceding section.

$$\begin{aligned} 2x + 6y + 2z &= 4 && (1) \\ 3x + 7y + z &= 4 && (2) \\ x - 2y - 2z &= 1 && (3) \end{aligned}$$

$$\begin{aligned} x + 3y + z &= 2 && \text{[Divide (1) by 2]} && (4) \\ 3x + 7y + z &= 4 && && (2) \\ x - 2y - 2z &= 1 && && (3) \end{aligned}$$

$$\begin{aligned} x + 3y + z &= 2 && && (4) \\ -2y - 2z &= -2 && \text{[Subtract } 3 \times (4) \text{ from (2)]} && (5) \\ -5y - 3z &= -1 && \text{[Subtract (4) from (3)]} && (6) \end{aligned}$$

$$\begin{aligned} x + 3y + z &= 2 && && (4) \\ y + z &= 1 && \text{[Divide (5) by } -2\text{]} && (7) \\ -5y - 3z &= -1 && && (6) \end{aligned}$$

$$\begin{aligned} x + 3y + z &= 2 && && (4) \\ y + z &= 1 && && (7) \\ 2z &= 4 && \text{[Subtract } -5 \times (7) \text{ from (6)]} && (8) \end{aligned}$$

$$\begin{aligned} x + 3y + z &= 2 && && (4) \\ y + z &= 1 && && (7) \\ z &= 2 && \text{[Divide (8) by 2].} && (9) \end{aligned}$$

Now it is necessary only to substitute for z in (7) to obtain $y = -1$ and substitute for y and z in (4) to obtain $x = 3$. Although this method appears to be difficult, it is very easily programmed on a digital computer.

A TABULAR FORM OF SOLUTION. Actually, the Gauss method can be performed by working only with coefficients, as shown below:

$$\begin{array}{rrr|r} 2 & 6 & 2 & 4 \\ 3 & 7 & 1 & 4 \\ 1 & -2 & -2 & 1 \end{array} \qquad \begin{array}{r} (1) \\ (2) \\ (3) \end{array}$$

By comparing rows (1), (2), and (3), we see that the numbers to the left of the vertical line correspond to the coefficients of Equations (1), (2), and (3), and the column of numbers to the right corresponds to the constant terms. If we keep in mind that the first column is associated with x, the second with y, and the third with z, we can operate on this system just as we did on the previous one. Note that row numbers in the following method correspond to equation numbers in the preceding development.

1	3	1	2	[Divide (1) by (2)]	(4)
3	7	1	4		(2)
1	−2	−2	1		(3)

1	3	1	2		(4)
0	−2	−2	−2	[Subtract 3 × (4) from (2)]	(5)
0	−5	−3	−1	[Subtract (4) from (3)]	(6)

1	3	1	2		(4)
0	1	1	1	[Divide (5) by −2]	(7)
0	−5	−3	−1		(6)

1	3	1	2		(4)
0	1	1	1		(7)
0	0	2	4	[Subtract −5 × (7) from (6)]	(8)

1	3	1	2		(4)
0	1	1	1		(7)
0	0	1	2	[Divide (8) by 2]	(9)

Comparing this result to (4), (7), and (9) of the preceding section shows that the end results are identical. Actually the process may be carried a few steps further with very interesting results:

1	3	0	0	[Subtract (9) from (4)]	(10)
0	1	0	−1	[Subtract (9) from (7)]	(11)
0	0	1	2		(9)

1	0	0	3	[Subtract 3 × (11) from (10)]	(12)
0	1	0	−1		(11)
0	0	1	2		(9)

The column to the right of the vertical line is the solution $(3, -1, 2)$ since, if we re-attach the original coefficients, we have

$$\begin{aligned} x + 0 + 0 &= 3 \\ 0 + y + 0 &= -1 \\ 0 + 0 + z &= 2. \end{aligned}$$

This method (which is basically the Gauss-Jordan method) for solving linear systems borders on the subject of matrix methods (Chapter 10). Furthermore, we shall see more of this tabular procedure in our study of linear programming (Chapter 11).

Although the various methods of solution have been illustrated using the same sets of equations, they are completely general and may be used to solve any system which is not dependent or inconsistent. Nor are the techniques limited to systems of three equations in three variables. They may also be applied to systems of four equations in four variables, five equations in five variables, and so on.

EXERCISE 9.3

Solve the following sets of equations, using either the method of substitution or linear combination:

1. $x+y+z=1$
   $x-2y+2z=-3$
   $-2x+3y+z=0$
2. $2x-y+3z=11$
   $3x+y-z=-6$
   $2y+z=6$
3. $3x-y+2z=3$
   $3x+3z=0$
   $2x+y-2z=0$
4. $x-y+3z=2$
   $2x+4y+z=1$
   $5x-3y-2z=0$
5. $x-y+2z=2$
   $2x-y+3z=5$
   $-2x+2y-4z=4$
6. $2x-y-z=0$
   $-x+2y-4z=0$
   $3x-2z=0$
7. $2x-3y+z=2$
   $9x-11y-4z=9$
   $x-3y=0$
8. $4x+y-3z=5$
   $2x-2y+z=0$
   $-6x+y-2z=-1$
9. $3x+2y-z=0$
   $x+4y-z=1$
   $-6x-4y+2z=2$
10. $2x+2y-4z=1$
    $x+3y-z=-2$
    $3x+y-7z=0$
11. $4x-y+3z=2$
    $x+2y-6z=\frac{1}{2}$
    $-2x+3y-12z=-2$
12. $ax+y-az=a$
    $x-ay+az=a$
    $-ax+ay+z=a$
13. $5x+3y-z=1$
    $2x+4y-6z=2$
    $5x+3y-2z=1$
14. $ax=1$
    $ax+by=2$
    $ax+by+cz=3$
15. Solve the system of problem 1 using the method of Gauss.
16. Solve the system of problem 11 using the method of Gauss. The order of the equations may be changed if convenient.
17. Solve the system of problem 8 using the method of Gauss.
18. Solve the system of problem 5 using the method of Gauss.
19. Solve the system of problem 10 using the method of Gauss.
20. Solve the system of problem 1 using the tabular method.
21. Solve the system of problem 11 using the tabular method. The order of the equations may be changed if convenient.
22. Solve the system of problem 8 using the tabular method.

23. Solve the system of problem 5 using the tabular method.
24. Solve the system of problem 10 using the tabular method.

9.4 APPLICATIONS OF LINEAR SYSTEMS

The objective in learning to solve systems of equations is to acquire a tool which can be used in solving practical problems. As is so often the case, the primary stumbling block is usually translating the problem into a set of equations. It is important to read the problem carefully, designate each variable by a symbol, and list each condition. Once the equations are obtained, the solution, if one exists, is worked out in fairly routine fashion.

EXAMPLE 9.1

If two numbers have a sum of 12, and twice the first is 6 greater than four times the second, what are the numbers?

SOLUTION

Since we are concerned with two numbers, let us refer to them as

$$x = \text{first number}$$
$$y = \text{second number}.$$

Their sum is 12, thus

$$x + y = 12. \tag{1}$$

Twice the first is six greater than four times the second, so

$$2x - 4y = 6. \tag{2}$$

Equations 1 and 2 are readily solved, and have the solution (9, 3).

Checking the solution with the word statement of the problem is the final step. Their sum is $9 + 3 = 12$. Twice the first ($2 \times 9 = 18$) is 6 greater ($18 - 12 = 6$) than four times the second ($4 \times 3 = 12$).

EXAMPLE 9.2

Find a set of coefficients a, b, and c for the linear equation $ax + by = c$ if its graph passes through the points (1, 2) and (4, 7).

SOLUTION

Since the graph passes through these points, the equation is satisfied when the given values are substituted for x and y. That is,

$$a(1) + b(2) = c$$
$$a(4) + b(7) = c,$$

or

$$a + 2b = c \tag{1}$$
$$4a + 7b = c. \tag{2}$$

Since the two equations (1) and (2) contain three unknowns, it is necessary to solve for two of them in terms of the third. Let us solve for a and b in terms of c, and consider the implications of this result:

$$a = -2b + c \qquad (3)$$
$$4(-2b + c) + 7b = c \qquad [\textit{Substitute (3) in (2)}] \qquad (4)$$
$$b = 3c \qquad [\textit{Solve for } b] \qquad (5)$$
$$a = -2(3c) + c \qquad [\textit{Substitute (5) in (3)}] \qquad (6)$$
$$a = -5c \qquad (7)$$

The required linear equation has the form

$$-5cx + 3cy = c. \qquad (8)$$

Since c occurs in each term, it is possible to multiply both sides of (8) by $1/c$, thus eliminating c:

$$-5x + 3y = 1. \qquad (9)$$

The system is independent of the value of c. That the graph of (9) passes through the points (1, 2) and (4, 7) can be verified simply by substituting these values in (9).

EXAMPLE 9.3

A soft-drink vendor has two flavor mixes which he desires to close out. He has 7 gallons of flavor A and 12 gallons of flavor B. He sells two standard drinks that contain both of these flavors; drink I uses 1 part of A and 2 parts of B; drink II uses 2 parts of A and 3 parts of B. How much of each flavor A and B should he use for drink I and how much of A and B should he use in drink II?

SOLUTION

A problem of this nature can be extremely confusing unless a step-by-step approach is used in defining the variables and the various relationships. If a is gallons of flavor A in drink I, then $7 - a$ will be gallons of flavor A in drink II. If b is gallons of flavor B in drink I, then $12 - b$ will be gallons of flavor B in drink II. The ratio of A and B in drink I is 1 to 2, twice as much B as A.

$$2a = b. \qquad (1)$$

The ratio of A to B in drink II is 2 to 3. Thus

$$3(7 - a) = 2(12 - b). \qquad (2)$$

Equations 1 and 2 represent a linear system in two variables, which is readily solved by substitution to yield the solution (3, 6). In terms of the original problem, this means that the vendor should use 3 gallons of A and 6 gallons of B for drink I; he should also use 4 gallons of A and 6 gallons of B for drink II. Note that the algebra of finding the solution to (1) and (2) is almost trivial. However, the original problem statement appeared to be anything but trivial.

EXERCISE 9.4

1. The sum of two numbers is 99 and their difference is 50. What are the numbers?
2. The digits of a three-digit number have the following relationships: (a) The sum of the digits is 9. (b) Twice the tens digit equals the sum of the other two. (c) The sum of twice the hundreds digit and the tens digit minus the units digit is zero. Find the number.
3. If one number is divided by another, the quotient is 4 and the remainder 5. If 1 is added to the numerator and 10 to the denominator, the quotient is 2. What are the numbers?
4. The graph of a linear equation of the form $y = mx + b$ passes through the points $(2, -3)$ and $(-6, -7)$. Find the coefficients m and b.
5. Is it possible to determine an equation of the form $y = mx + b$ which passes through the points $(1, 1)$, $(-2, 7)$, and $(7, -2)$? If so, what are the coefficients; if not, why is this an impossible task?
6. Find the parabola, $y = ax^2 + bx + c$, which passes through the points $(1, 8)$, $(2, -11)$, and $(4, -11)$.
7. In Chapter 8 we interpolated the logarithm function by assuming that two points of the function are connected by a straight line segment. A common method of interpolation readily adaptable to computers involves three points and the assumption that they are connected by a parabola (refer to problem 6). Find the equation (of form $y = ax^2 + bx + c$) which passes through the points $(1, -7)$, $(2, -6)$, and $(3, -1)$.
8. Using the equation developed in problem 7, make a table of values for x and y, in the interval $1 < x < 3$, at intervals of 0.2. Compare this to a table obtained by linearly interpolating between the points $(1, -7)$ and $(2, -6)$.
9. A young man seeking employment as a salesman is informed that he can work on a straight commission or on a base salary plus a smaller commission. The two methods (both paid monthly) are (a) a commission of 10% on gross sales, or (b) a guaranteed minimum of \$300 plus 6% on gross sales. He feels that his monthly gross sales would be about \$7000; should he accept plan *a* or plan *b*? What is the "break-even" point for gross sales?

9.5 SOLUTION BY DETERMINANTS

SYSTEMS IN TWO VARIABLES. Let us consider again the general case of a linear system in two variables in the following form:

$$a_1x + b_1y = c_1 \tag{9.6}$$
$$a_2x + b_2y = c_2. \tag{9.7}$$

In the following development we will obtain an expression for y in terms of a, b, and c by using the method of linear combination:

Multiply (9.6) by $-a_2$:

$$-a_1a_2x - a_2b_1y = -a_2c_1. \tag{1}$$

Multiply (9.7) by a_1:

$$a_1a_2x + a_1b_2y = a_1c_2. \tag{2}$$

Add (1) to (2):

$$a_1b_2y - a_2b_1y = a_1c_2 - a_2c_1. \tag{3}$$

Factor y:

$$(a_1b_2 - a_2b_1)y = a_1c_2 - a_2c_1. \tag{4}$$

Solve for y:

$$y = \frac{a_1c_2 - a_2c_1}{a_1b_2 - a_2b_1}. \tag{5}$$

Note that (5) is true only if $a_1b_2 - a_2b_1 \neq 0$.

We will now consider the following symbolism:

$$\begin{vmatrix} p_1 & q_1 \\ p_2 & q_2 \end{vmatrix}$$

and define it by the equation

$$\begin{vmatrix} p_1 & q_1 \\ p_2 & q_2 \end{vmatrix} = p_1q_2 - p_2q_1.$$

This symbol is called a **determinant**, and consists of two rows and two columns; it is therefore referred to as a **second-order** or two-by-two (2×2) determinant. The arbitrary constants p_1, p_2, q_1, and q_2 are called the **elements** of the determinant, and the number represented by $p_1q_2 - p_2q_1$ is the **value** of the determinant. For example, if $p_1 = 4$, $p_2 = 3$, $q_1 = -1$, and $q_2 = 5$, the determinant would be

$$\begin{vmatrix} 4 & -1 \\ 3 & 5 \end{vmatrix}$$

The value of the determinant is determined from the defining equation as

$$\begin{aligned} p_1q_2 - p_2q_1 &= (4)(5) - (3)(-1) \\ &= 20 - (-3) \\ &= 23. \end{aligned}$$

Referring again to (5), we can see that the expression for y can be replaced with equivalent determinants as shown in (6).

$$y = \frac{\begin{vmatrix} a_1 & c_1 \\ a_2 & c_2 \end{vmatrix}}{\begin{vmatrix} a_1 & b_1 \\ a_2 & b_2 \end{vmatrix}} \tag{6}$$

It is left to the student to solve (9.6) and (9.7) for x by multiplying by b_2 and $-b_1$, respectively, and to show that

$$x = \frac{\begin{vmatrix} c_1 & b_1 \\ c_2 & b_2 \end{vmatrix}}{\begin{vmatrix} a_1 & b_1 \\ a_2 & b_2 \end{vmatrix}} \tag{7}$$

The importance of forms (6) and (7) becomes apparent by reference to Eqs. 9.6 and 9.7 repeated below:

$$a_1x + b_1y = c_1 \tag{9.6}$$
$$a_2x + b_2y = c_2 \tag{9.7}$$

The denominators of (6) and (7) are identical determinants whose elements are the coefficients on the variables of the original equations. On the other hand, we find in the numerator of (6) that the coefficients for y have been replaced by the constant terms of the original equations. That is, b_1 and b_2 have been replaced by c_1 and c_2. Similarly, the coefficients of x have been replaced by the constant terms in the numerator of (7). Rather than write out the complete array that comprises a determinate, it is common practice to use the designation D for the denominator, D_y for the numerator of (6), and D_x for the numerator of (7). Equations 6 and 7 can then be written in the more convenient form

$$y = \frac{D_y}{D} \tag{8}$$

$$x = \frac{D_x}{D}. \tag{9}$$

In using this technique (known as Cramer's rule) to solve linear systems, it is frequently difficult to remember the proper order of the variables. The expression can be remembered by considering the evaluation of a determinant as the difference of the products of the opposite diagonal elements. The process is shown schematically as

$$\begin{vmatrix} p_1 & q_1 \\ p_2 & q_2 \end{vmatrix} = p_1q_2 - p_2q_1.$$

For example,

$$\begin{vmatrix} 1 & -5 \\ 3 & 7 \end{vmatrix} = (1)(7) - (3)(-5) = 22.$$

As an illustration of solving simultaneous linear equations with this method, consider

$$3x - 2y = -4$$
$$11x - 3y = 7$$

$$D = \begin{vmatrix} 3 & -2 \\ 11 & -3 \end{vmatrix} = (3)(-3) - (11)(-2) = 13$$

$$D_x = \begin{vmatrix} -4 & -2 \\ 7 & -3 \end{vmatrix} = (-4)(-3) - (7)(-2) = 26$$

$$D_y = \begin{vmatrix} 3 & -4 \\ 11 & 7 \end{vmatrix} = (3)(7) - (11)(-4) = 65$$

$$x = \frac{D_x}{D} = \frac{26}{13} = 2$$

$$y = \frac{D_y}{D} = \frac{65}{13} = 5.$$

The solution to this system is (2, 5), which can be verified by substitution in the original equations.

EXERCISE 9.5-1

Solve the following systems, using determinants:

1. $2x + y = 7$
   $x - 3y = -7$
2. $5y + z = 5$
   $2y + 3z = 15$
3. $-3x + 4z = 7$
   $2x - 1 = -3z$
4. $3x - z = -10$
   $3z + 2x = 8$
5. $-3y + x = 13$
   $-2x - 7y = 10$
6. $7x - 9y = -3$
   $-5x + 4y = 7$
7. $-4z - 6y = 9$
   $-5z + 7y = 10$
8. $5x + 8y = 11$
   $11x + 12y = -15$
9. $7x - 3y = 21$
   $-14x + 6y = 42$
10. $3x - 4 = 8z$
    $9x - 24z = 12$
11. $-2x + 9z = 2$
    $6x - 27z = 3$
12. $x = 5y - 2$
    $5x - 25y + 4 = 5$
13. Illustrate by example what results when a solution by determinants is attempted of a dependent system; of an inconsistent system. What condition rules out the solution of dependent and inconsistent equations by using determinants?

SYSTEMS IN THREE VARIABLES. In solving the second-order system

$$a_1x + b_1y = c_1$$
$$a_2x + b_2y = c_2$$

we use the determinant

$$\begin{vmatrix} a_1 & b_1 \\ a_2 & b_2 \end{vmatrix}.$$

Similarly, in solving the third-order system

$$a_1x + b_1y + c_1z = d_1$$
$$a_1x + b_2y + c_2z = d_2$$
$$a_3x + b_3y + c_3z = d_3,$$

we use the determinant

$$\begin{vmatrix} a_1 & b_1 & c_1 \\ a_2 & b_2 & c_2 \\ a_3 & b_3 & c_3 \end{vmatrix}.$$

However, higher order determinants, such as this 3×3, cannot be evaluated by merely considering the product of diagonal elements. On the other hand, it can be shown (using the technique of mathematical induction, which is beyond the scope of this text) that a third-order determinant can be represented in the following way:

$$\begin{vmatrix} a_1 & b_1 & c_1 \\ a_2 & b_2 & c_2 \\ a_3 & b_3 & c_3 \end{vmatrix} = a_1 \begin{vmatrix} b_2 & c_2 \\ b_3 & c_3 \end{vmatrix} - a_2 \begin{vmatrix} b_1 & c_1 \\ b_3 & c_3 \end{vmatrix} + a_3 \begin{vmatrix} b_1 & c_1 \\ b_2 & c_2 \end{vmatrix}. \tag{1}$$

But the expression on the right may be expanded, using techniques for second-order determinants, giving

$$\begin{vmatrix} a_1 & b_1 & c_1 \\ b_2 & b_2 & c_2 \\ a_3 & b_3 & c_3 \end{vmatrix} = a_1b_2c_3 - a_1b_3c_2 - a_2b_1c_3 + a_2b_3c_1 + a_3b_1c_2 - a_3b_2c_1. \tag{2}$$

This technique for evaluating determinants is called expansion by **minors**. A minor of an element is the determinant remaining after deleting the entire row and column in which the element appears. Thus the minor of a_1 is

$$\begin{vmatrix} b_2 & c_2 \\ b_3 & c_3 \end{vmatrix}$$

and so on for a_2 and a_3. Similarly, the minor for b_2 is

$$\begin{vmatrix} a_1 & c_1 \\ a_3 & c_3 \end{vmatrix}.$$

Generally speaking, a determinant may be evaluated by expanding on any row or column. The resulting products of elements and their minors are then combined by alternate plus and minus signs, as shown in (1). A convenient means for remembering whether to add or subtract a given term is to associate the element whose minor is of interest with the corresponding sign in the following array:

$$\begin{vmatrix} + & - & + \\ - & + & - \\ + & - & + \end{vmatrix}.$$

Note that the upper left-hand sign is plus, and the rest of them alternate; there are no two adjacent signs which are identical. (The justification of this sign configuration is beyond the scope of this text; the interested student is referred to references listed at the end of this chapter.)

To illustrate this technique, we will evaluate the following determinant:

$$\begin{vmatrix} 3 & -1 & 2 \\ 1 & 0 & -4 \\ 2 & 3 & 3 \end{vmatrix}.$$

Since the middle element is zero, it will be convenient to expand this determinant by the second row (the second column probably would be just as convenient).

$$\begin{vmatrix} 3 & -1 & 2 \\ 1 & 0 & -4 \\ 2 & 3 & 3 \end{vmatrix} = -(1)\begin{vmatrix} -1 & 2 \\ 3 & 3 \end{vmatrix} + (0)\begin{vmatrix} 3 & 2 \\ 2 & 3 \end{vmatrix} - (-4)\begin{vmatrix} 3 & -1 \\ 2 & 3 \end{vmatrix}$$

$$\begin{aligned} &= -(1)[(-1)(3) - (3)(2)] + 0[(3)(3) - (2)(2)] \\ &\quad -(-4)[(3)(3) - (2)(-1)] \\ &= 9 + 0 + 44 \\ &= 53. \end{aligned}$$

SOLVING LINEAR EQUATIONS. A third-degree system may be solved by using determinants in much the same manner as second-degree systems are solved. Considering the general case, we have

$$a_1x + b_1y + c_1z = d_1 \qquad (9.8)$$

$$a_2x + b_2y + c_2z = d_2 \qquad (9.9)$$

$$a_3x + b_3y + c_3z = d_3. \qquad (9.10)$$

The solution may be represented as

$$x = \frac{D_x}{D} = \frac{\begin{vmatrix} d_1 & b_1 & c_1 \\ d_2 & b_2 & c_2 \\ d_3 & b_3 & c_3 \end{vmatrix}}{\begin{vmatrix} a_1 & b_1 & c_1 \\ a_2 & b_2 & c_2 \\ a_3 & b_3 & c_3 \end{vmatrix}} \quad (1)$$

$$y = \frac{D_y}{D} = \frac{\begin{vmatrix} a_1 & d_1 & c_1 \\ a_2 & d_2 & c_2 \\ a_3 & d_3 & c_3 \end{vmatrix}}{\begin{vmatrix} a_1 & b_1 & c_1 \\ a_2 & b_2 & c_2 \\ a_3 & b_3 & c_3 \end{vmatrix}} \quad (2)$$

$$z = \frac{D_z}{D} = \frac{\begin{vmatrix} a_1 & b_1 & d_1 \\ a_2 & b_2 & d_2 \\ a_3 & b_3 & d_3 \end{vmatrix}}{\begin{vmatrix} a_1 & b_1 & c_1 \\ a_2 & b_2 & c_2 \\ a_3 & b_3 & c_3 \end{vmatrix}}. \quad (3)$$

Solving third-order linear equations by Cramer's rule is somewhat mechanical, although other techniques (which we shall not consider) are often introduced to simplify the process. The following example illustrates the process:

$$\begin{aligned} x &= 2y + 3z - 15 \\ y - 2x &= 3z \\ 3x + 7y &= -5. \end{aligned}$$

In solving equations with determinants, it is important to arrange the equations in the general form of (9.8), (9.9), and (9.10) before using the determinant form. This gives

$$\begin{aligned} x - 2y - 3z &= -15 \\ -2x + y - 3z &= 0 \\ 3x + 7y + 0 \cdot z &= -5. \end{aligned}$$

For D we have

$$D = \begin{vmatrix} 1 & -2 & -3 \\ -2 & 1 & -3 \\ 3 & 7 & 0 \end{vmatrix} = +(3)\begin{vmatrix} -2 & -3 \\ 1 & -3 \end{vmatrix} - (7)\begin{vmatrix} 1 & -3 \\ -2 & -3 \end{vmatrix} + (0)\begin{vmatrix} 1 & -2 \\ -2 & 1 \end{vmatrix}$$

$$= 3(9) - (7)(-9) + 0 = 90.$$

Here D was expanded by the third row in order to include the zero element, which reduces the number of expansions. Similarly, for D_x, D_y, and D_z we have

$$D_x = \begin{vmatrix} -15 & -2 & -3 \\ 0 & 1 & -3 \\ -5 & 7 & 0 \end{vmatrix} = (-15)\begin{vmatrix} 1 & -3 \\ 7 & 0 \end{vmatrix} - (0)\begin{vmatrix} -2 & -3 \\ 7 & 0 \end{vmatrix} + (-5)\begin{vmatrix} -2 & -3 \\ 1 & -3 \end{vmatrix}$$

$$= (-15)(21) - (0) + (-5)(9) = -360$$

$$D_y = \begin{vmatrix} 1 & -15 & -3 \\ -2 & 0 & -3 \\ 3 & -5 & 0 \end{vmatrix} = (3)\begin{vmatrix} -15 & -3 \\ 0 & -3 \end{vmatrix} - (-5)\begin{vmatrix} 1 & -3 \\ -2 & -3 \end{vmatrix} + (0)\begin{vmatrix} 1 & -15 \\ -2 & 0 \end{vmatrix}$$

$$= (3)(45) + 5(-9) = 90$$

$$D_z = \begin{vmatrix} 1 & -2 & -15 \\ -2 & 1 & 0 \\ 3 & 7 & -5 \end{vmatrix} = -(-2)\begin{vmatrix} -2 & -15 \\ 7 & -5 \end{vmatrix} + (1)\begin{vmatrix} 1 & -15 \\ 3 & -5 \end{vmatrix} - (0)\begin{vmatrix} 1 & -2 \\ 3 & 7 \end{vmatrix}$$

$$= -(-2)(115) + (1)(40) - 0 = 270.$$

These give

$$x = \frac{D_x}{D} = \frac{-360}{90} = -4$$

$$y = \frac{D_y}{D} = \frac{90}{90} = 1$$

$$z = \frac{D_z}{D} = \frac{270}{90} = 3.$$

The solution (−4, 1, 3) may be shown to be correct by substitution into the original equations.

In using Cramer's rule for solving simultaneous systems, it is not possible to get something for nothing. That is, we cannot obtain solutions for dependent or inconsistent systems, since D will always be zero under these conditions and division by zero is impossible.

This method, which was devised in the 18th century by the Swiss mathematician Gabriel Cramer, may be expanded to include systems of more than three equations by successive expansion by minors. Each such expansion will reduce the order by one until a series of second-order determinants is obtained.

EXERCISE 9.5-2

Solve the following sets of equations, using determinants:

1. $-2x + 2y + z = 0$
   $x + 4y - 3z = 5$
   $x - 6y - 2z = -1$
2. $x + 3y - z = -6$
   $2x + z = 6$
   $-x + 2y + 3z = 11$
3. $x - 2y + 2z = -3$
   $x + y + z = 1$
   $-2x + 3y + z = 0$
4. $2x - y - z = 0$
   $-x - 4y + 2z = 0$
   $3x - 2z = 0$
5. Solve Eqs. 9.8-9.10 (which represent the general form of a third-order system) for x, then solve for x in Eq. 1 by expanding the determinants and show that the results are identical.

REFERENCES

DeAngelo, S. and Jorgensen, P., *Mathematics for Data Processing.* New York, McGraw-Hill, Inc., 1970.

Dubisch, R., Howes, V. E. and Bryant, S. J., *Intermediate Algebra,* 2d ed. New York, John Wiley & Sons, 1969.

Fowler, F. P. and Sandberg, E. W., *Basic Mathematics for Administration.* New York, John Wiley & Sons, 1962.

Wooten, W. and Drooyan, I., *Intermediate Algebra,* 2d alt. ed. Belmont, Calif., Wadsworth Publishing Company, Inc., 1968.

10 MATRICES

10.1 THE NOTION OF A MATRIX / **322**
Definition of a Matrix / **322**
Equality of Matrices / **323**
Multiplication of a Matrix by a Number / **323**
Addition of Matrices / **324**
Notation / **324**
Exercise 10.1 / **325**

10.2 MATRIX MULTIPLICATION / **327**
Multiplication / **327**
The Identity Matrix / **330**
Elementary Transformation Matrices / **331**
Exercise 10.2 / **332**

10.3 SOLVING SIMULTANEOUS SYSTEMS WITH MATRICES / **334**
Inverse Matrices / **334**
Matrices Stored in the Computer / **336**
Simultaneous Equations in Matrix Form / **336**
Applying Matrix Relationships / **337**
Exercise 10.3 / **339**

10.4 FINDING THE INVERSE OF A MATRIX / **342**
Inverse of a 2×2 *Matrix* / **342**
Inverse of an $n \times n$ *Matrix* / **343**
Exercise 10.4 / **345**

10.5 MATRIX MULTIPLICATION ON THE COMPUTER / **346**
The Product of Square Matrices / **346**
Summation Notation / **347**
Multiplication of $n \times n$ *Matrices* / **348**
A Use of the Notation / **348**

10.1 THE NOTION OF A MATRIX

Solving two linear equations in two variables (a 2×2 linear system) is a relatively simple task which can be accomplished by any of several techniques, as we discovered in Chapter 9. These same methods can be applied to 3×3 systems with only a little more "bookkeeping" and manipulation required. In fact, the methods are general in nature, and may be used to solve higher order systems as required. However, if we consider a 10×10 system (10 equations and 10 variables) the mere writing of the original equations is a cumbersome task, not to mention the process of solving them. However, because of the importance of linear systems, necessary means have been devised for more easily manipulating higher order systems. These means are called matrix algebra.

DEFINITION OF A MATRIX. In solving a 3×3 system using Gauss' method, we employed a convenient tabular form and omitted writing the variables. For example, the linear system in Chapter 9

$$2x - 3y - 2z = 6 \quad (1)$$
$$-3x + y + 2z = -1 \quad (2)$$
$$7x - 2y - 6z = -7 \quad (3)$$

was expressed as

$$\begin{array}{rrr|r} 2 & -3 & -2 & 6 \\ -3 & 1 & 2 & -1 \\ 7 & -2 & -6 & -7. \end{array}$$

Here the array of numbers is meaningful only in a given context. That is, the 3×3 array

$$\begin{array}{rrr} 2 & -3 & -2 \\ -3 & 1 & 2 \\ 7 & -2 & -6 \end{array}$$

must be associated with the variables x, y, and z in a particular manner. To consider the sum of the numbers in the first row $(2 - 3 - 2 = -3)$ is meaningless in this situation. Although we studied the determinant of the coefficients in Chapter 9, it is not meaningful to speak of the determinant in this context. In other words, this array cannot be considered to have a numerical value. Such a grouping or **array** is commonly called a **matrix**, and is distinguished from a determinant by enclosing it in braces:

$$\begin{bmatrix} 2 & -3 & -2 \\ -3 & 1 & 2 \\ 7 & -2 & -6 \end{bmatrix}.$$

This particular matrix has three rows and three columns, so it is called a 3×3

matrix, and its **dimensions** are said to be 3×3. On the other hand, if we consider the constant terms of (1), (2), and (3), we have a matrix consisting only of a single column:

$$\begin{bmatrix} 6 \\ -1 \\ -7 \end{bmatrix}.$$

This is a 3×1 matrix, since it consists of 3 rows and 1 column. Matrices which have only one column are called **column matrices** (or commonly **column vectors**).

Although these notions may at first appear to have little bearing on solving simultaneous equations, we must first consider the basic properties of matrices given by definition before applying them to obtaining solutions.

EQUALITY OF MATRICES. Two matrices, by definition, are said to be equal if and only if they are identical. Thus the matrices must have the same dimensions and their corresponding elements must all be equal. Of the following four matrices, only the first and second are equal:

$$\begin{bmatrix} 3 & 7 & -1 \\ 1 & 0 & 5 \\ -2 & 9 & -4 \end{bmatrix} \quad \begin{bmatrix} 3 & 7 & -1 \\ 1 & 0 & 5 \\ -2 & 9 & -4 \end{bmatrix} \quad \begin{bmatrix} 3 & 7 & 1 \\ 1 & 0 & 5 \\ -2 & 9 & -4 \end{bmatrix} \quad \begin{bmatrix} 3 & 7 \\ 1 & 0 \\ -2 & 9 \end{bmatrix}.$$

Furthermore, by this definition of equality, if two matrices such as

$$\begin{bmatrix} x \\ y \\ z \end{bmatrix} = \begin{bmatrix} -3 \\ 5 \\ -1 \end{bmatrix}$$

are equal, then $x = -3$, $y = 5$, and $z = -1$.

MULTIPLICATION OF A MATRIX BY A NUMBER. Another important characteristic of matrices involves multiplication of a matrix by a number. The product of a matrix and a number is defined as a matrix whose elements are the product of the number and the corresponding elements of the original matrix. For example,

$$5 \times \begin{bmatrix} 2 & 1 \\ 5 & 3 \\ -1 & 2 \end{bmatrix} = \begin{bmatrix} 10 & 5 \\ 25 & 15 \\ -5 & 10 \end{bmatrix}$$

and

$$-\frac{1}{2} \times \begin{bmatrix} a & -b \\ c & d \end{bmatrix} = \begin{bmatrix} -\dfrac{a}{2} & \dfrac{b}{2} \\ -\dfrac{c}{2} & -\dfrac{d}{2} \end{bmatrix}.$$

ADDITION OF MATRICES. The sum of two matrices of the same dimensions is, by definition, a matrix whose elements consist of the sums of the corresponding elements of the corresponding matrices. For example,

$$\begin{bmatrix} 2 & 1 & 5 \\ 3 & 0 & -2 \end{bmatrix} + \begin{bmatrix} 4 & -1 & 7 \\ -5 & 1 & 3 \end{bmatrix} = \begin{bmatrix} 6 & 0 & 12 \\ -2 & 1 & 1 \end{bmatrix}.$$

Similarly, for matrices whose elements are represented by letters, we have

$$\begin{bmatrix} a_1 & b_1 \\ a_2 & b_2 \end{bmatrix} + \begin{bmatrix} m_1 & n_1 \\ m_2 & n_2 \end{bmatrix} = \begin{bmatrix} a_1 + m_1 & b_1 + n_1 \\ a_2 + m_2 & b_2 + n_2 \end{bmatrix}.$$

In this case it is important to note that the sum $a_1 + m_1$ is the element in the first row and first column. Likewise, the sum $b_1 + n_1$ is the element in the first row and second column, and so on for each of the other two elements. Concerning the addition of matrices we might now ask the following questions:

1. Are matrices commutative under the operation of addition?
2. Are matrices associative under the operation of addition?

If

$$\begin{bmatrix} a_1 & b_1 \\ c_1 & d_1 \end{bmatrix} + \begin{bmatrix} a_2 & b_2 \\ c_2 & d_2 \end{bmatrix} = \begin{bmatrix} a_1 + a_2 & b_1 + b_2 \\ c_1 + c_2 & d_1 + d_2 \end{bmatrix}$$

$$\begin{bmatrix} a_2 & b_2 \\ c_2 & d_2 \end{bmatrix} + \begin{bmatrix} a_1 & b_1 \\ c_1 & d_1 \end{bmatrix} = \begin{bmatrix} a_2 + a_1 & b_2 + b_1 \\ c_2 + c_1 & d_2 + d_1 \end{bmatrix}.$$

then obviously the corresponding elements of the two sums are equal, since the elements are real numbers, which are commutative. For example, $a_1 + a_2 = a_2 + a_1$; therefore, the two matrices are equal. Using a similar approach, it can be readily shown that matrices are associative under the operation of addition. Although we have considered only 2×2 matrices, the proof may be generalized to $m \times n$ dimensioned matrices.

NOTATION. Before going on to the next important definition regarding matrices, we shall consider a new notation. The elements in a matrix are commonly numbered with a double subscript to more easily identify a particular element. For example,

$$A = \begin{bmatrix} a_{11} & a_{12} & a_{13} \\ a_{21} & a_{22} & a_{23} \\ a_{31} & a_{32} & a_{33} \end{bmatrix}$$

The element in the second row and third column of this matrix is called a_{23};

note that the first digit of the subscript indicates the row and the second indicates the column. Thus in the matrix

$$\begin{bmatrix} 1 & 4 & 3 \\ 5 & -2 & 6 \\ 2 & 0 & 4 \end{bmatrix}$$

we can say $a_{22} = -2$, $a_{31} = 2$, and so on. Uppercase letters such as A are used to represent matrices, whenever it is convenient to do so, to avoid writing the complete matrix each time. The choice of the letters (A and a in this case) is as arbitrary as in the algebra of real numbers.

In Chapter 7 we saw that single subscripted variables have their counterpart in subscripted variables of Fortran. But Fortran, like algebra, involves quantities with two or more subscripts. (Most versions of Fortran allow up to three subscripts on a variable while some allow even more.) Thus the general element of A could be represented in Fortran as A(I,J) and any specific element, say a_{23}, as A(2,3). As a basic illustration of how such quantities would be incorporated into a program, let us consider the simple task of reading data into a Fortran array. We will assume that the array consists of 15 rows and 10 columns and that each data point is on a separate card in the order a_{11}, a_{12}, a_{13}, $\cdots$ a_{21}, a_{22}, $\cdots$ (that is, the data is arranged in row order). The flowchart and two program segments to perform this operation are illustrated in Figure 10.1. Note the convenience in using the subscripted variable in conjunction with the DO statement as illustrated in the alternate program segment of 10.1(c).

EXERCISE 10.1

What are the dimensions of each of the following matrices? Identify the element in the second row and third column where appropriate:

1. $\begin{bmatrix} 2 \\ 3 \end{bmatrix}$

2. $\begin{bmatrix} 1 & -5 \\ 3 & 0 \end{bmatrix}$

3. $\begin{bmatrix} 1 & 4 & -5 \\ 1 & 0 & 1 \end{bmatrix}$

4. $\begin{bmatrix} 1 & 5 & 3 \end{bmatrix}$

5. $\begin{bmatrix} 2 & 1 & -2 \\ 4 & 1 & 5 \\ 3 & 0 & 4 \end{bmatrix}$

6. $\begin{bmatrix} 7 & -5 & 1 \\ 2 & 0 & -5 \\ 1 & 0 & 1 \\ 9 & -3 & -8 \end{bmatrix}$

7. $\begin{bmatrix} 2 & -4 & -8 & 1 \\ 9 & 0 & 5 & 6 \end{bmatrix}$

8. $\begin{bmatrix} a+b & c-d \\ c+d & a-b \end{bmatrix}$

9. $\begin{bmatrix} a_1+b_1 \\ a_2-b_2 \end{bmatrix}$

10. $\begin{bmatrix} x+z & b-c \\ b+c & x-z \\ x-c & b+z \end{bmatrix}$

What are the implications in each of the following equations?

11. $\begin{bmatrix} a & b \\ c & d \end{bmatrix} = \begin{bmatrix} 1 & 5 \\ 3 & -1 \end{bmatrix}$

12. $\begin{bmatrix} x \\ y \\ z \end{bmatrix} = \begin{bmatrix} 1 \\ 2 \\ 3 \end{bmatrix}$

13. $\begin{bmatrix} x_1 + x_2 \\ x_1 - x_2 \end{bmatrix} = \begin{bmatrix} 3 \\ 1 \end{bmatrix}$

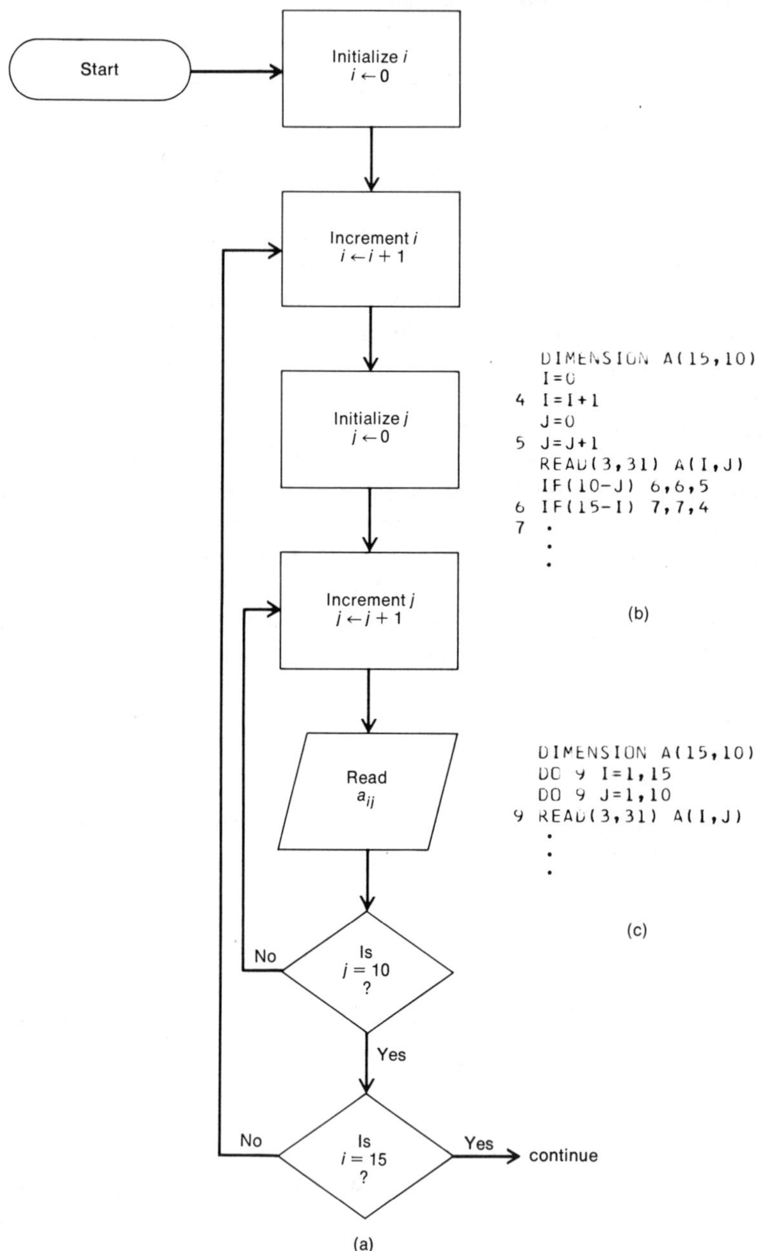

FIGURE 10.1 (a) The flowchart; (b) a program segment; (c) Using the DO

14. $\begin{bmatrix} x-y & a-b \\ a+b & x+y \end{bmatrix} = \begin{bmatrix} 5 & 5 \\ -1 & 7 \end{bmatrix}$

15. $\begin{bmatrix} 2x_1 - x_2 \\ -x_1 + 4x_2 \end{bmatrix} = \begin{bmatrix} 5 \\ 8 \end{bmatrix}$

Determine the equivalent matrix:

16. $\begin{bmatrix} 1 & 2 & 3 \\ 3 & 1 & -5 \end{bmatrix} + \begin{bmatrix} 1 & 4 & -5 \\ 2 & -1 & 5 \end{bmatrix}$

17. $\begin{bmatrix} 1 & 2 \\ 8 & 3 \\ 7 & 2 \end{bmatrix} + \begin{bmatrix} -5 & 4 \\ 2 & -3 \\ 0 & 1 \end{bmatrix}$

18. $\begin{bmatrix} 3 & 1 \\ -1 & 2 \end{bmatrix} + 2 \times \begin{bmatrix} 1 & 1 \\ 1 & 1 \end{bmatrix}$

19. $\begin{bmatrix} 2 & -5 & 6 \\ 0 & 1 & 0 \\ 4 & 0 & 3 \end{bmatrix} - \begin{bmatrix} 7 & 1 & 8 \\ 9 & 1 & 2 \\ 4 & 1 & 3 \end{bmatrix}$

20. $\begin{bmatrix} x & y \\ 2x & 3y \end{bmatrix} + \begin{bmatrix} 1 & 2 \\ -1 & 1 \end{bmatrix}$

21. $\begin{bmatrix} a_1 & b_1 \\ a_2 & b_2 \end{bmatrix} + \begin{bmatrix} 4 & 3 \\ 2 & -1 \end{bmatrix}$

22. $\begin{bmatrix} x_1 & y_1 \\ x_2 & z_2 \end{bmatrix} + \begin{bmatrix} x_1 & -z_1 \\ -x_2 & z_2 \end{bmatrix}$

23. $\begin{bmatrix} 1 & 4 \\ 2 & -1 \\ 0 & -3 \end{bmatrix} + \begin{bmatrix} -1 & -4 \\ -2 & 1 \\ 0 & 3 \end{bmatrix}$

24. $\begin{bmatrix} 2 & -5 \\ -3 & 8 \end{bmatrix} + \begin{bmatrix} -2 & 5 \\ 3 & -8 \end{bmatrix}$

25. $\begin{bmatrix} 1 & 3 \\ 5 & 9 \end{bmatrix} + \begin{bmatrix} 0 & 0 \\ 0 & 0 \end{bmatrix}$

26. $\begin{bmatrix} 0 & 0 \\ 0 & 0 \end{bmatrix} + \begin{bmatrix} -1 & 5 \\ -2 & 0 \end{bmatrix}$

27. $\begin{bmatrix} 1 & -2 & -3 \\ 3 & -2 & 1 \\ 5 & 2 & 7 \end{bmatrix} + \begin{bmatrix} 0 & 0 & 0 \\ 0 & 0 & 0 \\ 0 & 0 & 0 \end{bmatrix}$

28. What is the significance of the results of problems 23–27 relative to the basic properties of real numbers?

29. Referring to the matrix of Figure 10.1, write two program segments to subtract the fourth column from the first column, with the results to replace the first column. Write the first program segment without using the DO statement; utilize the DO in the second.

30. Referring to the matrix of Figure 10.1, write two program segments to find the sum of the third and seventh rows; store the results in a new array C. Write this segment both with and without using the DO statement.

10.2 MATRIX MULTIPLICATION

MULTIPLICATION. The product of two matrices is not intuitive, but rather it is defined to relate it to linear systems of equations. Consider the following product:

$$\begin{bmatrix} 3 & 4 \\ -1 & 7 \end{bmatrix} \times \begin{bmatrix} 5 & -2 \\ -6 & -3 \end{bmatrix} = \begin{bmatrix} a_{11} & a_{12} \\ a_{21} & a_{22} \end{bmatrix}.$$

The element a_{11} is determined by multiplying the individual elements of the first row of the first matrix by the corresponding elements of the first column of the second matrix, and taking their sum. That is,

$$\begin{aligned} a_{11} &= 3 \times 5 + 4(-6) \\ &= 15 - 24 \\ &= -9. \end{aligned}$$

Similarly, a_{12} involves the first row of the first matrix and the second column of the second matrix:

$$a_{12} = 3(-2) + 4(-3) = -18.$$

The elements a_{21} and a_{22} are

$$\begin{aligned} a_{21} &= (-1) \times 5 + 7(-6) = -47 \\ a_{22} &= (-1) \times (-2) + 7(-3) = -19. \end{aligned}$$

The resulting product is

$$\begin{bmatrix} 3 & 4 \\ -1 & 7 \end{bmatrix} \times \begin{bmatrix} 5 & -2 \\ -6 & -3 \end{bmatrix} = \begin{bmatrix} -9 & -18 \\ -47 & -19 \end{bmatrix}$$

Matrix multiplication is not limited to 2×2 systems or to square matrices, as we can see by the following example:

$$\begin{bmatrix} 4 & 1 \\ 7 & -3 \\ -2 & 1 \end{bmatrix} \times \begin{bmatrix} -1 & 3 & 2 & 1 \\ 5 & -2 & 0 & 4 \end{bmatrix}$$

The product of the first row (first matrix) and first column (second matrix) is

$$a_{11} = 4(-1) + 1 \times 5 = 1.$$

Continuing in this fashion, we have

$$\begin{aligned} a_{12} &= 4 \times 3 + 1(-2) = 10 \\ a_{13} &= 4 \times 2 + 1 \times 0 = 8 \\ a_{14} &= 4 \times 1 + 1 \times 4 = 8. \end{aligned}$$

Now, using the second row, we have

$$\begin{aligned} a_{21} &= 7(-1) + (-3) \times 5 = -22 \\ a_{22} &= 7 \times 3 + (-3)(-2) = 27 \\ a_{23} &= 7 \times 2 + (-3) \times 0 = 14 \\ a_{24} &= 7 \times 1 + (-3) \times 4 = -5. \end{aligned}$$

Finally, performing these operations with the third row gives

$$\begin{aligned} a_{31} &= (-2)(-1) + 1 \times 5 = 7 \\ a_{32} &= (-2) \times 3 + 1(-2) = -8 \\ a_{33} &= (-2) \times 2 + 1 \times 0 = -4 \\ a_{34} &= (-2) \times 1 + 1 \times 4 = 2. \end{aligned}$$

The required product is

$$\begin{bmatrix} 4 & 1 \\ 7 & -3 \\ -2 & 1 \end{bmatrix} \times \begin{bmatrix} -1 & 3 & 2 & 1 \\ 5 & -2 & 0 & 4 \end{bmatrix} = \begin{bmatrix} 1 & 10 & 8 & 8 \\ -22 & 27 & 14 & -5 \\ 7 & -8 & -4 & 2 \end{bmatrix}.$$

Much is to be learned from this example. It is apparent that matrices need not have the same dimensions in order to be multiplied. However, it is necessary that the first matrix have as many columns as the second matrix has rows. The product of a matrix with dimensions $m \times n$ and one with dimensions $n \times p$ is a third matrix with dimensions $m \times p$. In the above example, the product of a 3×2 matrix and a 2×4 matrix is a 3×4 matrix. Furthermore, it is evident that matrix multiplication is not commutative, since the following product is not even defined:

$$\begin{bmatrix} -1 & 3 & 2 & 1 \\ 5 & -2 & 0 & 4 \end{bmatrix} \begin{bmatrix} 4 & 1 \\ 7 & -3 \\ -2 & 1 \end{bmatrix}.$$

In other words, if A and B are two matrices, as a general rule,

$$AB \neq BA.$$

Try to perform this multiplication and note the results.

In both of the preceding examples, the product of two matrices was a third matrix whose dimensions depended upon the dimensions of the original matrices. Somewhat of a special case results when a row matrix multiplies a column matrix. For example,

$$[3 \quad 5 \quad 4] \begin{bmatrix} 6 \\ 1 \\ -3 \end{bmatrix} = 3 \times 6 + 5 \times 1 + 4(-3) = 11.$$

Here the product of two matrices results in a single number rather than another matrix. In vector analysis this is commonly referred to as the *scalar* product. Matrix products similar to this and the preceding examples are very useful in Chapter 11 (Linear Programming). However, in this chapter we shall concentrate on the products of square matrices ($n \times n$) and on the product of a square matrix and a column matrix. The latter is illustrated by the following example:

$$\begin{bmatrix} 1 & 3 \\ 2 & -5 \end{bmatrix} \begin{bmatrix} x \\ y \end{bmatrix} = \begin{bmatrix} x + 3y \\ 2x - 5y \end{bmatrix}.$$

Note that a 2×2 matrix multiplied by a 2×1 matrix yields a 2×1 (column) matrix.

THE IDENTITY MATRIX. Basic to the algebra of matrices is the multiplicative identity. In fact, this identity plays such an important role in solving linear systems that it is referred to as the *identity matrix* (commonly represented by I), with the operation of multiplication being understood. For 2×2 matrices, the identity element is

$$I = \begin{bmatrix} 1 & 0 \\ 0 & 1 \end{bmatrix}$$

since

$$\begin{bmatrix} 1 & 0 \\ 0 & 1 \end{bmatrix} \begin{bmatrix} a_{11} & a_{12} \\ a_{21} & a_{22} \end{bmatrix} = \begin{bmatrix} a_{11} & a_{12} \\ a_{21} & a_{22} \end{bmatrix}$$

or

$$IA = A.$$

The identity for 3×3 matrices is similar in structure, as shown by the following example:

$$\begin{bmatrix} 1 & 0 & 0 \\ 0 & 1 & 0 \\ 0 & 0 & 1 \end{bmatrix} \begin{bmatrix} b_{11} & b_{12} & b_{13} \\ b_{21} & b_{22} & b_{23} \\ b_{31} & b_{32} & b_{33} \end{bmatrix} = \begin{bmatrix} b_{11} & b_{12} & b_{13} \\ b_{21} & b_{22} & b_{23} \\ b_{31} & b_{32} & b_{33} \end{bmatrix}$$

or

$$IB = B.$$

The nonzero elements in the identity are said to constitute the **main diagonal** (in B the main diagonal consists of the elements b_{11}, b_{22}, and b_{33}). Generalizing, we can see that the identity for an $n \times n$ matrix is another $n \times n$ matrix whose main diagonal consists of ones and whose other elements are all zero.

In the previous section we learned that matrices do not commute under multiplication; in fact, in many instances, the product is not even defined. Even for square matrices, commutivity does not hold. For example,

$$\begin{bmatrix} 2 & 1 \\ -3 & 6 \end{bmatrix} \begin{bmatrix} 5 & 4 \\ 1 & -2 \end{bmatrix} = \begin{bmatrix} 11 & 6 \\ -9 & -24 \end{bmatrix}$$

but

$$\begin{bmatrix} 5 & 4 \\ 1 & -2 \end{bmatrix} \begin{bmatrix} 2 & 1 \\ -3 & 6 \end{bmatrix} = \begin{bmatrix} -2 & 29 \\ 8 & -11 \end{bmatrix}.$$

Thus it is extremely important in matrix algebra to avoid commuting factors for the sake of convenience. However, the identity matrix introduces an exception to this rule:

$$\begin{bmatrix} a_{11} & a_{12} \\ a_{21} & a_{22} \end{bmatrix} \begin{bmatrix} 1 & 0 \\ 0 & 1 \end{bmatrix} = \begin{bmatrix} a_{11} & a_{12} \\ a_{21} & a_{22} \end{bmatrix}$$

and

$$\begin{bmatrix} 1 & 0 \\ 0 & 1 \end{bmatrix} \begin{bmatrix} a_{11} & a_{12} \\ a_{21} & a_{22} \end{bmatrix} = \begin{bmatrix} a_{11} & a_{12} \\ a_{21} & a_{22} \end{bmatrix}$$

In other words, the product of a square matrix and its identity is commutative, which may be expressed as

$$IA = AI = A.$$

However, that which is true for a square matrix is not true for a column matrix:

$$\begin{bmatrix} 1 & 0 & 0 \\ 0 & 1 & 0 \\ 0 & 0 & 1 \end{bmatrix} \begin{bmatrix} c_1 \\ c_2 \\ c_2 \end{bmatrix} = \begin{bmatrix} c_1 \\ c_2 \\ c_3 \end{bmatrix}$$

but the product

$$\begin{bmatrix} c_1 \\ c_2 \\ c_3 \end{bmatrix} \begin{bmatrix} 1 & 0 & 0 \\ 0 & 1 & 0 \\ 0 & 0 & 1 \end{bmatrix}$$

is not defined.

ELEMENTARY TRANSFORMATION MATRICES. In solving linear systems, we employ the basic operations of (1) interchanging two equations (for convenience), (2) multiplying an equation by a constant, and (3) adding one equation to another.

In matrix algebra it is possible to interchange two rows, multiply a row by a constant, and add one row to another, by using so-called **elementary transformation matrices**. To illustrate these operations, we shall consider the matrix

$$\begin{bmatrix} a_{11} & a_{12} & a_{13} \\ a_{21} & a_{22} & a_{23} \\ a_{31} & a_{32} & a_{33} \end{bmatrix}.$$

Note the effect of multiplying this matrix by an identity matrix in which two rows have been interchanged:

$$\begin{bmatrix} 1 & 0 & 0 \\ 0 & 0 & 1 \\ 0 & 1 & 0 \end{bmatrix} \begin{bmatrix} a_{11} & a_{12} & a_{13} \\ a_{21} & a_{22} & a_{23} \\ a_{31} & a_{32} & a_{33} \end{bmatrix} = \begin{bmatrix} a_{11} & a_{12} & a_{13} \\ a_{31} & a_{32} & a_{33} \\ a_{21} & a_{22} & a_{23} \end{bmatrix}.$$

Multiplication on the left (premultiplication) by an identity matrix in which the second and third rows have been interchanged results in a product which is identical to the original matrix, except that corresponding rows have been interchanged.

To illustrate multiplication of a given row by a constant, consider premultiplication by an identity matrix in which one of the diagonal elements has been multiplied by a constant:

$$\begin{bmatrix} 1 & 0 & 0 \\ 0 & 1 & 0 \\ 0 & 0 & 3 \end{bmatrix} \begin{bmatrix} a_{11} & a_{12} & a_{13} \\ a_{21} & a_{22} & a_{23} \\ a_{31} & a_{32} & a_{33} \end{bmatrix} = \begin{bmatrix} a_{11} & a_{12} & a_{13} \\ a_{21} & a_{22} & a_{23} \\ 3a_{31} & 3a_{32} & 3a_{33} \end{bmatrix}.$$

Here the elementary transformation matrix contains 3 as one of the diagonal elements. The result is that the third row of the original matrix is multiplied by 3.

To cause one row to be added to another, it is necessary to employ a matrix with ones in the diagonal and a one in the proper off-diagonal position:

$$\begin{bmatrix} 1 & 1 & 0 \\ 0 & 1 & 0 \\ 0 & 0 & 1 \end{bmatrix} \begin{bmatrix} a_{11} & a_{12} & a_{13} \\ a_{21} & a_{22} & a_{23} \\ a_{31} & a_{32} & a_{33} \end{bmatrix} = \begin{bmatrix} a_{11}+a_{21} & a_{12}+a_{22} & a_{13}+a_{23} \\ a_{21} & a_{22} & a_{23} \\ a_{31} & a_{32} & a_{33} \end{bmatrix}.$$

(What is the precise relationship involved in locating the off-diagonal 1?)

Although these elementary transformation matrices are basic to the subject of matrix algebra, we will not explore them much further. However, this brief discussion should serve to introduce the matrix algebra manipulations required to solve linear systems by using matrix methods. Furthermore, it will provide the necessary background to completely justify some of the techniques used in later sections.

EXERCISE 10.2

Find the product of the given matrices below.

1. $\begin{bmatrix} 1 & 2 \\ -4 & 3 \\ 0 & -2 \end{bmatrix} \begin{bmatrix} 1 & 5 \\ -2 & 3 \end{bmatrix}$

2. $\begin{bmatrix} 1 & 0 & 5 \end{bmatrix} \begin{bmatrix} 3 & 2 \\ 1 & 6 \\ -4 & 5 \end{bmatrix}$

3. $\begin{bmatrix} 1 & -2 & 4 \\ -3 & 0 & 1 \\ 5 & 4 & -1 \end{bmatrix} \begin{bmatrix} 7 \\ -5 \\ -2 \end{bmatrix}$

4. $\begin{bmatrix} 2 & -3 \\ 1 & 5 \end{bmatrix} \begin{bmatrix} 1 & 0 \\ 5 & -4 \end{bmatrix}$

5. $\begin{bmatrix} 6 & 1 \\ 1 & 2 \end{bmatrix} \begin{bmatrix} 4 & -3 \\ 0 & 1 \end{bmatrix}$

6. $\begin{bmatrix} 2 & 5 \\ -1 & 3 \end{bmatrix} \begin{bmatrix} 1 & 1 \\ 2 & 1 \end{bmatrix}$

7. $\begin{bmatrix} 5 & -1 \\ 6 & 2 \end{bmatrix} \begin{bmatrix} 1 & 4 \\ 3 & -7 \end{bmatrix}$

8. $\begin{bmatrix} 8 & -1 \\ -1 & 7 \end{bmatrix} \begin{bmatrix} 2 & 7 \\ 3 & -5 \end{bmatrix}$

9. $\begin{bmatrix} 8 & -1 \\ 2 & 2 \end{bmatrix} \begin{bmatrix} 1 & 0 \\ 0 & -1 \end{bmatrix}$

10. $\begin{bmatrix} -5 & 3 \\ 2 & -7 \end{bmatrix} \begin{bmatrix} x \\ 7 \end{bmatrix}$

11. $\begin{bmatrix} 6 & 8 \\ -2 & 3 \end{bmatrix} \begin{bmatrix} m \\ z \end{bmatrix}$

12. $\begin{bmatrix} 7 & 0 \\ 3 & -1 \end{bmatrix} \begin{bmatrix} m \\ n \end{bmatrix}$

13. $\begin{bmatrix} 8 & 6 \\ 1 & 5 \end{bmatrix} \begin{bmatrix} x \\ z \end{bmatrix}$

14. $\begin{bmatrix} 3 & 1 & -1 \\ 0 & 5 & -2 \\ 1 & 5 & -7 \end{bmatrix} \begin{bmatrix} -3 & 6 & 2 \\ 4 & 1 & -2 \\ 3 & 0 & 1 \end{bmatrix}$

15. $\begin{bmatrix} 17 & 5 & -9 \\ 6 & -12 & 3 \\ 18 & 2 & 1 \end{bmatrix} \begin{bmatrix} 1 & 5 & 1 \\ 0 & 1 & -2 \\ 3 & 9 & 7 \end{bmatrix}$

16. $\begin{bmatrix} 3 & -1 & -1 \\ 4 & -1 & 2 \\ -19 & 0 & 3 \end{bmatrix} \begin{bmatrix} 1 & 5 & 1 \\ -2 & 16 & 3 \\ 12 & -13 & 7 \end{bmatrix}$

17. $\begin{bmatrix} 1 & 0 & -1 \\ 5 & 0 & 3 \\ 0 & 1 & 0 \end{bmatrix} \begin{bmatrix} 1 & 2 & 0 \\ 8 & -9 & 1 \\ 0 & 0 & -5 \end{bmatrix}$

18. $\begin{bmatrix} 6 & -3 & 8 \\ 4 & 5 & -1 \\ 7 & 9 & 3 \end{bmatrix} \begin{bmatrix} x \\ y \\ z \end{bmatrix}$

19. $\begin{bmatrix} 2 & 0 & 5 \\ 3 & -1 & 0 \\ 5 & 2 & -1 \end{bmatrix} \begin{bmatrix} x \\ y \\ z \end{bmatrix}$

20. $\begin{bmatrix} 4 & 5 & -1 \\ 3 & 0 & 1 \\ 0 & 0 & 2 \end{bmatrix} \begin{bmatrix} a \\ b \\ c \end{bmatrix}$

21. $\begin{bmatrix} 1 & 0 & 0 \\ 0 & 5 & 0 \\ 0 & 0 & 1 \end{bmatrix} \begin{bmatrix} 1 & 2 & 4 \\ 3 & -2 & 5 \\ 4 & 7 & -5 \end{bmatrix}$

22. $\begin{bmatrix} -3 & 0 & 0 \\ 0 & 1 & 0 \\ 0 & 0 & 1 \end{bmatrix} \begin{bmatrix} 1 & 2 & 4 \\ 3 & -2 & 5 \\ 4 & 7 & -5 \end{bmatrix}$

23. $\begin{bmatrix} 1 & 0 & 0 \\ 0 & 1 & 0 \\ 0 & 0 & -1 \end{bmatrix} \begin{bmatrix} 1 & 2 & 4 \\ 3 & -2 & 5 \\ 4 & 7 & -5 \end{bmatrix}$

24. $\begin{bmatrix} 1 & 0 & 0 \\ 0 & 0 & 1 \\ 0 & 1 & 0 \end{bmatrix} \begin{bmatrix} 1 & 2 & 4 \\ 3 & -2 & 5 \\ 4 & 7 & -5 \end{bmatrix}$

25. $\begin{bmatrix} 0 & 0 & 1 \\ 0 & 1 & 0 \\ 1 & 0 & 0 \end{bmatrix} \begin{bmatrix} 1 & 2 & 4 \\ 3 & -2 & 5 \\ 4 & 7 & -5 \end{bmatrix}$

26. $\begin{bmatrix} 1 & 0 & 1 \\ 0 & 1 & 0 \\ 0 & 0 & 1 \end{bmatrix} \begin{bmatrix} 1 & 2 & 4 \\ 3 & -2 & 5 \\ 4 & 7 & -5 \end{bmatrix}$

27. $\begin{bmatrix} 1 & 0 & 0 \\ 1 & 1 & 0 \\ 0 & 0 & 1 \end{bmatrix} \begin{bmatrix} 1 & 2 & 4 \\ 3 & -2 & 5 \\ 4 & 7 & -5 \end{bmatrix}$

28. $\begin{bmatrix} 1 & 0 & -2 \\ 0 & 1 & 0 \\ 0 & 0 & 1 \end{bmatrix} \begin{bmatrix} 1 & 2 & 4 \\ 3 & -2 & 5 \\ 4 & 7 & -5 \end{bmatrix}$

29. $\begin{bmatrix} a_{11} & a_{12} & a_{13} \\ a_{21} & a_{22} & a_{23} \\ a_{31} & a_{32} & a_{33} \end{bmatrix} \begin{bmatrix} 1 & 0 & 0 \\ 0 & 0 & 1 \\ 0 & 1 & 0 \end{bmatrix}$

30. $\begin{bmatrix} a_{11} & a_{12} & a_{13} \\ a_{21} & a_{22} & a_{23} \\ a_{31} & a_{32} & a_{33} \end{bmatrix} \begin{bmatrix} 1 & 0 & 0 \\ 0 & -4 & 0 \\ 0 & 0 & 1 \end{bmatrix}$

31. $\begin{bmatrix} a_{11} & a_{12} & a_{13} \\ a_{21} & a_{22} & a_{23} \\ a_{31} & a_{32} & a_{33} \end{bmatrix} \begin{bmatrix} 1 & 0 & 0 \\ 1 & 1 & 0 \\ 0 & 0 & 1 \end{bmatrix}$

32. In problems 29-31, matrices are postmultiplied by elementary transformation matrices. How does the end result differ from premultiplying?

33. Find the values for a, b, c, and d if

$$\begin{bmatrix} -3 & -7 \\ 2 & 5 \end{bmatrix} \begin{bmatrix} a & b \\ c & d \end{bmatrix} = \begin{bmatrix} 1 & 0 \\ 0 & 1 \end{bmatrix}$$

or if

$$\begin{bmatrix} a & b \\ c & d \end{bmatrix} \begin{bmatrix} -3 & -7 \\ 2 & 5 \end{bmatrix} = \begin{bmatrix} 1 & 0 \\ 0 & 1 \end{bmatrix}$$

34. Find the value for p, q, r, and s if

$$\begin{bmatrix} p & q \\ r & s \end{bmatrix} \begin{bmatrix} 18 & 5 \\ 11 & 3 \end{bmatrix} = \begin{bmatrix} 1 & 0 \\ 0 & 1 \end{bmatrix}$$

or if

$$\begin{bmatrix} 18 & 5 \\ 11 & 3 \end{bmatrix} \begin{bmatrix} p & q \\ r & s \end{bmatrix} = \begin{bmatrix} 1 & 0 \\ 0 & 1 \end{bmatrix}.$$

Three basic considerations in the study of ordinary algebra are (*a*) for any two real numbers a and b, $ab = ba$; (*b*) if $ab = 0$, either $a = 0$, $b = 0$, or $a = b = 0$; (*c*) if $ab = ac$ and $a \neq 0$, $b = c$. We have seen that matrices are not commutative under multiplication. Consider (*b*) and (*c*) in view of the following exercises. Determine the indicated matrix products.

35. $\begin{bmatrix} 3 & -3 \\ -2 & 2 \end{bmatrix} \begin{bmatrix} 1 & 1 \\ 1 & 1 \end{bmatrix}$

36. $\begin{bmatrix} -2 & -8 \\ 1 & 4 \end{bmatrix} \begin{bmatrix} 4 & 4 \\ -1 & -1 \end{bmatrix}$

37. $\begin{bmatrix} 1 & 3 & 0 \\ 5 & -2 & 0 \\ 6 & 3 & 0 \end{bmatrix} \begin{bmatrix} 1 & 4 & -7 \\ 3 & -1 & 2 \\ 1 & 3 & 0 \end{bmatrix}$ and $\begin{bmatrix} 1 & 3 & 0 \\ 5 & -2 & 0 \\ 6 & 3 & 0 \end{bmatrix} \begin{bmatrix} 1 & 4 & -7 \\ 3 & -1 & 2 \\ 5 & 2 & 7 \end{bmatrix}$

38. $\begin{bmatrix} 9 & -3 & 7 \\ -16 & 8 & 1 \\ 12 & 3 & -7 \end{bmatrix} \begin{bmatrix} 0 & 0 & 0 \\ 3 & -2 & 7 \\ 5 & -1 & -4 \end{bmatrix}$ and $\begin{bmatrix} 8 & -3 & 7 \\ 9 & 8 & 1 \\ 0 & 3 & -7 \end{bmatrix} \begin{bmatrix} 0 & 0 & 0 \\ 3 & -2 & 7 \\ 5 & -1 & -4 \end{bmatrix}$

39. Demonstrate that 2×2 matrices are associative under multiplication, using

$$\begin{bmatrix} a_{11} & a_{12} \\ a_{21} & a_{22} \end{bmatrix} \quad \begin{bmatrix} b_{11} & b_{12} \\ b_{21} & b_{22} \end{bmatrix} \quad \begin{bmatrix} c_{11} & c_{12} \\ c_{21} & c_{22} \end{bmatrix}.$$

Argue that, in general, $n \times n$ matrices are associative.

40. Demonstrate that 2×2 matrices are distributive using the matrices of problem 39. Argue that, in general, $n \times n$ matrices are distributive.

10.3 SOLVING SIMULTANEOUS SYSTEMS WITH MATRICES

INVERSE MATRICES. In studying multiplication of real numbers, the concept of inverse was associated with the identity. That is, since $a \times 1 = 1 \times a$, 1 is called the identity for multiplication. (Furthermore, since $a \times (1/a) = 1$, $1/a$ is defined as the inverse of a.) In matrix algebra a similar identity exists, which we refer to as I. Now we might ask if, for every square matrix A, an inverse exists (which we shall call A^{-1}) such that $A^{-1} \times A = I$. Proof that such an inverse will always exist (except for certain types of matrices) is beyond the scope of this text, but we shall use inverses, and in a later section we shall study a general method of finding them.

As an illustration of the use of inverses, let us begin with the 2×2 matrix A, which is

$$A = \begin{bmatrix} 8 & 5 \\ 3 & 2 \end{bmatrix}.$$

Let us multiply A by the matrix

$$B = \begin{bmatrix} 2 & -5 \\ -3 & 8 \end{bmatrix}$$

$$\begin{bmatrix} 2 & -5 \\ -3 & 8 \end{bmatrix} \begin{bmatrix} 8 & 5 \\ 3 & 2 \end{bmatrix} = \begin{bmatrix} 1 & 0 \\ 0 & 1 \end{bmatrix}$$

or

$$BA = I.$$

Thus B is the inverse of A or $B = A^{-1}$. Unlike matrix multiplication in general, the product of a matrix and its inverse is commutative:

$$A^{-1} \times A = A \times A^{-1} = I.$$

This is confirmed by the preceding example, since

$$\begin{bmatrix} 8 & 5 \\ 3 & 2 \end{bmatrix} \begin{bmatrix} 2 & -5 \\ -3 & 8 \end{bmatrix} = \begin{bmatrix} 1 & 0 \\ 0 & 1 \end{bmatrix}.$$

For 3×3 and larger matrices, the criterion is the same; that is, the product of a matrix and its inverse is the identity matrix. For example, if

$$B = \begin{bmatrix} 2 & -3 & -2 \\ -3 & 1 & 2 \\ 7 & -2 & -6 \end{bmatrix}$$

then

$$B^{-1} = \begin{bmatrix} -\frac{1}{5} & -\frac{7}{5} & -\frac{2}{5} \\ -\frac{2}{5} & \frac{1}{5} & \frac{1}{5} \\ -\frac{1}{10} & -\frac{17}{10} & -\frac{7}{10} \end{bmatrix}$$

since

$$\begin{bmatrix} -\frac{1}{5} & -\frac{7}{5} & -\frac{2}{5} \\ -\frac{2}{5} & \frac{1}{5} & \frac{1}{5} \\ -\frac{1}{10} & -\frac{17}{10} & -\frac{7}{10} \end{bmatrix} \begin{bmatrix} 2 & -3 & -2 \\ -3 & 1 & 2 \\ 7 & -2 & -6 \end{bmatrix} = \begin{bmatrix} 1 & 0 & 0 \\ 0 & 1 & 0 \\ 0 & 0 & 1 \end{bmatrix}$$

or

$$B^{-1}B = I.$$

Show that $BB^{-1} = I$.

MATRICES STORED IN THE COMPUTER. Whenever the elements of a matrix or its inverse are in fractional form, we can carry out hand computations using the fractional forms. However, when such a number is processed by the computer it is normally carried in decimal form. For example, the computer might use 0.33333333 in place of 1/3. As a result, the product of a matrix and the computer's version of its inverse will not always result in precisely the identity matrix. As an example, consider

$$A = \begin{bmatrix} 7 & 4 \\ 8 & 5 \end{bmatrix} \qquad A^{-1} = \begin{bmatrix} \frac{5}{3} & -\frac{4}{3} \\ -\frac{8}{3} & \frac{7}{3} \end{bmatrix}.$$

Within the computer, A^{-1} might appear as

$$A^{-1} = \begin{bmatrix} 1.666 & -1.333 \\ -2.666 & 2.333 \end{bmatrix}$$

and its product of A and A^{-1} would be

$$\begin{bmatrix} 7 & 4 \\ 8 & 5 \end{bmatrix} \begin{bmatrix} 1.666 & -1.333 \\ -2.666 & 2.333 \end{bmatrix} = \begin{bmatrix} 0.998 & 0.001 \\ -0.002 & 1.001 \end{bmatrix}$$

Although the result is not the identity, it is sufficiently close that the computer form of A^{-1} can be considered as the inverse.

In each of these examples, it is apparent from the results that the selected matrices are the appropriate inverses. Means for determining these matrices is quite another problem, as we shall see in Section 10.4.

SIMULTANEOUS EQUATIONS IN MATRIX FORM. In Section 10.1, equality of matrices was defined to imply that the matrices are identical in every respect; in Section 10.2, a means for multiplying one matrix by another was defined. Keeping in mind both of these notions, let us study the following matrix equation:

$$\begin{bmatrix} 2 & -3 & -2 \\ -3 & 1 & 2 \\ 7 & -2 & -6 \end{bmatrix} \begin{bmatrix} x \\ y \\ z \end{bmatrix} = \begin{bmatrix} 6 \\ -1 \\ -7 \end{bmatrix}. \tag{1}$$

Performing the indicated matrix multiplication on the left gives

$$\begin{bmatrix} 2x - 3y - 2z \\ -3x + y + 2z \\ 7x - 2y - 6z \end{bmatrix} = \begin{bmatrix} 6 \\ -1 \\ -7 \end{bmatrix}. \tag{2}$$

Note that the matrix product in (2) is a 3×1 matrix, and that each element consists of three algebraic terms. However, in order that two matrices of equal dimensions be equal, corresponding elements must be equal. Thus, by equating corresponding elements, we have

$$\begin{aligned} 2x - 3y - 2z &= 6 \\ -3x + y + 2z &= -1 \\ 7x - 2y - 6z &= -7. \end{aligned} \tag{3}$$

In other words, the simultaneous system of (3) may be represented by the matrix equation of (1) which, in turn, may be represented in a compact form:

$$A = \begin{bmatrix} 2 & -3 & -2 \\ -3 & 1 & 2 \\ 7 & -2 & -6 \end{bmatrix} \qquad X = \begin{bmatrix} x \\ y \\ z \end{bmatrix} \qquad B = \begin{bmatrix} 6 \\ -1 \\ -7 \end{bmatrix}.$$

Remembering that A, X, and B are matrices rather than real numbers, we have

$$AX = B. \tag{4}$$

This may be solved for the matrix X by employing the inverse A^{-1}:

$$A^{-1}AX = A^{-1}B \tag{5}$$

$$IX = A^{-1}B$$

$$X = A^{-1}B. \tag{6}$$

It is important to recognize that division by the matrix A to solve for X in (4) is not possible, since division is not defined in matrix algebra. Furthermore, we must take great care in performing the multiplication by A^{-1} that gives (5), since

$$A^{-1}AX \neq BA^{-1}. \tag{7}$$

Compare (6) and (7) and determine why (7) is not an equality.

APPLYING MATRIX RELATIONSHIPS. Let us apply the relationships of (4)-(6) to solve (3). It can be seen that A is the matrix whose inverse was examined previously:

$$A^{-1} = \begin{bmatrix} -\frac{1}{5} & -\frac{7}{5} & -\frac{2}{5} \\ -\frac{2}{5} & \frac{1}{5} & \frac{1}{5} \\ -\frac{1}{10} & -\frac{17}{10} & -\frac{7}{10} \end{bmatrix}.$$

Thus (5) becomes

$$\begin{bmatrix} -\frac{1}{5} & -\frac{7}{5} & -\frac{2}{5} \\ -\frac{2}{5} & \frac{1}{5} & \frac{1}{5} \\ -\frac{1}{10} & -\frac{17}{10} & -\frac{7}{10} \end{bmatrix} \begin{bmatrix} 2 & -3 & -2 \\ -3 & 1 & 2 \\ 7 & -2 & -6 \end{bmatrix} \begin{bmatrix} x \\ y \\ z \end{bmatrix} = \begin{bmatrix} -\frac{1}{5} & -\frac{7}{5} & -\frac{2}{5} & 6 \\ -\frac{2}{5} & \frac{1}{5} & \frac{1}{5} & -1 \\ -\frac{1}{10} & -\frac{17}{10} & -\frac{7}{10} & -7 \end{bmatrix} \tag{8}$$

But $A^{-1}A = I$, and $A^{-1}B$ yields

$$\begin{bmatrix} -\frac{1}{5} & -\frac{7}{5} & -\frac{2}{5} \\ -\frac{2}{5} & \frac{1}{5} & \frac{1}{5} \\ -\frac{1}{10} & -\frac{17}{10} & -\frac{7}{10} \end{bmatrix} \begin{bmatrix} 6 \\ -1 \\ -7 \end{bmatrix} = \begin{bmatrix} 3 \\ -4 \\ 6 \end{bmatrix}.$$

The result is that (8) reduces to

$$\begin{bmatrix} 1 & 0 & 0 \\ 0 & 1 & 0 \\ 0 & 0 & 1 \end{bmatrix} \begin{bmatrix} x \\ y \\ z \end{bmatrix} = \begin{bmatrix} 3 \\ -4 \\ 6 \end{bmatrix}. \tag{9}$$

But this, by the nature of the identity matrix, gives

$$\begin{bmatrix} x \\ y \\ z \end{bmatrix} = \begin{bmatrix} 3 \\ -4 \\ 6 \end{bmatrix}. \tag{10}$$

Since the equality of matrices implies that corresponding elements are equal,

$$x = 3 \qquad y = -4 \qquad z = 6$$

which is the solution to the original system (3). Thus, by putting the system in matrix form, and then multiplying both sides of the matrix equation by the inverse of the coefficient matrix, we obtain the solution.

EXERCISE 10.3

In each of the following, determine whether or not $B = A^{-1}$; that is, whether or not B is the inverse of A.

1. $A = \begin{bmatrix} 5 & 7 \\ 3 & 4 \end{bmatrix} \qquad B = \begin{bmatrix} -4 & 7 \\ 3 & -5 \end{bmatrix}$

2. $A = \begin{bmatrix} -2 & 7 \\ 3 & 1 \end{bmatrix} \qquad B = \begin{bmatrix} 10 & 7 \\ 3 & 2 \end{bmatrix}$

3. $A = \begin{bmatrix} -5 & 9 \\ 6 & -11 \end{bmatrix} \qquad B = \begin{bmatrix} -11 & -9 \\ -6 & -5 \end{bmatrix}$

4. $A = \begin{bmatrix} -5 & -11 \\ 2 & 4 \end{bmatrix} \qquad B = \begin{bmatrix} 2 & \frac{11}{2} \\ -1 & -\frac{5}{2} \end{bmatrix}$

5. $A = \begin{bmatrix} 3 & 4 \\ 5 & 6 \end{bmatrix} \qquad B = \begin{bmatrix} -3 & 2 \\ \frac{5}{2} & -\frac{3}{2} \end{bmatrix}$

6. $A = \begin{bmatrix} 11 & -4 \\ 6 & 1 \end{bmatrix} \qquad B = \begin{bmatrix} 3 & 1 \\ 8 & -5 \end{bmatrix}$

7. $A = \begin{bmatrix} -5 & -7 \\ 3 & 4 \end{bmatrix} \qquad B = \begin{bmatrix} -4 & -7 \\ 3 & 5 \end{bmatrix}$

8. $A = \begin{bmatrix} 7 & -1 \\ -6 & 1 \end{bmatrix} \qquad B = \begin{bmatrix} 1 & 1 \\ 6 & 7 \end{bmatrix}$

9. $A = \begin{bmatrix} 1 & 2 \\ 1 & -1 \end{bmatrix} \qquad B = \begin{bmatrix} \frac{1}{3} & \frac{2}{3} \\ \frac{1}{3} & -\frac{1}{3} \end{bmatrix}$

10. $A = \begin{bmatrix} -5 & 2 \\ 8 & -4 \end{bmatrix} \qquad B = \begin{bmatrix} -1 & -0.5 \\ -2 & -1.25 \end{bmatrix}$

11. $A = \begin{bmatrix} 5 & 3 \\ 4 & 3 \end{bmatrix} \qquad B = \begin{bmatrix} 1 & -1 \\ -1.333 & 1.667 \end{bmatrix}$

12. $A = \begin{bmatrix} 2 & 3 \\ -1 & 2 \end{bmatrix} \qquad B = \begin{bmatrix} 0.286 & -0.429 \\ 0.143 & 0.286 \end{bmatrix}$

Identify which of the preceding matrices may be identified with each of the following systems, and use the corresponding inverse to solve them using matrix methods. (It may be necessary to rearrange the equations in some instances.)

13. $7x - y = -1$
$-6x + y = 2$

14. $3x + 4y = 11$
$5x + 6y = 15$

15. $3x + 4y = -4$
$5x + 7y = -9$

16. $5x - 9y = 6$
$6x - 11y = 7$

17. $-x + y = -7$
$x + 2y = 22$

18. $5x + 11y = 2$
$x + 2y = 0$

19. $x - 2y = 3$
$2x + 3y = 7$

20. $3x + 4y = 7$
$3x + 5y = 14$

21. $x - 2y = 5$
$2x - 5y = 14$

In each of the following, determine whether or not $B = A^{-1}$:

22. $A = \begin{bmatrix} 1 & 0 & 3 \\ 1 & 1 & 3 \\ 0 & 2 & 2 \end{bmatrix} \qquad B = \begin{bmatrix} -2 & 3 & -\frac{3}{2} \\ -1 & 1 & 0 \\ 1 & -1 & \frac{1}{2} \end{bmatrix}$

23. $A = \begin{bmatrix} 1 & -3 & -1 \\ -14 & 45 & 16 \\ -9 & 29 & 10 \end{bmatrix} \qquad B = \begin{bmatrix} 14 & -1 & 3 \\ 4 & -1 & 2 \\ 1 & 2 & -3 \end{bmatrix}$

24. $A = \begin{bmatrix} 4 & 2 & 1 \\ 2 & 1 & 1 \\ 1 & 2 & 8 \end{bmatrix} \qquad B = \begin{bmatrix} 1 & 1 & 0 \\ -1 & -1 & 2 \\ -1 & 2 & -4 \end{bmatrix}$

25. $A = \begin{bmatrix} 4 & 4 & -6 \\ -4 & -6 & 8 \\ 2 & 4 & -4 \end{bmatrix} \qquad B = \begin{bmatrix} 1 & 1 & \frac{1}{2} \\ 0 & \frac{1}{2} & 1 \\ \frac{1}{2} & 1 & 1 \end{bmatrix}$

26. $A = \begin{bmatrix} 2 & -3 & \frac{3}{2} \\ 1 & -1 & 0 \\ -1 & 1 & -\frac{1}{2} \end{bmatrix}$ $B = \begin{bmatrix} 1 & 0 & 3 \\ 1 & 1 & 3 \\ 0 & 2 & 2 \end{bmatrix}$

27. $A = \begin{bmatrix} 0 & 1 & 1 & 1 \\ -1 & 0 & 1 & 1 \\ -1 & -1 & 0 & 1 \\ -1 & -1 & -1 & 0 \end{bmatrix}$ $B = \begin{bmatrix} 0 & -1 & 1 & -1 \\ 1 & 0 & -1 & 1 \\ -1 & 1 & 0 & -1 \\ 1 & -1 & 1 & 0 \end{bmatrix}$

28. $A = \begin{bmatrix} 6 & -1 & 0 \\ 0 & 1 & 3 \\ 4 & 0 & -9 \end{bmatrix}$ $B = \begin{bmatrix} \frac{3}{22} & \frac{3}{22} & \frac{1}{22} \\ -\frac{2}{11} & \frac{9}{11} & \frac{3}{11} \\ \frac{2}{33} & \frac{2}{33} & -\frac{1}{11} \end{bmatrix}$

29. $A = \begin{bmatrix} 2 & -1 & 6 \\ 3 & -5 & -9 \\ 3 & 0 & 3 \end{bmatrix}$ $B = \begin{bmatrix} -0.1562 & 0.0312 & 0.4062 \\ -0.3749 & -0.1250 & 0.3749 \\ 0.1562 & -0.0312 & -0.0729 \end{bmatrix}$

30. $A = \begin{bmatrix} 5 & -2 & -3 \\ 8 & 9 & -1 \\ -1 & 10 & 1 \end{bmatrix}$ $B = \begin{bmatrix} -0.1202 & 0.1772 & -0.1835 \\ 0.0443 & -0.0126 & 0.1202 \\ -0.5632 & 0.3037 & -0.3860 \end{bmatrix}$

Identify which of the preceding matrices may be identified with each of the following systems, and use the corresponding inverse to solve them using matrix methods. (It may be necessary to rearrange the equations in some instances.)

31. $x + 3z = 4$
$x + y + 3z = 3$
$2y + 2z = 0$

32. $2x + 2y - 3z = 12$
$-2x - 3y + 4z = -14$
$x + 2y - 2z = 7$

33. $14x - y + 3z = 3$
$4x - y + 2z = -5$
$x + 2y - 3z = 17$

34. $2x - y + 6z = 10$
$3x - 5y - 9z = -10$
$3x + 3z = 12$

35. $-x + 10y + z = 34$
$5x - 2y - 3z = -24$
$8x + 9y - z = 2$

36. If $A = \begin{bmatrix} -3 & 3 & -3 & 2 \\ 3 & -4 & 4 & -2 \\ -3 & 4 & -5 & 3 \\ 2 & -2 & 3 & -2 \end{bmatrix}$ show that $A^{-1} = \begin{bmatrix} 0 & 1 & 2 & 2 \\ 1 & 1 & 2 & 3 \\ 2 & 2 & 2 & 3 \\ 2 & 3 & 3 & 3 \end{bmatrix}$

37. Solve the following system, using the matrices of problem 36.

$$\begin{aligned} -3w + 3x - 3y + 2z &= -11 \\ 3w - 4x + 4y - 2z &= 13 \\ -3w + 4x - 5y + 3z &= -15 \\ 2w - 2x + 3y - 2z &= 9. \end{aligned}$$

10.4 FINDING THE INVERSE OF A MATRIX

The solution to a linear system can be found by simple multiplication—if the inverse of the coefficient matrix is known. However, the problem of finding the inverse for a matrix is much more complex than finding the inverse of a real number. Because of the importance of the inverse, much attention has been devoted by mathematicians to devising techniques for determining it. In fact, within the last several years, many very sophisticated methods have been evolved for use on computers; but the mathematics of these techniques is considerably beyond the scope of this text.

INVERSE OF A 2 × 2 MATRIX. A simple technique can be used to obtain the inverse of a 2 × 2 matrix. It consists of interchanging the two main diagonal elements, changing the sign of the off-diagonal elements, and dividing each element by the determinant of the matrix. For example,

$$2x - 5y = 1 \qquad 3x - 9y = -3.$$

In matrix form this system is

$$\begin{bmatrix} 2 & -5 \\ 3 & -9 \end{bmatrix} \begin{bmatrix} x \\ y \end{bmatrix} = \begin{bmatrix} 1 \\ -3 \end{bmatrix}.$$

The required determinant of the coefficient matrix is

$$\begin{vmatrix} 2 & -5 \\ 3 & -9 \end{vmatrix} = -18 - (-15) = -3.$$

Interchanging the main diagonal elements, changing the sign on the off-diagonal elements, and dividing each element in the coefficient matrix, gives the inverse, which is

$$\begin{bmatrix} \frac{-9}{-3} & \frac{5}{-3} \\ \frac{-3}{-3} & \frac{2}{-3} \end{bmatrix} = \begin{bmatrix} 3 & -\frac{5}{3} \\ 1 & -\frac{2}{3} \end{bmatrix}.$$

Thus

$$\begin{bmatrix} 3 & -\frac{5}{3} \\ 1 & -\frac{2}{3} \end{bmatrix} \begin{bmatrix} 2 & -5 \\ 3 & -9 \end{bmatrix} \begin{bmatrix} x \\ y \end{bmatrix} = \begin{bmatrix} 3 & -\frac{5}{3} \\ 1 & -\frac{2}{3} \end{bmatrix} \begin{bmatrix} 1 \\ -3 \end{bmatrix}$$

$$\begin{bmatrix} 1 & 0 \\ 0 & 1 \end{bmatrix} \begin{bmatrix} x \\ y \end{bmatrix} = \begin{bmatrix} 8 \\ 3 \end{bmatrix}$$

$$\begin{bmatrix} x \\ y \end{bmatrix} = \begin{bmatrix} 8 \\ 3 \end{bmatrix}.$$

The solution to the original equation is $x = 8$, $y = 3$.

Although we have used a pragmatic approach for finding the inverse of a 2×2 matrix, the derivation of this method is a relatively easy process. It can be found in the first reference listed at the end of this chapter. To justify it, refer to problem 18, at the end of this section.

INVERSE OF AN N × N *MATRIX.* The method for finding the inverse of an $n \times n$ matrix is basically a variation of the Gauss-Jordan method for solving linear equations (see Chapter 9). It involves operating on rows of the matrix to reduce it to the identity matrix; the same operations performed on an identity matrix will transform it to the inverse of the original matrix. To illustrate it, we shall consider the following linear system:

$$\begin{aligned} 2x - 3y - 2z &= 6 \\ -3x + y + 2z &= -1 \\ 7x - 2y - 6z &= -7. \end{aligned}$$

This system has the coefficient matrix

$$\begin{bmatrix} 2 & -3 & -2 \\ -3 & 1 & 2 \\ 7 & -2 & -6 \end{bmatrix}.$$

Now we shall operate on its rows to reduce it to the identity matrix, and simultaneously perform the same operations on an identity matrix. For convenience, we will write the two arrays within the same set of parentheses, but separated by a vertical line:

$$\left[\begin{array}{rrr|rrr} 2 & -3 & -2 & 1 & 0 & 0 \\ -3 & 1 & 2 & 0 & 1 & 0 \\ 7 & -2 & -6 & 0 & 0 & 1 \end{array}\right] \begin{array}{r} (1) \\ (2) \\ (3) \end{array}$$

$$\left[\begin{array}{rrr|rrr} 1 & -\frac{3}{2} & -1 & \frac{1}{2} & 0 & 0 \\ -3 & 1 & 2 & 0 & 1 & 0 \\ 7 & -2 & -6 & 0 & 0 & 1 \end{array}\right] \quad \begin{array}{ll} [\textit{Multiply } (1) \textit{ by } 1/2] & (4) \\ & (2) \\ & (3) \end{array}$$

$$\left[\begin{array}{rrr|rrr} 1 & -\frac{3}{2} & -1 & \frac{1}{2} & 0 & 0 \\ 0 & -\frac{7}{2} & -1 & \frac{3}{2} & 1 & 0 \\ 0 & \frac{17}{2} & 1 & -\frac{7}{2} & 0 & 1 \end{array}\right] \quad \begin{array}{ll} & (4) \\ [\textit{Add } 3 \textit{ times } (4) \textit{ to } (2)] & (5) \\ [\textit{Add } -7 \textit{ times } (4) \textit{ to } (3)] & (6) \end{array}$$

$$\left[\begin{array}{rrr|rrr} 1 & -\frac{3}{2} & -1 & \frac{1}{2} & 0 & 0 \\ 0 & 1 & \frac{2}{7} & -\frac{3}{7} & -\frac{2}{7} & 0 \\ 0 & \frac{17}{2} & 1 & -\frac{7}{2} & 0 & 1 \end{array}\right] \quad \begin{array}{ll} & (4) \\ [\textit{Multiply } (5) \text{ by } -2/7] & (7) \\ & (6) \end{array}$$

$$\left[\begin{array}{rrr|rrr} 1 & 0 & -\frac{4}{7} & -\frac{1}{7} & -\frac{3}{7} & 0 \\ 0 & 1 & \frac{2}{7} & -\frac{3}{7} & -\frac{2}{7} & 0 \\ 0 & 0 & -\frac{10}{7} & \frac{1}{7} & \frac{17}{7} & 1 \end{array}\right] \quad \begin{array}{ll} [\textit{Add } 3/2 \textit{ times } (7) \textit{ to } (4)] & (8) \\ & (7) \\ [\textit{Add } -17/2 \textit{ times } (7) \textit{ to } (6)] & (9) \end{array}$$

$$\left[\begin{array}{rrr|rrr} 1 & 0 & -\frac{4}{7} & -\frac{1}{7} & -\frac{3}{7} & 0 \\ 0 & 1 & \frac{2}{7} & -\frac{3}{7} & -\frac{2}{7} & 0 \\ 0 & 0 & 1 & -\frac{1}{10} & -\frac{17}{10} & -\frac{7}{10} \end{array}\right] \quad \begin{array}{ll} & (8) \\ & (7) \\ [\textit{Multiply } (9) \textit{ by } -7/10] & (10) \end{array}$$

$$\left[\begin{array}{rrr|rrr} 1 & 0 & 0 & -\frac{1}{5} & -\frac{7}{5} & -\frac{2}{5} \\ 0 & 1 & 0 & -\frac{2}{5} & \frac{1}{5} & \frac{1}{5} \\ 0 & 0 & 1 & -\frac{1}{10} & -\frac{17}{10} & -\frac{7}{10} \end{array}\right] \quad \begin{array}{ll} [\textit{Add } 4/7 \textit{ times } (10) \textit{ to } (8)] & (11) \\ [\textit{Add } -2/7 \textit{ times } (10) \textit{ to } (7)] & (12) \\ & (10) \end{array}$$

The required inverse is the matrix on the right, which easily may be verified by multiplying it by the original matrix, or by referring to the same example in Section 10.3.

Compare this to the example of the Gauss-Jordan method in Chapter 9. Although this approach is also pragmatic, it seems to have an intuitive justification when viewed in light of the Gauss-Jordan method. Attempt to justify it, using the elementary transformation matrices discussed in Section 10.2 (begin with $AA^{-1} = I$ and multiply both sides by elementary transformation matrices to reduce A to I).

EXERCISE 10.4

In each of the following, first find the inverse of the coefficient matrix, then determine the solution, using matrix methods:

1. $5x + 7y = 3$
   $3x + 4y = 2$
2. $8x - 5y = 7$
   $-5x + 3y = -4$
3. $-2x + 3y = 11$
   $x - 2y = -10$
4. $3x - 5y = 36$
   $-5x + 9y = -64$
5. $x + 4y = -3$
   $-2x + 3y = -5$
6. $2x - 4y = -8$
   $3x + y = -5$
7. $4x - y = 23$
   $3x + 2y = 9$
8. $3x + 5y = 3$
   $4x - 7y = -37$
9. $x + 2y - 3z = -4$
   $4x - y + 2z = 8$
   $14x - y + 3z = 21$
10. $4x - 9z = -11$
    $y + 3z = -7$
    $6x - y = 26$
11. $2x - y + 6z = 5$
    $3x - 5y - 9z = -7$
    $3x + 3z = 0$
12. $5x - 2y - 5z = -2$
    $8x + 3y - z = 10$
    $-3x - z = -4$
13. $4x - 2y + 8z = 4$
    $3x + 4y - 2z = 33$
    $-6x + y - 3z = -23$
14. $20x + y + z = 18$
    $2x + 30y + 3z = -31$
    $x + 5y + 40z = -44$
15. $w + x + y + z = 2$
    $w + 3x + 3y + 2z = -1$
    $2w + 4x + 3y + 3z = 3$
    $w + 2x + 3y + z = -3$
16. Using the inverse from the preceding exercise, find the solution of

$$\begin{aligned} w + x + y + z &= 9 \\ w + 3x + 3y + 2z &= 16 \\ 2w + 4x + 3y + 3z &= 19 \\ w + 2x + 3y + z &= 18. \end{aligned}$$

17. Attempt to solve the following system using matrix methods, and then explain the peculiar result, based on the nature of the system of equations

$$\begin{aligned} x - 3y + 2z &= 7 \\ -4x + 8y - 2z &= -3 \\ 2x - 4y + z &= 1. \end{aligned}$$

18. Show that the inverse of

$$A = \begin{bmatrix} a & b \\ c & d \end{bmatrix} \quad \text{is} \quad A^{-1} = \frac{1}{|A|} \begin{bmatrix} d & -b \\ -c & a \end{bmatrix}$$

where $|A|$ = determinant of $A = (ad - cb)$.

10.5 MATRIX MULTIPLICATION ON THE COMPUTER

THE PRODUCT OF SQUARE MATRICES. Earlier in this chapter the use of double subscripts to indicate elements of a matrix was mentioned. For example, the general form of a 3 × 3 matrix, call it A, may be represented as

$$A = \begin{bmatrix} a_{11} & a_{12} & a_{13} \\ a_{21} & a_{22} & a_{23} \\ a_{31} & a_{32} & a_{33} \end{bmatrix}.$$

Recall that the first subscript denotes the row and the second the column; thus a_{23} is in the second row and third column. This notation also makes it convenient to identify an entire row or column. The first row of A is

$$a_{11} \quad a_{12} \quad a_{13}$$

and the third column is

$$\begin{matrix} a_{13} \\ a_{23} \\ a_{33}. \end{matrix}$$

For a particular row, the first subscript is the same for each element but the second varies from 1 to 3. Similarly, for a particular column, the first subscript varies from 1 to 3 but the second remains the same.

Let us consider the matrix product $AB = C$ or

$$\begin{bmatrix} a_{11} & a_{12} & a_{13} \\ a_{21} & a_{22} & a_{23} \\ a_{31} & a_{32} & a_{33} \end{bmatrix} \begin{bmatrix} b_{11} & b_{12} & b_{13} \\ b_{21} & b_{22} & b_{23} \\ b_{31} & b_{32} & b_{33} \end{bmatrix} = \begin{bmatrix} c_{11} & c_{12} & c_{13} \\ c_{21} & c_{22} & c_{23} \\ c_{31} & c_{32} & c_{33} \end{bmatrix}.$$

Here individual rows of A are matched with individual columns of B to give the elements of C. For example, the result of matching the second row of A with the third column of B produces the elements in the second row and third column of C:

$$c_{23} = a_{21}b_{13} + a_{22}b_{23} + a_{23}b_{33} \tag{1}$$

For convenience, in the product $a_{21}b_{13}$, we will consider 2 and 3 as the outer subscripts and the 1 and 1 as the inner subscripts. Note that in each term of (1), the outer subscripts are 2 and 3, corresponding to the row and column of c_{23}. The inner subscripts in each term are identical, due to the manner in which the elements are matched in matrix multiplication. These results are not by coincidence, as can be demonstrated by further multiplication.

SUMMATION NOTATION. Due to the special nature of matrix sums, a more concise representation is possible through the use of the summation symbol Σ. Although we shall study this notion in more detail in Chapter 12 (Section 12.4), we will consider it briefly for use here. The symbol Σ (uppercase Greek letter **sigma**) simply indicates that a sequence of terms should be added. For example, the sum

$$S = a_1 + a_2 + a_3 + a_4 + a_5$$

can be represented as

$$S = \sum_{i=1}^{5} a_i = a_1 + a_2 + a_3 + a_4 + a_5.$$

This is read "S equals the sum of a_i as i goes from 1 to 5." The symbolism indicates that values of 1-5 should successively replace the subscript i and that the resulting terms should be summed as shown. Summation of variables with double subscripts is also possible, but only the subscript indicated under the summation sign (the **index**) is to be replaced:

$$T = \sum_{j=1}^{4} b_{ij} = b_{i1} + b_{i2} + b_{i3} + b_{i4}.$$

Of significance in this discussion is the fact that

$$c_{23} = a_{21}b_{13} + a_{22}b_{23} + a_{23}b_{33}$$

may be replaced with

$$c_{23} = \sum_{k=1}^{3} a_{2k}b_{k3}.$$

Note that when k is replaced by 1, 2, or 3, the subscripts k of both a and b are affected. Thus the elements of the first row of C may be represented as

$$c_{11} = \sum_{k=1}^{3} a_{1k}b_{k1}$$

$$c_{12} = \sum_{k=1}^{3} a_{1k}b_{k2}$$

$$c_{13} = \sum_{k=1}^{3} a_{1k}b_{k3}.$$

The second row is

$$c_{21} = \sum_{k=1}^{3} a_{2k}b_{k1}$$

$$c_{22} = \sum_{k=1}^{3} a_{2k}b_{k2}$$

$$c_{23} = \sum_{k=1}^{3} a_{2k} b_{k3}.$$

The result for the third row will be similar.

MULTIPLICATION OF n × n *MATRICES.* As the matrices become larger (n increases), the number of terms in each element of the product matrix also increases. As an illustration, consider the following matrix C, obtained from the product of two 5×5 matrices A and B ($C = AB$). For the purpose of illustration we will consider only the entry c_{54} (fifth row and fourth column). These portions of A and B are shaded in the following:

$$\begin{bmatrix} a_{11} & a_{12} & a_{13} & a_{14} & a_{15} \\ a_{21} & a_{22} & a_{23} & a_{24} & a_{25} \\ a_{31} & a_{32} & a_{33} & a_{34} & a_{35} \\ a_{41} & a_{42} & a_{43} & a_{44} & a_{45} \\ a_{51} & a_{52} & a_{53} & a_{54} & a_{55} \end{bmatrix} \begin{bmatrix} b_{11} & b_{12} & b_{13} & b_{14} & b_{15} \\ b_{21} & b_{22} & b_{23} & b_{24} & b_{25} \\ b_{31} & b_{32} & b_{33} & b_{34} & b_{35} \\ b_{41} & b_{42} & b_{43} & b_{44} & b_{45} \\ b_{51} & b_{52} & b_{53} & b_{54} & b_{55} \end{bmatrix}$$

The element c_{54} is then

$$c_{54} = \sum_{k=1}^{5} a_{5k} b_{k4}. \tag{1}$$

Had this been a 25×25 matrix ($n = 25$), the same entry would have been expressed almost identically to (1); that is,

$$c_{54} = \sum_{k=1}^{25} a_{5k} b_{k4}. \tag{2}$$

Note that it is only necessary to change the upper limit of the summation in going from (1) to (2), but (1) contains only 5 terms, while (2) contains 25.

The general element of C is usually indicated as C_{ij}, where i represents the row and j the column. Then any element of C may be represented by

$$c_{ij} = \sum_{k=1}^{n} a_{ik} b_{kj}.$$

In the first row of C, j varies from 1 to n but i remains 1; in the second column, i varies from 1 to n but j remains 2. However, in calculating each entry for C, the sum of n products still must be determined.

A USE OF THE NOTATION. What is the value of this notation since multiplication of two matrices still requires the same effort, regardless of representation? If a computer is used to perform the multiplication, this notation becomes a vital tool. Not only is it possible to clearly identify each number

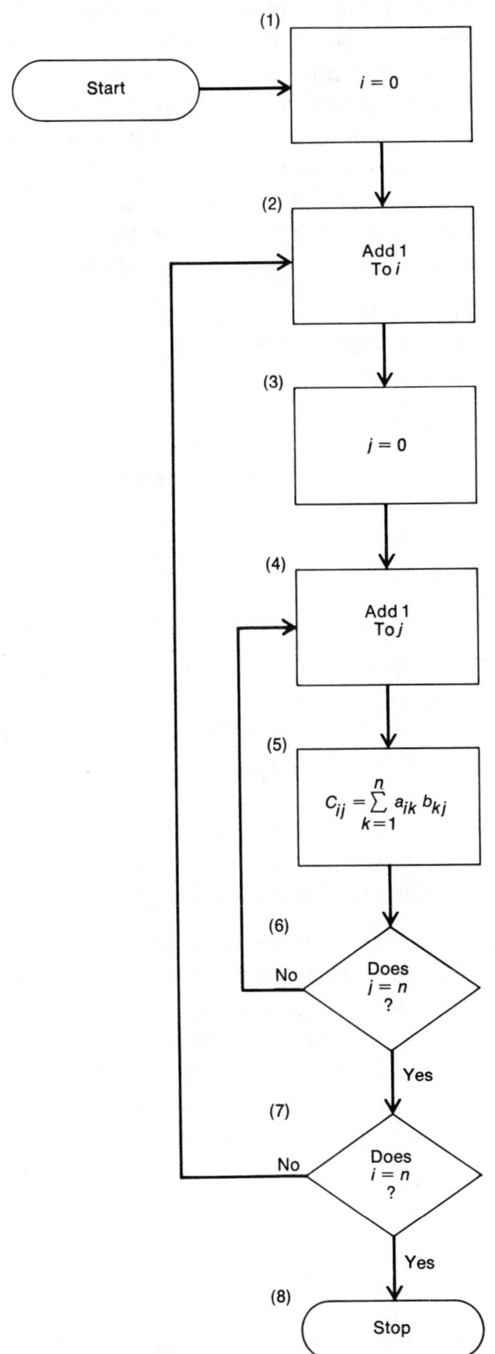

FIGURE 10.2 Flowchart for matrix product

as it is entered into the computer, but it also relates the resulting product elements with the corresponding component rows and columns.

In using a computer to perform the multiplication, it would be necessary to outline a systematic procedure before beginning the programming. This is best illustrated by a flowchart, a commonly used device in displaying the sequence of steps in programming. Figure 10.2 is a flowchart for this problem. For convenience of reference, each of the blocks is numbered above and to the left. Before beginning at block 1, the value of n (which we shall assume to be 25 for this discussion) and the values of the elements in both A and B must be loaded into the computer. The subsequent sequence of steps would be

1. Set $i=0$ (block 1).
2. Increase i by 1, thus $i=1$ (block 2).
3. Set $j=0$ (block 3).
4. Increase j by 1, thus $j=1$ (block 4).
5. Calculate c_{11} (block 5)
6. Check the value of j; since $j \neq 25$, return to block 4 (block 6).
7. Increase j by 1, thus $j=2$ (block 4).
8. Calculate c_{12} (block 5).
9. Check the value of j; since $j \neq 25$, return to block 4 (block 6).

This process would be repeated until $j=25$; then the following would result:

10. Calculate $c_{1,25}$ (block 5).
11. Check the value of j; since $j=25$, go on to block 7 (block 6).
12. Check the value of i; since $i \neq 25$, go back to block 2 (block 7).
13. Increase i by 1, thus $i=2$ (block 2).
14. Set $j=0$ (block 3).
15. Increase j by 1, thus $j=1$ (block 4).
16. Calculate c_{21} (block 5).
17. Check the value of j; since $j \neq 25$, return to block 4 (block 6).

This process would continue until both i and j reached 25, at which time the final entry of C would be determined and the process would halt at block 8. Note that there is nothing mysterious about the flowchart. It is simply a graphical means for illustrating a necessary sequence of events. Whether the product of two matrices is to be calculated by hand or on a computer, some such orderly procedure is highly desirable.

A Fortran program to perform these calculations (less important initialization statements) is shown in Figure 10.3. The entire set of calculations is

```
      .
      .
      .
      DO 10 I=1,N
      DO 10 J=1,N
      DO 10 K=1,N
   10 C(I,J) = C(I,J) + A(I,K)*B(K,J)
      .
      .
      .
```

FIGURE 10.3 Segment of Fortran program

summarized in these four lines. The astonishing simplicity of this program is possible because of the proper choice of notation. With this program, the product of any two $n \times n$ matrices could be calculated, subject only to the size of the particular computer being used.

REFERENCES

Allendoerfer, C. B. and Oakley, C. O., *Fundamentals of College Algebra.* New York, McGraw-Hill, Inc., 1967.

DeAngelo, S. and Jorgenson, P., *Mathematics for Data Processing.* New York, McGraw-Hill, Inc., 1970.

Dodes, A. and Greitzer, B., *Numerical Analysis.* New York, Hayden Book Companies, 1964.

Fowler, F. P. and Sandberg, E. W., *Basic Mathematics for Administration.* New York, John Wiley & Sons, 1962.

Hohn, F. E., *Elementary Matrix Algebra.* New York, The Macmillan Company, 1958.

11 LINEAR PROGRAMMING

11.1 A GRAPHICAL SOLUTION / **353**
A Classical Problem: Maximizing Profit / **353**
Estimating the Solution / **354**
Graphing the Problem / **355**
Profit Lines / **357**
The Maximum Feasible Profit Line / **358**
A Change in the Objective Function / **359**
Exercise 11.1 / **360**

11.2 FINDING A MINIMUM COST / **361**
Vitamin Supplements / **361**
The Equal Condition / **363**
The Feasible Region / **363**
Including the Cost Function / **364**
Exercise 11.2 / **366**

11.3 AN ALGEBRAIC METHOD / **368**
Slack Variables / **368**
The First Solution / **369**
Key Variable and Pivotal Equation / **369**
The Second Solution / **370**
An Improved Solution / **371**
Summarizing the Algebraic Method / **372**
Another Example of Maximizing Profit / **372**
Exercise 11.3 / **375**

11.4 THE ALGEBRAIC METHOD IN MATRIX FORM / **375**
The Key Column and Pivotal Row / **377**
Exercise 11.4 / **380**

11.5 THE SIMPLEX METHOD / **380**
A Tabular Form / **381**
An Example Containing Four Variables / **382**
Exercise 11.5 / **385**

During World War II, the application of scientific methods to certain tactical problems led to a new field of endeavor called operations research. The successful solution of a number of military problems was followed by attempts to apply new techniques to the solution of industrial problems. Systematic methods were applied to problems in government and business operations, such as minimizing production costs and determining the best product allocation or the best use of limited resources.

An important tool in coping with these large-scale problems is a mathematical technique called **linear programming**. This technique involves finding an optimum value of a linear function subject to certain given restrictions. Basic resources, such as labor, time, and raw materials, are usually available in limited supply. These restrictions or constraints must be taken into consideration when minimizing costs or maximizing profits.

In order to provide some insight into the nature of a linear programming problem and its solution, a graphical method will be presented first. However, as the number of restrictions increases so does the number of variables, requiring that the graphical technique be replaced by an algebraic method. Although we can use several algebraic methods in finding a solution, we shall study a particular format which eventually explains the manipulations involved in the basic simplex method (the subject of Section 11.5). Hence the algebraic method is the key development in this chapter.

11.1 A GRAPHICAL SOLUTION

A CLASSICAL PROBLEM: MAXIMIZING PROFIT. To illustrate the notion of linear programming and a graphical technique for finding a solution, let us consider Example 11.1.

EXAMPLE 11.1

A company manufactures two types of electric knives, a standard and a deluxe model. For greater efficiency, the required production and assembly operations are performed in two separate shops. The standard model requires 5 hours of work in shop 1 and 7 hours in shop 2, while the deluxe model requires 8 hours in shop 1 and 6 hours in shop 2. Because of a scarcity of skilled labor, only 400 hours of work per week can be performed in shop 1 and 420 hours in shop 2. The company can sell its total production and realize a profit of \$3 on the standard model and \$4 on the deluxe model. How many of each type should the company produce in order to achieve the greatest possible profit?

SOLUTION

Let $P =$ the total profit, $S =$ the number of standard models produced, and $D =$ the number of deluxe models produced. In shop 1, the S standard knives require $5S$ hours of labor, while the D deluxe models require $8D$ hours. Since this shop provides only 400 hours of labor, the sum, $5S + 8D$, must not exceed the weekly total of 400 hours. This **constraint** or **restriction** is expressed by the inequality

$$5S + 8D \leq 400. \tag{1}$$

A similar limitation exists in shop 2, where the S standard knives require $7S$ hours and the D deluxe models, $6D$ hours. The 420 hour restriction for shop 2 leads to the second constraint,

$$7S + 6D \leq 420. \tag{2}$$

Since the number of knives produced must be non-negative (zero or positive), we also have

$$S \geq 0 \qquad \text{and} \qquad D \geq 0. \tag{3}$$

The profit on S standard knives is $3S$ dollars and for D deluxe models, $4D$ dollars. Hence, the total profit is

$$P = 3S + 4D. \tag{4}$$

Our objective is to maximize profit subject to the constraints described by (1), (2), and (3). For this reason, (4) is frequently called the **objective function**.

ESTIMATING THE SOLUTION. Before we describe a systematic technique for finding the best solution, we try an educated guess. Since the profit is greater on the deluxe knives, it may seem better to produce only this model. To check this possibility, we let S equal zero and determine how many deluxe models can be handled by both shops. The second constraint (2) indicates that shop 2 can handle 70 deluxe knives per week. However, the first constraint (1) indicates that shop 1 can handle only 50 knives. The combined time restrictions of both shops will allow a maximum production of just 50 deluxe models. Thus the company will realize a profit of $50 \times 4 = \$200$.

The use of row and column matrices (vectors) and the property of matrix multiplication can result in a convenient form here. For example, the number of knives to be produced (none of the standard model and 50 of the deluxe) can be represented as

$$[0 \quad 50]$$

and the profit on each model ($3 on the standard and $4 on the deluxe) can be represented as

$$\begin{bmatrix} 3 \\ 4 \end{bmatrix}.$$

Now the profit is the matrix product

$$[0 \quad 50] \begin{bmatrix} 3 \\ 4 \end{bmatrix} = 0 + 200 = \$200.$$

Recall from Chapter 10 that the product of two such matrices is simply a number (that is, a scalar).

Pursuing this matter further, we see that this is not the greatest possible profit. If 16 standard and 40 deluxe models are produced, the profit will be larger, since

$$\begin{bmatrix}16 & 40\end{bmatrix}\begin{bmatrix}3\\4\end{bmatrix} = 48 + 160 = \$208.$$

However, the original constraints (no more than 400 hours in shop 1 and 420 in shop 2) must be satisfied. This may also be represented in matrix form, by using

$$\begin{bmatrix}5\\8\end{bmatrix}$$

to represent required manufacturing time for each model. That is,

$$\begin{bmatrix}16 & 40\end{bmatrix}\begin{bmatrix}5\\8\end{bmatrix} = 80 + 320 = 400.$$

Also, for shop 2,

$$\begin{bmatrix}16 & 40\end{bmatrix}\begin{bmatrix}7\\6\end{bmatrix} = 112 + 240 = 352 < 420.$$

As we can see, this combination of standard and deluxe models satisfies the basic constraints. Hence the profit is not maximized by producing only the deluxe model.

But we cannot be certain that the second choice is optimum. The best combination can be found by a limited graphical method and a more general algebraic method.

GRAPHING THE PROBLEM. The graphical solution, governed by the basic constraints, will give a better indication of the possible combinations that can be handled by both shops. Using the techniques described in Chapter 7. the graph of the line

$$5S + 8D = 400$$

is shown in Figure 11.1(a). Because S and D must be non-negative, we need consider only the first quadrant portion of the line.

The shaded triangle in Fjgure 11.1(b) contains all solutions of inequality (1). For example, the points (30 20) and (40 10) are in the shaded area, since

$$\begin{bmatrix}30 & 20\end{bmatrix}\begin{bmatrix}5\\8\end{bmatrix} = 150 + 160 = 310 < 400$$

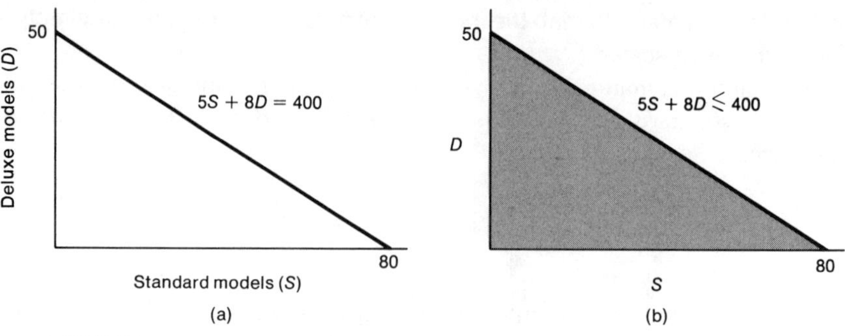

FIGURE 11.1

and

$$[40 \quad 10]\begin{bmatrix}5\\8\end{bmatrix} = 200 + 80 = 280 < 400.$$

This indicates that shop 1 could handle 30 standard and 20 deluxe knives or 40 standard and 10 deluxe knives. On the other hand, the point (40 30) is not in the shaded region, since

$$[40 \quad 30]\begin{bmatrix}5\\8\end{bmatrix} = 200 + 240 = 440 > 400.$$

This combination requires more hours in shop 1 than are available.

The graph of

$$7S + 6D = 420$$

is combined with the first line in Figure 11.2(a). The combinations of standard and deluxe models which satisfy inequality (2) are represented by points below the second line. However, to find solutions of both inequalities, we must stay below both lines. The result is shaded in Figure 11.2(b). Each point in this

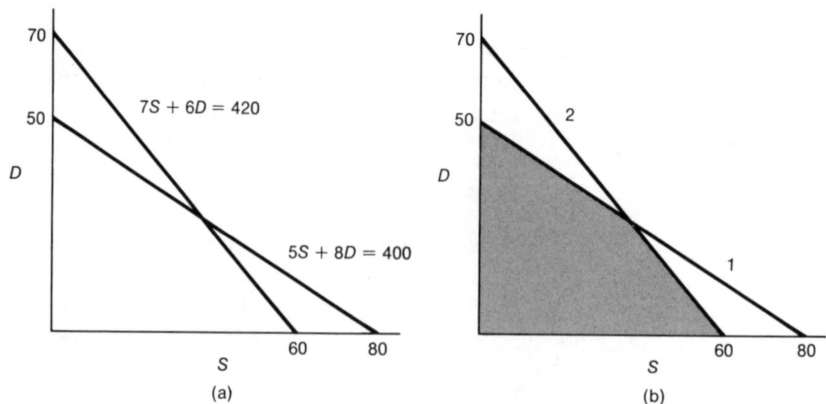

FIGURE 11.2 (a) Both lines; (b) feasible region

feasible region represents a combination of standard and deluxe models that can be handled by shop 1 and also by shop 2.

In Figure 11.2(a), we can readily verify that shop 2 can handle as many as 70 deluxe knives but shop 1 only 50. On the other hand, shop 1 can handle up to 80 standard models while shop 2 is limited to 60. In both cases, the smaller number must be selected to stay within the limitations of both shops.

PROFIT LINES. We now return to our main goal of determining which combination will yield the greatest profit. To this end, we use the profit relationship

$$P = 3S + 4D$$

and graph **profit lines**, shown in Figure 11.3, using values for P of 150, 200, and 250 (for example, when $P = 200$, the linear function $3S + 4D = 200$ or $D = -\frac{3}{4}S + 50$ results). The three profit lines are parallel because the slope is $-\frac{3}{4}$ in each case. As P varies, the coefficients of S and D remain constant; hence, the slope is constant.

Several combinations that will produce a profit of \$150 include 50 standard and 0 deluxe models, 38 standard and 9 deluxe models, 18 standard and 24 deluxe models, and 2 standard and 36 deluxe models. This is verified by

$$[50 \quad 0]\begin{bmatrix} 3 \\ 4 \end{bmatrix} = 150 \qquad [38 \quad 9]\begin{bmatrix} 3 \\ 4 \end{bmatrix} = 114 + 36 = 150$$

$$[18 \quad 24]\begin{bmatrix} 3 \\ 4 \end{bmatrix} = 54 + 96 = 150 \qquad [2 \quad 36]\begin{bmatrix} 3 \\ 4 \end{bmatrix} = 6 + 144 = 150.$$

Figure 11.3 indicates that each of these points is in the feasible region and hence within the time limitation of the two shops.

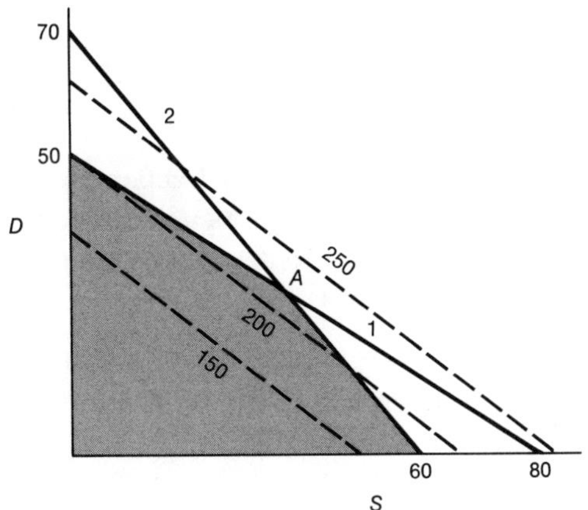

FIGURE 11.3

A profit of $200 is realized by any of the following combinations: 64 standard and 2 deluxe models, 56 standard and 8 deluxe models, and 0 standard and 50 deluxe models. However, (64 2) and (56 8) are not in the feasible region, since

$$[64 \quad 2]\begin{bmatrix}7\\ \\6\end{bmatrix} = 448 + 12 = 460$$

and

$$[56 \quad 8]\begin{bmatrix}7\\ \\6\end{bmatrix} = 392 + 48 = 440.$$

In both cases, the capacity of shop 2 is exceeded. The graph in Figure 11.3 indicates that there are other combinations in the feasible region which produce a $200 profit.

A comparison of the $150 and $200 profit lines indicates that the line representing the higher profit is farther from the origin. As the profit increases, the profit line moves away from the origin but retains the same slope. The $250 profit line in Figure 11.3, however, fails to intersect the feasible region at a single point. The combinations of standard and deluxe models which would produce a $250 profit are beyond the capacities of the two shops.

THE MAXIMUM FEASIBLE PROFIT LINE. Our problem is to find the profit line that is as far as possible from the origin, but still intersects the feasible region. If we visualize moving the $200 profit line away from the origin in Figure 11.3, we note that it can be shifted to point A. This point is the intersection of the lines

$$5S + 8D = 400 \qquad 7S + 6D = 420.$$

Solving these equations simultaneously, we have

$$S = 36\frac{12}{13} \qquad D = 26\frac{12}{13}.$$

Since it is not realistic to produce a fractional part of a knife, we consider integral solutions only. Two possibilities that are within the limitations of the two shops are 37 standard and 26 deluxe models, or 35 standard and 28 deluxe models. These values satisfy (1) and (2), since

$$[37 \quad 26]\begin{bmatrix}5\\ \\8\end{bmatrix} = 185 + 208 = 393 < 400 \qquad \textit{(maximum hours, shop 1)}$$

$$[37 \quad 26]\begin{bmatrix}7\\ \\6\end{bmatrix} = 259 + 156 = 415 < 420 \qquad \textit{(maximum hours, shop 2)}$$

and

$$[35 \quad 28]\begin{bmatrix}5\\8\end{bmatrix} = 175 + 224 = 399 < 400$$

$$[35 \quad 28]\begin{bmatrix}7\\6\end{bmatrix} = 245 + 168 = 413 < 420.$$

The profit on each combination is

$$[37 \quad 26]\begin{bmatrix}3\\4\end{bmatrix} = 111 + 104 = 215$$

and

$$[35 \quad 28]\begin{bmatrix}3\\4\end{bmatrix} = 105 + 112 = 217.$$

Hence, the company should produce 35 standard and 28 deluxe models to realize the maximum profit of $217.

A CHANGE IN THE OBJECTIVE FUNCTION. In the preceding graphical solution, the slopes of the constraints (shop limitations) and objective function (profit relationship) were important factors. A change in one or more of these slopes could result in a different solution. For example, suppose that competition required that the profit on the standard model be reduced from $3 to $2. Then the total profit would be

$$P = 2S + 4D.$$

The slope of this line for any value of P would be $-\frac{2}{4}$ or $-\frac{1}{2}$, which results in a flatter profit line. If the time limitations of the two shops are the same, the feasible region will not change, but the new profit line through point A of the feasible region will appear as shown in Figure 11.4.

It is evident from the graph that this profit line can be shifted further from the origin and still intersect the feasible region. In this case, the profit line can

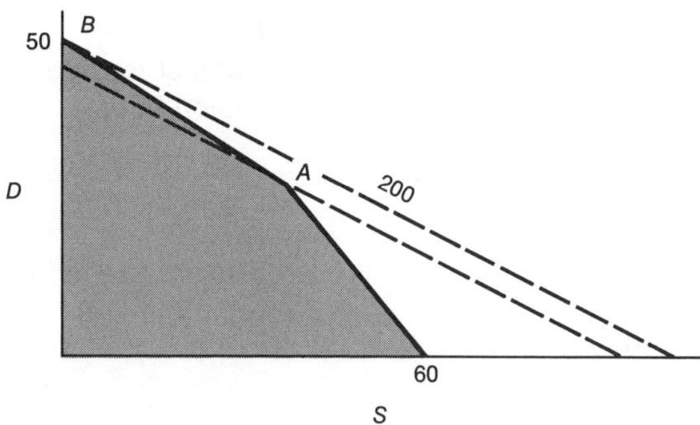

FIGURE 11.4

be shifted to point B, indicating that total profit will be maximized by producing deluxe models only. The profit on standard knives has decreased to the point where their production can no longer be justified. Hence, the production of 50 deluxe knives yields the maximum profit of $200.

EXERCISE 11.1

1. Determine whether or not the following points are in the feasible region shown in Figure 11.2(b):

   (5 45) (10 45) (20 38) (35 25) (40 20) (50 15) (55 6)

2. Find the profit corresponding to each point given in problem 1, if (a) $P = 3S + 4D$, (b) $P = 2S + 4D$.

3. Find the number of hours of unused time in each shop for shop 1 and shop 2, if

   (a) $S = 10$, $D = 10$
   (b) $S = 16$, $D = 40$
   (c) $S = 0$, $D = 50$
   (d) $S = 54$, $D = 7$
   (e) $S = 60$, $D = 0$
   (f) $S = 36$, $D = 27$

4. Find the maximum profit in each of the following situations by first determining the feasible region graphically and then drawing various profit lines:

   (a) $2S + 7D = P$
   Constraints:
   $3S + 5D \leq 300$
   $6S + 2D \leq 216$

   (b) $2S + 7D = P$
   Constraints:
   $3S + 5D \leq 300$
   $6S + 10D \leq 600$

   (c) $2S + 7D = P$
   Constraints:
   $3S + 5D \leq 300$
   $2S + 6D \leq 264$

   (d) $2S + 7D = P$
   Constraints:
   $3S + 5D \leq 300$
   $9S + 15D \leq 300$

Find the maximum value of P in problems 5-8.

5. (a) $3x + 2y = P$
   Subject to:
   $x - y \leq 50$
   $x \leq 70$
   $2x + y \leq 200$
   $x + y \leq 170$

   (b) $2x + 3y = P$
   Subject to the feasible region in 5(a).

6. (a) $6x + 3y = P$
   Subject to:
   $-x + y \leq 40$
   $x + y \leq 80$
   $3x + y \leq 220$

   (b) $8x + 2y = P$
   Subject to the feasible region in 6(a).

7. (a) $4x + 3y = P$
   Subject to:
   $x + 7y \leq 420$
   $3x + 9y \leq 660$
   $4x + 4y \leq 560$
   $5x + 3y \leq 640$

   (b) $2x + 8y = P$
   Subject to the feasible region in 7(a).

8. (a) $7x + 2y = P$
Subject to:
$x + 6y \le 300$
$x + 4y \le 220$
$x - y \le 120$

(b) $7x + 2y = P$
Subject to the feasible region in 8(a) with the additional constraint $4x + 13y \le 650$

9. In the manufacture of two types of lamps, a company uses three machines which we call *A*, *B*, and *C*. The time requirements for each lamp on each machine are listed below:

Machine	Lamp *I*	Lamp *II*	Maximum time available
A	3	4	190
B	4	2	150
C	6	2	202

The profit on lamp *I* is $5.00 and on lamp *II* is $3.00. Find the production combination that would maximize total profit.

10. In problem 9, suppose the time requirements are as follows:

Machine	Lamp *I*	Lamp *II*	Maximum time available
A	3	2	120
B	5	7	310
C	1	4	153

Find the maximum total profit, if the profit on each lamp is as given in problem 9.

11. Maximize $4x + 5y$ subject to

$$7x + 6y \le 19$$
$$x + 4y \le 9$$
$$2x + 3y \le 13.$$

12. Maximize $x + 5y$ subject to

$$5x + 4y \le 35$$
$$2x + 3y \le 21$$
$$x + 6y \le 24.$$

11.2 FINDING A MINIMUM COST

VITAMIN SUPPLEMENTS. In certain cases, linear programming techniques are used to find the minimum value of a function rather than the maximum. A good example involves a function which represents a cost of some type. While a company would like its profit to be as large as possible, it would also like its costs to be as small as possible. Graphical techniques for finding minimum cost are very similar to those used in the preceding section, so long as only two variables are involved. This is perhaps best demonstrated by the following example:

EXAMPLE 11.2

A dairy farmer decides to provide the minimum requirements of three vitamins for his cows by adding a supplement to the feed he normally uses. The vitamins are available in differing amounts in product A which costs 9¢ an ounce, and in product B which costs 7¢ an ounce. Product A contains 15 units of vitamin 1, 20 units of vitamin 2, and 15 units of vitamin 3. Product B contains 10 units of vitamin 1, 5 units of vitamin 2, and 25 units of vitamin 3. The minimum requirements are 60 units of vitamin 1, 40 units of vitamin 2, and 75 units of vitamin 3. The farmer wishes to find the mixture that will supply not only the minimum requirements but also will be the least expensive to buy.

SOLUTION

If a ounces of product A and b ounces of product B are used, the cost or objective function is

$$C = 9a + 7b.$$

Ideally, he would like a mixture providing the exact minimum requirements of each vitamin. However, if this is not possible, an excess rather than a deficiency of one or two vitamins is acceptable. This possible excess requires the use of inequalities involving the "greater than" symbol. In particular, the following inequalities must be satisfied:

$$15a + 10b \geq 60 \tag{1}$$
$$20a + 5b \geq 40 \tag{2}$$
$$15a + 25b \geq 75 \tag{3}$$
$$a \geq 0 \quad \text{and} \quad b \geq 0 \tag{4}$$

The inequalities (1), (2), and (3) can be expressed in matrix form as

$$\begin{bmatrix} 15 & 10 \\ 20 & 5 \\ 15 & 25 \end{bmatrix} \begin{bmatrix} a \\ b \end{bmatrix} \geq \begin{bmatrix} 60 \\ 40 \\ 75 \end{bmatrix}.$$

Although this is just a different way of writing the inequalities, the coefficient matrix contains essential information in a convenient form. By indicating the product by row and vitamin by column, we have

$$\begin{array}{c} \\ \text{Vitamin} \\ 1 \\ 2 \\ 3 \end{array} \quad \begin{array}{c} \text{Product} \\ \begin{array}{cc} A & B \end{array} \\ \begin{bmatrix} 15 & 10 \\ 20 & 5 \\ 15 & 25 \end{bmatrix} \end{array}$$

Each entry in the first column relates to product A; similarly, each entry in the second row relates to vitamin 2. This suggests a possible organization of facts that facilitates the translation to a mathematical form, even though inequalities are involved.

THE EQUAL CONDITION. In determining the solution by graphical techniques, we first consider only the equal signs in (1), (2), and (3), which leads to three linear equations whose graphs are shown in Figure 11.5. Each line represents various combinations that supply the minimum requirements of one vitamin. For example, the points (4,0) and (0,6) are on the line whose equation is

$$15a + 10b = 60.$$

Four ounces of product A or 6 ounces of product B will provide the exact minimum requirement of vitamin 1. However, 4 ounces of product A provides 80 units of vitamin 2 and 60 units of vitamin 3. This is a 40-unit excess of vitamin 2 and a 15-unit deficiency of vitamin 3. The 6 ounces of product B will provide more than enough of vitamin 3 but an insufficient amount of vitamin 2. Hence each combination provides the exact minimum requirement of vitamin 1 but fails to provide enough of other vitamins. A similar situation exists on the other two lines in Figure 11.5.

THE FEASIBLE REGION. These results suggest a feasible region in this case as indicated in Figure 11.6. Any point in the feasible region represents a combination of the two products that will supply at least the minimum require-

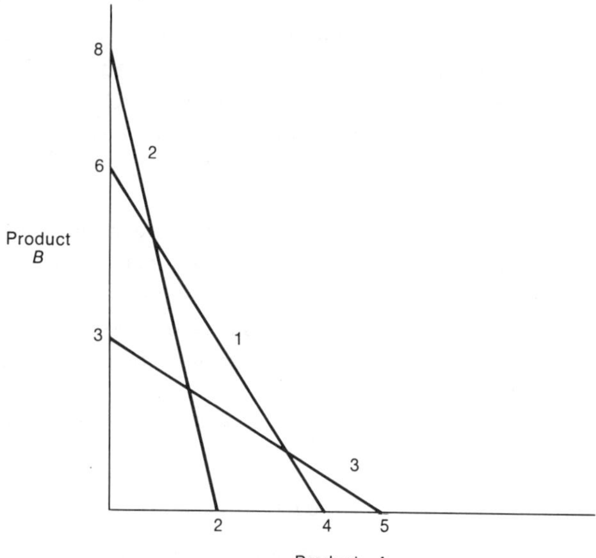

FIGURE 11.5

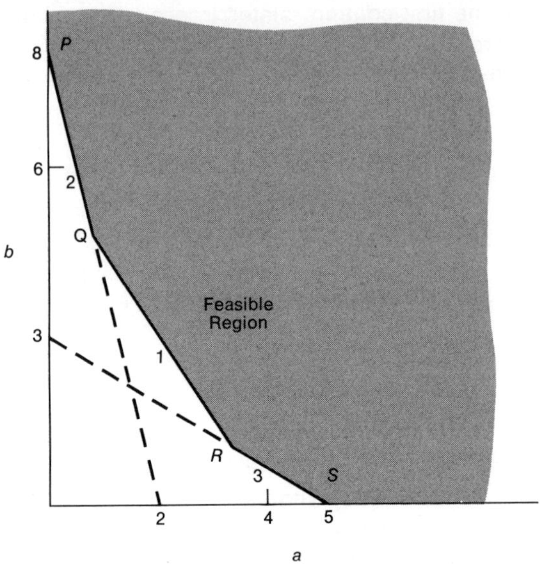

FIGURE 11.6

ments of each vitamin. For example, the point (3,5) within the feasible region represents a combination of 3 ounces of product *A* and 5 ounces of product *B*. This will provide 95 units of vitamin 1, 85 units of vitamin 2, and 170 units of vitamin 3, which is more than enough of each vitamin. The same is true for any point in the feasible region shown in Figure 11.6.

Part of the boundary of the feasible region in Figure 11.6 consists of the line segments *PQ, QR,* and *RS*. As indicated previously, each combination on *QR* provides the exact minimum requirements of vitamin 1. Similarly, the exact minimum requirement of vitamin 2 is supplied on the segment *PQ* and of vitamin 3 on the segment *RS*. The point *Q* is on both *PQ* and *QR*, and hence satisfies the exact minimum requirements of vitamin 1 and vitamin 2. By the same reasoning, point *R* represents a combination that provides the exact minimum requirements of 1 and 3.

Since any excess over minimum requirements will increase total cost, it appears reasonable that the minimum cost should occur at point *Q* or point *R*. At point *Q*, an excess of vitamin 3 is provided; at point *R*, an excess of vitamin 2 is included. The least expensive excess should determine the combination that minimizes total cost. We will verify this by graphing the cost function.

INCLUDING THE COST FUNCTION. In Figure 11.7, the graph of the cost function

$$C = 9a + 7b$$

is shown for *C* equal to 25, 37, and 48. No point on the 25¢ cost line is in the feasible region; hence, it would not supply any of the minimum requirements.

For instance, 2 ounces of product *A* and 1 ounce of product *B* would cost 25¢,

$$[2 \quad 1] \begin{bmatrix} 9 \\ 7 \end{bmatrix} = 18 + 7 = 25,$$

but provide only 55 of the required 75 units of vitamin 3. On the other hand, many points on the 48¢ cost line are in the feasible region but they provide a large excess over the minimum requirements. For example, (3 3) is on the 48¢ cost line:

$$[3 \quad 3] \begin{bmatrix} 9 \\ 7 \end{bmatrix} = 27 + 21 = 48.$$

However, 3 ounces of product *A* and 3 ounces of product *B* provide an excess of 15, 35, and 45 units of the three vitamins.

It is evident from Figure 11.7 that the least expensive mixture of the two products will cost more than 25¢ and less than 48¢. In fact, the 37¢ cost line touches the feasible region at only one point. Any cost line that is closer to the origin will not intersect the feasible region. Hence the minimum cost is 37¢.

The least expensive mixture, then, is represented by the coordinates of point *R*. At this point, the exact requirements of vitamin 1 and vitamin 3 are supplied. This indicates that the coordinates of *R* satisfy the equations

$$15a + 10b = 60 \qquad 15a + 25b = 75.$$

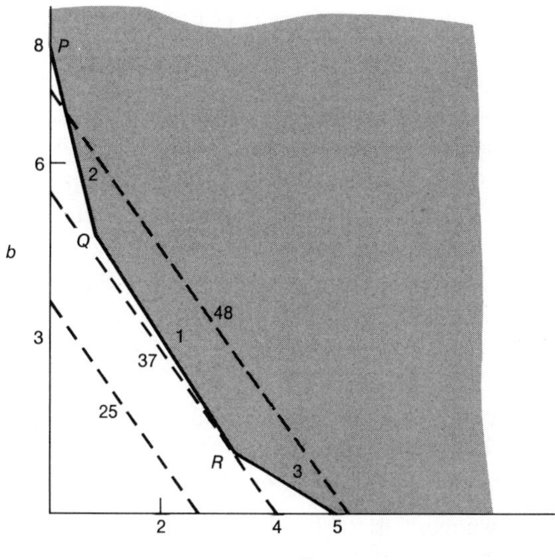

FIGURE 11.7

Solving these equations simultaneously, we have

$$a = \frac{10}{3} \qquad b = 1.$$

In this case we need not round off the solution; the least expensive mixture consists of $3\frac{1}{3}$ ounces of product A and 1 ounce of product B. The cost of this mixture is

$$\begin{bmatrix} \frac{10}{3} & 1 \end{bmatrix} \begin{bmatrix} 9 \\ 7 \end{bmatrix} = 30 + 7 = 37.$$

In summary,

Amount of vitamin 1:

$$\begin{bmatrix} \frac{10}{3} & 1 \end{bmatrix} \begin{bmatrix} 15 \\ 10 \end{bmatrix} = 50 + 10 = 60.$$

Amount of vitamin 2:

$$\begin{bmatrix} \frac{10}{3} & 1 \end{bmatrix} \begin{bmatrix} 20 \\ 5 \end{bmatrix} = \frac{200}{3} + 5 = 71\frac{2}{3}.$$

Amount of vitamin 3:

$$\begin{bmatrix} \frac{10}{3} & 1 \end{bmatrix} \begin{bmatrix} 15 \\ 25 \end{bmatrix} = 50 + 25 = 75.$$

These calculations indicate that the only excess supplied by this combination is $31\frac{2}{3}$ units of vitamin 2. The graphs in Figure 11.7 assure us that this is the least expensive combination that supplies the minimum requirements.

EXERCISE 11.2

1. (a) Find the coordinates of point Q in Figure 11.6.
   (b) Verify that the exact minimum requirements of vitamins 1 and 2 are supplied at point Q.
   (c) Find the excess of vitamin 3 at Q.
   (d) Find the cost of the mixture represented by point Q.
2. (a) Find the coordinates of points P and S in Figure 11.6.
   (b) If only one product is used, determine which would be the least expensive.
3. (a) Determine which mixture would supply the exact minimum requirements of vitamins 2 and 3.
   (b) Does the mixture in (a) provide an excess or deficiency of vitamin 1?
   (c) Find the cost of this mixture.
4. For the feasible region in Figure 11.6, determine the least expensive mixture if the cost function is
   (a) $C = 2a + 5b$ (b) $C = 6a + 3b$ (c) $C = 5a + b$ (d) $C = 4a + 4b$
5. A gardener discovers that his lawn will appear best if he uses at least 15 units of nitrogen and 8 units of phosphate to fertilize it. He considers two products, A and B.

Product	Number of units (per lb) Nitrogen	Phosphate	Cost (per lb)
A	4	1	7¢
B	3	5	5¢

Find the least expensive mixture that would satisfy the minimum requirements he has established.

6. In problem 5, find the least expensive mixture if product *A* costs 6¢ and product *B* 2¢.
7. In problem 5, find the least expensive mixture if product *A* costs 4¢ and product *B* 12¢.
8. Minimize $2x + 3y$ subject to the constraints:

$$\begin{aligned} 7x + y &\geq 54 \\ 3x + 3y &\geq 36 \\ x + 6y &\geq 67. \end{aligned}$$

9. Minimize $4x + y$ subject to the constraints:

$$\begin{aligned} 10x + 2y &\geq 68 \\ 9x + 3y &\geq 66 \\ 8x + 4y &\geq 72 \\ 2x + 6y &\geq 88. \end{aligned}$$

10. Minimize $5x + 2y$ subject to the constraints given in problem 9.
11. Minimize $3x + 3y$ subject to the constraints given in problem 9.
12. Minimize $3x + y$ subject to the constraints given in problem 8.
13. Minimize $3x + 2y$ subject to the constraints:

$$\begin{aligned} 11x + 3y &\geq 29 \\ 5x + 2y &\geq 17 \\ 6x + 10y &\geq 28. \end{aligned}$$

14. A druggist wishes to provide for his customers the minimum daily requirements of three vitamins at a minimum cost. He narrows his choice to two products, *A* and *B*.

Vitamin	Product *A*	*B*	Minimum daily requirements
1	4	5	37
2	3	6	39
3	4	3	22

Find the least expensive combination if product *A* costs 7¢ per ounce and product *B* costs 6¢ per ounce.

15. In problem 14, find the least expensive combination if (a) product *A* costs 6¢ per ounce and product *B* costs 10¢ per ounce, (b) product *A* costs 5¢ per ounce and product *B* costs 3¢ per ounce, (c) product *A* costs 3¢ per ounce and product *B* costs 9¢ per ounce.

16. Given the constraints

$$\begin{aligned} 13x + 12y &\geq 62 \\ 5x + 6y &\geq 28 \\ 2x + 3y &\geq 12. \end{aligned}$$

find the minimum value for each of the following:

(a) $2x + 2y$ (c) $x + 3y$
(b) $3x + y$ (d) $3x + 4y$

17. Minimize $5x + 2y$ subject to the constraints

$$\begin{aligned} 3x + y &\geq 11 \\ 4x + 2y &\geq 18 \\ x + 5y &\geq 9. \end{aligned}$$

11.3 AN ALGEBRAIC METHOD

The use of the graphical method is limited to those problems that involve only two variables. By graphing certain linear equations in two dimensions, we are able to determine the best solution for the given circumstance. In a problem involving three variables, a graphical solution would require graphing linear relationships in three dimensions. This, of course, is possible but it becomes more difficult to visualize the shape of the feasible region.

A graphical technique cannot be used for problems involving more than three variables. Solutions are provided in these cases by algebraic methods, which are not limited by the number of variables. We will demonstrate the use of an algebraic method by solving the problem discussed in Section 11.1. Since it has been solved graphically, we can compare the two methods.

SLACK VARIABLES. The mathematical formulation of the problem was as follows:

Objective function: $3S + 4D = P$

Constraints:

$$5S + 8D \leq 400 \quad (1)$$

$$7S + 6D \leq 420 \quad (2)$$

$$S \geq 0 \text{ and } D \geq 0. \quad (3)$$

In constraint 1, the sum $5S + 8D$ represents the number of hours required in shop 1 to produce S standard and D deluxe knives. Whenever this sum is less than 400, the shop is not being used to full capacity. The unused time in this shop is the difference $400 - (5S + 8D)$. If

$$x_1 = 400 - (5S + 8D)$$

then

$$5S + 8D + x_1 = 400. \quad (4)$$

The variable x_1, called a **slack variable**, enables us to write constraint 1 as an equation.

A second slack variable can be introduced to convert constraint 2 to the equation

$$7S + 6D + x_2 = 420. \tag{5}$$

In this case, the variable x_2 represents the unused time in shop 2. Note that both x_1 and x_2 can equal zero but cannot be negative.

After introducing the two slack variables x_1 and x_2, we have two equations in four variables. Since the number of variables exceeds the number of equations, more than one solution exists. These solutions can be determined by arbitrarily assigning values to two of the four variables, and solving the resulting two equations in two variables.

It so happens that we need consider only those solutions in which two of the four variables are zero. In a particular solution, these will be referred to as either the zero variables or the nonzero variables. This distinction will be important in what follows.

In general, the algebraic method consists of shifting from one solution to another in such a way that total profit always increases. The objective function is expressed in terms of the zero variables (variables which equal zero) of each solution as it is encountered. Then, a simple check of coefficients will determine how to find a better solution.

Although there may be considerable algebraic manipulation in changing the form of equations, there are two basic decisions which give direction to the whole process. These will be emphasized in the following demonstration.

For convenience, we rewrite the three basic equations. (Note that the first of these is the objective function.)

$$3S + 4D = P$$

$$5S + 8D + x_1 = 400 \tag{4}$$

$$7S + 6D + x_2 = 420. \tag{5}$$

THE FIRST SOLUTION. We begin with the solution which yields the smallest profit:

$$S = 0 \qquad D = 0 \qquad x_1 = 400 \qquad x_2 = 420 \qquad P = 0.$$

In effect, both shops are closed, no knives are produced, and there is no profit. Since S and D are the zero variables in this solution, the objective function is already expressed in terms of the zero variables.

KEY VARIABLE AND PIVOTAL EQUATION. It now seems reasonable that profit will be increased most by assigning a positive value to D, since it has the larger positive coefficient in the objective function. At the same time, one of the nonzero variables x_1 or x_2 is assigned a zero value. We describe this change by indicating that D is brought into the solution (allowed to assume a nonzero value) and either x_1 or x_2 is removed (assigned a zero value). The variable that has the largest positive coefficient in the objective function will be called the **key variable** (D in this case, which has a coefficient of 4).

The decision as to which variable should be removed depends on determining the largest possible value for the key variable D. In (4), if S and x_1 are zero, D would equal 50. On the other hand, if S and x_2 are zero in (5), D would equal 70. However, a value of 70 for D in (4) would require that S or x_1 be a negative number. Since this is impossible, D cannot exceed 50. Equation (4) then determines the maximum value of D and also indicates that x_1 should be assigned a value of zero. Thus (4) is called the **pivotal equation**.

The selection of the key variable and the pivotal equation are basic and very important decisions in using the algebraic method. For this reason, it is well to note the essential steps in determining the pivotal equation. In (4), the constant term 400 was divided by 8, the coefficient of the key variable D. The similar quotient from (5) was 420/6. Since $400/8 < 420/6$, (4) is the pivotal equation. The smaller quotient is selected to avoid negative values for S or x_1 in the next solution (for example, see problem 1 in Exercise 11.3).

THE SECOND SOLUTION. At this point, we have enough information to continue from the first solution (which was no production and no profit) to the second solution. However, a change in form of the basic equations is necessary to determine subsequent action. The change in form will involve eliminating the D terms from the objective function and (5).

First, we divide each side of (4), the pivotal equation, by the coefficient of the key variable D.

$$\frac{5}{8}S + D + \frac{1}{8}x_1 = 50. \tag{6}$$

Then we multiply (6) by -4 and add the resulting equation to the objective function. In addition, we multiply (6) by -6 and add the resulting equation to (5). These changes will produce the following equations:

$$\frac{1}{2}S - \frac{1}{2}x_1 = P - 200$$

$$\frac{5}{8}S + D + \frac{1}{8}x_1 = 50 \tag{6}$$

$$\frac{13}{4}S - \frac{3}{4}x_1 + x_2 = 120. \tag{7}$$

Note that the key variable has been eliminated in the second objective function and (7).

Now, by recalling that S and x_1 should equal zero, the values for D, x_2, and P can be found by inspection.

$$S = 0 \qquad D = 50 \qquad x_1 = 0 \qquad x_2 = 120 \qquad P = 200$$

This solution indicates a production of 50 deluxe models at a profit of \$200.

It is well at this point to note the general structure of the new equations relative to the second solution. After zero values are assigned to the zero variables, each equation contains only one other variable. Equation 6, for example, becomes

$$\frac{5}{8}\times 0 + D + \frac{1}{8}\times 0 = 50$$

or $D = 50$. Thus, because of the general form, each equation can be solved without reference to the other equations. In fact, we shall see that the constant term in each equation will be a value of a nonzero variable in each solution. For this reason, the general form, which is a result of the algebraic changes, is very convenient.

AN IMPROVED SOLUTION. The shift from the first to the second solution has improved total profit from \$0 to \$200. We know, of course, from the graphical solution that further improvement is possible, but wish to use only algebraic techniques to determine the best solution. Hence, we check the transformed objective function for positive coefficients. In this case, S has the only positive coefficient and becomes the key variable. As before, we reason that if S is assigned a positive value, total profit will increase. Thus, in the next solution, S becomes a nonzero variable. (x_1, however, remains zero.)

The pivotal equation is determined by dividing the constant term in (6) and (7) by the corresponding coefficient of S. This yields ($50 \div 5/8$) or 80 in (6) and ($120 \div 13/4$) or 36 12/13 in (7). The smaller quotient indicates the largest possible value for S without forcing x_2 in (7) to be negative (recall that $x_1 = 0$). Equation 7 is the new pivotal equation and the slack variable x_2 is assigned a zero value, to permit the largest possible increase in S.

Next, we eliminate the key variable S from the objective function and Equation 6. This is accomplished by first changing the coefficient of S in the pivotal equation to 1. Multiplying each side of (7) by 4/13 produces

$$S - \frac{3}{13}x_1 + \frac{4}{13}x_2 = 36\frac{12}{13}. \tag{9}$$

The coefficient of 1 in the S term is very convenient in eliminating the key variable in the objective function and (6). We multiply (9) by $-1/2$ and add the result to the objective function, and multiply (9) by $-5/8$ and add it to (6). This results in the following three equations:

$$-\frac{5}{13}x_1 - \frac{2}{13}x_2 = P - 218\frac{6}{13}$$

$$D + \frac{7}{26}x_1 - \frac{5}{26}x_2 = 26\frac{12}{13} \tag{8}$$

$$S - \frac{3}{13}x_1 + \frac{4}{13}x_2 = 36\frac{12}{13}. \tag{9}$$

Again, these three equations present a very convenient form. The slack variables x_1 and x_2 are zero, and hence the values for S, D, and P can be determined by inspection:

$$S = 36\frac{12}{13} \qquad D = 26\frac{12}{13} \qquad x_1 = 0 \qquad x_2 = 0 \qquad P = 218\frac{6}{13}.$$

In this case, total profit includes a profit for fractions of two knives which, of course, is unrealistic. Before correcting this oversight, we note that the last objective function has only negative coefficients. This indicates that total profit cannot be increased further. If the slack variables are assigned positive values, total profit will decrease.

The solution that has been determined by the algebraic method is exactly the same as the graphical solution. Since it is unrealistic to produce a fraction of a knife, the final solution is

$$S = 35 \qquad D = 28 \qquad x_1 = 1 \qquad x_2 = 7 \qquad P = 217.$$

As predicted above, positive values for x_1 and x_2 result in a decrease of total profit. However, circumstances require 1 hour of slack time in shop 1 and 7 hours in shop 2.

SUMMARIZING THE ALGEBRAIC METHOD. In comparing the graphical and algebraic methods, the latter is probably more difficult to follow. The steps in the graphical method are easier to visualize because they can be described in terms of appropriate graphs. The same degree of understanding of an algebraic technique may require much practice before we can see why it produces a desired result.

The algebraic manipulations may seem rather tedious, so it is important to realize why they are performed. The following brief outline of the method may be of some value:

1. The largest positive coefficient in the objective function determines the key variable.
2. The constant terms in each constraint are divided by the nonzero coefficients of the key variable. The smallest quotient determines the pivotal equation.
3. The coefficient of the key variable in the pivotal equation is changed to 1.
4. The transformed pivotal equation is used to eliminate the key variable in all other equations.
5. The values of the nonzero variables in the new solution will be the constant terms on the right sides.

ANOTHER EXAMPLE OF MAXIMIZING PROFIT. As a second illustration of the algebraic method, consider the following example:

EXAMPLE 11.3

In the manufacture of two types of radio components, a firm uses a drill and a sander. Component X requires 2 min for drilling and 1 min for sanding; component Y requires 1 min for drilling and 3 min for sanding.

The drill and sander are also used to manufacture other products, so that in a 30-min period, the drill is available for 7 min and the sander for 11 min.

The firm makes a profit of \$1.00 on component X and \$2.00 on component Y. If the firm can sell all the components it can produce, find the number of each it should produce to maximize profit.

SOLUTION

If $x =$ the number of X components which are produced, and $y =$ the number of Y components which are produced,

$$x + 2y = P \qquad 2x + y \leq 7 \qquad x + 3y \leq 11.$$

By introducing two slack variables, the inequalities are transformed into equations:

$$x + 2y = P$$
$$2x + y + x_1 = 7 \qquad (1)$$
$$x + 3y + x_2 = 11. \qquad (2)$$

SOLUTION 1

$$x = 0 \quad y = 0 \quad x_1 = ? \quad x_2 = ? \quad P = ?$$
$$x = 0 \quad y = 0 \quad x_1 = 7 \quad x_2 = 11 \quad P = 0$$

$x + 2y = P$ [*Largest positive coefficient is* 2 (*y is key variable*)]

$2x + y + x_1 = 7$ [7/1 *dividing constant term of* (1) *by coefficient of key variable*] (1)

$x + 3y + x_2 = 11$ [11/3 *dividing constant term of* (2) *by coefficient of key variable*] (2)

Since $11/3 < 7/1$, (2) is the pivotal equation. Therefore, y is nonzero, x_2 is zero.

$$x + 2y = P$$

$$2x + y + x_1 = 7 \qquad (1)$$

$$\frac{1}{3}x + y + \frac{1}{3}x_2 = \frac{11}{3}. \qquad (4)$$

Eliminate key variable from objective function and (1) by adding $[-2 \times (4)]$ to the objective function and $[-1 \times (4)]$ to (1).

$$\frac{1}{3}x - \frac{2}{3}x_2 = P - \frac{22}{3}$$

$$\frac{5}{3}x + x_1 - \frac{1}{3}x_2 = \frac{10}{3} \qquad (3)$$

$$\frac{1}{3}x + y + \frac{1}{3}x_2 = \frac{11}{3}. \qquad (4)$$

SOLUTION 2

$x = 0 \quad y = ? \quad x_1 = ? \quad x_2 = 0 \quad P = ?$

$x = 0 \quad y = \frac{11}{3} \quad x_1 = \frac{10}{3} \quad x_2 = 0 \quad P = \frac{22}{3}$

$$\frac{1}{3}x - \frac{2}{3}x_2 = P - \frac{22}{3}$$ [*x is new key variable*]

$$\frac{5}{3}x + x_1 - \frac{1}{3}x_2 = \frac{10}{3} \qquad \frac{10}{3} \div \frac{5}{3} = 2$$

$$\frac{1}{3}x + y + \frac{1}{3}x_2 = \frac{11}{3} \qquad \frac{11}{3} \div \frac{1}{3} = 11$$

Since $2 < 11$, (3) is the new pivotal equation, x is nonzero, x_1 is zero.

$$\frac{1}{3}x - \frac{2}{3}x_2 = P - \frac{22}{3}$$

$$x + \frac{3}{5}x_1 - \frac{1}{5}x_2 = 2 \qquad (5)$$

$$\frac{1}{3}x + y + \frac{1}{3}x_2 = \frac{11}{3}. \qquad (4)$$

Eliminate key variable in the objective function and (4):

$$-\frac{1}{5}x_1 - \frac{3}{5}x_2 = P - 8$$

$$x + \frac{3}{5}x_1 - \frac{1}{5}x_2 = 2 \qquad (5)$$

$$y - \frac{1}{5}x_1 + \frac{2}{5}x_2 = 3. \qquad (6)$$

SOLUTION 3

$x = ? \quad y = ? \quad x_1 = 0 \quad x_2 = 0 \quad P = ?$

$x = 2 \quad y = 3 \quad x_1 = 0 \quad x_2 = 0 \quad P = 8$

Since there are no positive coefficients in the objective function, the best solution has been found. The maximum profit is \$8, with $x = 2$ and $y = 3$.

EXERCISE 11.3

Use the algebraic method to find the solutions for problems 1-8:

1. Maximize: $2x + 3y = P$
   Constraints: $4x + y \leq 17$
   $2x + 2y \leq 16$
2. Maximize: $x + 2y = P$
   Constraints: $3x + y \leq 25$
   $x + 4y \leq 23$
3. Maximize: $5x + 4y = P$
   Constraints: $7x + y \leq 28$
   $5x + 5y \leq 50$
4. Maximize: $2x + 5y = P$
   Constraints: $4x + 4y \leq 40$
   $x + 6y \leq 15$
5. Maximize: $4x + 3y = P$
   Constraints: $2x + 7y \leq 55$
   $5x + 3y \leq 36$
6. Maximize: $4x + 3y = P$
   Constraints: $3x + 2y \leq 39$
   $2x + 4y \leq 50$
7. Maximize: $3x + 8y = P$
   Constraints: $x + 2y \leq 25$
   $3x + 9y \leq 84$
8. Maximize: $9x + 5y = P$
   Constraints: $x + 7y \leq 23$
   $4x + 2y \leq 14$
9. Solve problem 1 by using appropriate graphs.
10. Solve problem 2 by using appropriate graphs.
11. Solve problem 4 by using appropriate graphs.
12. Solve problem 7 by using appropriate graphs.
13. The following problem can be solved by introducing three slack variables:
    Maximize: $2x + y = P$
    Constraints: $x + y \leq 7$
    $x + 3y \leq 17$
    $x + 4y \leq 20$
14. Maximize: $x + 3y = P$
    Constraints: $x + y \leq 13$
    $2x + y \leq 18$
    $3x + y \leq 19$
15. Maximize: $3x + 2y = P$
    Constraints: $x + y \leq 10$
    $2x + y \leq 13$
    $x \leq 6$
16. Solve problem 13 by using appropriate graphs.
17. Solve problem 14 by using appropriate graphs.
18. Solve problem 15 by using appropriate graphs.

11.4 THE ALGEBRAIC METHOD IN MATRIX FORM

In Chapter 10 the concept of matrices was used to represent a system of equations in a different form. One advantage of this matrix form was the convenient display of the coefficients of the variables involved. Since the algebraic method discussed in the last section requires considerable changing of coefficients, the matrix form can be used as an aid in keeping track of these changes. We will demonstrate the use of the matrix form by reviewing the solution of the second example in the preceding section.

Before converting to matrix form we note that the objective function, (1), and (2) can be written as follows:

$$\begin{aligned} x + 2y + 0x_1 + 0x_2 &= P \\ 2x + y + x_1 + 0x_2 &= 7 \\ x + 3y + 0x_1 + x_2 &= 11. \end{aligned}$$

In matrix form, these three equations can be represented as

$$\begin{bmatrix} 1 & 2 & 0 & 0 \\ 2 & 1 & 1 & 0 \\ 1 & 3 & 0 & 1 \end{bmatrix} \begin{bmatrix} x \\ y \\ x_1 \\ x_2 \end{bmatrix} = \begin{bmatrix} P \\ 7 \\ 11 \end{bmatrix}. \tag{1}$$

In a similar manner, the second objective function, (3), and (4) can be written

$$\begin{bmatrix} \frac{1}{3} & 0 & 0 & -\frac{2}{3} \\ \frac{5}{3} & 0 & 1 & -\frac{1}{3} \\ \frac{1}{3} & 1 & 0 & \frac{1}{3} \end{bmatrix} \begin{bmatrix} x \\ y \\ x_1 \\ x_2 \end{bmatrix} = \begin{bmatrix} P - \frac{22}{3} \\ \frac{10}{3} \\ \frac{11}{3} \end{bmatrix} \tag{2}$$

and the third objective function, (5), and (6) as

$$\begin{bmatrix} 0 & 0 & -\frac{1}{5} & -\frac{3}{5} \\ 1 & 0 & \frac{3}{5} & -\frac{1}{5} \\ 0 & 1 & -\frac{1}{5} & \frac{2}{5} \end{bmatrix} \begin{bmatrix} x \\ y \\ x_1 \\ x_2 \end{bmatrix} = \begin{bmatrix} P - 8 \\ 2 \\ 3 \end{bmatrix}. \tag{3}$$

A significant part of the algebraic method is to effect the change of matrix equation (1) to matrix equation (2). As an illustration of the operations that produced the change, we recall that in the preceding section the third row of (1) was divided by 3. In matrix form, the result could be written

$$\begin{bmatrix} 1 & 2 & 0 & 0 \\ 2 & 1 & 1 & 0 \\ \frac{1}{3} & 1 & 0 & \frac{1}{3} \end{bmatrix} X = \begin{bmatrix} P \\ 7 \\ \frac{11}{3} \end{bmatrix} \tag{1a}$$

where

$$X = \begin{bmatrix} x \\ y \\ x_1 \\ x_2 \end{bmatrix}$$

Since X represents the variables, it does not change.

The coefficient 1 in the third row and second column is then used to change the other entries in the second column to zeros. This is accomplished in part by multiplying the third row by -2 and adding the result to the first row. In turn the third row is multiplied by -1 and the result added to the second row. The indicated operations are

$$\begin{bmatrix} 1-2\left(\frac{1}{3}\right) & 2-2(1) & 0-2(0) & 0-2\left(\frac{1}{3}\right) \\ 2-1\left(\frac{1}{3}\right) & 1-(1) & 1-(0) & 0-\left(\frac{1}{3}\right) \\ \frac{1}{3} & 1 & 0 & \frac{1}{3} \end{bmatrix} X = \begin{bmatrix} P-2\frac{11}{3} \\ 7-\frac{11}{3} \\ \frac{11}{3} \end{bmatrix}. \tag{1b}$$

Note that the row operations are performed on both sides of the matrix equations. When simplified, the first two entries in the second column are zero, corresponding to the elimination of the y terms in the first two equations.

The change from (2) to (3) is accomplished in a similar manner. The entry in the second row and first column is changed to 1 by multiplying by 3/5.

$$\begin{bmatrix} \frac{1}{3} & 0 & 0 & -\frac{2}{3} \\ 1 & 0 & \frac{3}{5} & -\frac{1}{5} \\ \frac{1}{3} & 1 & 0 & \frac{1}{3} \end{bmatrix} X = \begin{bmatrix} P-\frac{22}{3} \\ 2 \\ \frac{11}{3} \end{bmatrix}. \tag{2a}$$

Then $(-1/3)$ times the second row is added to the first and third rows. These operations are performed on both sides of the matrix equation and result in (3).

In general, the nonzero variables occur in the columns containing two zeros and a single 1. The other columns can then be ignored, since they contain coefficients of variables that are zero. In (2), for example, $x_1 = 10/3$, $y = 11/3$, and $P = 22/3$. The first and fourth columns can be ignored. From (3), we conclude that $P = 8$, $x = 2$, and $y = 3$ by blocking out the third and fourth columns.

THE KEY COLUMN AND PIVOTAL ROW. The key variable and the pivotal equation can also be determined from the matrix equation. It is more convenient, however, to speak of the key column and pivotal row. Since the coefficients of the objective function are in the first row, the column in which the largest positive entry occurs will be the key column. The entries in the key column (excluding the first row) are divided into the constants in the matrix on the right side. The smallest quotient determines the pivotal row.

EXAMPLE 11.4

Suppose the mathematical form of a linear programming problem is as follows:

$$\begin{aligned} 2x + 3y &= P \\ 4x + 5y &\leq 51 \\ x + 2y &\leq 18. \end{aligned}$$

SOLUTION

After introducing two slack variables, we have

$$\begin{aligned} 2x + 3y + 0x_1 + 0x_2 &= P \\ 4x + 5y + x_1 + 0x_2 &= 51 \\ x + 2y + 0x_1 + x_2 &= 18. \end{aligned}$$

In matrix form, these equations appear as follows:

$$\begin{matrix} & \begin{matrix} & \downarrow & & \end{matrix} \\ \begin{matrix} \\ \\ \rightarrow \end{matrix} & \begin{bmatrix} 2 & 3 & 0 & 0 \\ 4 & 5 & 1 & 0 \\ 1 & 2 & 0 & 1 \end{bmatrix} \end{matrix} \qquad X = \begin{bmatrix} P \\ 51 \\ 18 \end{bmatrix} \tag{4}$$

The key column can be conveniently indicated by placing an arrow above the largest positive coefficient in the first row. Then, since $18/2 < 51/5$, the arrow on the left in (4) indicates that the third row is the pivotal row.

Next, the entry in the key column and pivotal row, sometimes called the **pivot**, becomes the center of attention. By circling this entry as in (5), the key column and pivotal row are marked simultaneously.

$$\begin{bmatrix} 2 & 3 & 0 & 0 \\ 4 & 5 & 1 & 0 \\ 1 & \textcircled{2} & 0 & 1 \end{bmatrix} \qquad X = \begin{bmatrix} P \\ 51 \\ 18 \end{bmatrix}. \tag{5}$$

This convenient notation will be used in the rest of this chapter.

Once determined, the pivot is always changed to one; in this case, by dividing each entry in the third row by two. The result is

$$\begin{bmatrix} 2 & 3 & 0 & 0 \\ 4 & 5 & 1 & 0 \\ \frac{1}{2} & 1 & 0 & \frac{1}{2} \end{bmatrix} \qquad X = \begin{bmatrix} P \\ 51 \\ 9 \end{bmatrix}. \tag{6}$$

The new third row is used to change the other entries in the key column to zero, as in (7):

$$\begin{bmatrix} \frac{1}{2} & 0 & 0 & -\frac{3}{2} \\ \frac{3}{2} & 0 & 1 & -\frac{5}{2} \\ \frac{1}{2} & 1 & 0 & \frac{1}{2} \end{bmatrix} \qquad X = \begin{bmatrix} P-27 \\ 6 \\ 9 \end{bmatrix}. \tag{7}$$

At this point, the profit is \$27, $y = 9$, and $x_1 = 6$.

The second cycle begins by noting the single positive coefficient in the first row. This determines the key column, and the inequality

$$\left(6 \div \frac{3}{2}\right) < \left(9 \div \frac{1}{2}\right)$$

determines the pivotal row. Accordingly, the pivot 3/2 is circled in (8):

$$\begin{bmatrix} \frac{1}{2} & 0 & 0 & -\frac{3}{2} \\ \boxed{\frac{3}{2}} & 0 & 1 & -\frac{5}{2} \\ \frac{1}{2} & 1 & 0 & \frac{1}{2} \end{bmatrix} \qquad X = \begin{bmatrix} P-27 \\ 6 \\ 9 \end{bmatrix}. \tag{8}$$

The pivot could have been circled in (7), but (8) is included to indicate the proper sequence.

Next, the pivot is changed to one by multiplying the second row by 2/3:

$$\begin{bmatrix} \frac{1}{2} & 0 & 0 & -\frac{3}{2} \\ 1 & 0 & \frac{2}{3} & -\frac{5}{3} \\ \frac{1}{2} & 1 & 0 & \frac{1}{2} \end{bmatrix} \qquad X = \begin{bmatrix} P-27 \\ 4 \\ 9 \end{bmatrix} \tag{9}$$

In turn, the other coefficients in the first column are changed to zero:

$$\begin{bmatrix} 0 & 0 & -\frac{1}{3} & -\frac{2}{3} \\ 1 & 0 & \frac{2}{3} & -\frac{5}{3} \\ 0 & 1 & -\frac{1}{3} & \frac{4}{3} \end{bmatrix} \qquad X = \begin{bmatrix} P-29 \\ 4 \\ 7 \end{bmatrix} \tag{10}$$

Now, the first row does not contain any positive entries, indicating that the best solution has been determined. The maximum profit is \$29, with $x = 4$ and $y = 7$.

EXERCISE 11.4

In problems 1-6, write a matrix equation and solve:

1. Maximize: $2x + y = P$
   Constraints: $6x + 2y \leq 50$
   $x + 4y \leq 23$
2. Maximize: $4x + 7y = P$
   Constraints: $3x + 3y \leq 30$
   $x + 6y \leq 15$
3. Maximize: $2x + 3y = P$
   Constraints: $x + y \leq 15$
   $2x + y \leq 22$
4. Maximize: $6x + 3y = P$
   Constraints: $x + y \leq 80$
   $3x + y \leq 220$
5. Maximize: $3x + y = P$
   Constraints: $2x + y \leq 9$
   $4x + y \leq 11$
6. Maximize: $x + 4y = P$
   Constraints: $x + 2y \leq 14$
   $x + 6y \leq 38$

Solve problems 7-10 by the graphical method, the algebraic method, and by using a matrix form.

7. Maximize: $x + 2y = P$
   Constraints: $x + y \leq 13$
   $2x + y \leq 18$
8. Maximize: $5x + 4y = P$
   Constraints: $2x + y \leq 22$
   $x + 2y \leq 26$
9. Maximize: $3x + 2y = P$
   Constraints: $x + y \leq 10$
   $2x + y \leq 13$
   $x \leq 6$
10. Maximize: $5x + 4y = P$
    Constraints: $7x + y \leq 28$
    $x + y \leq 10$
    $x + 6y \leq 15$

11.5 THE SIMPLEX METHOD

The method discussed in the preceding section can be changed slightly so that total profit appears as a positive number. If both sides of the objective function are multiplied by (-1) and the letter P is omitted, each solution will be listed directly on the right side. In effect, this is one form of the well-known simplex method for solving a linear programming problem. We will demonstrate the method by considering again the second example in Section 11.4.

Recall that the matrix equation

$$\begin{bmatrix} 2 & 3 & 0 & 0 \\ 4 & 5 & 1 & 0 \\ 1 & 2 & 0 & 1 \end{bmatrix} \qquad X = \begin{bmatrix} P \\ 51 \\ 18 \end{bmatrix}$$

was used to represent the basic system of equations. The top row of the coefficient matrix relates to the objective function, which is to be maximized; the second and third rows relate to the constraints.

A TABULAR FORM. In the basic simplex method, this same information is displayed in table form. The arrangement, however, differs, as is shown in Table 11.1. This table, called the **initial simplex tableau**, contains in its

TABLE 11.1 INITIAL SIMPLEX TABLEAU

	x	y	x_1	x_2	
x_1	4	5	1	0	51
x_2	1	(2)	0	1	18
	−2	−3	0	0	0

first two rows the coefficients and constants from the constraints. In the last row of this table, the coefficients of the objective function have been multiplied by (−1) and P has been replaced by zero. The variables that are related to each column are listed at the top of the tableau. In addition, the variables which assume nonzero values in the first solution are listed at the left of the table. When x and y are zero, the values 51, 18, and 0 are the initial values of x_1, x_2, and P. Hence, the initial simplex tableau contains all the information displayed in the matrix equation and also clearly indicates the first solution.

The second solution is determined by the same arithmetic operations as before except for one difference. Since the coefficients of the variables in the objective function are negative, the key column is determined by the **smallest** negative coefficient—in this case $-3 < -2$—so the second column is the key column.

The pivotal row and pivot (circled in Table 11.1) are determined as before. However, in the simplex method, the pivot indirectly indicates which variables enter and leave the solution. The nonzero variable (listed on the left), which is in the pivotal row, is replaced by the key variable.

After changing the pivot to 1 and the other entries in the key column to zero, the second solution can be read directly from Table 11.2, the second simplex tableau. Note that y has replaced x_2 on the left as a nonzero variable. With x and x_2 zero, the values for x_1, y, and P are 6, 9, and 27, respectively.

The negative coefficient in the last row of the second simplex tableau indicates that further improvement is possible. Since $(6 \div 3/2) < (9 \div 1/2)$, the pivot is 3/2 (circled in Table 11.2). In the next solution, x_1 will be replaced by x on the left, so that x and y will be the nonzero variables.

After effecting the appropriate changes in the first column, the third and final simplex tableau is shown in Table 11.3. Since all entries in the last row are positive, the optimum solution is

$$x = 4 \qquad y = 7 \qquad x_1 = 0 \qquad x_2 = 0 \qquad P = 29.$$

TABLE 11.2 SECOND SIMPLEX TABLEAU

	x	y	x_1	x_2	
x_1	$\frac{3}{2}$ (circled)	0	1	$-\frac{5}{2}$	6
y	$\frac{1}{2}$	1	0	$\frac{1}{2}$	9
	$-\frac{1}{2}$	0	0	$\frac{3}{2}$	27

TABLE 11.3 THIRD SIMPLEX TABLEAU

	x	y	x_1	x_2	
x	1	0	$\frac{2}{3}$	$-\frac{5}{3}$	4
y	0	1	$-\frac{1}{3}$	$\frac{4}{3}$	7
	0	0	$\frac{1}{3}$	$\frac{2}{3}$	29

As noted above, the simplex method is essentially the same as the algebraic method considered previously. The coefficients of the objective function appear in the last row with opposite signs, the profit corresponding to a particular solution is now positive, and the optimum solution has been found when all entries in the last row are positive. Moreover, the listing of the variables that assume nonzero values in a particular solution make it possible to determine a solution directly from the table.

AN EXAMPLE CONTAINING FOUR VARIABLES. In an earlier section, we noted that an algebraic method was necessary to solve a linear programming problem involving more than three variables (not including the slack variables). We shall now consider an example that requires four basic variables and five additional variables. It is significant that, with the simplex method, the coefficients of nine variables can be handled quite efficiently.

EXAMPLE 11.5

An individual with \$4000 to invest considers the following investments at the stated interest rate: bonds, 3%; stocks, 4%; loans, 5%; second mortgages, 8%.

A broker advised him to invest no more than \$1500 in mortgages and loans, at least \$2000 in bonds and stocks, but less than \$2500 in stocks and mortgages. Find the most profitable investment combination.

SOLUTION

Let B, M, S, and L represent the amounts invested in bonds, mortgages, stocks, and loans, respectively. The objective function is

$$0.03B + 0.08M + 0.04S + 0.05L = P. \tag{1}$$

The restrictions or constraints lead to the following inequalities:

$$M + L \leq 1500 \tag{2}$$
$$B + S \geq 2000 \tag{3}$$
$$M + S \leq 2500. \tag{4}$$

In addition,

$$B + M + S + L = 4000. \tag{5}$$

In this problem we have included two situations that require special treatment. Inequality (3) can be converted to an equation by subtracting a non-negative slack variable from the sum $B + S$. This slack variable represents the amount by which $B + S$ exceeds \$2000. However, if B and S are zero, this sum is also zero, forcing the slack variable to be negative. For this reason another non-negative variable is introduced into the equation. The "greater than" inequality is then written

$$B + S - x_1 + A_1 = 2000.$$

x_1 is the slack variable and A_1 is called an *artificial* variable. A second artificial variable A_2 is also added to (5). It then becomes

$$B + M + S + L + A_2 = 4000.$$

In the initial solution, the artificial variable A_2 will allow B, M, S, and L to be zero. Without A_2 at least one of these four, variables would have to be nonzero.

Now, introducing slack variables for the other inequalities, the entire system can be written as follows:

$$\begin{aligned}
B + 0 \times M + S + 0 \times L - x_1 + A_1 \qquad\qquad &= 2000 \\
0 \times B + M + 0 \times S + L \qquad + x_2 \qquad\qquad &= 1500 \\
0 \times B + M + S + 0 \times L \qquad\qquad + x_3 \qquad &= 2500 \\
B + M + S + L \qquad\qquad\qquad + A_2 &= 4000 \\
0.03B + 0.08M + 0.04S + 0.05L \qquad\qquad &= P
\end{aligned}$$

TABLE 11.4 INITIAL SIMPLEX TABLEAU

	B	M	S	L	x_1	A_1	x_2	x_3	A_2	
A_1	1	0	1	0	−1	1	0	0	0	2000
x_2	0	(1)	0	1	0	0	1	0	0	1500
x_3	0	1	1	0	0	0	0	1	0	2500
A_2	1	1	1	1	0	0	0	0	1	4000
	−0.03	−0.08	−0.04	−0.05	0	0	0	0	0	0

The initial simplex tableau is shown in Table 11.4. The list of variables on the left indicates that B, M, S, L, and x_1 are zero, which of course accounts for the zero profit (lower right-hand entry). The smallest entry in the last row is -0.08, and since

$$\frac{1500}{1} < \frac{2500}{1} < \frac{4000}{1}$$

the pivot is as shown in Table 11.4. The key variable M will replace x_2 in the next solution. Using the pivotal row to change all other entries in the second column to zero, we have the second simplex tableau, as shown in Table 11.5.

Note that, even though the second row did not change, the variable M now appears on the left. From this, we observe that $1500 has been invested in mortgages. The last entry in the right-hand column shows a return of $120 on this investment.

TABLE 11.5 SECOND SIMPLEX TABLEAU

	B	M	S	L	x_1	A_1	x_2	x_3	A_2	
A_1	1	0	1	0	−1	1	0	0	0	2000
M	0	1	0	1	0	0	1	0	0	1500
x_3	0	0	(1)	−1	0	0	−1	1	0	1000
A_2	1	0	1	0	0	0	−1	0	1	2500
	−0.03	0	−0.04	0.03	0	0	0.08	0	0	120

The smallest entry in the last row is now -0.04. Hence S is the key variable and will replace x_3 in the next solution. After the pivot is used to change all other entries in the third column to zero, we have the tableau of Table 11.6.

TABLE 11.6 THIRD SIMPLEX TABLEAU

	B	M	S	L	x_1	A_1	x_2	x_3	A_2	
A_1	(1)	0	0	1	−1	1	1	−1	0	1000
M	0	1	0	1	0	0	1	0	0	1500
S	0	0	1	−1	0	0	−1	1	0	1000
A_2	1	0	0	1	0	0	0	−1	1	1500
	−0.03	0	0	−0.01	0	0	0.04	0.04	0	160

The presence of negative coefficients in the last row indicates that further improvement is possible. At this point the return on $1500 in mortgages and $1000 in stocks is $160. Since $-0.03 < -0.01$, profit will show the greatest increase by investing in bonds. Hence B replaces A_1 in the next tableau (Table 11.7).

The investment now consists of $1000 in bonds, $1500 in mortgages, and $1000 in stocks. The net return has increased to $190, but the last row still has a negative entry. In selecting 1 or (−1) as the pivot, only positive coefficients can be considered. If the first row were the pivotal row, x_1 would replace B and be forced to assume a negative value. Hence the final simplex tableau is found by using the pivot 1 to eliminate the other x_1 terms. The result is shown in Table 11.8.

TABLE 11.7 FOURTH SIMPLEX TABLEAU

	B	M	S	L	x_1	A_1	x_2	x_3	A_2	
B	1	0	0	1	−1	1	1	−1	0	1000
M	0	1	0	1	0	0	1	0	0	1500
S	0	0	1	−1	0	0	−1	1	0	1000
A_2	0	0	0	0	(1)	−1	−1	0	1	500
	0	0	0	0.02	−0.03	0.03	0.07	0.01	0	190

TABLE 11.8 FINAL SIMPLEX TABLEAU

	B	M	S	L	x_1	A_1	x_2	x_3	A_2	
B	1	0	0	1	0	0	0	−1	1	1500
M	0	1	0	1	0	0	1	0	0	1500
S	0	0	1	−1	0	0	−1	1	0	1000
x_1	0	0	0	0	1	−1	−1	0	1	500
	0	0	0	0.02	0	0	0.04	0.01	0.03	205

All negative coefficients have been eliminated from the last row, indicating that the best return is \$205. The \$4000 should be divided so that \$1500 is invested in bonds, \$1500 in mortgages, and \$1000 in stocks. The slack variable x_1 is nonzero, since the investment in bonds and stocks exceeds \$2000 by \$500. The interest rates in this problem were selected so that the solution could be easily verified (see problem 1, Exercise 11.3).

This has been a necessarily brief introduction to the simplex method of linear programming. The main emphasis has been on the algebraic manipulations that change one tableau to another. Some of the refinements, variations, and special difficulties are discussed in the references listed at the end of this chapter.

EXERCISE 11.5

1. A solution to the investment problem could have been determined by the following reasoning: Since the return on mortgages was so high, the maximum amount (\$1500) would naturally be invested in them. From (2), this determines the amount invested in loans, and from (4) the maximum amount in stocks. Since the return on stocks is higher than on bonds, a complete solution can be determined. Indicate in sequence the amount invested in loans, stocks, and bonds, justifying each answer.
2. In the investment problem, use the simplex method to find the most

profitable combination if the interest rates are: for bonds, 4%; mortgages, 8%; stocks, 3%; loans, 5%.

3. Repeat problem 2, with the interest rates for bonds, 3%; mortgages, 5%; stocks, 4%; loans, 8%.

In problems 4-21, use the simplex method to determine the best solution.

4. Maximize: $2x + 3y + z$
   Subject to:
   $$\begin{aligned} x + 3y + 2z &\leq 11 \\ x + 2y + 5z &\leq 19 \\ 3x + y + 4z &\leq 25. \end{aligned}$$

5. Maximize: $4x + y + 5z$
   Subject to:
   $$\begin{aligned} x \quad\;\; + z &\leq 8 \\ x + y \quad\;\; &\leq 3 \\ x + y + z &\leq 5. \end{aligned}$$

6. Maximize $x + y + 2z$, subject to the constraints given in problem 5.
7. Maximize $2x + 3y + z$, subject to the constraints given in problem 5.
8. Maximize $3x + y + z$, subject to the constraints given in problem 5.
9. Maximize: $2x + y + z$
   Subject to:
   $$\begin{aligned} y + z &\leq 15 \\ x + y \quad\;\; &\leq 12 \\ x \quad\;\; + z &\leq 7 \end{aligned}$$

10. Maximize $x + 2y + z$, subject to the constraints given in problem 9.
11. Maximize $x + y + 2z$, subject to the constraints given in problem 9.
12. Maximize $3x + 2y + z$, subject to the constraints given in problem 9.
13. Maximize: $2x + y + u + v$
    Subject to:
    $$\begin{aligned} x + y \quad\;\; + v &\leq 9 \\ x \quad\;\; + u \quad\;\; &\leq 10 \\ + y \quad\;\; + v &\leq 7 \\ x + y + u + v &\leq 16 \end{aligned}$$

14. Maximize $x + 2y + u + v$, subject to the constraints given in problem 13. In this problem, there will be two different solutions associated with the maximum profit.
15. Maximize $x + y + 3u + 2v$, subject to the constraints given in problem 13.
16. Maximize $x + 2y + 3u + 4v$, subject to the constraints given in problem 13.
17. Maximize: $2x + 3y$
    Subject to:
    $$\begin{aligned} 4x + y &\leq 47 \\ 6x + 3y &\leq 87 \\ 3x + 5y &\leq 117 \end{aligned}$$

18. Maximize: $5x + 4y$
    Subject to:
    $$\begin{aligned} 7x + y &\leq 28 \\ 5x + 5y &\leq 50 \\ x + 6y &\leq 15 \end{aligned}$$

19. Maximize: $3x + 8y$
    Subject to:
    $$\begin{aligned} 3x + 2y &\leq 39 \\ 2x + 4y &\leq 50 \\ 3x + 9y &\leq 84 \end{aligned}$$

20. Maximize: $9x + 5y$
    Subject to:
    $$\begin{aligned} 2x + 14y &\leq 46 \\ x + 6y &\leq 21 \\ 3x + 4y &\leq 49 \end{aligned}$$

21. Maximize: $4x + 7y$
    Subject to:
    $$\begin{aligned} 6x + 5y &\leq 80 \\ 2x + 7y &\leq 48 \\ x + 9y &\leq 57 \end{aligned}$$

REFERENCES

Campbell, H. G., *An Introduction to Matrices, Vectors, and Linear Programming.* New York, Appleton-Century-Crofts, 1965.

Hadley, G., *Linear Programming.* Reading, Mass., Addison-Wesley Publishing Company, Inc., 1962.

Kaufmann, A., *Methods and Models of Operations Research.* Englewood Cliffs, N.J., Prentice-Hall, Inc., 1963.

Kemeny, J., Snell, J., Thompson, G., and Schleifer, A. Jr., *Finite Mathematics with Business Applications.* Englewood Cliffs, N. J., Prentice-Hall, Inc., 1962.

Levin, R. I. and Kirkpatrick, C. A., *Quantitative Approaches to Management.* New York, McGraw-Hill, Inc., 1965.

Metzger, R. W., *Elementary Mathematical Programming.* New York, John Wiley & Sons, 1968.

12 SERIES

12.1 SEQUENCES / **389**
What is Meant by Sequence / **389**
Money and Interest / **390**
Other Sequences / **391**
Determining a Formula / **393**
Exercise 12.1 / **394**

12.2 PROGRESSIONS / **395**
Arithmetic Progressions / **395**
The Recursive Form of Arithmetic Progressions / **396**
Geometric Progressions / **397**
Exercise 12.2 / **398**

12.3 CONVERGENCES OF SEQUENCES / **400**
Bounds and Limits / **400**
Convergence of Progressions / **401**
Calculation of Pi / **402**
Exercise 12.3 / **404**

12.4 SUMMATIONS / **404**
Exercise 12.4 / **407**

12.5 SUMS OF ARITHMETIC AND GEOMETRIC PROGRESSIONS / **408**
Arithmetic Progressions / **408**
Geometric Progressions / **410**
Exercise 12.5 / **411**

12.6 INFINITE GEOMETRIC SERIES / **412**
Achilles and the Tortoise / **412**
Repeating Decimals / **413**
Exercise 12.6 / **414**

The words *series* and *sequence* are used in everyday English in a variety of ways; in fact, they are sometimes used interchangeably. However, when used in mathematics, they have a much more specific meaning and, although directly related, can never be used as synonyms. Not only are the notions that we shall study in this chapter a foundation for the study of numerical methods (Chapter 13), but they are basic to the powerful DO loop in Fortran.

12.1 SEQUENCES

WHAT IS MEANT BY SEQUENCE. In Chapter 7 the term *function* was defined as a correspondence between two sets, such that each element of the first set is associated with exactly one element of the second set. The linear function defined by the equation $y = x/2$ is valid for x as any real number; in other words, the domain of the function consists of the real numbers. Similarly, the nonlinear function $y = 1/x$ also has a domain, consisting of the real numbers (except zero).

On the other hand, we could limit the domains in both cases to the positive integers, resulting in Table 12.1. Whenever a function is so limited,

TABLE 12.1

$y = x/2$		$y = 1/x$	
x	y	x	y
1	$\frac{1}{2}$	1	1
2	1	2	$\frac{1}{2}$
3	$\frac{3}{2}$	3	$\frac{1}{3}$
4	2	4	$\frac{1}{4}$
⋮	⋮	⋮	⋮

it is called a **sequence function**, or simply a **sequence**. More specifically, we will define the term *sequence* as *a function whose domain consists of consecutive nonnegative integers larger than a given integer.*

A sequence is said to be finite if its domain has a finite number of members, and infinite if it has an infinite number of members. In order to clearly indicate a sequence, rather than a function whose domain is not so restricted, we will use the notation

$$a_n = \frac{n}{2} \qquad (n = 1, 2, 3, \cdots) \tag{1}$$

and

$$a_n = \frac{1}{n} \qquad (n = 1, 2, 3, \cdots) \tag{2}$$

rather than the previous formulas $y = x/2$ and $y = 1/x$. Furthermore, for the sake of simplicity, we shall always assume that the domain of our sequence is the positive integers, unless explicitly indicated otherwise. Thus the range elements of the sequence defined by (1) are

$$a_1 = \frac{1}{2}(1) = \frac{1}{2}$$

$$a_2 = \frac{1}{2}(2) = 1$$

$$a_3 = \frac{1}{2}(3) = \frac{3}{2}$$

$$\vdots$$

$$a_n = \frac{1}{2}(n)$$

$$\vdots$$

Here the original equation is included to represent the general or the nth **member of the sequence.** The dots are used to indicate that the sequence continues indefinitely.

In a similar manner, the range elements of the sequence defined by (2) are

$$a_1 = \frac{1}{1} = 1$$

$$a_2 = \frac{1}{2}$$

$$a_3 = \frac{1}{3}$$

$$\vdots$$

$$a_n = \frac{1}{n}$$

$$\vdots$$

MONEY AND INTEREST. Another example of a sequence, this one very meaningful in our modern world of credit buying and saving, involves compound interest, and the formula

$$A = P\left(1 + \frac{i}{n}\right)^{nt},$$

where n is the number of times per year the interest is compounded. What is the effect of compounding on a savings of \$100 over a period of one year at a 4% rate? The formula may be written in the familiar form representing a sequence:

$$A_n = 100\left(1 + \frac{0.04}{n}\right)^n. \tag{3}$$

If it were necessary to calculate the amount A_n for values of n ranging from 1 (interest compounded yearly) through 365 (interest compounded daily), we would have a problem well suited to the computer. The problem logic is illustrated in the flowchart of Figure 12.1. Results of the computation would, technically speaking, form a sequence; the first four and the 365th elements of this sequence are

$$A_1 = 100\left(1 + \frac{0.04}{1}\right)^1 = 104.000$$

$$A_2 = 100\left(1 + \frac{0.04}{2}\right)^2 \simeq 104.040$$

$$A_3 = 100\left(1 + \frac{0.04}{3}\right)^3 \simeq 104.053$$

$$A_4 = 100\left(1 + \frac{0.04}{4}\right)^4 \simeq 104.060$$

$$\vdots$$

$$A_{365} = 100\left(1 + \frac{0.04}{365}\right)^{365} \simeq 104.081.$$

It is interesting to note that a bank which advertises its interest to be compounded daily ($n = 365$) is actually offering very little additional money over a bank which compounds quarterly.

OTHER SEQUENCES. An example of a sequence with an oscillating effect because of alternating signs on one of the terms is

$$a_n = 1 + \frac{(-1)^n}{n} \tag{4}$$

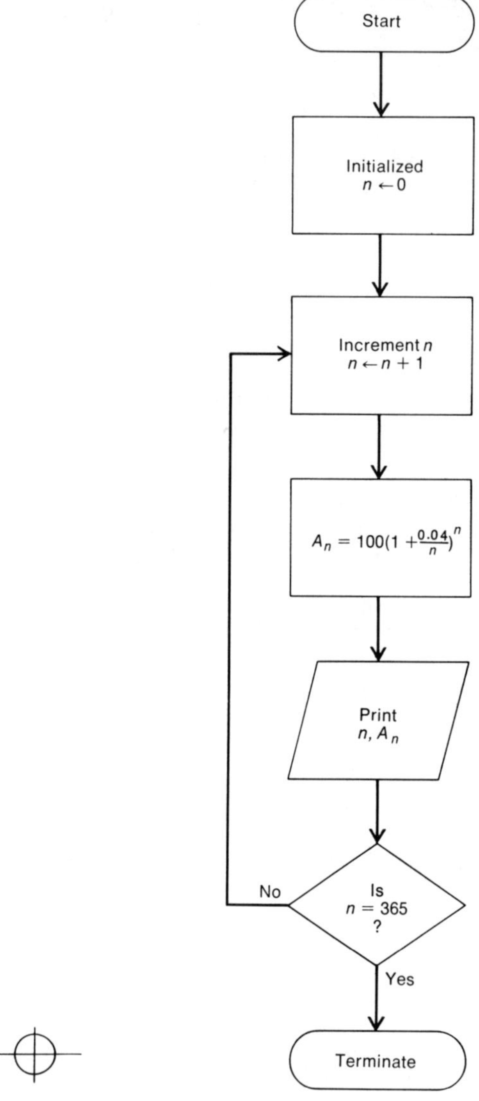

FIGURE 12.1 Calculating interest

whose first six terms are 0, 3/2, 2/3, 5/4 and 4/5; the fiftieth and fifty-first terms are

$$a_{50} = 1 + \frac{(-1)^{50}}{50} = \frac{51}{50}$$

$$a_{51} = 1 + \frac{(-1)^{51}}{51} = \frac{50}{51}.$$

Another example is the **Fibonacci sequence**, defined by the following formula:

$$a_n = a_{n-1} + a_{n-2} \qquad a_1 = a_2 = -1. \tag{5}$$

Here the values for a_1 and a_2 are defined as -1; a_3 can be calculated by using $n=3$, giving

$$a_3 = a_{3-1} + a_{3-2}$$

or

$$\begin{aligned} a_3 &= a_2 + a_1 = -1 - 1 = -2 \\ a_4 &= a_3 + a_2 = -2 - 1 = -3 \\ a_5 &= a_4 + a_3 = -3 - 2 = -5 \\ a_6 &= a_5 + a_4 = -5 - 3 = -8. \end{aligned}$$

Since (5) defines a given term as a function of previous terms (evident by the above calculations), it is necessary to determine all previous terms to obtain any desired one. Thus (5) is commonly referred to as a **recursive** formula.

DETERMINING A FORMULA. The elements of the sequences in Exercise 12.1 are relatively simple to calculate. On the other hand, the process of obtaining the general expression, given several members of the sequence, is usually much more difficult. If the first five elements of a sequence are 1, 4, 9, 16, 25, we can see by inspection that they are all perfect squares. The general term is

$$a_n = n^2. \tag{6}$$

The sequence whose first four elements are 2/1, 4/3, 6/5, 8/7 is not as obvious at first glance, but the general term can be obtained by noting that the numerator can be represented by $2n$ and the denominator, being one less in each case, by $2n-1$. Thus

$$a_n = \frac{2n}{2n-1}. \tag{7}$$

In both of these examples, the general expression chosen to represent the sequence is the obvious one. However, in neither case is the choice unique. Consider the sequence whose first three terms are 6, 3, and 2. The most straightforward sequence including these terms is

$$a_n = \frac{6}{n}. \tag{8}$$

However, each of the following expressions leads to the same first three terms, yet do not generate the same infinite sequence:

$$a_n = n^2 - 6n + 11 \tag{9}$$

$$a_n = \frac{n^4 - 6n^3 + 11n^2 - 6n + 6}{n} \tag{10}$$

$$a_n = \frac{6}{n} + (n-1)(n-2)(n-3)(n-4). \tag{11}$$

EXERCISE 12.1

Find the first five terms of each of the following sequences:

1. $a_n = 2n$
2. $a_n = 2n - 1$
3. $a_k = k^2$
4. $a_j = \frac{j^2}{2}$
5. $a_n = n^2 - 1$
6. $a_n = \frac{n^2 - 1}{n^2}$
7. $a_m = (-1)^m$
8. $a_n = (-1)^{n-1}$
9. $a_n = 1 + (-1)^n$
10. $a_n = \left(-\frac{1}{2}\right)^n$
11. $a_i = \left(1 + \frac{1}{i}\right)^i$
12. $a_n = 1 - \frac{(-1)^n}{n}$
13. $a_n = n \log 10^{-n}$
14. $a_k = \log k$
15. $a_{i+1} = a_i + 2 \quad a_1 = 2$
16. $a_{j+1} = a_j - 4 \quad a_1 = 1$
17. $a_{m+1} = a_m + \frac{1}{3} \quad a_1 = -\frac{1}{3}$
18. $a_{n+1} = a_n - 2 \quad a_1 = -1$
19. $a_{n+1} = 3a_n \quad a_1 = -1$
20. $a_{k+1} = -\frac{1}{3}a_k \quad a_1 = -3$
21. $a_n = na_{n-1} \quad a_1 = 1$
22. $a_{n+1} = na_n \quad a_1 = 1$
23. $a_{n+1} = \frac{a_n}{n} \quad a_1 = 1$
24. $a_{n+2} = a_{n+1} + a_n \quad a_1 = a_2 = \frac{1}{2}$

Find the first eight terms of the sequences of problems 25 and 26. Round off to one decimal in each calculaton.

25. $a_{n+1} = \frac{16/a_n + a_n}{2} \quad a_1 = 1$
26. $a_{n+1} = \frac{16/a_n + a_n}{2} \quad a_1 = 2$
27. Based on the results in problems 25 and 26, predict a_{20}. How is this result related to the constant 16 used in the recursive formula? (This formula will be studied in detail in Chapter 13.)

Find the general expression for a_n in each of the following sequences:

28. 3, 4, 5, 6, 7, ⋯
29. 2, 4, 6, 8, 10, ⋯
30. 0, 2, 4, 6, 8, ⋯
31. 1, 3, 5, 7, 9, ⋯
32. 1, 2, 4, 8, 16, ⋯
33. 1, 1, 2, 3, 5, 8, 13, ⋯
34. −1, 1, −1, 1, −1, 1, ⋯
35. 1, −1, 1, −1, 1, ⋯
36. $-\frac{1}{2}, \frac{1}{4}, -\frac{1}{8}, \frac{1}{16}, \cdots$
37. $-\frac{2}{1}, \frac{3}{2}, -\frac{4}{3}, \frac{5}{4}, \cdots$
38. 0.1, 0.01, 0.001, 0.0001, ⋯
39. 0.9, 0.99, 0.999, 0.9999, ⋯
40. An ancient fable concerns a bright serf who agreed to work for the king for 30 days if he were paid one grain of wheat for the first day, two for

the second, four for the third, eight for the fourth, and so on. How many grains would he receive on the tenth day? The sixteenth day? The thirtieth day? The nth day?

41. In the fission of uranium, each neutron that combines with a uranium atom causes, on the average, 2.5 neutrons to be released. Thus, after two fissions, 6.25 neutrons (on the average) will be emitted. How many neutrons will be emitted after 4 fissions? After 20 fissions? After 1000 fissions? Use logarithms for these calculations.

42. In physics, the *coefficient of restitution* is the distance an object will rebound, divided by the distance from which it was dropped. An object with a coefficient of 2/3 will rebound (on the first bounce) to a height of 6 ft if dropped from a distance of 9 ft. If an object with a coefficient of 1/4 is dropped from 16 ft, how far will it rebound on the first bounce? On the fifth bounce? On the nth bounce?

12.2 PROGRESSIONS

ARITHMETIC PROGRESSIONS. If the domain of the linear function $y = mx + b$ is restricted to zero and the positive integers, the result is a linear sequence, and may be written as

$$a_n = dn + a_0 \qquad (n = 0, 1, 2, 3, \cdots). \tag{1}$$

In order to restrict the domain to the positive integers, as was the practice in the preceding section, we will modify (1) and use it in the form

$$a_n = a_1 + (n - 1)d \qquad (n = 1, 2, 3, \cdots). \tag{2}$$

Note that relationship (2) for the sequence of odd numbers remains basically the same as (1). That is, the first term is now called a_1 rather than a_0, and it is now necessary to multiply d (corresponding to the slope of a linear function) by $n - 1$ rather than n. Substituting $n = 1, 2, 3, 4$, and 5 in (2) gives the first five terms, which can be expressed as

$$\begin{aligned} a_1 &= a_1 \\ a_2 &= a_1 + d \\ a_3 &= a_1 + 2d = a_2 + d \\ a_4 &= a_1 + 3d = a_3 + d \\ a_5 &= a_1 + 4d = a_4 + d. \end{aligned} \tag{3}$$

Note that each term may be represented as a sum of the preceding term and a common difference d. A sequence that exhibits this characteristic is commonly referred to as an **arithmetic progression**. For example, the sequence of odd numbers, $1, 3, 5, \cdots$ is an arithmetic progression with $a_1 = 1$ and $d = 2$. It can be expressed, in the form of (2), as

$$a_n = 1 + 2(n - 1) \qquad (n = 1, 2, 3, \cdots)$$

or, in the form of (1), as

$$a_n = 2n - 1 \qquad (n = 1, 2, 3, \cdots).$$

The sequence 6, 2, -2, -6, -10, $\cdots$ is also an arithmetic progression, whose first term is $a_1 = 6$ and whose common difference is $d = -4$. Thus the general term is

$$a_n = 6 - 4(n - 1) = 10 - 4n \qquad (n = 1, 2, 3, \cdots).$$

From this form, it is possible to obtain any specified term. For example, a_{50} is

$$a_{50} = 10 - 4(50) = -190.$$

By the use of a little ingenuity, it is possible to find the general term of any arithmetic progression, given two terms. For example, substituting $a_7 = 9$ and $a_{16} = 15$ in (2) gives .

$$9 = a_1 + (7 - 1)d \qquad \text{and} \qquad 15 = a_1 + (16 - 1)d.$$

This reduces to the linear system

$$a_1 + 6d = 9 \qquad a_1 + 15d = 15.$$

Since the solution to this system is $a_1 = 5$ and $d = 2/3$, the general term is

$$a_n = 5 + \frac{2}{3}(n - 1)$$

from which any desired term may be obtained.

THE RECURSIVE FORM OF ARITHMETIC PROGRESSIONS. Referring once again to the final form of (3), we note that they are suggestive of the recursive-type equations of Section 12.1. In fact, it is common to define an arithmetic progression as a sequence whose elements can be expressed in the form

$$a_{n+1} = a_n + d. \tag{4}$$

Using this recursive form, the preceding examples of arithmetic progressions take the form

$$a_{n+1} = a_n + 2 \qquad a_1 = 1 \tag{5}$$

and

$$a_{n+1} = a_n - 4 \qquad a_1 = 6. \tag{6}$$

The odd numbers may be easily generated by using (5):

$$\begin{aligned} a_1 &= 1 \\ a_2 &= a_1 + 2 = 3 \\ a_3 &= a_2 + 2 = 5 \\ a_4 &= a_3 + 2 = 7 \\ &\cdot \\ &\cdot \\ &\cdot \end{aligned}$$

Similarly, the second arithmetic progression can be generated by using (6):

$$\begin{aligned} a_1 &= 6 \\ a_2 &= a_1 - 4 = 2 \\ a_3 &= a_2 - 4 = -2 \\ a_4 &= a_3 - 4 = -6 \\ &\cdot \\ &\cdot \\ &\cdot \end{aligned}$$

Although the recursive form appears to offer little significant advantage over the form of (2), many types of sequences can be represented only by recursive equations. In fact, the primary emphasis in Chapter 13 (Numerical Methods) is in developing recursive forms that can be solved by iterative (repetitive) methods on a computer.

GEOMETRIC PROGRESSIONS. Although the nonlinear exponential function was studied only briefly as an introduction to logarithms in Chapter 8, its form leads to an especially important class of sequences. If the domain of the function $y = a \times b^x$ is limited to the positive integers, a sequence results, with the general term

$$a_n = a_1 r^{n-1} \qquad (n = 1, 2, 3, \cdots). \tag{1}$$

Insight into the nature of this sequence can be obtained by studying its terms:

$$\begin{aligned} a_1 &= a_1 \\ a_2 &= a_1 r \\ a_3 &= a_1 r^2 = a_2 r \\ a_4 &= a_1 r^3 = a_3 r \\ &\cdot \\ &\cdot \\ &\cdot \\ a_{n+1} &= a_n r. \end{aligned} \tag{2}$$

A sequence of this type is characterized by the fact that any term can be represented as the product of the previous term and a common constant. Such a sequence is commonly called a **geometric progression**, and may be defined by (1), or by the general term of (2), which is $a_{n+1} = a_n r$. Since the general term of (2) may be solved for r, giving $r = a_{n+1}/a_n$ (that is, the ratio of consecutive terms), r is called the **common ratio**.

The sequence 5, 25, 125, 625, $\cdots$ is a geometric progression with a ratio of 5 (since $25/5 = 5$) and with a first term of 5. Thus the general term is

$$a_n = 5 \times 5^{n-1}.$$

This sequence may also be represented by the recursive form as

$$a_{n+1} = 5a_n \qquad a_1 = 5.$$

As another example, a geometric progression whose terms have alternating signs is readily recognized as having a negative ratio. For example, the progression 3, −2, 4/3, −8/9, $\cdots$ has a ratio $r = -2/3$ and a first term $a_1 = 3$; consequently, the general term is

$$a_n = 3\left(-\frac{2}{3}\right)^{n-1}$$

or

$$a_{n+1} = -\frac{2}{3}a_n \qquad a_1 = 3.$$

It is also possible to find the general term, and consequently any particular term, if given any two terms of a geometric progression. For example, the progression with the terms $a_3 = 8$ and $a_6 = 1$ may be found by substituting in (1) as shown below:

$$8 = a_1 r^{3-1} = a_1 r^2 \tag{3}$$
$$1 = a_1 r^{6-1} = a_1 r^5. \tag{4}$$

Solve (3) for a_1:

$$a_1 = \frac{8}{r^2}. \tag{5}$$

Solve (4) for a_1:

$$a_1 = \frac{1}{r^5}. \tag{6}$$

Equate (5) and (6):

$$\frac{8}{r^2} = \frac{1}{r^5}. \tag{7}$$

Multiply (7) by r^2:

$$r^3 = \frac{1}{8}. \tag{8}$$

Solve for r:

$$r = \frac{1}{2}. \tag{9}$$

Substitute in (5) for r:

$$a_1 = \frac{8}{(1/2)^2} = 32 \tag{10}$$

The general term of the progression is

$$a_n = 32\left(\frac{1}{2}\right)^{n-1}$$

which can be verified by checking for $n = 3$ and $n = 6$ to obtain the original two terms.

EXERCISE 12.2

Identify each of the following sequences as an arithmetic or geometric progression, determine the general term, and calculate the indicated term:

1. 2, 4, 6, 8, $\cdots$; ninth term

2. 2, 4, 8, 16, ⋯; sixth term
3. 1, 4, 7, 10, ⋯; seventh term
4. 19, 13, 7, 1, ⋯; eighth term
5. 27, −18, 12, −8, ⋯; sixth term
6. $\frac{1}{2}, \frac{7}{16}, \frac{3}{8}, \frac{5}{16}, \cdots$; seventh term
7. $\frac{1}{2}, \frac{1}{4}, \frac{1}{8}, \frac{1}{16}, \cdots$; ninth term
8. $-3, \frac{3}{5}, -\frac{3}{25}, \frac{3}{125}, \cdots$; sixth term
9. −13, 0, 13, ⋯; seventh term
10. 1, −1, 1, −1, ⋯; twelfth term
11. In Exercise 12.1, which of the first 14 sequences are arithmetic progressions and which are geometric progressions?
12. In Exercise 12.1 (problems 15-24), which of the recursive forms define arithmetic progressions and which define geometric progressions?
13. The second and sixth terms of a sequence are 16 and 256, respectively. Find the general term (a) assuming this to be an arithmetic progression, (b) assuming it to be a geometric progression.
14. A geometric progression has a first term of 1, and an arithmetic progression a first term of 0. The third terms are equal, but the fifth term of the geometric progression is twice the fifth term of the arithmetic progression. Determine the general term of each progression.
15. The salary policy of an office firm is to start each employee at $5000 per year and grant an increase of $600 each year. Is the sequence of salaries an arithmetic or geometric progression? Express this relationship as an equation. How much will an employee be earning after 10 years?
16. Another employer pays at the rate of $5000 per year but grants a 10% increase each year based on the current salary. Is this sequence of salaries an arithmetic or geometric progression? Express the relationship as an equation. How much will an employee be earning after 10 years?
17. Compile a table of yearly salaries for problems 15 and 16 and determine the year that the employee in 16 would be earning more than the employee in 15. Show how the same result could have been obtained by solving the equations simultaneously.
18. A businessman decides to build a new warehouse at a cost of $10,000. Not only will it benefit his business, but it will increase the GNP as well. Since he studies economics as a hobby, he concludes that building contractors, building materials producers, and their various employees will receive the $10,000. Assuming that they will all spend 70% of the $10,000 on other so-called consumption goods, the producers of these will receive $7000. If, in turn, the second group spends 70% on further consumption goods, a third group will receive 70% of $7000 or $4900, of which they will spend 70%, and so on. Does this represent an arith-

metic progression or a geometric progression? Represent the relationship as a recursive equation.

12.3 CONVERGENCES OF SEQUENCES

In Section 12.1 we considered the following five sequences:

$$a_n = \frac{1}{2}n \tag{1}$$

$$a_n = \frac{1}{n} \tag{2}$$

$$A_n = 100\left(1 + \frac{0.04}{n}\right)^n \tag{3}$$

$$a_n = 1 + \frac{(-1)^n}{n} \tag{4}$$

$$a_n = a_{n-1} + a_{n-2} \qquad a_1 = a_2 = -1. \tag{5}$$

The elements of some of these sequences appear to increase without limit as n increases. This is true in (1) and (5). On the other hand (2), (3), and (4) appear to level off as n becomes very large. These characteristics are illustrated in Figure 12.2, which shows graphs of these five sequences.

BOUNDS AND LIMITS. The sequence of (1) increases without limit as n increases; in other words, a_n goes to infinity as n goes to infinity. This is commonly written as $a_n \to \infty$ as $n \to \infty$; similarly in (5), $a_n \to \infty$ as $n \to \infty$. These sequences are said to be **unbounded**, since it is not possible to find some positive number M such that $|a_n| \leqslant M$ for all values of n.*

On the other hand, a bound can be found for each of the sequences (2), (3), and (4). For example, in (2) $0 < a_n \leqslant 1$, so we can use $M = 1$, giving $|a_n| \leq 1$ for all values of n. Similarly (4) is bounded by $M = 3/2$, and (3) is bounded by an irrational number, whose approximate value is $M = 104.0811$ (determined by means other than those which we are studying).

Another characteristic illustrated by these examples is that of **convergence**, an extremely important consideration in many computer applications. In lay terms, a sequence is convergent if respective values of a_n tend to "zero in" on some given value. In other words, a_n approaches some **limit** L as n approaches infinity ($a_n \to L$ as $n \to \infty$). In (2) $a_n \to 0$ as $n \to \infty$; in (3) $A_n \to 104.0811$ (approximate value) as $n \to \infty$; in (4) $a_n \to 1$ as $n \to \infty$. Thus the sequence of (2) is said to converge to 0, the sequence of (3) to 104.0811, and the sequence of (4) to 1. Notice that in (4) the limit is different from the bound. On the other

*Although this definition is adequate for our purposes, the concept is studied in much more detail in advanced courses. That is, a sequence may be bounded from above or below or both. Furthermore, such concepts of the smallest upper bound (least upper bound) and greatest lower bound are also important.

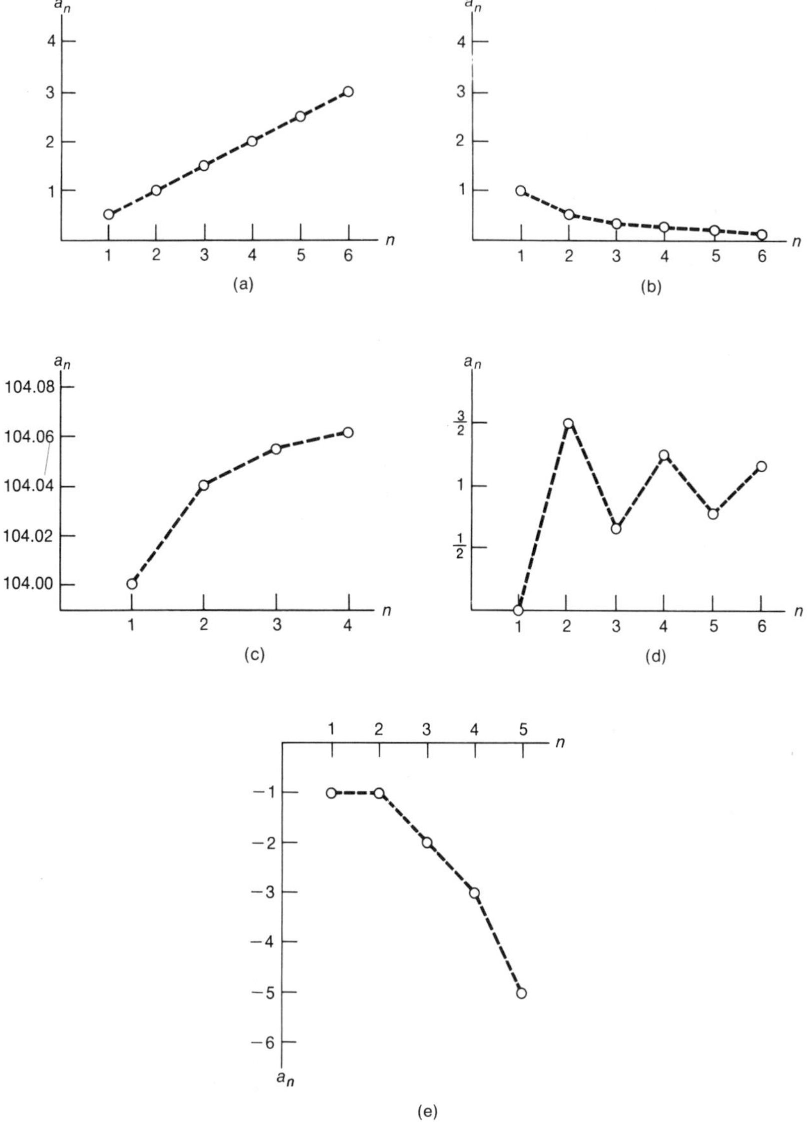

FIGURE 12.2 (a) $a_n = n/2$; (b) $a_n = 1/n$; (c) $A_n = 100(1 + 0.04/n)^n$; (d) $a_n = 1 + (-1)^n/n$; (e) $a_n = a_{n-1} + a_{n-2}$

hand, since sequences (1) and (5) do not approach some finite limits as $n \to \infty$, they are said to **diverge**.

CONVERGENCE OF PROGRESSIONS. By relating arithmetic progression to the linear function $y = mx + b$, we can quickly see that all arithmetic progressions diverge (with the exception of arithmetic progressions that have a common difference of zero). That is, as n becomes larger without limit, a_n also becomes larger without limit, due to the continual addition of the common difference d. This characteristic is especially well illustrated by the recursive form of the arithmetic progression

$$a_{n+1} = a_n + d.$$

On the other hand, if we relate the geometric progression to the exponential function $y = ab^x$, we might tend to conclude (incorrectly) that a geometric progression always diverges. Certainly if the base b is greater than 1 ($b > 1$), y will increase without limit as x increases without limit ($y \to \infty$ as $x \to \infty$). For example, the function $y = 2^x$ has the values

x	y
1	2
2	4
3	8
4	16
.	.
.	.
.	.

However, if the absolute value of the base is between 0 and 1 ($|b| < 1$), y will approach 0 as x becomes larger. For example, if $y = -\frac{1}{2}x$ then the function has the values

x	y
1	−1/2
2	1/4
3	−1/8
4	1/16
.	.
.	.
.	.

Since the base b in the exponential function corresponds to the common ratio r in the geometric progression $a_n = a_1 r^{n-1}$, we can conclude that

$$a_n \to \infty \text{ as } n \to \infty \text{ for } |r| > 1$$
$$a_n \to 0 \text{ as } n \to \infty \text{ for } |r| < 1.$$

In other words, the geometric progression diverges if $|r| > 1$, and converges to 0 if $|r| < 1$.

CALCULATION OF PI. The last sequence we shall study in this section leads to means for determining the irrational number π, and can be represented by

$$a_n = \frac{(2n)^2}{(2n)^2 - 1}.$$

The first four terms are 4/3, 16/15, 36/35, 64/63. It is easy to see that the numerators will all be squares of the positive integers and the denominators will be 1 less than the numerators; thus a_n will converge to 1 as a limit. If we consider the sequence formed by taking the product of the first n terms (and call its terms A_n), we have

$$A_1 = \frac{4}{3} \simeq 1.3333$$

$$A_2 = \frac{4}{3} \cdot \frac{16}{15} \simeq 1.4222$$

$$A_3 = \frac{4}{3} \cdot \frac{16}{15} \cdot \frac{36}{35} \simeq 1.4629$$

$$A_4 = \frac{4}{3} \cdot \frac{16}{15} \cdot \frac{36}{35} \cdot \frac{64}{63} \simeq 1.4861.$$

To represent the general term of this sequence, it is necessary to introduce a new symbolism. The product $x_1 \times x_2 \times x_3 \times x_4 \times x_5 \times x_6$ can be represented as

$$\prod_{n=1}^{6} x_n = x_1 \times x_2 \times x_3 \times x_4 \times x_5 \times x_6,$$

where the symbol Π (the upper case Greek letter *pi*) is used to represent a product, and x_n the variable. The letter n is called the *index*, and can range from 1 through 6, as indicated by

$$\prod_{n=1}^{6}$$

(The choice of pi for this representation has nothing to do with the use of pi to relate the diameter to the circumference of a circle.) The expression

$$\prod_{i=1}^{4} (x_i + 1)$$

tells us to multiply the expression obtained when $i = 1$, by that obtained when $i = 2$, and so on to $i = 4$, or

$$\prod_{i=1}^{4} (x_i + 1) = (x_1 + 1)(x_2 + 1)(x_3 + 1)(x_4 + 1).$$

Note that the index used in this case is i rather than n. The choice is arbitrary, but the same letter should be used in both places.

Now the general term of the original sequence can be represented by

$$A_n = \prod_{i=1}^{n} \frac{(2i)^2}{(2i)^2 - 1}. \qquad (6)$$

Does this sequence converge? That is, does there exist some finite quantity L such that

$$L = \prod_{i=1}^{\infty} \frac{(2i)^2}{(2i)^2 - 1} \tag{7}$$

or

$$L = \frac{4}{3} \times \frac{16}{15} \times \frac{36}{35} \times \frac{64}{63} \cdots.$$

We can see that succeeding factors approach closer and closer to 1 and thus have less and less effect on the product; however, there are an infinite number of such factors to consider. On the other hand, it can be shown (by methods beyond the scope of this text) that $L = \pi/2$.

Actually, (7) was one of the first formulas developed for calculation of the irrational number π. The mathematician John Wallis proved the relationship in 1655.

EXERCISE 12.3

1-12. Referring to Exercise 12.1 (problems 1-12), state (a) whether each sequence appears to converge or diverge, (b) if convergent, the limit to which the function appears to converge (if obvious), (c) the bounds on the sequence (if obvious).

13. What are the first four terms of the geometric progression $a_n = r^{n-1}$ if (a) $r = \frac{1}{2}$, (b) $r = \frac{1}{4}$, (c) $r = \frac{1}{8}$?
To what limit does each converge? Which sequence converges the most quickly?

12.4 SUMMATIONS

Suppose a small company which leases a computer keeps a monthly running time log. Such a record might appear as in Table 12.2. By the definition of

TABLE 12.2

Month	Running Time
1	63.26
2	59.87
3	75.05
4	82.59
5	41.37
6	52.23
7	60.28
8	71.04
9	80.16
10	82.95
11	76.82
12	73.41

sequence (a function whose domain consists of the positive integers), we can regard Table 12.2 as a sequence. Further, if m is used to denote month, a_m can be used to denote the hourly running time. In other words,

$$\begin{aligned} a_1 &= 63.26 \\ a_2 &= 59.87 \\ &\vdots \\ a_{12} &= 73.41. \end{aligned}$$

Of particular interest might be the average monthly running time that would be calculated by adding all of a_m and dividing by 12. It would be necessary to calculate the sum of these terms:

$$S_{12} = a_1 + a_2 + a_3 + a_4 + a_5 + a_6 + a_7 + a_8 + a_9 + a_{10} + a_{11} + a_{12}. \tag{1}$$

Here, associated with the original sequence, is the sum obtained by adding the terms of the sequence; we will refer to (1) as a **series**. A more rigorous definition of the term *series* is usually given in advanced mathematics, but it will suffice here to call a series *a sum of the terms of a sequence*. Since series are so commonly encountered in mathematics, the following summation form (which was briefly used in Chapter 10) is commonly used in place of (1).

$$S_{12} = \sum_{m=1}^{12} a_m.$$

Here the upper case Greek sigma (Σ) is used to indicate the sum of the terms a_m, as the index of summation m ranges from 1 to 12 (similar to the use of Π in Section 12.3). Although the letters *i*, *j*, and *k* are most commonly used for the index, any letter may be employed so long as the index and the corresponding subscript are the same. The expression is read as "the sum of a_m as m ranges from 1 to 12." The sum of the first four terms would be represented as

$$S_4 = \sum_{m=1}^{4} a_m = a_1 + a_2 + a_3 + a_4.$$

It is possible to represent the desired monthly average in the convenient form

$$\text{Average} = \frac{S_{12}}{12} = \frac{1}{12} \sum_{m=1}^{12} a_m.$$

To illustrate how closely Fortran parallels mathematical usage, a short program in Fortran to calculate average is shown in Figure 12.3. This illustrates the powerful DO loop of Fortran.

In the first line, the variable SUM is set equal to 0. The second line requires that all calculations indicated by statements down to and including

```
          SUM = 0.0
          DO 100  M = 1,12
    100   SUM = SUM + A(M)
          AVE = SUM/12.0
```

FIGURE 12.3

the statement numbered 100 be executed for M = 1, 2, 3, ..., 12. In this example, the designated statement is the very next one; however, many such intermediate statements are often included.

In considering the third line (statement 100) we must remember that the equals sign is used in Fortran to indicate "replace the current value of the variable on the left by the calculated value from the expression on the right." In executing this statement twelve times, the computer will substitute 1, 2, 3, ..., 12 for M and thus calculate the total. That is, the first calculation will give

$$\text{SUM} = 0.0 + \text{A}(1) = 0.0 + 63.26 = 63.26.$$

The second will give

$$\text{SUM} = 63.26 + \text{A}(2) = 63.26 + 59.87 = 123.13.$$

The third will give

$$\text{SUM} = 123.13 + \text{A}(3) = 123.13 + 75.05 = 198.18.$$

The process continues until the twelfth (and final) summation will give

$$\text{SUM} = 819.03.$$

Since the loop has been executed the required 12 times, the machine will continue to line four and calculate the average by dividing SUM by 12.0.

In the preceding illustration, it would be necessary to provide the computer with the values for a_n in tabular form, since they do not appear to be easily related by an equation. However, neither the DO loop nor the sigma notation are limited to tabulated sequences. Each and every one of the sequences we have discussed in this chapter has an associated series. For example, the series defined by the sequence $a_i = 2i - 1$ may be represented as

$$S_n = \sum_{i=1}^{n} (2i - 1)$$

where S_n is the sum of the first n terms. Specifically, for $n = 7$ we have

$$\begin{aligned} S_7 &= \sum_{i=1}^{7} (2i - 1) \\ &= 1 + 3 + 5 + 7 + 9 + 11 + 13 \\ &= 49. \end{aligned}$$

More complex forms are handled in much the same manner as the sequences of Exercise 12.1; for example,

$$\sum_{n=2}^{5} \left(1 - \frac{(-1)^n}{n}\right).$$

Here the summation is indicated over the range $n=2$ to $n=5$. Thus, when

$$n=2 \qquad a_2 = 1 - \frac{(-1)^2}{2} = 1 - \frac{1}{2} = \frac{1}{2}$$

$$n=3 \qquad a_3 = 1 - \frac{(-1)^3}{3} = 1 + \frac{1}{3} = \frac{4}{3}$$

$$n=4 \qquad a_4 = 1 - \frac{(-1)^4}{4} = 1 - \frac{1}{4} = \frac{3}{4}$$

$$n=5 \qquad a_5 = 1 - \frac{(-1)^5}{5} = 1 + \frac{1}{5} = \frac{6}{5}$$

and the series is

$$\sum_{n=2}^{5}\left(1 - \frac{(-1)^n}{n}\right) = \frac{1}{2} + \frac{4}{3} + \frac{3}{4} + \frac{6}{5}$$

$$= \frac{227}{60}.$$

EXERCISE 12.4

Calculate the indicated sums:

1. $\sum_{i=1}^{5} i$
2. $\sum_{j=1}^{4} j^2$
3. $\sum_{m=2}^{6} (3m-1)$
4. $\sum_{k=1}^{4} (k-7)$
5. $\sum_{n=4}^{9} (-n)$
6. $\sum_{i=1}^{4} \left(\frac{1}{i}\right)$
7. $\sum_{m=1}^{4} \frac{m+1}{m}$
8. $\sum_{n=1}^{6} n(-1)^n$
9. $\sum_{j=1}^{6} j(-1)^{j+1}$
10. $\sum_{k=1}^{6} k(-1)^{k-1}$
11. $\sum_{i=1}^{6} 2^i$
12. $\sum_{n=1}^{4} \frac{1}{2^n}$
13. $\sum_{i=1}^{3} 2^{2i}$
14. $\sum_{i=1}^{3} 2^{2i-1}$
15. $\sum_{j=1}^{4} \frac{(-1)^{j-1}(j^2-1)}{j}$
16. Calculate the sum

$$\sum_{i=1}^{n} (2i-1)$$

when (a) $n=1$, (b) $n=2$, (c) $n=3$, (d) $n=4$, (e) $n=5$, (f) $n=6$, (g) $n=7$, (h) $n=8$.

The result of this exercise is a new sequence whose elements are the squares of the positive integers. Prove that the sum of the first n odd integers is n^2.

17. Argue that

$$\sum_{i=1}^{5} 3 = 5 \times 3 = 15$$

by considering the form

$$\sum_{i=1}^{5} \frac{3i}{i}.$$

18. In chemistry, the *half life* of a radioactive material is the time required for one half of a given material to decay. For an element with a half life of 1 year, one half of the element will decay the first year, one half of the remaining one half (1/4) will decay the next year, and so on. Express the amount that decays each year for 7 years as a sequence. Represent the sum of this sequence in sigma notation.

12.5 SUMS OF ARITHMETIC AND GEOMETRIC PROGRESSIONS

ARITHMETIC PROGRESSIONS. Let us consider the task of computing the sum of the first hundred positive integers, that is, the sum of the sequence

$$1, 2, 3, 4, 5, \cdots, 96, 97, 98, 99, 100.$$

The obvious reaction to such a task would likely be "dull and time-consuming." However, the sum may be calculated very quickly by noting that a partial sum obtained by adding the first and last terms is 101 (that is, $1 + 100 = 101$); similarly, the second and next to last terms also total 101, since $2 + 99 = 101$. In fact, all of the remaining numbers may be grouped in a similar manner, yielding partial sums of 101. Since there are 100 terms in the sequence, there will be 50 such pairs. Thus the sum is $50 \times 101 = 5050$ (Gauss is said to have first presented this technique at the age of eight). Using sigma notation, this may be represented as

$$S_n = \sum_{n=1}^{100} n$$

$$S_{100} = \frac{100}{2}(1 + 100) \tag{1}$$

$$S_{100} = 5050.$$

Although this technique might appear to be in the category of a "parlor game," (1) represents the basic form of the sum of the first n terms of any arithmetic progression. To show this, consider the general form of the arithmetic progression

$$a_n = a_1 + (n - 1)d. \tag{2}$$

The series associated with the first n terms is

$$S_n = a_1 + (a_1 + d) + (a_1 + 2d) + \cdots + (a_n - 2d) + (a_n - d) + a_n. \tag{3}$$

Since each of these terms represents a real number, the terms are commutative, and the series may be rewritten as

$$S_n = a_n + (a_n - d) + (a_n - 2d) + \cdots + (a_1 + 2d) + (a_1 + d) + a_1. \tag{4}$$

Addition of (3) and (4) gives

For the first pair of terms: $a_1 + a_n$
For the second pair of terms: $(a_1 + d) + (a_n - d) = a_1 + a_n$

For the third pair of terms: $(a_1 + 2d) + (a_n - 2d) = a_1 + a_n$

.
.
.

For the $(n-1)$th pair of terms: $(a_n - d) + (a_1 + d) = a_1 + a_n$
For the nth pair of terms: $a_n + a_1 = a_1 + a_n$

Note that this process is basically the same one used in summing the first 100 positive integers. Now, since there are n such terms, their sum $2S_n$ is

$$2S_n = n(a_1 + a_n)$$

or

$$S_n = \frac{n(a_1 + a_n)}{2}. \tag{5}$$

EXAMPLE 12.1
Determine the sum of the first 11 terms of the arithmetic progression

$$a_n = 2n - 5.$$

SOLUTION
In order to use (5), it is necessary to determine the first and last terms:

$$a_1 = 2(1) - 5 = -3$$
$$a_{11} = 2(11) - 5 = 17$$

$$S_{11} = \frac{11(-3 + 17)}{2}$$
$$= 77.$$

EXAMPLE 12.2
A man sets up the following savings plan for himself: the first month he will save \$20; each month thereafter he will increase the amount by \$1. How much will he have saved (excluding interest) after five years (60 months)?

SOLUTION
First it is necessary to determine the form for the nth term:

$$a_n = 20 + (n-1)(1)$$
$$= 20 + (n-1).$$

Thus

$$a_1 = 20$$
$$a_{60} = 20 + (60 - 1) = 79$$

$$S_{60} = \frac{60(20 + 79)}{2}$$
$$= 2970.$$

GEOMETRIC PROGRESSIONS. By careful manipulation of the terms of a series, (5) was obtained, which yields the sum of the first n terms of an arithmetic progression. Similar manipulation, although not as intuitive in nature, will lead to a representation for the sum of the first n terms of a geometric progression. Recall that the general term of a geometric progression may be represented by $a_n = a_1 r^{n-1}$. The sum of the first n terms is

$$S_n = a_1 + a_1 r + a_1 r^2 + \cdots + a_1 r^{n-2} + a_1 r^{n-1}. \tag{6}$$

Multiplying both sides of the equation by r will give

$$rS_n = a_1 r + a_1 r^2 + a_1 r^3 + \cdots + a_1 r^{n-1} + a_1 r^n. \tag{7}$$

Subtracting (7) from (6) and pairing identical terms gives

$$S_n - rS_n = a_1 + (a_1 r - a_1 r) + (a_1 r^2 - a_1 r^2) + \cdots + (a_1 r^{n-1} - a_1 r^{n-1}) - a_1 r^n$$

or

$$S_n - rS_n = a_1 - a_1 r^n.$$

Solving for S_n yields

$$S_n = \frac{a_1 - a_1 r^n}{1 - r}. \tag{8}$$

EXAMPLE 12.3

Find the sum of the first eight terms of the geometric progression whose first four terms are

6, 12, 24, 48.

SOLUTION

Note that the first terms is $a_1 = 6$, and the common ratio is $r = 12/6 = 2$. The required sum is

$$\begin{aligned} S_8 &= \frac{a_1 - a_1 r^8}{1 - r} \\ &= \frac{6 - 6 \cdot 2^8}{1 - 2} \\ &= 1530. \end{aligned}$$

EXAMPLE 12.4

A student deposited \$100 in a savings account in January 1970, and \$100 each January thereafter. If the bank pays 4% interest compounded annually, how much would be in his account in January 1990?

SOLUTION

The amount that will accumulate may be calculated by considering each deposit separately and summing them as the terms of a sequence:

$$S = 100(1.04)^{20} + 100(1.04)^{19} + \cdots + 100(1.04)^2 + 100(1.04) + 100.$$

Note that there will be 21 terms and that the last deposit (made in January

1990) will not yet have earned any interest. Inspection of the terms of this series shows that they can be represented as a sequence of 21 terms whose first term is $a_1 = 100$ and whose common ratio is $r = 1.04$. The sum may be computed from (8) as

$$S_{21} = \frac{100(1 - 1.04^{21})}{1 - 1.04}.$$

Using logarithms for computational convenience gives

$S_{21} = \$3196$ (correct to the nearest dollar).

EXERCISE 12.5

Determine whether each of the following is an arithmetic or geometric progression, and then determine the sum of the indicated number of terms:

1. $-1, 3, 7, 11, \cdots$; 12 terms
2. $2, 4, 6, 8, \cdots$; 20 terms
3. $1, 3, 9, 27, \cdots$; 8 terms
4. $1, 2, 4, 8, \cdots$; 10 terms
5. $1, -1, 1, -1, \cdots$; 10 terms
6. $64, 16, 4, 1, \cdots$; 8 terms
7. $-5, 0, 5, \cdots$; 10 terms
8. $-1, 2, -4, 8, \cdots$; 10 terms
9. $2, 1, \frac{1}{2}, \frac{1}{4}, \cdots$; 8 terms
10. $3, 5, 7, 9, \cdots$; 10 terms
11. Show that the sum

$$S_n = \frac{n(a_1 + a_n)}{2}$$

may be represented as

$$S_n = \frac{n}{2}[2a_1 + (n - 1)d].$$

12. Show that the sum

$$S_n = \frac{a_1 + a_1 r^n}{1 - r}$$

may be represented as

$$S_n = \frac{a_1 + ra_n}{1 - r}.$$

13. The serf who agreed to work for the king for 30 days given 1 grain of wheat the first day, 2 the second, 4 the third, and so on, was wise indeed (problem 40, Exercise 12.1). How many grains would he have accumulated by the fifteenth day? How many grains by the thirtieth day? Compare the total received for 30 days to the number of grains he would have received on the thirty-first day.
14. An object starting from rest falls approximately 16 ft during the first second, 32 ft during the second, 48 ft during the third, and so on. How far will it fall during the tenth second? How far will it have fallen in 10 seconds?
15. A store holds a sale in which a given item is priced at \$100 for the first week. Each succeeding week the price is reduced by 5%. What will be the discount during the eighth week? What will be the selling price during the eighth week?
16. A contractor finishes a job 15 days later than required by contract. If he

must pay a penalty of \$30 the first day, \$32 the second day, \$34 the third day, and so on, what will be the total penalty?

17. A ball, when dropped, rebounds to $\frac{2}{3}$ the height from which it last fell. If it is initially dropped from a height of 9 ft, how far will it have traveled by the time it strikes the ground for the eighth time? Consider as two separate series the distance traveled in moving upward and the distance traveled in moving downward.

18. In Exercise 12.2, problems 15 and 16 described two employee salary policies: a yearly starting salary of \$5000 with annual increases of \$600, and a starting salary of \$5000 with annual increases of 10% current salary. How much will an employee have earned after 10 years of employment under each pay schedule? At the beginning of which year will the total income under the second schedule exceed the total income under the first schedule?

12.6 INFINITE GEOMETRIC SERIES

ACHILLES AND THE TORTOISE. The following story involving the summation of an infinite number of terms of a geometric progression provided puzzlement for mathematicians and logicians for many centuries. Achilles, the hero of Greek mythology, gave chase to a tortoise 100 yards away. Upon traveling the 100 yards, he noted that the tortoise was 10 yards ahead, having traveled that distance while Achilles covered the 100 yards. Upon continuing 10 yards farther, Achilles noted that the tortoise had gone 1 yard. Similarly, upon covering 1 yard, the tortoise traveled 1/10 yard, and so on. Thus, Achilles could never catch the tortoise, since the tortoise would always remain ahead of him, even though by only an extremely small margin. This conclusion, which is obviously false, appears to follow directly from the argument which precedes it. Using basic relationships from physics, it is simple to show that Achilles will be abreast of the tortoise after running 111 1/9 yards.

In order to show that this result is consistent with the preceding argument, it is helpful to study each of the distances traveled as a term of a geometric progression, and then determine the sum of an infinite number of terms. In other words, this problem involves an **infinite geometric progression**. Since the sequence is 100, 10, 1, 1/10, $\cdots$, we have $a_1 = 100$ and $r = 1/10$. In order to calculate the sum of the first n terms, we may use the formula

$$S_n = \frac{a_1(1 - r^n)}{1 - r} \tag{1}$$

$$S_1 = \frac{100[1 - (1/10)]}{1 - 1/10} = \frac{100(0.9)}{0.9} = 100$$

$$S_2 = \frac{100[1 - (1/10)^2]}{1 - 1/10} = \frac{100(0.99)}{0.9} = 110$$

$$S_3 = \frac{100[1 - (1/10)^3]}{1 - 1/10} = \frac{100(0.999)}{0.9} = 111$$

$S_4 = 111.1$
$S_5 = 111.11$
.
.
.

The pattern becomes apparent if we carefully note that the quantity appearing within the bracket approaches the value 1 as n becomes larger. These results may be interpreted as a sequence of partial sums, which approaches the limit $(100)(1)/0.9 = 111\ 1/9$. In Section 12.3 it was pointed out that, for geometric progressions in general, terms of the sequence converge to 0 whenever $|r| < 1$. That an infinite geometric progression will converge under the same conditions becomes apparent with reference to (1). If $|r| < 1$, $r^n \to 0$; consequently, $(1 + r^n) \to 1$ as $n \to \infty$, as illustrated by the preceding example. Thus we find that, as $n \to \infty$,

$$S_n = \frac{a_1(1 - r^n)}{1 - r} \to \frac{a_1}{1 - r} \qquad |r| < 1.$$

This is commonly written as

$$\lim_{n \to \infty} S_n = \frac{a_1}{1 - r}$$

and is read "the limit, as n approaches infinity, of S_n is $a_1/(1 - r)$." However, whenever the sum of an infinite geometric series is to be calculated, we will use the form

$$S = \frac{a_1}{1 - r} \qquad |r| < 1. \tag{2}$$

Checking back on Achilles, we find that he would indeed have overtaken the tortoise after traveling 111 1/9 yards, since

$$S = \frac{100}{1 - (1/10)} = 111\frac{1}{9}$$

REPEATING DECIMALS. In Chapter 3, it was pointed out that a rational number is a number that can be represented in the form a/b. Furthermore, any rational number can be written as a repeating decimal; for example, $1/37 = 0.027027027 \cdots$. Can all repeating decimals be converted to the form a/b, and if so, how? To answer this, let us consider the decimal $0.333 \cdots$ (which we know to be 1/3) as

$$0.3333 \cdots = 0.3 + 0.03 + 0.003 + \cdots.$$

This is a geometric series, with the first term $a_1 = 0.3$ and a ratio 1/10. The sum may be calculated from (2) giving

$$S = \frac{0.3}{1 - 0.1} = \frac{0.3}{0.9} = \frac{1}{3}.$$

This technique is further illustrated by the not so obvious decimal form:

$$2.081081081 \cdots = 2 + 0.081 + 0.000081 + 0.000000081 + \cdots.$$

Ignoring the 2, we have a geometric progression, with a first term $a_1 = 0.081$ and a ratio $r = 1/1000 = 0.001$. The series yields

$$S = \frac{0.081}{1 - 0.001}$$

$$= \frac{0.0081}{0.999}$$

$$= \frac{3}{37}.$$

Therefore,

$$2.081081 \cdots = 2 + \frac{3}{37} = \frac{77}{37}.$$

EXERCISE 12.6

Determine the sum of each of the following geometric series, whenever possible. Where a series does not converge, so indicate.

1. $1 + \frac{1}{2} + \frac{1}{4} + \cdots$
2. $4 + 2 + 1 + \cdots$
3. $1 + \frac{1}{4} + \frac{1}{16} + \cdots$
4. $1 - \frac{1}{4} + \frac{1}{16} - \frac{1}{64} + \cdots$
5. $\frac{1}{64} + \frac{1}{32} + \frac{1}{16} + \cdots$
6. $-9 + 6 - 4 + \cdots$
7. $128 - 64 + 32 - 16 + \cdots$
8. $\frac{7}{36} - \frac{7}{30} + \frac{7}{25} - \cdots$
9. $1 + x + x^2 + x^3 + \cdots \quad (0 < x < 1)$
10. $1 - x + x^2 - x^3 + \cdots \quad (0 < x < 1)$
11. $\frac{6}{10} + \frac{6}{100} + \frac{6}{1000} + \cdots$
12. $\frac{3}{10} + \frac{14}{10^3} + \frac{14}{10^5} + \frac{14}{10^7} + \cdots$

Convert each of the following repeating decimals to its fraction equivalent.

13. $0.5555 \cdots$
14. $0.05555 \cdots$
15. $0.081081081 \cdots$
16. $0.008108108 \cdots$
17. $0.1272727 \cdots$
18. $0.1027027 \cdots$
19. $0.09999 \cdots$
20. $1.49999 \cdots$
21. State the fallacy of the following solution: The sequence $1, -1, 1, -1, \cdots$ is a geometric progression with $a_1 = 1$ and $r = -1$. The sum of the corresponding infinite series is

$$S = \frac{a_1}{1 - r} = \frac{1}{1 - (-1)} = \frac{1}{2}.$$

22. Referring to Exercise 12.5, problem 17, how far will the ball travel in coming to rest?
23. One method for depreciating equipment and buildings for income tax purposes is called the declining balance method. To illustrate this tech-

nique, consider a machine costing \$1000, which has an expected useful life of 10 years. A businessman may then depreciate 1/10 of the undepreciated value each year. For example, the first year he may deduct $1000/10 = \$100$ for depreciation, the second year $(1000 - 100)/10 = \$90$, the third year, $(900 - 90)/10 = \$81$, and so on. Express the next three terms of the sequence whose terms are the yearly deductions for depreciation. Show that after an infinite number of years, the machine will be totally depreciated.

24. In Exercise 12.2, problem 18 describes the overall effect of the investment of \$10,000 on the economy. In theory, what will be the total amount spent if the process is carried out an infinite number of times?

REFERENCES

Apostle, H. G., *Survey of Basic Mathematics.* Boston, Little, Brown & Company, 1960.

Barnett, R. A., *Intermediate Algebra: Structure and Use.* New York, McGraw-Hill, Inc., 1971.

Groza, V. S. and Shelley, S., *Modern Intermediate Algebra for College Students.* New York, Holt, Rinehart and Winston, Inc., 1969.

Wooten, W. and Drooyan, I., *Intermediate Algebra,* 2d alt. ed. Belmont, Calif., Wadsworth Publishing Company, Inc., 1968.

13 NUMERICAL METHODS

13.1 SQUARE ROOT / **417**
Estimating the Side of a Square / **417**
Formulating the Equations / **418**
Convergence of the Solution / **418**
Computer Results / **420**
Exercise 13.1 / **421**

13.2 ROOTS BY THE BOLZANO METHOD / **421**
Refining the Approximation / **422**
Graphical Interpretation / **423**
Formalizing the Procedure / **423**
A Flowchart / **424**
Computer Results / **424**
Additional Comments / **424**
Exercise 13.2 / **426**

13.3 ROOTS BY THE METHOD OF FALSE POSITION / **426**
Improving the Bolzano Method / **426**
Graphical Interpretation / **427**
Formalizing the Procedure / **428**
Computer Results / **429**
General Comments / **429**
Exercise 13.3 / **429**

13.4 SOLUTION OF LINEAR SYSTEMS BY THE JACOBI METHOD / **429**
A Simple Example / **430**
A Diverging Result from the Jacobi Method / **431**
Conditions for Convergence / **432**
Applying the Convergence Test / **433**
Formalizing the Procedure / **434**
Computer Results / **434**
Exercise 13.4 / **434**

13.5 SOLUTION OF LINEAR SYSTEMS BY THE GAUSS-SEIDEL METHOD / **437**
Improving the Jacobi Method / **437**
Computer Results / **437**
Geometric Interpretation of a Converging Case / **438**
Geometric Interpretation of a Diverging Case / **439**
Application to Equations in Several Variables / **439**
Convergence for Equations in Several Variables / **441**
Exercise 13.5 / **443**

Numerical techniques for solving equations and systems of equations were of interest to mathematicians long before the electronic computer was developed. However, with the development of the computer, numerical analysis has become a large and important branch of mathematics. This single chapter is intended only to introduce the so-called **iterative** techniques, in which an estimate of the solution is made and then continuously refined by calculation. Although the five numerical methods in this chapter are primarily of only historical significance, they will serve to illustrate the basic nature of the subject.

13.1 SQUARE ROOT

Many methods have been devised for finding the square root of a number. The one which we shall consider here is actually a special case of a method for finding roots of polynomials. It is not generally used for obtaining roots on a computer, but it is a fascinating illustration of a simple iterative process that displays unusually rapid convergence characteristics.

ESTIMATING THE SIDE OF A SQUARE. The familiar formula for the area of a square is $A = x^2$, where A is the area and x is the length of a side, as shown in Figure 13.1. Suppose we have a square whose area is 120 and we wish to know the length of the side (that is, we have A and want x). If we had the good fortune to make a correct estimate for x, the result obtained by dividing the estimate into A would be equal to the estimate itself.

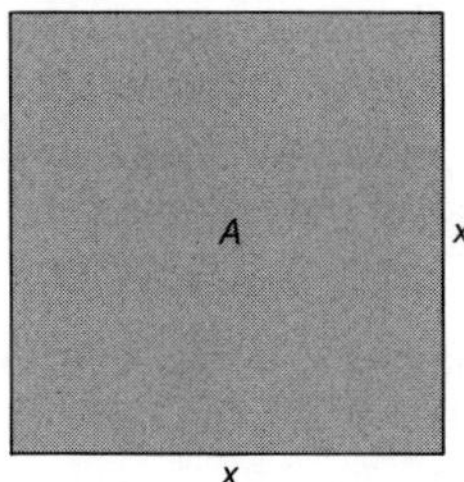

FIGURE 13.1

If, on the other hand, the estimate were not correct, the quotient would differ. For example, a first estimate of 10 gives

$$\frac{130}{10} = 13.$$

Geometrically, this represents a rectangle whose area is equal to the area of the desired square, but whose sides are 10 and 13, as illustrated by Figure 13.2. Intuitively we can see that the desired value of x will be bounded by 10 and 13; that is, $10 < x < 13$.

A simple means for determining the second estimate is to average these two numbers, which gives $(10 + 13)/2 = 11.5$. Dividing 130 by this estimate gives $130/11.5 = 11.3$. Now the rectangle takes on proportions very nearly those of a square. Averaging once more gives (11.5 + 11.3)/2 = 11.4; using 11.4 as the fourth estimate gives 130/11.4 = 11.4. On the fourth estimate the rectangle becomes a square, as shown by Figure 13.3. The length of the side is the desired value of $\sqrt{130}$; in other words, we have $\sqrt{130} = 11.4$ correct to the nearest tenth. The steps we have followed to obtain $x = \sqrt{A}$ are as follows:

1. Obtain an initial approximation for x.
2. Calculate a new estimate for x, using the previous estimate and involving the following steps:

   (a) Divide A by the estimate for x.
   (b) Calculate the average of the estimate and the quotient from (a).
   (c) Call the average obtained in (b) the new estimate.

3. Compare the new estimate to the previous estimate. If they differ, go back to step 2; otherwise, continue to step 4.
4. The final result is $x = \sqrt{A}$.

These steps are clearly illustrated by the flowchart of Figure 13.4.

FORMULATING THE EQUATIONS. Let us now consider an explicit mathematical relationship for the intuitive approach we have been using. In this problem we use four estimates, 10, 11.5, 11.3, and 11.4 (the final value). The first value, 10, differs from the others in that it is a sheer guess; the other two are calculated refinements. The first estimate is often called the **zero**th approximation; 11.5 is called the first approximation; 11.3 the second; and 11.4 the third. Using the notation of Chapter 12, these may be considered a sequence that converges on A, and represented as

$$x_0 = 10 \qquad x_1 = 11.5 \qquad x_2 = 11.3 \qquad x_3 = 11.4.$$

Furthermore, this sequence may be defined by the following recursive formula (see problem 18, Exercise 12.1):

$$x_{n+1} = \frac{A/x_n + x_n}{2} \qquad (x_0 = 10).$$

Using this nomenclature, the flowchart of Figure 13.4 can be made much more meaningful, as shown in Figure 13.5. In the last rectangular block, the representation $x_n \leftarrow x_{n+1}$ indicates that, for a new calculation, the newly calculated value for x replaces the previous value.

CONVERGENCE OF THE SOLUTION. A convenient format for compiling results illustrates how this technique converges, regardless of how poor the initial guess might be. Using the same number, 130, we will assume that its square root is 1, with the results shown in Table 13.1. In spite of a very poor first estimate, the solution rapidly converges to the proper value.

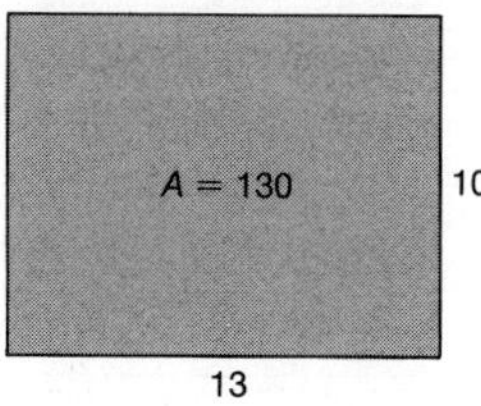

FIGURE 13.2

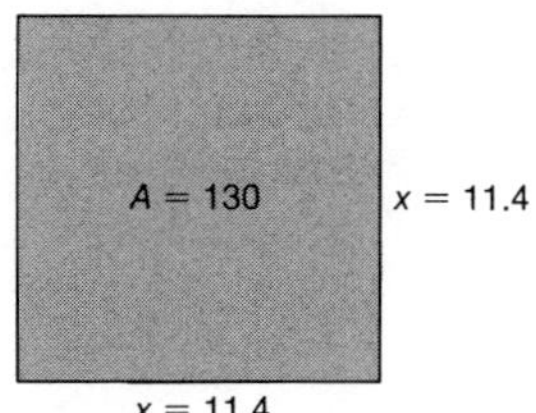

FIGURE 13.3

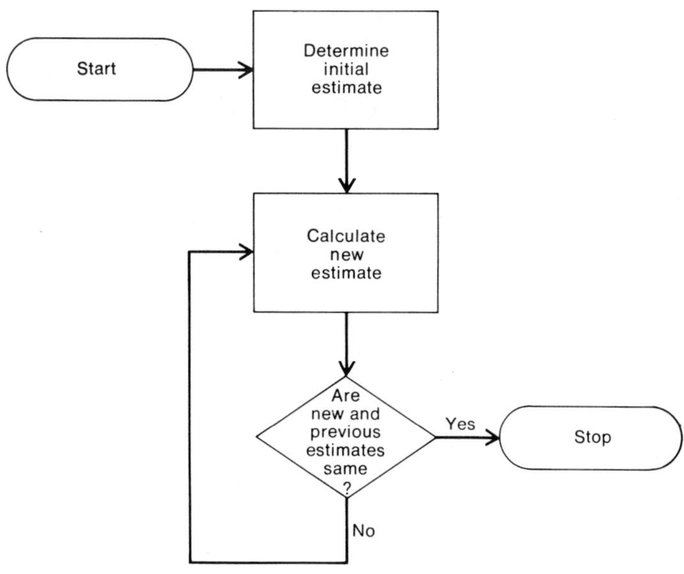

FIGURE 13.4

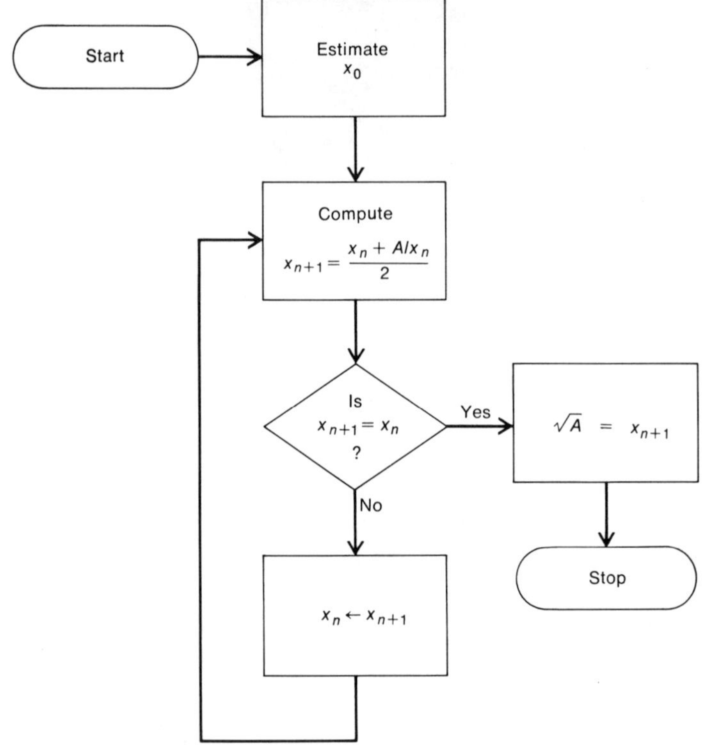

FIGURE 13.5

Not all numerical methods result in such rapid convergence; in fact, some techniques converge so slowly that they have no practical value. In some instances, results even diverge; that is, they are continually further and further from the correct result. This characteristic is illustrated by some of the other techniques we shall study.

TABLE 13.1

n	x_n	$\frac{A/x_n + x_n}{2}$
0	1	65.5
1	65.5	33.8
2	33.8	18.8
3	18.8	12.9
4	12.9	11.5
5	11.5	11.3
6	11.3	11.4
7	11.4	11.4

COMPUTER RESULTS. The foregoing problem was programmed in Fortran, and the output for each iteration printed according to the format of

```
N        X(N)          X(N+1)

0          1.00        321.00
1        321.00        161.50
2        161.50         82.73
3         82.73         45.24
4         45.24         29.70
5         29.70         25.64
6         25.64         25.32
7         25.32         25.32

   THE SQUARE ROOT OF      641.
          IS   25.32
```

FIGURE 13.6

Table 13.1. The results are shown in Figure 13.6, where the square root of 641 has been calculated to the nearest one hundredth.

EXERCISE 13.1

1. Using the recursive form defined in Section 13.1, calculate, rounded to the nearest whole number, the square roots of 7, 8, 9, and 10. Use as the initial estimate, $x_0 = 1$.
2. Repeat problem 1 but determine each root to the nearest tenth.
3. Calculate the square root of 23,600, corrected to the nearest whole number. Make your initial estimate as close as possible, to minimize the number of computations.
4. Calculate the square root of 236,000, corrected to the nearest whole number. Make your initial estimate as close as possible to minimize the number of computations.
5. Devise a simple rule for obtaining a first square root estimate based on the number of digits in a number.
6. Newton's method for finding the roots of polynomials (which we have not studied here) yields the following recursive formula for finding the cube root of N:

$$x_{n+1} = \frac{2x_n + N/x_n^2}{3}$$

Using an initial estimate of $x_0 = 1$, show that this form yields 3 for the cube root of 27. Round off all values of x_n to the nearest whole number.

13.2 ROOTS BY THE BOLZANO METHOD

For a computer program to calculate roots of a second degree polynomial, the quadratic formula developed in Chapter 6 is ideal. However, if roots of higher degree polynomials are required, no such convenient formula exists (it can be shown that no formula of any type exists for polynomials of degree five or greater). Thus it is necessary to employ other techniques similar to those for finding square root. The problem is often simplified by the fact that approximate values of the desired root or roots are usually known or may be obtained by other means (for example, by graphing).

REFINING THE APPROXIMATION. Determine the root between $x = 3$ and $x = 4$ of the polynomial

$$y = x^3 - 2x^2 + x - 16.$$

We see by the following table that a root must lie in this interval; that is, the graph must cross the x axis:

x	y
3	−4
4	20

Since y goes from negative to positive between $x = 3$ and $x = 4$, it must be zero somewhere in the interval $3 < x < 4$. If we bisect the interval-that is, consider $x = (3 + 4)/2 = 3.5$-and test, we have

$$\begin{aligned} y(3.5) &= (3.5)^3 - 2(3.5)^2 + 3.5 - 16 \\ &= 42.88 - 25.50 + 3.5 - 16 \\ &= 4.88. \end{aligned}$$

Now we have

x	y
3.0	−4.00
3.5	4.88
4.0	20.00

Since the root is between the first and second entries in the table, we will use the average of these two points to obtain a more refined estimate:

$$x = \frac{3 + 3.5}{2} = 3.25,$$

$$\begin{aligned} y(3.25) &= (3.25)^3 - 2(3.25)^2 + 3.25 - 16 \\ &= 34.33 - 21.12 + 3.25 - 16 \\ &= 0.46. \end{aligned}$$

Repeating the process again, we try $x = (3 + 3.25)/2 = 3.13$ and find that $y(3.13) = -1.80$. The table now appears as

x	y
3.00	−4.00
3.13	−1.80
3.25	0.46
3.50	4.88
4.00	20.00

Note that the root now lies between the second and third entries, which are used to obtain the new estimate, $x = (3.13 + 3.25)/2 = 3.19$. Correspondingly, we have $y(3.19) = -0.70$. The process may be continued until the desired accuracy is obtained.

GRAPHICAL INTERPRETATION. Figure 13.7 gives a good insight into the nature of the foregoing process. The desired root r lies between $x = a$ and $x = b$, where a corresponds to the lower bound 3 and b corresponds to the upper bound 4. Bisecting the interval between a and b yields x_1 (corresponding to a value of 3.5), which is closer to the root than b, so b is replaced by x_1. Now a and x_1 bracket the root. Bisecting this interval gives x_2, which in turn replaces x_1 as the upper bracket. Bisecting this new interval yields the point x_3, which is on the other side of r; now the root is bracketed by x_3 on the low side and x_2 on the high side. This process may be continued until the exact value of the root is found or until a desired degree of accuracy is attained.

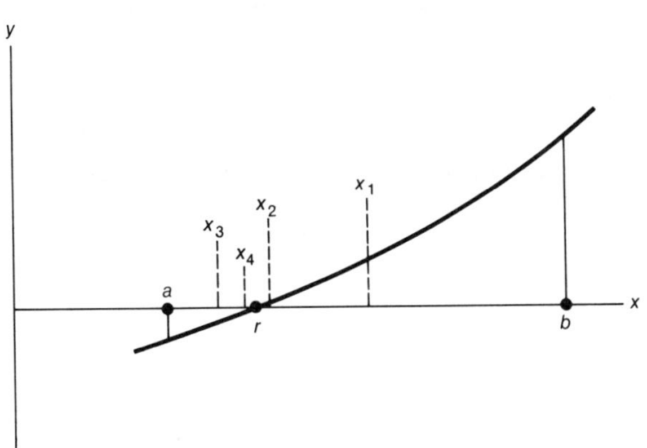

FIGURE 13.7

FORMALIZING THE PROCEDURE. Refer back to the techniques of Chapter 12 and consider the following set of numbers as a sequence converging on the desired root. In subscript notation we would have

$$x_1 = 3.50$$
$$x_2 = 3.25$$
$$x_3 = 3.13$$
$$x_4 = 3.19$$
$$\vdots$$

This technique may be generalized by the following recursive formula:

$$x_{n+1} = \frac{a_n + b_n}{2},$$

where a_n and b_n are the values of x which most closely bracket the desired root, as illustrated in Figure 13.7. Thus the procedure is

$$a_0 = 3 \qquad b_0 = 4$$

$$x_1 = \frac{a_0 + b_0}{2} = \frac{3 + 4}{2} = 3.5$$

$$y(3.5) = 4.88.$$

Therefore,

$$a_1 = 3,\ b_1 = 3.5,$$

$$x_1 = \frac{a_1 + b_1}{2} = \frac{3 + 3.5}{2} = 3.25,$$

$$y(3.25) = 0.46,$$

and so on.

In the interest of simplifying the following discussion, we shall assume that the choices for a_0 and b_0 are made such that $y(a_0) < 0$ and $y(b_0) > 0$.

A FLOWCHART. The logic of this problem is basically the same as that of the preceding problem. The steps involved are as follows:

1. Substitute predetermined values of x to bracket the root for

$$a_0 \qquad \text{and} \qquad b_0.$$

2. Calculate an estimated value for x, using the formula

$$x_{n+1} = \frac{a_n + b_n}{2}.$$

3. Calculate y, using the new value of x.
4. Compare x_{n+1} to x_n, and if $x_{n+1} = x_n$ to the desired degree of accuracy; stop.
5. Compare y to zero. (a) If $y > 0$, substitute x_{n+1} for b_n and return to step 2. (b) If $y < 0$, substitute x_{n+1} for a_n and return to step 2.

These steps are summarized by the flowchart of Figure 13.8.

COMPUTER RESULTS. The output of the computer is shown in Figure 13.9. Note that, in addition to x_n, values for y_n, a_n, and b_n are also printed. These quantities are printed only for the purpose of illustration and would not normally be included.

ADDITIONAL COMMENTS. The problem we have studied and flowcharted here involves the calculation of a particular root for a given third-degree polynomial. Writing a computer program to calculate this root would require considerably more time than performing the calculation by hand.

The program would normally be written to include arbitrary coefficients for the polynomial. Then the numeric values for these coefficients would be

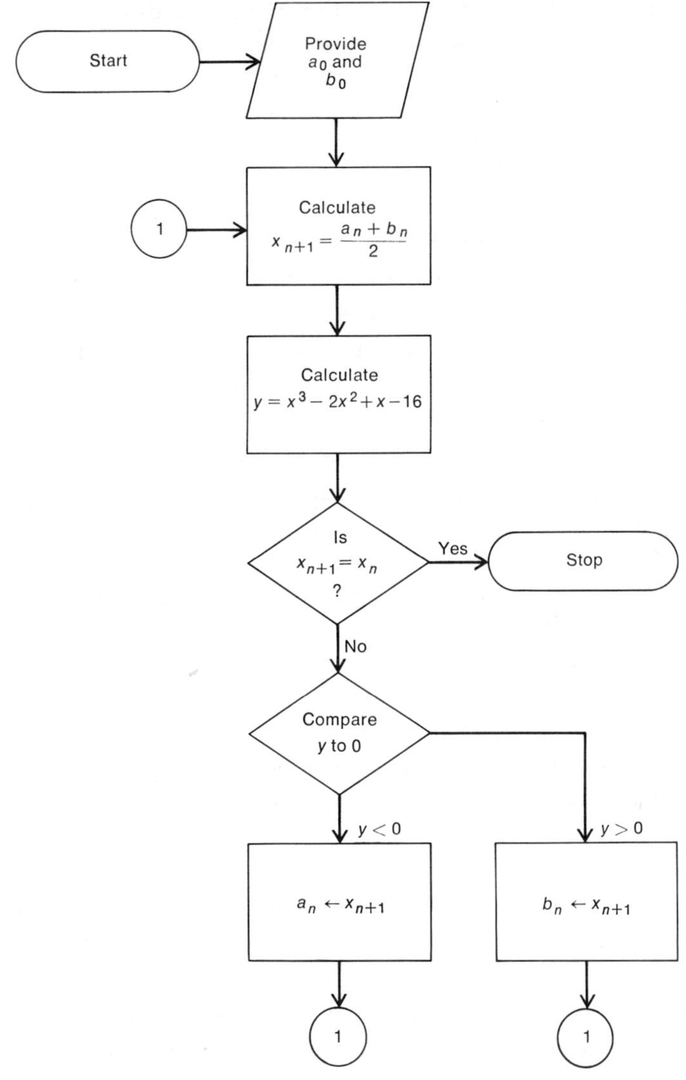

FIGURE 13.8

N	XN	YN	AN	BN
1	3.50	5.88	3.00	4.00
2	3.25	.45	3.00	3.50
3	3.13	-1.79	3.00	3.25
4	3.19	-.69	3.13	3.25
5	3.22	-.12	3.19	3.25
6	3.24	.26	3.22	3.25
7	3.23	.06	3.22	3.24
8	3.23	.06	3.22	3.23

THE ROOT IS 3.23

FIGURE 13.9

read into the computer whenever a root (or roots) was desired. Thus the program would be general in nature and, if properly written, could be used to determine roots of polynomials of degree 2, 3, 4, or any magnitude within the capacity of the computer.

EXERCISE 13.2

In each of the following, use the Bolzano method to find the indicated root(s) corrected to the nearest hundredth:

1. $y = x^2 - 4x - 8$ (a) root in interval $5 < x < 6$; (b) root in interval $-2 < x < -1$.
2. $y = x^2 - 5x + 3$ (a) root in interval $4 < x < 5$; (b) root in interval $0 < x < 1$.
3. $y = x^2 - 2x - 9$ (a) root in vicinity of $x = 5$; (b) root in vicinity of $x = -1$.
4. $y = x^3 - 12x - 9$; root in interval $-1 < x < 0$.
5. $y = x^3 - x + 1$; root in vicinity of $x = -1.4$.
6. $y = x^3 + 10x^2 - 81$; root in vicinity of $x = -3.5$.
7. If $x^2 - 29 = 0$, the solution to the equation is $x = \pm\sqrt{29}$. Similarly, the positive root of the function $y = x^2 - 29$ is the square root of 29. The square root of any number may be obtained by using the Bolzano method. Find the square root of 29, correct to the nearest tenth (assuming that it is between 5 and 6).

13.3 ROOTS BY THE METHOD OF FALSE POSITION

IMPROVING THE BOLZANO METHOD. In Section 13.2, a better approximation for x_1 could have been obtained by employing linear interpolation. For example, the table of $y = x^3 - 2x^2 + x - 16$:

x	y
3	−4
4	.20

suggests that the root will be closer to 3 than 4, since −4 is closer to 0 than 20. Thus we can employ the technique used in interpolating logarithm tables, and obtain

x			y		
3	$x_1 - 3$	1	−4	4	24
x_1			0		
4			20		

Thus,

$$\frac{x_1 - 3}{1} = \frac{4}{24}$$

$$x_1 = 3.17.$$

The corresponding value for y is

$$y = (3.17)^3 - 2(3.17)^2 + 3.17 - 16 = -1.09.$$

This process is then repeated to give

x			y		
3.17	$x_2 - 3.17$		-1.09	1.09	
x_2		.83	0		21.09
4.00			20.00		

$$\frac{x_2 - 3.17}{0.83} = \frac{1.09}{21.09}$$

$$x_2 = 3.21,$$

and so on.

Note the very rapid convergence of this method as opposed to the Bolzano method; using the former, $x_4 = 3.19$; with the latter, $x_1 = 3.17$. This technique is commonly referred to as **regula falsi**, or *the method of false position*, and was used by the Egyptians over 3000 years ago. The name is derived from the graphical interpretation.

GRAPHICAL INTERPRETATION. A segment of a polynomial function similar to the one we are studying here is shown in Figure 13.10. The points $[a, f(a)]$ and $[b, f(b)]$ correspond to $(3, -4)$ and $(4,20)$ of the problem here. Furthermore, r represents the desired root and x_1 represents the first approximation to it. The approximation x_1 is determined by the intersection of the x axis

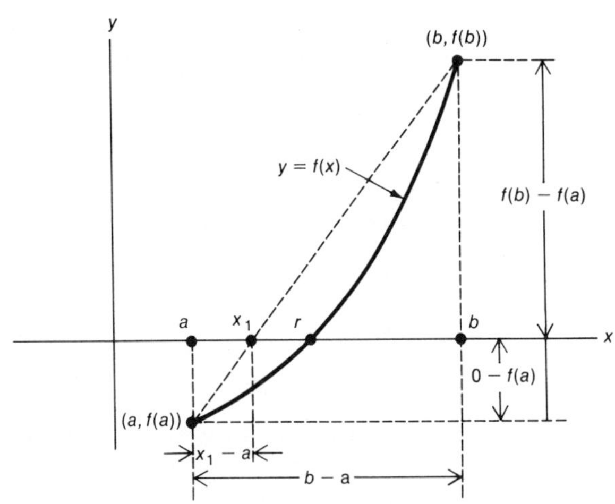

FIGURE 13.10

and the straight line connecting the two points. The value is calculated by employing techniques identical with those used with logarithms:

$$\frac{x_1 - a}{b - a} = \frac{0 - f(a)}{f(b) - f(a)}, \tag{1}$$

which may be solved for x_1, giving

$$x_1 = a + \frac{-f(a)}{f(b) - f(a)}(b - a). \tag{2}$$

It should be noted that (1) corresponds to the equation

$$\frac{x_1 - 3}{1} = \frac{4}{24}$$

which we used earlier in this section.

The newly calculated value for x_1 may then be substituted for a (or b as the case may be) and the calculation continued.

FORMALIZING THE PROCEDURE. For calculation of roots on a computer, it is convenient to change (2) to a recursive form, which is easily done by use of subscripts:

$$x_{n+1} = a_n + \frac{-f(a_n)}{f(b_n) - f(a_n)}(b_n - a_n).$$

Use of the recursive form is illustrated by finding the root between $x = 3$ and $x = 4$ of $y = x^3 - 2x^2 + x - 16$:

$$a_0 = 3, \qquad b_0 = 4,$$

$$f(a_0) = -4, \qquad f(b_0) = 20,$$

$$x_1 = a_0 + \frac{-f(a_0)}{f(b_0) - f(a_0)}(b_0 - a_0)$$

$$= 3 + \frac{-(-4)}{20 - (-4)}(4 - 3)$$

$$= 3 + \frac{4}{24}(1)$$

$$= 3.17;$$

$$y(3.17) = 3.17^3 - 2 \times 3.17^2 + 3.17 - 16$$

$$= -1.09;$$

$$a_1 = 3.17 \qquad b_1 = 4$$

$$f(a_1) = -1.09 \qquad f(b_1) = 20$$

$$x_2 = a_1 + \frac{-f(a_1)}{f(b_1) - f(a_1)}(b_1 - a_1)$$

$$= 3.17 + \frac{-(-1.09)}{20-(-1.09)}$$

$$= 3.21;$$

and so on. Compare these calculations, step by step, with those developed from the intuitive approach at the beginning of this section.

The flowchart for this problem is almost identical to that of Figure 13.8. It is only necessary to change the equation in the second block to (2) here.

COMPUTER RESULTS. Output for a computer run, in which the root has been calculated to 3 decimals, is included as Figure 13.11. Output quantities correspond to those of Section 13.2.

N	XN	YN	AN	BN
1	3.167	-1.128	3.000	4.000
2	3.211	-.303	3.167	4.000
3	3.223	-.073	3.211	4.000
4	3.226	-.015	3.223	4.000
5	3.227	.004	3.226	4.000
6	3.227	.004	3.226	3.227

THE ROOT IS 3.227

FIGURE 13.11

GENERAL COMMENTS. Both the method of false position and Bolzano's method are straightforward and intuitive in nature. Generally speaking, however, they do not provide the simplest means for determining roots of equations. As a result, many other means have been devised for determining roots of equations (not necessarily limited to polynomials) using methods of calculus and other higher mathematics.

In evaluating any such numerical method for use on a computer, it is necessary to consider factors such as speed of convergence and number of arithmetic operations required for each iteration. However, the methods we have studied represent fundamental techniques and illustrate well the process of determining roots by iteration. In following sections we shall apply iterative techniques to solving systems of linear equations.

EXERCISE 13.3

1-7. For problems 1-7 of Exercise 13.2, find the indicated root(s), corrected to the nearest hundredth, using the method of false position.

8. Using the polynomial function $y = x^3 - 25$, find the cube root of 25, corrected to the nearest hundredth. Use as starting points $x = 2$ and $x = 3$.

9. Using the polynomial function $y = x^5 - 500$, find the fifth root of 500.

13.4 SOLUTION OF LINEAR SYSTEMS BY THE JACOBI METHOD

In Chapter 9, systems of simultaneous linear equations were solved by using algebraic methods, determinants, and the Gauss-Jordan method. Chapter 10 was devoted to solving linear systems using matrices. All of these, except the algebraic methods, are basically numerical processes. However, they differ in one important respect. The methods studied earlier lead directly to the exact solution, whereas the methods of this chapter for finding roots yield approximations to the solution, which may be refined to any desired degree. We will now consider the first of two iterative techniques for solving systems of linear equations.

A SIMPLE EXAMPLE. The linear system

$$5x + y = 6 \tag{1}$$
$$2x - 3y = 16 \tag{2}$$

may be solved readily by the method of substitution. However, let us attempt to determine the solution by using an iterative method which is adaptable to the computer. Solving (1) for x and (2) for y gives the equivalent system

$$x = \frac{1}{5}(-y + 6) \tag{3}$$

$$y = \frac{1}{3}(2x - 16). \tag{4}$$

As a matter of convention, we shall always solve the first equation for the first variable (x in this case) and the second equation for the second variable (y in this case). If we make an estimate for the solution, we can substitute the value for y in (3) and calculate a new value for x. Similarly, if we substitute the value for x in (4) we can calculate a new value for y, yielding a new estimate for the solution. For example, let us assume the solution $x = 0$, $y = 0$ and calculate better values for x and y:

$$x = \frac{1}{5}(0 + 6) = 1.2$$

$$y = \frac{1}{3}(0 - 16) = -5.33.$$

An improved estimate for the solution is $(1.2, -5.33)$, which is in turn substituted into (3) and (4) to further refine the solution:

$$x = \frac{1}{5}(5.33 + 6) = 2.27$$

$$y = \frac{1}{3}(2 \times 1.2 - 16) = -4.53.$$

This process may be continued, with the results shown in the following table:

Estimate	x	y
0	0	0
1	1.20	−5.33
2	2.27	−4.53
3	2.11	−3.82
4	1.96	−3.93
5	1.99	−4.03
6	2.01	−4.00
7	2.00	−4.00

The first line is referred to as the zeroth estimate (consistent with Section 13.2) since it is an arbitrary guess. However, the entries following are calculated estimates that converge, by the seventh iteration, to the correct solution $(2, -4)$. In spite of a poor initial guess, this process, called the **Jacobi method of iteration**, appears to be self-correcting in converging to the proper solution. However, we must not assume that the Jacobi method will always converge so readily.

A DIVERGING RESULT FROM THE JACOBI METHOD. The original decision to solve the first equation for the first unknown and the second equation for the second unknown in no way restricts us, since we can reorder the equations, giving

$$2x - 3y = 16$$
$$5x + y = 6.$$

Let us solve these for x and y, and apply the Jacobi method, beginning with the estimate (0,0).

$$x = \frac{1}{2}(3y + 16) \tag{5}$$

$$y = -5x + 6. \tag{6}$$

Estimate	x	y
0	0	0
1	8	6
2	17	−34
3	−43	−79
4	−110.5	221
5	339.5	558.5

It appears that the values for x and y continually become larger (in magnitude), diverging from the correct solution (2, −4). In other words, it is possible to solve (1) and (2) if the equivalent system consisting of (3) and (4) is used, but not if the equivalent system of (5) and (6) is used. Results such as these can be most discouraging (especially on a computer).

CONDITIONS FOR CONVERGENCE. In studying the conditions for convergence of a linear system, we shall refer to (1) and (2), repeated here for convenience:

$$5x + y = 6 \tag{1}$$
$$2x - 3y = 16. \tag{2}$$

In the first attempt (successful) to obtain a solution by the Jacobi method, (1) was solved for x and (2) for y. In the second attempt (unsuccessful), (1) was solved for y and (2) for x. Note that the magnitude of the coefficient on x is greater than the magnitude of the coefficient on y in (1); that is, $|5| > |1|$. In (2), the converse is true; that is, $|-3| > |2|$. The solution will converge if (1) is solved for x and (2) for y, as was done in (3) and (4).

Although this method of comparing coefficients is simple to use, it does not always tell the complete story. Refer to the general case for a 2 × 2 system, using the notation of Chapter 10. If it is possible to order the pair of equations

$$a_{11}x + a_{12}y = b_1 \tag{7}$$
$$a_{21}x + a_{22}y = b_2 \tag{8}$$

such that

$$|a_{12}\, a_{21}| < |a_{11}\, a_{22}|, \tag{9}$$

the Jacobi method will converge* if (7) is solved for x and (8) for y. In other words, the product of the main diagonal elements must be greater in magnitude than the product of the minor diagonal elements. For example,

$$5x + y = 6 \tag{1}$$
$$2x - 3y = 16 \tag{2}$$

satisfied the criterion of (9), since

$$a_{11} = 5 \qquad a_{12} = 1$$
$$a_{21} = 2 \qquad a_{22} = -3$$

and

$$|a_{12}\, a_{21}| = |(1)\,(2)| = 2$$
$$|a_{11}\, a_{22}| = |(5)\,(-3)| = 15$$
$$2 < 15.$$

On the other hand, in the unsuccessful attempt to solve this system, we con-

**The proof found on pages 255 and 256 of the fourth reference can be readily adjusted to the Jacobi method.*

sidered the equations in the following order (and by convention solved the first for x and the second for y):

$$2x - 3y = 16$$
$$5x + y = 6.$$

The coefficients were

$$a_{11} = 2 \qquad a_{12} = -3$$
$$a_{21} = 5 \qquad a_{22} = 1$$

and

$$|a_{12}\, a_{21}| = |(-3)(5)| = 15$$
$$|a_{11}\, a_{22}| = |(2)(1)| = 2$$
$$15 \not< 2.$$

APPLYING THE CONVERGENCE TEST. In the following system, note that both coefficients on x are larger than the corresponding coefficients on y:

$$9x + 2y = 15$$
$$12x - y = 9.$$

However, by reordering the equations, we have

$$a_{11} = 12 \qquad a_{12} = -1$$
$$a_{21} = 9 \qquad a_{22} = 2.$$

Thus

$$|a_{12}\, a_{21}| = |(-1)(9)| = 9$$
$$|a_{11}\, a_{22}| = |(12)(2)| = 24$$
$$9 < 24$$

and the iteration will converge as shown in the following.

Iteration	x	y
0	0	0
1	0.75	7.50
2	1.38	4.13
3	1.09	1.29
4	0.86	2.69
5	0.97	3.63
6	1.05	3.14
7	1.01	2.78
8	0.98	2.96
9	1.00	3.09
10	1.01	3.05
11	1.00	2.96
12	1.00	3.00

Although this solution converged slowly, which is often a problem when solving problems on a computer, it did converge to the solution (1,3).

FORMALIZING THE PROCEDURE. Although we have applied the Jacobi method only to systems of two equations in two unknowns, it can be readily generalized to solve systems of three or more equations. In this section we shall formalize the procedures for the case of two equations, and reserve the study of larger systems to the next section. However, these procedures could be generalized to any number of equations.

We begin with the general form of the linear system in two variables:

$$a_{11}x + a_{12}y = b_1$$
$$a_{21}x + a_{22}y = b_2.$$

Assuming that the convergence conditions have been met, we solve the first for x and the second for y, giving

$$x = \frac{1}{a_{11}}(-a_{12}y + b_1)$$

$$y = \frac{1}{a_{22}}(-a_{21}x + b_2).$$

The overall procedure is then as follows:

1. Order the equations such that $|a_{12}\, a_{21}| < |a_{11}\, a_{22}|$. If this is impossible, stop; the solution will not converge.
2. For the initial estimate, set $x_0 = 0$ and $y_0 = 0$.
3. Calculate

$$x_{n+1} = \frac{1}{a_{11}}(-a_{12}y_n + b_1)$$

$$y_{n+1} = \frac{1}{a_{22}}(-a_{21}x_n + b_2).$$

4. Compare x_{n+1} to x_n and y_{n+1} to y_n. If they are equal, this is the solution. If not, let $x_n \leftarrow x_{n+1}$ and $y_n \leftarrow y_{n+1}$ and return to step 3.

The flowchart of this procedure is shown in Figure 13.12.

COMPUTER RESULTS. Two computer runs are included here to illustrate the Jacobi method. The output shown in Figure 13.13 consists of the equations and the iterative solution, which converges to the correct solution (to the nearest tenth). In Figure 13.14 the equations have purposely been incorrectly ordered to produce a system that does not converge. The output diverges from the solution (4,5) with each iteration.

EXERCISE 13.4

Test each of the following systems for convergence (it may be necessary to reorder some of the equations). For each system that can be made to converge, use the Jacobi method to obtain the solution (corrected to two decimals).

1. $3x + y = -13$
   $x - 5y = 17$
2. $3x - 2y = -12$
   $x + 5y = 13$
3. $3x - 4y = -1$
   $5x + y = 6$
4. $x - 4y = 3$
   $2x + 8y = 5$
5. $2x - 3y = -5$
   $-6x + 9y = 15$
6. $3x + 2y = 3$
   $6x + y = -3$
7. $4x - y = -1$
   $3x - 2y = -7$
8. $5x + y = 3$
   $2x - 3y = -26$
9. $2x - y = 1$
   $x + 3y = -17$
10. $3x - y = 0$
    $-x + 5y = 14$
11. What is the significance (or geometric interpretation) of a pair of equations that cannot be made to satisfy the convergence criterion? That is, equations in which

$$a_{12}\, a_{21} = a_{11}\, a_{22}.$$

12. The system

$$2x + 3y = 5$$
$$2x - 3y = -1$$

does not satisfy the condition for convergence, so would be expected to diverge. Apply the Jacobi method to this system and note the results after the fourth iteration.

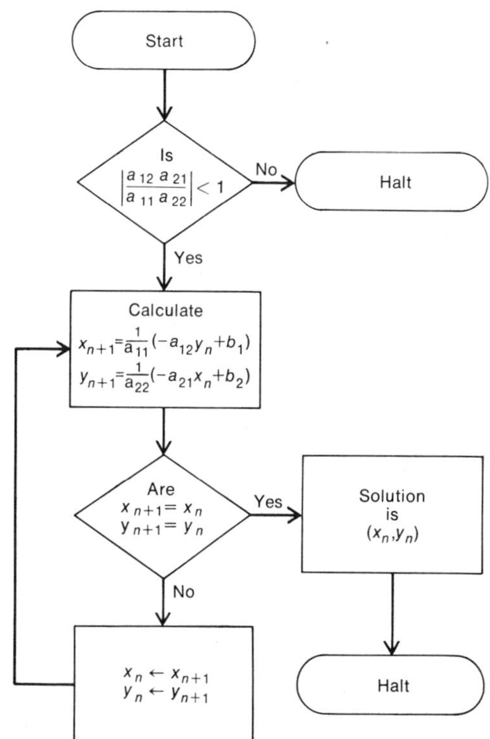

FIGURE 13.12

Apply the Jacobi method to the following systems of three equations. The first equation should be solved for x, the second for y, and the third for z.

13. $7x+2y-3z=1$
$x-5y-2z=6$
$-x+y+4z=8$

14. $6x-y+4z=18$
$x-10y+z=2$
$-4x+y-6z=-2$

SOLUTION OF THE FOLLOWING
SYSTEM BY THE JACOBI METHOD

+5X +1Y = +25
+3X -2Y = +2

ITERATION	X	Y
0	.00	.00
1	5.00	-1.00
2	5.20	6.50
3	3.70	6.80
4	3.64	4.55
5	4.09	4.46
6	4.11	5.14
7	3.97	5.17
8	3.97	4.96
9	4.01	4.96
10	4.01	5.02
11	4.00	5.02
12	4.00	5.00

FIGURE 13.13

SOLUTION OF THE FOLLOWING
SYSTEM BY THE JACOBI METHOD

+3X -2Y = +2
+5X +1Y = +25

ITERATION	X	Y
0	.00	.00
1	.67	25.00
2	17.33	21.65
3	15.10	-61.65
4	-40.43	-50.50
5	-33.00	227.15
6	152.10	190.00
7	127.33	-735.50
8	-489.67	-611.65
9	-407.10	2473.35
10	1649.57	2060.50
11	1374.33	-8222.85
12	-5481.23	-6846.65
13	-4563.77	27431.15
14	18288.10	22843.85

FIGURE 13.14

13.5 SOLUTION OF LINEAR SYSTEMS BY THE GAUSS-SEIDEL METHOD

IMPROVING THE JACOBI METHOD. In solving a system of equations using the Jacobi method, a solution of (0,0) is used to begin the iteration. New values for x and y are determined, using the previous estimate of the solution. However, the newly calculated value for x can immediately replace the previous estimate, thus speeding up the convergence. For example, consider the following equations used in studying the Jacobi method:

$$5x + y = 6 \qquad x = \frac{1}{5}(-y + 6) \tag{1}$$

$$2x - 3y = 16 \qquad y = \frac{1}{3}(2x - 16). \tag{2}$$

Beginning with $y_0 = 0$, we have $x_1 = (-0 + 6)/5 = 1.20$ and $y_1 = (2 \times 1.20 - 16)/5 = -4.53$ (note that x_1, not x_0, is used in computing y_1). The complete results are tabulated below:

Iteration	x	y
0		0
1	1.20	−4.53
2	2.11	−3.91
3	1.99	−4.01
4	2.00	−4.00

Note that this technique, called the **Gauss-Seidel** method, converges after four iterations, as opposed to seven for the Jacobi method. The recursive formula for the Gauss-Seidel method is identical to that of the Jacobi method, except x_{n+1} instead of x_n is used to compute y_{n+1}. In other words,

$$x_{n+1} = \frac{1}{a_{11}}(-a_{12}y_n + b_1) \tag{3}$$

$$y_{n+1} = \frac{1}{a_{22}}(-a_{21}x_{n+1} + b_2). \tag{4}$$

Conditions for convergence of the Gauss-Seidel method are identical to those for the Jacobi method; that is, it must be possible to order the equations such that

$$|a_{12}\, a_{21}| < |a_{11}\, a_{22}|.$$

COMPUTER RESULTS. In order to contrast the results of the Gauss-Seidel method with those of the Jacobi method, the following equations have been solved:

$$5x + y = 25$$
$$3x - 2y = 2.$$

Figure 13.15(a) is the output using the Jacobi method, and Figure 13.15(b) is the output using the Gauss-Seidel method.

ITEPATION	X	Y
0	.00	.00
1	5.00	-1.00
2	5.20	6.50
3	3.70	6.80
4	3.64	4.55
5	4.09	4.46
6	4.11	5.14
7	3.97	5.17
8	3.97	4.96
9	4.01	4.96
10	4.01	5.02
11	4.00	5.02
12	4.00	5.00

(*a*)

ITERATION	X	Y
1	5.00	6.50
2	3.70	4.55
3	4.09	5.14
4	3.97	4.96
5	4.01	5.02
6	4.00	5.00

(*b*)

FIGURE 13.15

GEOMETRIC INTERPRETATION OF A CONVERGING CASE. As we know, the geometric interpretation of the solution to a pair of linear equations is the intersection of the lines representing each equation. Reference to the geometric significance of the Gauss-Seidel iterative procedure is equally enlightening. The graphs of (1) and (2) and the successive points in the iterative solutions are shown in Figure 13.16.

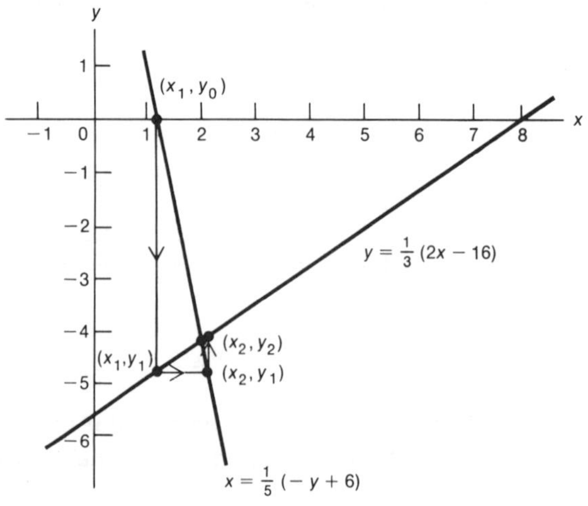

FIGURE 13.16

GEOMETRIC INTERPRETATION OF A DIVERGING CASE. Consider the application of the Gauss-Seidel method to the system

$$2x - 3y = -1 \qquad \text{or} \qquad x = \frac{1}{2}(3y - 1) \tag{5}$$

$$x + y = 2 \qquad \text{or} \qquad y = -x + 2. \tag{6}$$

Inspection of the results in the following table appear to indicate that this solution is diverging:

Iteration	x	y
0		0
1	−0.50	2.50
2	3.25	−1.25
3	−2.38	4.38
4	6.07	−4.07
5	−6.61	8.61

The graphical interpretation in Figure 13.17 confirms that the resulting process diverges and continually moves further from the solution (1,1). Determine the effect of reordering the original equations to satisfy the convergence conditions.

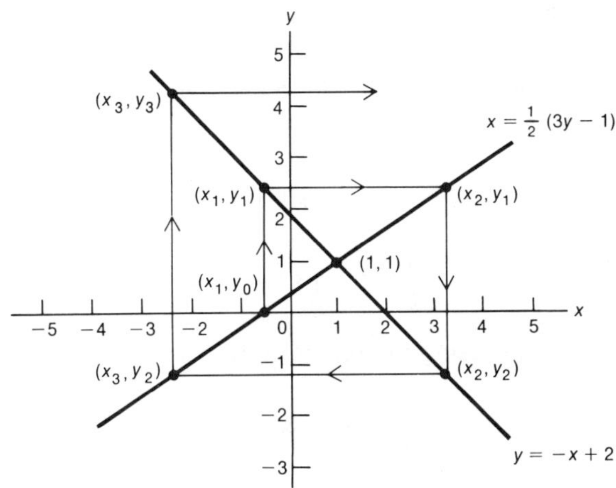

FIGURE 13.17

APPLICATION TO EQUATIONS IN SEVERAL VARIABLES. Both the Jacobi and the Gauss-Seidel methods may be used to solve simultaneous systems in three or more variables. For example, consider the application of the Gauss-Seidel method to the system

$$6x - y - 3z = 3 \tag{7}$$

$$-x + 3y + 2z = -6 \tag{8}$$

$$3x + y - 5z = -10. \tag{9}$$

Solving (7) for x, (8) for y, and (9) for z yields

$$x = \frac{1}{6}(y + 3z + 3)$$

$$y = \frac{1}{3}(x - 2z - 6)$$

$$z = \frac{1}{5}(3x + y + 10).$$

Here we begin with the assumption that $y_0 = 0$ and $z_0 = 0$, in order to begin the iteration. Thus

$$x_1 = \frac{1}{6}(0 + 0 + 3) = 0.5.$$

In calculating y_1, it is necessary to use z_0; however, the newly calculated x_1 may be used in place of x_0:

$$y_1 = \frac{1}{3}(x_1 - 2z_0 - 6)$$

$$= \frac{1}{3}(0.5 + 0 - 6) = -1.83.$$

Finally, in calculating z_1, it is possible to use both x_1 and y_1 rather than x_0 and y_0, with the result

$$z_1 = \frac{1}{5}(3x_1 + y_1 + 10)$$

$$= \frac{1}{5}(1.5 - 1.83 + 10) = 1.93.$$

In other words, we are using the recursive forms

$$x_{n+1} = \frac{1}{6}(y_n + 3z_n + 3)$$

$$y_{n+1} = \frac{1}{3}(x_{n+1} - 2z_n - 6) \tag{10}$$

$$z_{n+1} = \frac{1}{5}(3x_{n+1} + y_{n+1} + 10).$$

Continuing the iteration and using the above equations yields the following table, which converges to the solution (1,−3,2):

Iteration	x	y	z
0		0	0
1	0.50	−1.83	1.93
2	1.16	−2.90	2.12
3	1.08	−3.05	2.04
4	1.01	−3.02	2.00
5	1.00	−3.00	2.00

CONVERGENCE FOR EQUATIONS IN SEVERAL VARIABLES. Unfortunately it is usually more difficult to satisfy conditions for convergence when dealing with three or more equations than when dealing with two. Convergence of

$$6x - y - 3z = 3 \quad (7)$$
$$-x + 3y + 2z = -6 \quad (8)$$
$$3x + y - 5z = -10 \quad (9)$$

was assured by the following:

1. The absolute value of the coefficient on x in (7) is greater than the sums of the absolute values of the other coefficients; that is, $|6| > |-1| + |-3|$.
2. The absolute value of the coefficient on y in (8) is equal to the sums of the absolute values of the other coefficients; that is, $|3| = |-1| + |2|$.
3. The absolute value of the coefficient on z in (9) is greater than the sums of the absolute values of the other coefficients; that is, $|-5| > |3| + |1|$.

In general, convergence is assured if it is possible to order the equations

$$\begin{aligned} a_{11}x + a_{12}y + a_{13}z &= b_1 \\ a_{21}x + a_{22}y + a_{23}z &= b_2 \\ a_{31}x + a_{32}y + a_{33}z &= b_3 \end{aligned} \quad (11)$$

such that

$$\begin{aligned} |a_{11}| &\geq |a_{12}| + |a_{13}| \\ |a_{22}| &\geq |a_{21}| + |a_{23}| \\ |a_{33}| &\geq |a_{31}| + |a_{32}|. \end{aligned} \quad (12)$$

However, the convergence criterion of (12) *must be qualified* to the extent that at least *one* of the three equations be limited to "greater than." Although all systems that satisfy the convergence criterion will converge, it does not follow that all systems failing to satisfy it will not converge. Unfortunately there is no simple means for detecting all systems of equations that are convergent.

Application of the convergence criterion and of the Gauss-Seidel method to four simultaneous equations is shown as follows:

$$\begin{aligned} 2w + 6x + 3y &= 9 \\ 3w + 4x + 10z &= -11 \\ -2w - x + 9y &= 13 \\ 3w + y + 2z &= -10. \end{aligned} \tag{13}$$

Inspection of the equations indicates that they may be reordered to

$$3w + 0x + y + 2z = -10 \tag{14}$$
$$2w + 6x + 3y + 0z = 9 \tag{15}$$
$$-2w - x + 9y + 0z = 13 \tag{16}$$
$$3w + 4x + 0y + 10z = -11. \tag{17}$$

Thus the convergence criteria are satisfied, since

from (14)

$$|3| = |1| + |2|$$

from (15)

$$|6| > |2| + |3|$$

from (16)

$$|9| > |-2| + |-1|$$

from (17)

$$|10| > |3| + |4|.$$

Solving these equations for their respective variables yields

$$w = \frac{1}{3}(-y - 2z - 10)$$

$$x = \frac{1}{6}(-2w - 3y + 9)$$

$$y = \frac{1}{9}(2w + x + 13)$$

$$z = \frac{1}{10}(-3w - 4x - 11).$$

The iterative process can be started by assuming $x = y = z = 0$, with the following results.

Iteration	w	x	y	z
0		0	0	0
1	−3.33	2.61	0.99	−1.15
2	−2.90	1.97	1.02	−1.02
3	−2.99	1.99	1.00	−1.01
4	−3.00	2.00	1.00	−1.00

The rapid convergence of this solution is not generally typical of systems in several variables. These particular equations have been arranged to insure rapid convergence. Minor changes in one or two of the equations could easily require many more iterations. For example, see problem 13 below.

EXERCISE 13.5

Test each of the following systems for convergence. (It may be necessary to reorder some of the equations.) For each system that can be made to converge, use the Gauss-Seidel method to obtain the solution (correct to two decimals). Compare the results to those of the corresponding problems in Exercise 13.4.

1. $3x + y = -13$
   $x - 5y = 17$
2. $3x - 2y = -12$
   $x + 5y = 13$
3. $3x - 4y = -1$
   $5x + y = 6$
4. $x - 4y = 3$
   $2x + 8y = 5$
5. $2x - 3y = -5$
   $-6x + 9y = 15$
6. $3x + 2y = 3$
   $6x + y = -3$
7. $4x - y = -1$
   $3x - 2y = -7$
8. $5x + y = 3$
   $2x - 3y = -26$
9. $2x - y = 1$
   $x + 3y = -17$
10. $3x - y = 0$
    $-x + 5y = 14$

Apply the Gauss-Seidel method to the following systems of three equations. The first equation should be solved for x, the second for y, and the third for z.

11. $7x + 2y - 3z = 1$
    $x - 5y - 2z = 6$
    $-x + y + 4z = 8$
12. $6x - y + 4z = 18$
    $x - 10y + z = 2$
    $-4x + y - 6z = -2$
13. The following system is only slightly different from the equations of (13). Solve it, using the Gauss-Seidel method, and note that it requires more than twice as many iterations to converge.

$$\begin{aligned} 3w + y + 2z &= 18 \\ 2w + 6x + 3y &= 33 \\ -2w - x + 3y &= 0 \\ 3w + 4x + 7z &= 42 \end{aligned}$$

14. Occasionally a numeric solution will oscillate about the solution, and will not converge as required by the specified error criterion. This is illustrated by the following problem. Here it is of paramount importance that each value for x and y be rounded to two decimals in order to illustrate the problem. Carry this process out to 20 iterations.

$$\begin{aligned} 9x - 11y &= 65 \\ 5x + 8y &= 22. \end{aligned}$$

REFERENCES

Dodes, A. and Greitzer, B., *Numerical Analysis.* New York, Hayden Book Companies, 1964.

Gruenberger, F. J. and Jaffray, G., *Problems for Computer Solution,* New York, John Wiley & Sons, 1965.

Kovach, L., *Computer-Oriented Mathematics: An Introduction to Numerical Methods.* San Francisco, Holden-Day, Inc., 1964.

McCracken, D. D. and Dorn, W. S., *Numerical Methods and Fortran Programming.* New York, John Wiley & Sons, 1964.

14 BOOLEAN ALGEBRA

14.1 BASIC CIRCUITS / **446**
Parallel and Series Connections / **447**
Notation / **448**
Parallel Circuits / **449**
Series Circuits / **449**
Combination Circuits / **450**
Switches in Opposite States / **450**
Exercise 14.1 / **451**

14.2 ADDITIONAL CIRCUITS / **452**
A Simpler Circuit / **453**
The Value of Simpler Circuits / **455**
Exercise 14.2 / **455**

14.3 THE FORM $AB + A$ AND BOOLEAN PROPERTIES / **457**
Boolean Properties / **458**
Distributive Properties / **459**
Closed Switches / **459**
Simplification of AB + A / **460**
Distributive Properties / **460**
A Second Distributive Property / **461**
Exercise 14.3 / **463**

14.4 OTHER BOOLEAN PROPERTIES / **463**
Commutative Property / **463**
Associative Property / **464**
Identity Elements / **465**
Open or Closed Switches / **465**
Exercise 14.4 / **466**

14.5 SIMPLIFICATIONS / **466**
Basic Properties / **466**
Exercise 14.5 / **469**

14.6 AN APPLICATION OF BOOLEAN ALGEBRA / **471**
DeMorgan's Rules / **473**
Exercise 14.6 / **475**

In discussions of computers and programming, one may encounter the terms **Boolean operators, Boolean variables** or **Boolean algebra**. The logical connectives .AND. and .OR. may be referred to as Boolean operators and the logical forms as Boolean expressions. It is interesting to note the word Boolean refers to a man, George Boole, who lived more than one hundred years ago. In 1854, he published a book in which he related algebraic properties to logic. This set of properties has become known as Boolean algebra and will now be discussed in terms of switching circuits. The similarities between logic and sets which have been noted earlier will be extended to circuits involving electrical components. Since algebraic properties in the three situations are the same, the term *Boolean algebra* may relate to any of the three systems.

14.1 BASIC CIRCUITS

In this chapter we center our attention on a simple electrical switch (or its equivalent) and its ability to allow or prevent the passage of current through itself. When a light switch is turned on, current can pass through the switch to the light. Turned off, the switch creates a break or gap in the circuit so that electricity cannot flow. A symbolic representation of a switch in the OFF position is shown in Figure 14.1(a). Figure 14.1(b) represents the ON position. In Figure 14.1(a), current cannot jump across the break, but in Figure 14.1(b) current can pass through the switch. A more convenient representation of a switch is as shown in Figure 14.1(c), where the letter *A* represents the switch in either the ON or OFF position. Since there are only two possible states for the switch, 0 could represent the OFF position and 1 the ON position. The letter *A*, representing the state of the switch, acts as a variable in ordinary algebra, except that it can assume only two numerical values, 0 and 1.

Suppose we have two switches, the first represented by *A* and the second by *B*. Before considering possible paths between them, note that there are four different ON-OFF combinations for the two switches. Listing these in table form, we have

Case	*A*	*B*
1	0	0
2	0	1
3	1	0
4	1	1

In Case 1, both switches are OFF; in Case 4 they are both ON. In both the second and third cases, one switch is ON and the other is OFF.

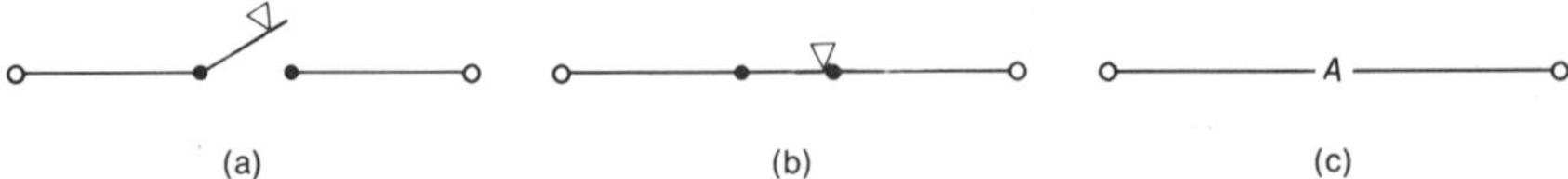

FIGURE 14.1 (a) Switch in OFF position; (b) Switch in ON position; (c) symbolic representation of switch

PARALLEL AND SERIES CONNECTIONS. When wires are used to connect the two switches, a *network* or *circuit* results. There are two possible ways in which two switches may be arranged. They may be side by side, wired in **parallel**; or they may be end to end, wired in **series**. These two possibilities are shown in Figure 14.2(a) and 14.2(b) respectively.

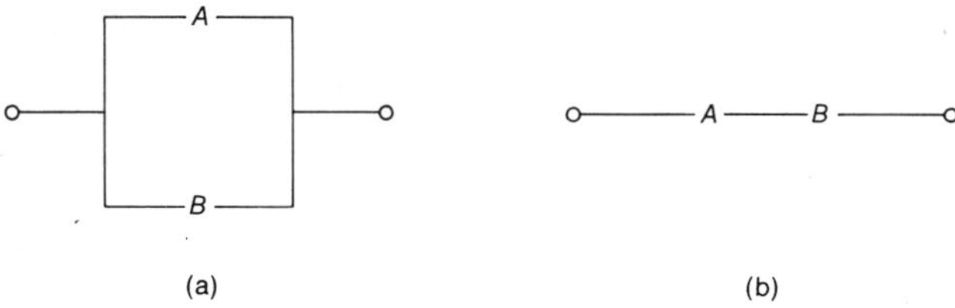

FIGURE 14.2 (a) Switches in parallel; (b) switches in series

The four possible ON-OFF combinations of two switches connected in parallel are shown in Figure 14.3. When switch *A* is closed, current can still pass between the end points, even though switch *B* is open; similarly, current flows if *B* is closed even though *A* is open. The ability of this circuit to pass current is represented by 1; its inability to pass current is represented by 0. This representation allows a convenient display of the four possibilities, as shown in Table 14.1.

TABLE 14.1

Case	*A*	*B*	Parallel Circuit
1	0	0	0
2	0	1	1
3	1	0	1
4	1	1	1

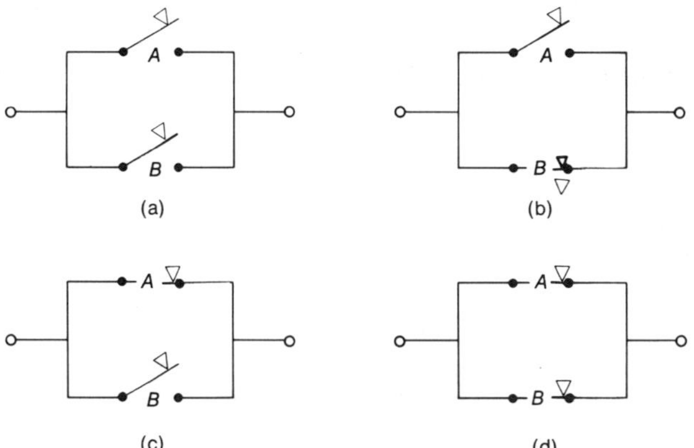

FIGURE 14.3 (a) $A = 0$, $B = 0$; (b) $A = 0$, $B = 1$; (c) $A = 1$, $B = 0$; (d) $A = 1$, $B = 1$

The table indicates that current will flow through the parallel circuit when either or both of the switches are closed. The only way of preventing passage of current through the circuit is to open both switches, as represented by Case 1.

In the series circuit of Figure 14.2(b), current will pass between the two end points only when both switches are closed. An open or OFF state for either or both switches is sufficient to prevent the flow of current through the circuit (Table 14.2). The four possibilities of Table 14.2 are shown in Figure 14.4.

TABLE 14.2

Case	A	B	Series Circuit
1	0	0	0
2	0	1	0
3	1	0	0
4	1	1	1

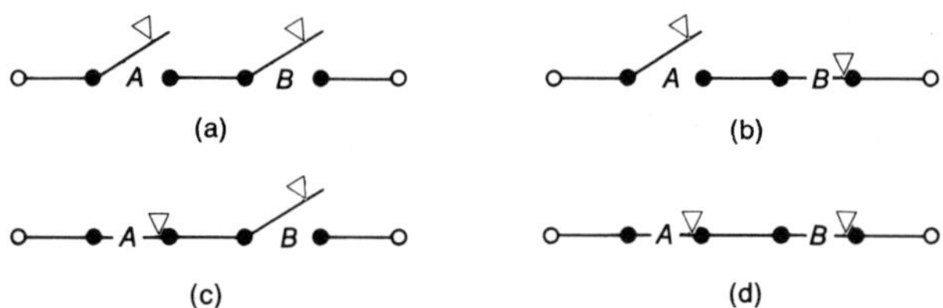

FIGURE 14.4 (a) $A = 0$, $B = 0$; (b) $A = 0$, $B = 1$; (c) $A = 1$, $B = 0$; (d) $A = 1, B = 1$

NOTATION. Although we could use the schematic representation shown in Figure 14.5(a) whenever we wish to discuss a parallel circuit, and the representation in Figure 14.5(b) whenever we wish to discuss a series circuit, it is more convenient to represent the parallel connection of A and B by

$$A + B$$

and the series connection by

$$A \cdot B \quad \text{or} \quad AB.$$

These representations are the same as those used in ordinary algebra for the sum and product of two numbers. However, since we are dealing with a system consisting of only the ON and OFF states (1 and 0), we cannot automatically use basic algebraic rules in this new system.

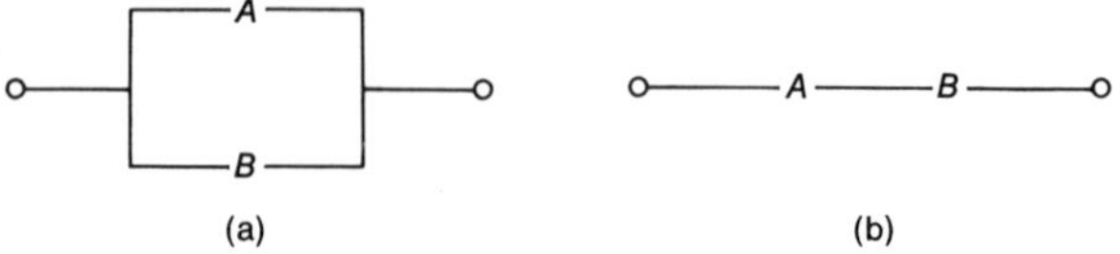

FIGURE 14.5

PARALLEL CIRCUITS. Rewriting Table 14.1 for a parallel circuit in terms of the new notation, some interesting facts come to light, as can be seen in Table 14.3.

TABLE 14.3

Case	A	B	$A+B$
1	0	0	0
2	0	1	1
3	1	0	1
4	1	1	1

Case 1 can also be represented by replacing A and B with zeros, since both switches are open. This results in $0+0$, which from the table "equals" 0. In Case 2 $A+B$ becomes $0+1$, which "equals" 1. If we indicate these two results as

$$0+0=0$$
$$0+1=1,$$

they have the appearance of ordinary addition. Again in Case 3, the results are consistent with ordinary arithmetic, since

$$1+0=1.$$

However, Case 4 differs from both ordinary decimal arithmetic and from binary arithmetic (recall that binary $1+1=10$), since

$$1+1=1.$$

This equation may seem somewhat strange until we recall what it actually represents. The left side, $1+1$, indicates that 2 switches are connected in parallel and that both switches are closed. In this circumstance we naturally expect current to pass through the circuit, indicated by 1 on the right side of the equation. As long as we are careful in interpreting our notation, the results will be reasonable.

SERIES CIRCUITS. In a similar manner, Table 14.2 for a series circuit becomes the "multiplication" table shown as Table 14.4. In this case, we

TABLE 14.4

Case	A	B	$A \cdot B$
1	0	0	0
2	0	1	0
3	1	0	0
4	1	1	1

have no inconsistency with normal decimal arithmetic, nor with binary, since

$$0 \cdot 0 = 0$$
$$0 \cdot 1 = 0$$
$$1 \cdot 0 = 0$$
$$1 \cdot 1 = 1.$$

COMBINATION CIRCUITS. More complex circuits occur when series and parallel connections occur in combination. As an example, consider the circuit in Figure 14.6(a). We know that the parallel portion of this circuit, shown in Figure 14.6(b), can be represented by $B + C$. Taking this one step further, the parallel portion of Figure 14.6(a) may be replaced by $B + C$, as shown in Figure 14.7.

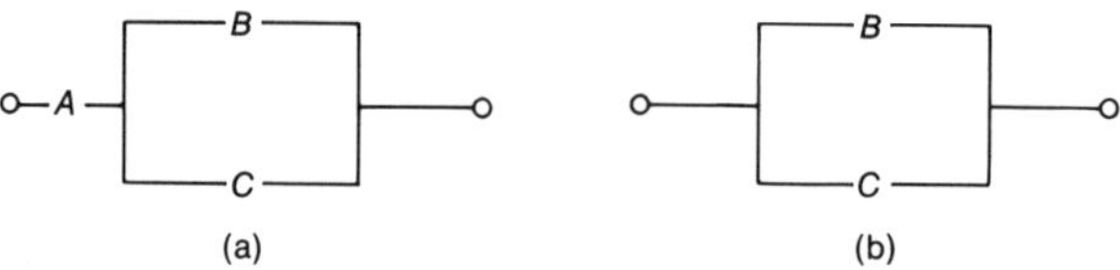

FIGURE 14.6 (a) Series-parallel circuit; (b) parallel portion

o——A——B + C——o

FIGURE 14.7

If we direct our attention to the circuit of Figure 14.7, we observe a series connection of A and $(B + C)$. Since a series circuit is represented by the product of the two components, we conclude that

$$A \cdot (B + C)$$

would represent Figure 14.7 and hence the original circuit of Figure 14.6(a).

In a similar manner, the circuit of Figure 14.8 would be represented by $A \cdot B + C$, since the series connection of A and B is connected in parallel with switch C.

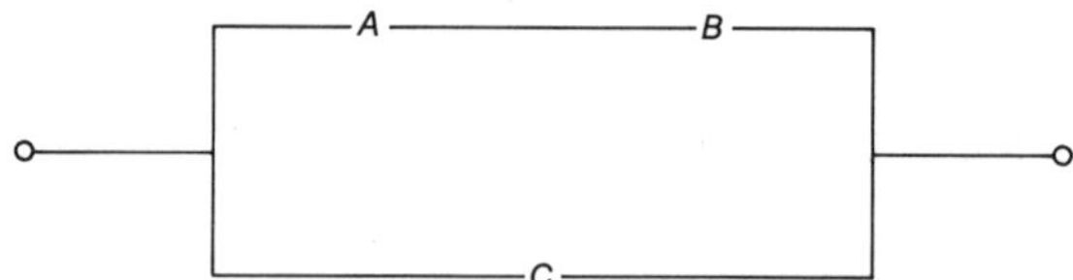

FIGURE 14.8

SWITCHES IN OPPOSITE STATES. In some cases, two separate switches in a circuit are always in opposite states. When the first is open the second is closed and when the first is closed the second is open. This relationship is indicated by representing the two switches as A and A'. Using a numerical representation for the state of the switch, we have

A	A'
0	1
1	0

The symbol A' is read *not A* or *A complement.*

The two switches A and A' may be connected in parallel [Figure 14.9(a)] or in series [Figure 14.9(b)]. In both cases we find an interesting result. In the parallel circuit, current will always flow, since one switch will always be closed. If A is open, A' is closed, and current will flow through A'; when A' is open, A is closed, and current will flow through A. Hence we have

$$A + A' = 1,$$

with the 1 representing the fact that current will always pass through this circuit. Note that, by definition, we can have neither the case $A = 0$, $A' = 0$, nor the case $A = 1$, $A' = 1$.

On the other hand, current will never pass through the series circuit $A \cdot A'$, shown in Figure 14.9(b). As before, the relationship between the two switches, A and A', requires that one always be open, which insures that current cannot flow through this series connection. Since the lack of flow is indicated by 0, we have

$$A \cdot A' = 0.$$

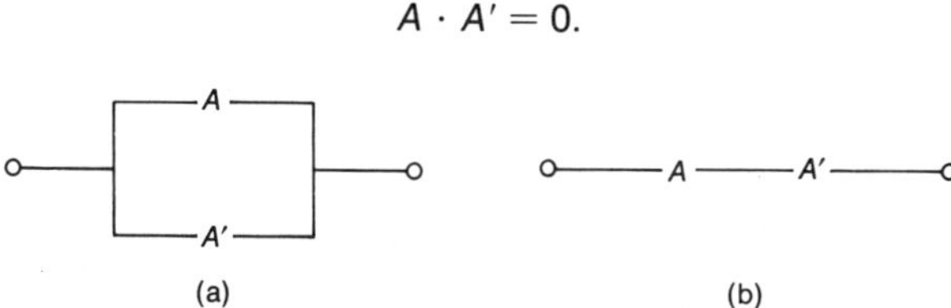

FIGURE 14.9 (a) $A + A'$; (b) $A \cdot A'$

EXERCISE 14.1

Determine if current will flow between the end points of the circuit in Figure 14.6(a) for the given switch settings:

1. $A = 1$, $B = 0$, $C = 0$
2. $A = 1$, $B = 0$, $C = 1$
3. $A = 0$, $B = 1$, $C = 0$
4. $A = 0$, $B = 1$, $C = 1$

Determine if current will flow between the end points of the circuit in Figure 14.10(a) for the given switch settings:

5. $A = 1$, $B = 0$, $C = 1$,
6. $A = 0$, $B = 1$, $C = 0$,
7. $A = 1$, $B = 1$, $C = 0$,
8. $A = 1$, $B = 1$, $C = 1$,

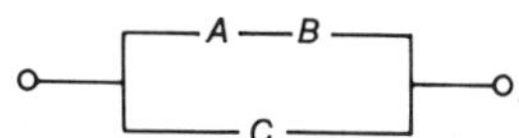

FIGURE 14.10(a)

Using the switch notation of Figure 14.10(b), draw the circuit in Figure 14.6(a) for the given switch settings:

9. $A = 0$, $B = 1$, $C = 1$
10. $A = 1$, $B = 0$, $C = 0$

FIGURE 14.10(b)

11. Draw the circuit represented by $A'B + A$. Construct a table showing current flow for the four possible switch settings.
12. Draw the circuit represented by $A(B + A')$ and construct a table showing current flow.
13. Construct a truth table for (.NOT.*P*.AND.*Q*).OR.*P* and compare it with the table in problem 11.
14. Construct a truth table for *P*.AND.(*Q*.OR..NOT.*P*) and compare it with the table in problem 12.
15. Write equations involving the sets A, A', U and ϕ which are similar to $A + A' = 1$ and $A \cdot A' = 0$.

14.2 ADDITIONAL CIRCUITS

Before more properties are developed, we shall demonstrate how more complex circuits can be represented with the notation that has been introduced. Suppose a circuit has the diagram shown in Figure 14.11. Taking advantage of

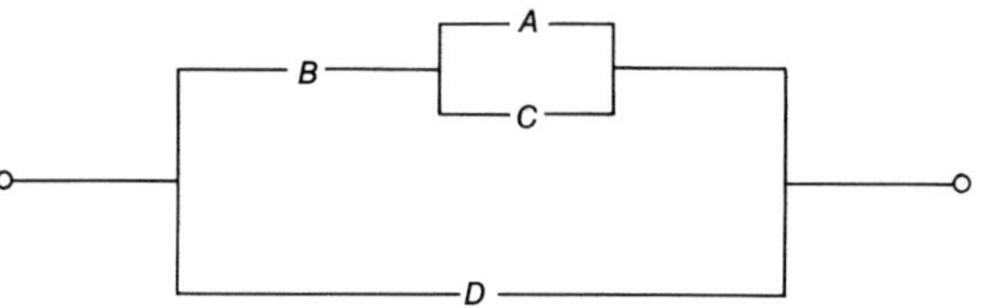

FIGURE 14.11

previous developments, we note that the top portion of this circuit (shown as Figure 14.12) is similar to Figure 14.6(a). Hence this circuit can be represented

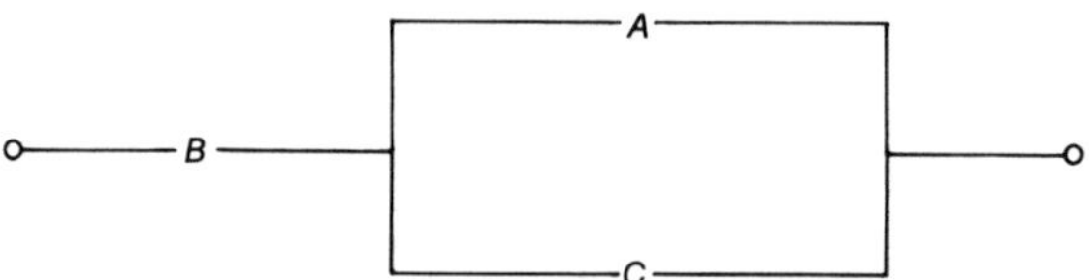

FIGURE 14.12

by $B(A + C)$. Switches A and C are in parallel and B is in series with $(A + C)$. Placing this representation, $B(A + C)$, in the appropriate position in the circuit, we have the result shown in Figure 14.13.

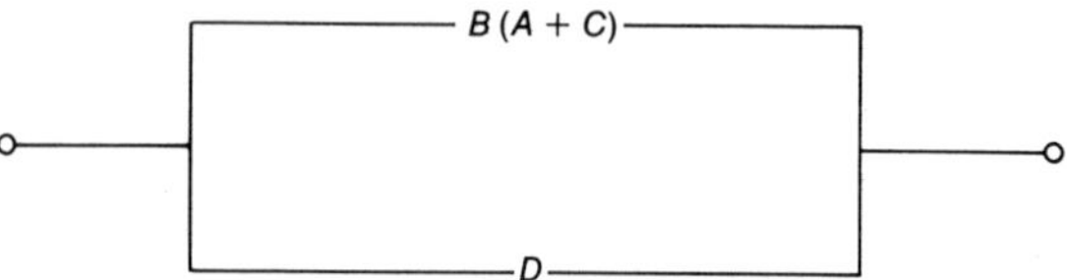

FIGURE 14.13

Recall that $B(A + C)$ will assume a value of 0 or 1, depending on current flow through the particular circuit. Since a variable Z representing the state of a single switch also assumes the values 0 and 1, we could replace $B(A + C)$ with Z. Switch Z, of course, must be controlled to allow current flow only when it would pass through $B(A + C)$. In other words, Z is 0 when $B(A + C)$ is 0, and Z is 1 when $B(A + C)$ is 1.

The use of Z to replace $B(A + C)$ in Figure 14.13 results in the circuit of Figure 14.14. This is represented by $Z + D$, and when Z is replaced by $B(A + C)$, the circuit of Figure 14.11 can be represented by

$$B(A + C) + D.$$

Note that switch Z is introduced only to emphasize the parallel arrangement of Z and D or of $B(A + C)$ and D. If the parallel connection is evident from Figure 14.13, this extra step can be eliminated.

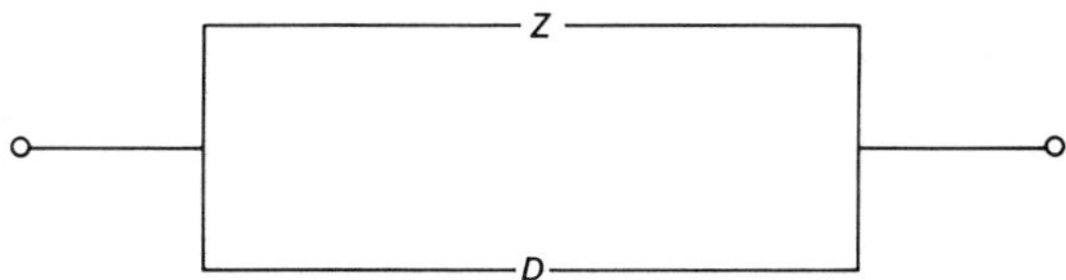

FIGURE 14.14

As another illustration, consider the circuit shown in Figure 14.15. Actually this is the circuit in Figure 14.11 with an additional switch A placed on the left. This results in two separate switches carrying the label A, which means

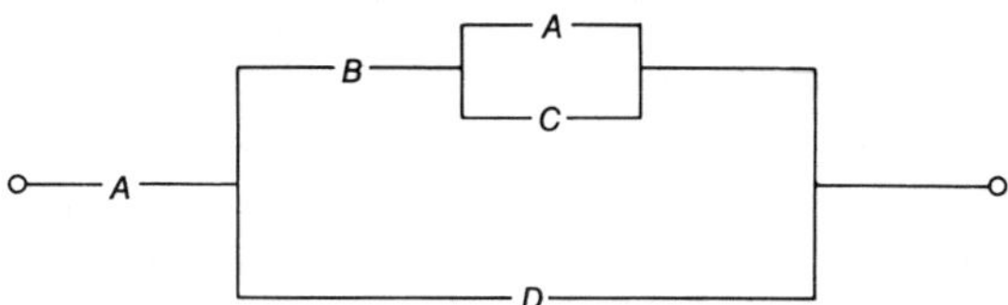

FIGURE 14.15

both are open or closed at the same time. Using the results of the previous example, an equivalent circuit is shown in Figure 14.16. If we imagine $B(A + C) + D$ as a single switch, we note that it is connected in series with A. Hence the "product" notation for a series arrangement results in

$$A[B(A + C) + D]$$

as a representation for the circuit in Figure 14.15.

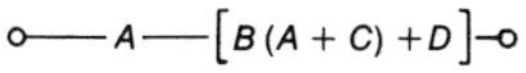

A — [B (A + C) + D]

FIGURE 14.16

A SIMPLER CIRCUIT. If we look at the circuit of Figure 14.15, an interesting result comes to light. If switch A is open, current passage is blocked immediately on the left (Figure 14.17). When switch A is closed, current can follow the upper or the lower path. The lower path can be used when switch D

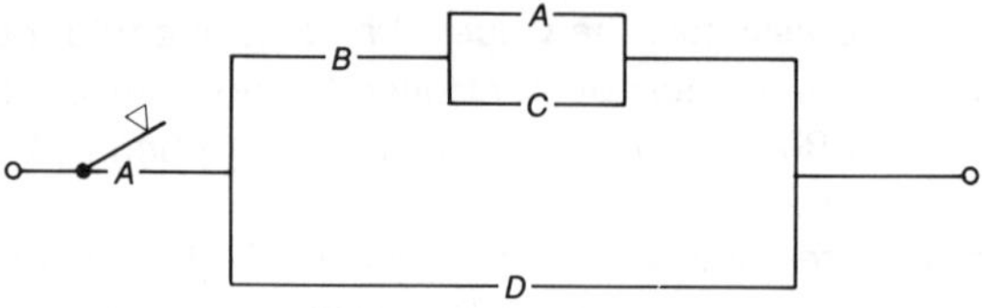

FIGURE 14.17

is closed, and the upper path when *B* is closed (Figure 14.18). When the leftmost switch *A* is closed, the other switch *A* must also be closed. Note that the position of switch *C* does not affect the passage of current along the upper path. Before current can pass through *C*, it must pass through *A* and *B*. However, if it can pass through one *A* switch it can pass through the other. For this reason, switch *C* does not affect the circuit—by itself it cannot prevent or allow current flow. When *A* and *B* are closed, current will flow regardless of the position of switch *C*; if *A* is open, current cannot flow and again switch *C* has no effect.

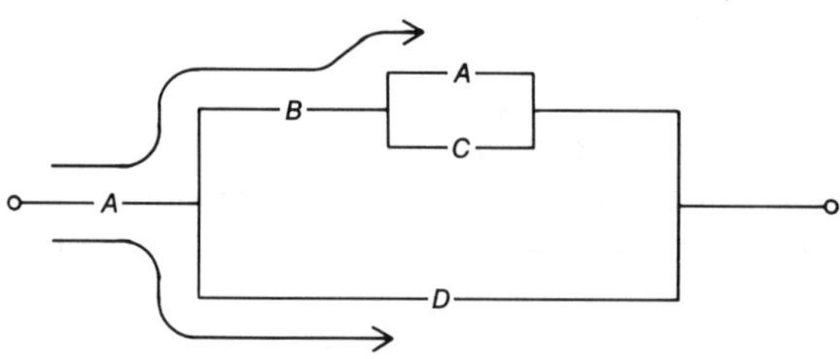

FIGURE 14.18

The above discussion indicates that current will flow through the simpler circuit shown in Figure 14.19 for the same switch settings as in the original circuit. For both circuits, current does not flow when *A* is open, but it will if *A* and *B* or *A* and *D* are closed.

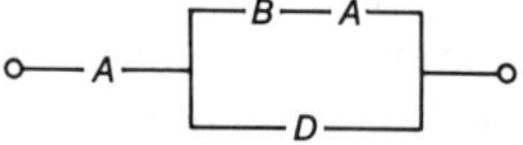

FIGURE 14.19

The representation for the new circuit can be found by considering, in sequence, the circuits shown in Figure 14.20. Note that *B* and *A* are in series, and *BA* is parallel with *D*. The sum $BA + D$, representing the parallel arrangement of *BA* and *D*, is inserted in the circuit in Figure 14.20(c). From this it is evident that *A* is in series with $BA + D$. Hence the representation is $A(BA + D)$.

B—A (a)

BA / D (b)

A—(BA + D) (c)

FIGURE 14.20 (a) BA; (b) $BA + D$; (c) $A(BA + D)$

THE VALUE OF SIMPLER CIRCUITS. The significance of finding a simpler circuit relates to the fact that one less switch is used. If these circuits were being mass produced, the elimination of one switch could result in a substantial savings. For instance, the removal of a 90-cent switch in 100,000 circuits would save $90,000. Hence, a systematic technique for simplifying a circuit is of great value.

In the last example the simplification was accomplished by studying the circuit diagram and determining when current would flow. This involved consideration of a number of different paths and switch settings. It is evident that if there is a large number of switches, the task of finding a simpler circuit would be very complex.

A better way of finding a simpler circuit uses the algebraic representation of the circuit. This representation can be simplified by using certain basic rules, which shall be developed later. The simplification is similar to the simplification of an ordinary algebraic expression. Recall that

$$\frac{ax + a}{x + 1}$$

can be written

$$\frac{a(x + 1)}{x + 1}$$

and then simplified to a by using the distributive property and division. Similar distributive properties hold in simplifying circuits. However, not all of the basic rules are the same, so a different name is used for this new algebra.

The term *Boolean algebra* distinguishes this new algebra from ordinary algebra, and circuit representations such as $AB + C$ are called *Boolean expressions* or *Boolean representations.* The rest of this chapter will be devoted to developing the rules of Boolean algebra and demonstrating how these can be used to simplify circuits.

EXERCISE 14.2

1. Find the Boolean representation for each of the circuits in Figure 14.21.
2. If $A = 1$, $B = 0$, $C = 1$, and $D = 1$, determine through which of the circuits in Figure 14.21 current will flow.
3. Repeat problem 2 for $A = 0$, $B = 1$, $C = 1$, and $D = 1$.
4. Draw the circuit for each of the following Boolean expressions:

(a) $AB(C + D)$	(e) $A[C(B + D) + A]$
(b) $A(C + D)B$	(f) $ABC + ABC'$
(c) $(A + B)(C + D)$	(g) $(A + B)\,C\,(A + D)$
(d) $AB + A'$	(h) $AB(C + D) + A'D$

5. In Figure 14.21(f), there are three possible paths through the circuit. One path goes through switches *A, B,* and *C,* which could be written *ABC.* Find the other two paths and their representation.
6. Find the number of paths and a representation for each path in the circuits of Figure 14.21(g), (h), (i), and (j).

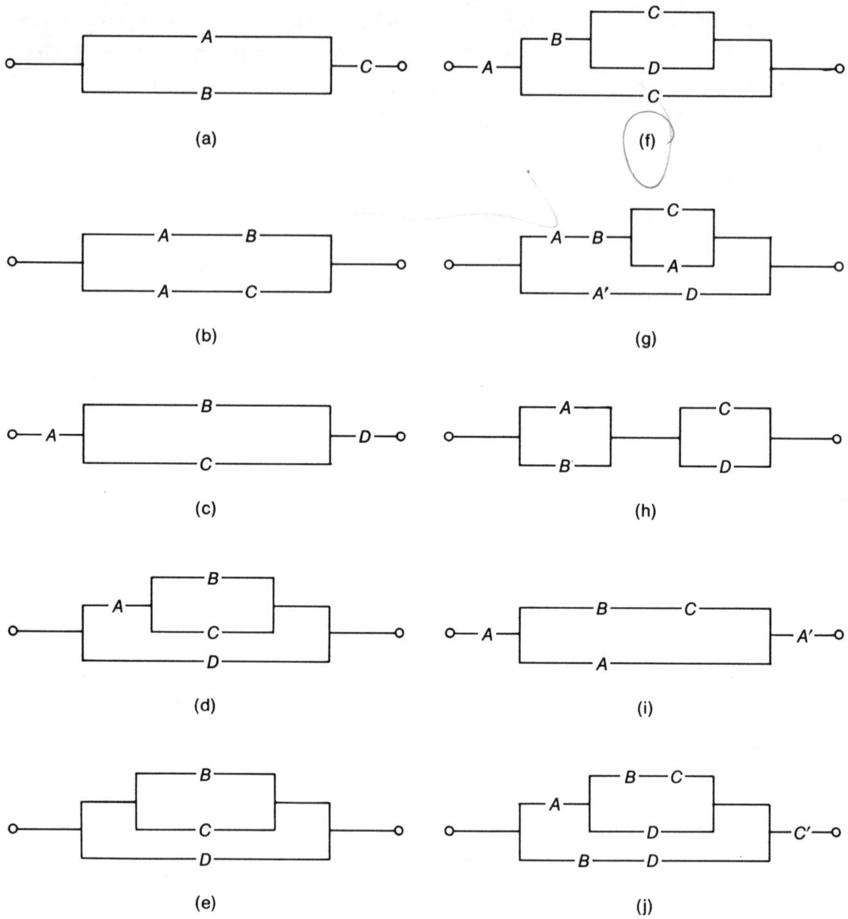

(a) (f)

(b) (g)

(c) (h)

(d) (i)

(e) (j)

FIGURE 14.21

7. A series connection of two switches, the intersection of two sets and an .AND. statement are similar. P.AND.Q is true only when both P and Q are true; X is in $C \cap D$ only when X is in both C and D; current flows through AB only when both A and B are closed. $A + B$, $C \cup D$, and P.OR.Q are also similar. Write the logical form and the set representation which is similar to each of the following:
(a) $A(B + C)$ (b) $A + BC$ (c) $AB + BC$

8. (a) Find a set representation for the shaded area

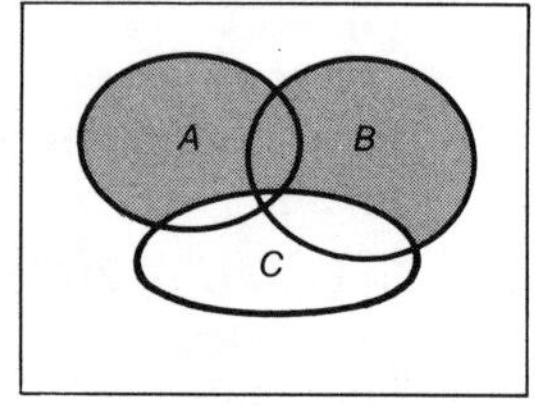

(b) Indicate the similar circuit representation (of part a) and draw the circuit.

(c) If 010 indicates X is in B but not in A or C, is X in the shaded area above?

(d) In part (c), 010 could indicate switch B is closed but A and C are open. Would current flow in this case?

9. (a) In the Venn diagram, 100 indicates "in A but not in B and not in C," while 011 indicates "not in A but in B and in C."

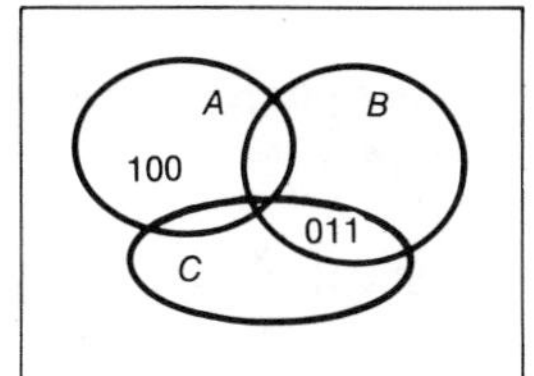

Fill in the six remaining areas with binary numbers from 000 to 111.

(b) If 100 indicates switch A is closed but B and C are open, current flows through the circuit represented by $A + BC$. For which other binary numbers would current flow in $A + BC$?

(c) For which binary numbers in the Venn diagram would current flow in $AB + AC$.

14.3 THE FORM $AB + A$ AND BOOLEAN PROPERTIES

As an illustration of simplification procedures, consider the circuit in Figure 14.22. Its Boolean representation is

$$AB + A.$$

FIGURE 14.22

To emphasize the fact that we are only concerned with the conditions under which current flows between the end points of the circuit, we connect an ordinary light bulb as shown in Figure 14.23. We wish to determine the switch settings that will allow current to pass through the circuit and light the bulb.

For this circuit, when switch A is closed, current can follow the lower path and the bulb will light. Leaving switch A closed and also closing switch B will produce no change, since the bulb was already illuminated. However, if we open switch A and close switch B, the bulb will not light. This indicates that switch B has no effect in the circuit. With switch A closed, switch B can be flipped back and forth with no change in the illumination of the bulb. Also, when switch A is open, flipping switch B will not illuminate the bulb. Hence switch B by itself has no effect so far as allowing or preventing current flow.

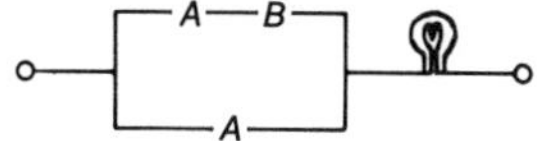

FIGURE 14.23

The circuit in Figure 14.24 will allow current to pass (or light the bulb) for the same switch settings as $AB + A$. Its representation, of course, is just A. The

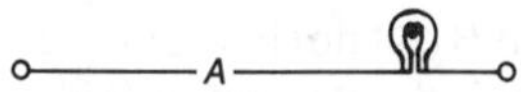

FIGURE 14.24

various switch settings and resulting current flow in both circuits are shown in Table 14.5

TABLE 14.5

Case	A	B	$AB + A$	A
1	0	0	0	0
2	0	1	0	0
3	1	0	1	1
4	1	1	1	1

In cases 1 and 2, switch A is open so that current cannot pass through $AB + A$ (Figure 14.23) or through A (Figure 14.24). This is indicated by the 0 entries in the last two columns. Switch A is closed for cases 3 and 4, and current will flow (indicated by entries of 1) in both circuits. The identical 0 and 1 entries for the two columns headed by $AB + A$ and A indicate that current will flow (or the light will illuminate) for the same switch positions. We then say the two circuits are **equivalent**, and indicate this by writing

$$AB + A = A.$$

We have now shown that the circuit represented by

$$AB + A$$

can be simplified to a circuit represented by

$$A.$$

However, we did this by looking at the circuits involved instead of their Boolean representations. We shall now demonstrate three basic properties of Boolean algebra that will allow us to carry out the simplification procedure in an algebraic manner.

BOOLEAN PROPERTIES. First, consider the Boolean expression

$$A(B + 1).$$

This expression is a representation for the circuit shown in Figure 14.25, where the 1 indicates that this particular switch is always *closed.* Because the lower switch is always closed, we need only close switch A to get current flow through the circuit. Switch B has no effect in the circuit. We shall show that this circuit is equivalent to $AB + A$.

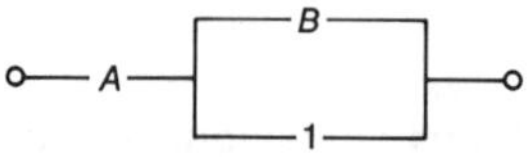

FIGURE 14.25

To show the equivalence, we indicate in table form (Table 14.6) current flow for the possible switch settings.

TABLE 14.6

Case	A	B	$A(B+1)$	$AB+A$
1	0	0	0	0
2	0	1	0	0
3	1	0	1	1
4	1	1	1	1

Current does not flow when switch A is open (cases 1 and 2), but it does flow when switch A is closed (cases 3 and 4). Therefore, we can write

$$A(B+1) = AB + A.$$

DISTRIBUTIVE PROPERTY. We are actually demonstrating that the ordinary distributive property holds in Boolean algebra. If we thought of $AB + A$ as an ordinary algebraic expression, we could use the distributive property to write

$$AB + A = A(B+1)$$

since A is a common factor. Using the same distributive property, we could also write

$$A(B+1) = AB + A$$

indicating that we are carrying out the multiplication $A(B+1)$. In either case, the equivalence shown in Table 14.6 indicates that the distributive property holds for this particular situation.

CLOSED SWITCHES. To complete the simplification, we consider the Boolean expressions

$$B+1 \qquad \text{and} \qquad A \cdot 1.$$

These represent the circuits shown in Figure 14.26. In Figure 14.26(a), the lower switch is always closed, as indicated by the entry 1; hence, current can always flow through the circuit. Since current flow is indicated by 1, we can write

$$B+1 = 1.$$

The circuit $B+1$ is equivalent to a closed switch.

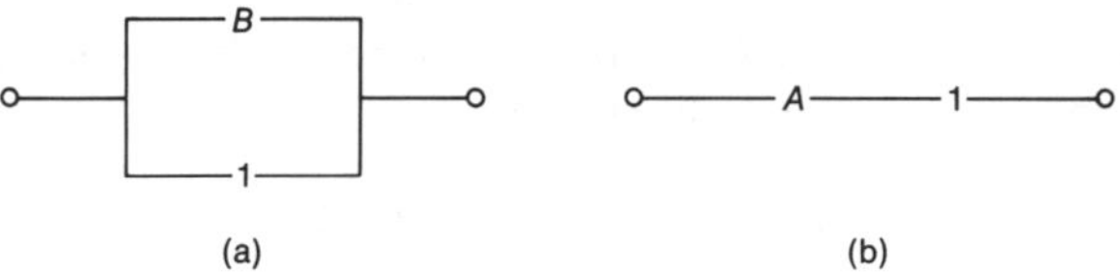

FIGURE 14.26 (a) $B+1$; (b) $A \cdot 1$

In Figure 14.26(b), A is connected in series with a switch that is always closed. The passage of current will depend entirely on the position of switch A, so the closed switch may as well not be in the circuit. This indicates that $A \cdot 1$ and A are equivalent, or

$$A \cdot 1 = A.$$

These two results are significant in Boolean algebra, even though they may seem relatively simple. The effect of a switch that is always closed is completely different in the two circuits. Connected in parallel, the closed switch cancels out the effect of the other switch; in series, the closed switch has no effect. The result

$$A \cdot 1 = A$$

is compatible with ordinary algebra, where the same algebraic equation can be written. However,

$$B + 1 = 1$$

is true only in Boolean algebra, where it is quite useful in simplifying Boolean expressions.

SIMPLIFICATION OF AB + A. Putting the foregoing results together allows us to simplify

$$A \cdot B + A.$$

Using the distributive property, this can be written

$$A \cdot B + A = A(B + 1)$$

as was shown in Table 14.6.

In turn $B + 1$ can be replaced by 1, or

$$A(B + 1) = A \cdot 1.$$

Finally,

$$A \cdot 1 = A.$$

Hence we conclude that

$$A \cdot B + A = A.$$

This type of simplification can be performed quite efficiently, after we become familiar with the basic properties of Boolean algebra. We shall show that many properties of ordinary algebra hold in Boolean algebra, and some others, such as

$$B + 1 = 1$$

are completely new.

DISTRIBUTIVE PROPERTIES. In the last section we demonstrated that the distributive property held for a particular case. The general distributive property as it occurs in ordinary algebra is

$$A \cdot (B + C) = A \cdot B + A \cdot C.$$

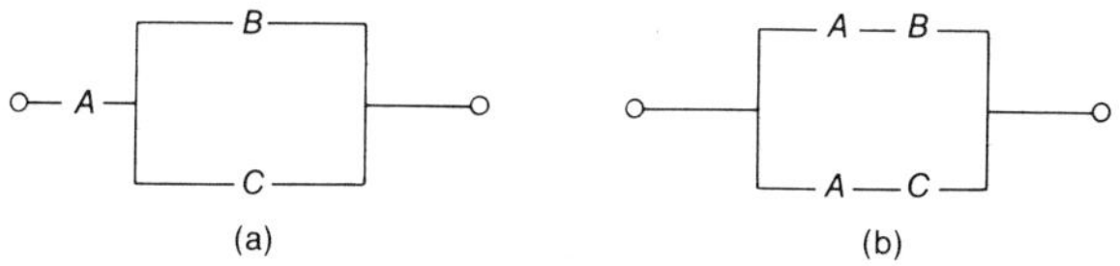

FIGURE 14.27 (a) $A(B+C)$; (b) $AB+AC$;

Consider each side as a Boolean expression. The circuits that they represent are shown in Figure 14.27.

The equivalence of these two circuits requires that current flow for identical switch settings in the two circuits. Since there are three switches in each circuit, there are eight possible combinations that must be checked. The most efficient way to do this is by indicating in table form the various combinations. Before doing this, however, we shall show that it is reasonable to expect the two circuits to be equivalent.

In Figure 14.27(a), when switch A is open, current is immediately blocked. The same situation holds in Figure 14.27(b), since both A switches will be open at the same time. When switches A and B or A and C are closed, current flows in both circuits. The single switch A in $A(B+C)$ has the same effect as the two A switches in $AB+AC$.

The eight different ON-OFF combinations for the three switches are indicated in Table 14.7. In each of the first four cases, switch A is open. From Figure 14.27, we note that this will prevent current flow.

For case 5, switch A is closed, but B and C are both open, which also prevents current flow. This accounts for the five 0 entries for $A(B+C)$ and $AB+AC$. Switch A is also closed in the last three cases, and since B or C is also closed, current can flow through both circuits. The **identical columns** for $A(B+C)$ and $AB+AC$ indicates their **equivalence**, so we can write

$$A(B+C) = AB + AC.$$

TABLE 14.7

Case	A	B	C	$A(B+C)$	$AB+AC$
1	0	0	0	0	0
2	0	0	1	0	0
3	0	1	0	0	0
4	0	1	1	0	0
5	1	0	0	0	0
6	1	0	1	1	1
7	1	1	0	1	1
8	1	1	1	1	1

A SECOND DISTRIBUTIVE PROPERTY. Some Boolean properties are similar to ordinary algebraic properties; other are quite different. One of the most marked differences involves a second distributive property. The equivalence is

$$A + BC = (A + B)(A + C).$$

This, of course, is simply not true if A, B, and C represent numbers. For instance, $2 + 3 \cdot 4 = 14 \neq (2 + 3)(2 + 4) = 5 \times 6 = 30$. However, we will show that the two circuits (Figure 14.28) represented by $A + BC$ and $(A + B)(A + C)$ will allow current to pass, for identical switch settings.

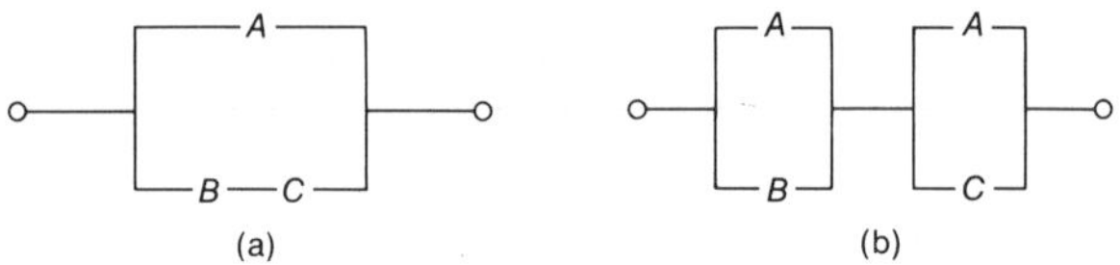

FIGURE 14.28 (a) $A + BC$; (b) $(A + B)(A + C)$

A table of the eight possible switch settings for A, B, and C would indicate current flow in both circuits when switch A is closed. In the other four cases, when A is open, a check of Figure 14.28(a) reveals that switches B and C must be closed for current to flow through $A + BC$. When B and C are closed, current will also flow in $(A + B)(A + C)$, as can be seen in Figure 14.28(b). Furthermore, when switch A is open, current will flow in $(A + B)(A + C)$ only when B and C are closed. Hence, the two circuits are equivalent, and

$$A + BC = (A + B)(A + C).$$

In showing an equivalence, there is always the danger of forgetting one or more combinations of switch settings. This type of omission can be eliminated by constructing a table that lists all possible combinations. A systematic check of each case will insure a correct conclusion. The construction of a table showing this equivalence is left as an exercise (problem 1, Exercise 14.3).

This distributive property is more difficult to use in simplifying Boolean expressions because it is not familiar from ordinary algebra. In fact, a careful check is required just to see why it is called a distributive property. To justify the name, we write the first distributive property in the form

$$A \cdot (B + C) = A \cdot B + A \cdot C.$$

The multiplication symbol is inserted with A to emphasize that A and the multiplication symbol always occur together. In $A \cdot (B + C)$, the $A \cdot$ is applied first to B and then to C. Writing the second distributive property in a similar manner, we have

$$A + BC = (A + B)(A + C).$$

Note that A and the addition symbol also always occur together, and that $A +$ is first applied to B and then to C. In this sense, the two properties are similar, with the multiplication and addition symbols reversed.

EXERCISE 14.3

1. Using Figure 14.28, construct a table showing the equivalence of $A + BC$ and $(A + B)(A + C)$.
2. Draw a circuit, and from the circuit construct a table for each of the following Boolean expressions:

   (a) $BC + 1$ (c) $A + B + 1$
   (b) $A + B \cdot 1$ (d) $(A + B) \cdot 1$
3. Each expression in problem 2 can be simplified by using the properties $B + 1 = 1$ or $A \cdot 1 = A$. Simplify each expression and determine which expressions are equivalent. Do the tables in problem 2 indicate these equivalences?
4. Draw the circuit, and from the circuit construct a table for each of the following Boolean expressions:

   (a) $B + BC$ (c) $A + AB + AC$
   (b) $AB + AB'$ (d) $A'A + A'$
5. The first distributive property $AB + AC = A(B + C)$ can be applied to each Boolean expression in problem 4. Write the resulting expression for each part. Recalling that $A + A' = 1$ and $A \cdot A' = 0$ (Figure 14.9) and that $B + 1 = 1$ and $A \cdot 1 = A$, simplify further each of the results. Are the final results consistent with the tables in problem 4?

Use one of the distributive properties to write a different form for each of the following:

6. $B(A + C)$
7. $B + AC$
8. $1 + AB$
9. $1(A + C)$
10. $CD + CB$
11. $(C + A)(C + D)$
12. $(A' + B)(A' + A)$
13. $A'(A + A')$
14. Write set equations similar to $B + 1 = 1$ and $B \cdot 1 = B$.
15. (a) Write a set expression similar to $A + AB$ (check problem 7, exercise 14.2).
    (b) In a Venn diagram, shade the set form of $A + AB$.
    (c) From the Venn diagram in (b), find a simpler form for $A + AB$.
16. Repeat problem 15 for $AB + AB'$.
17. Repeat problem 15 for $A + AB + AC$.

14.4 OTHER BOOLEAN PROPERTIES

Other properties from basic algebra that also hold in Boolean algebra include the commutative and associative properties. In addition, we will find equivalent circuits which demonstrate that certain switches act as identity elements.

COMMUTATIVE PROPERTY. The sum and product notations for parallel and series connections lead to the possibility of changing the order of the elements. For real numbers, the commutative property gives familiar results, such as

$$3 + 4 = 4 + 3$$

and

$$5 \times 7 = 7 \times 5.$$

For Boolean algebra, we check

$$A + B = B + A$$

and

$$AB = BA.$$

The circuits involved in these two equivalences are shown in Figure 14.29.

Although the positions of the switches have been reversed for $A + B$ and AB in Figure 14.29, it is evident that current flow is unaltered for the two pairs. Hence the two commutative properties are valid.

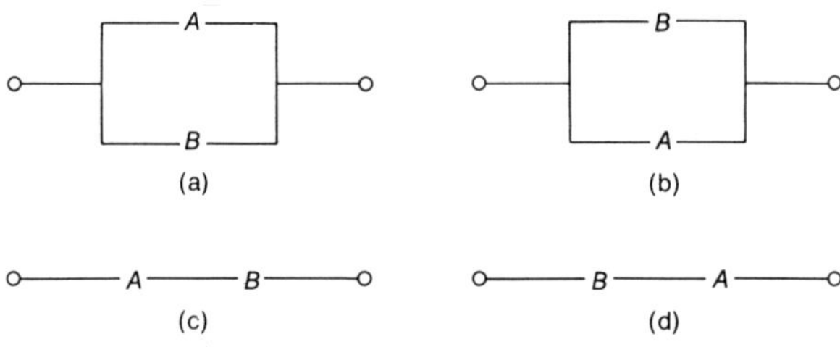

FIGURE 14.29 (a) $A + B$; (b) $B + A$; (c) AB; (d) BA

ASSOCIATIVE PROPERTY. For real numbers, we know that different groupings for addition and multiplication will give the same result. For instance,

$$(2 + 3) + 4 = 2 + (3 + 4)$$

since

$$5 + 4 = 2 + 7$$

or

$$9 = 9.$$

Also

$$(2 \times 3) \times 4 = 2 \times (3 \times 4)$$

because

$$6 \times 4 = 2 \times 12$$

or

$$24 = 24.$$

The associative property indicates that two groupings are possible when the same operation is to be performed twice.

The circuits that must be considered in checking associative properties are shown in Figure 14.30. In 14.30(a) and (b), current will be blocked **only** when all three switches are open. For the other seven cases, at least one switch is closed, which will allow current to flow through either circuit. Thus we have

$$(A + B) + C = A + (B + C).$$

(a) (b)

(c) (d)

FIGURE 14.30 (a) $(A+B)+C$; (b) $A+(B+C)$; (c) $(AB)C$; (d) $A(BC)$

For $(AB)C$ and $A(BC)$, the situation is reversed. Figure 14.30(c) and (d) indicates that current is blocked for seven of the eight cases and flows **only** when all of the switches are closed. This equivalence confirms the associative property

$$(AB)C = A(BC).$$

IDENTITY ELEMENTS. In the algebra of real numbers, 0 and 1 were the identity elements for addition and multiplication, respectively. The equations

$$a + 0 = a \qquad \text{and} \qquad a \times 1 = a$$

demonstrated the properties of each identity. Starting with any real number a, the operation (addition or multiplication) can be performed with the identity element, and the result is the same real number a.

We have already demonstrated that

$$A \cdot 1 = A$$

by considering the circuit of Figure 14.26(b). Since the 1 entry indicates that the second switch is always closed, we have a switch that acts as an identity element. The effect of switch A is not altered by a closed switch connected in series.

The corresponding equivalence for a parallel arrangement would be

$$A + 0 = A.$$

The 0 entry in $A + 0$ (Figure 14.31) is interpreted as a representation for a switch that is always open. With the bottom switch always open, current flow depends only on switch A so that $A + 0$ is equivalent to A.

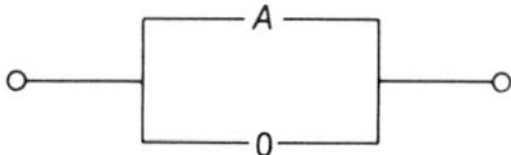

FIGURE 14.31

OPEN OR CLOSED SWITCHES. It may seem strange that a switch is always closed, since an unbroken wire would serve the same purpose. In a similar manner a broken wire would act as an open switch. However, earlier we have shown that

$$A + A' = 1$$

since the parallel arrangement, $A + A'$ always has one closed switch. In addition the equivalence

$$A \cdot A' = 0$$

results from the series arrangement of the same two switches. In simplifying a Boolean expression, either of these combinations may occur and be replaced by 1 or 0. Hence, the entry 1 in $A \cdot 1$ or $A + 1$ may represent some **combination** of switches which always allows current to flow. The purpose of the simplification process is to detect such combinations so that they can be replaced with less expensive components.

EXERCISE 14.4

Indicate which basic property justifies each of the following equivalences:

1. $A + (B + A') = A + (A' + B)$
2. $A + (A' + B) = (A + A') + B$
3. $(A + A') + B = 1 + B$
4. $1 + B = 1$
5. $A'(BA) = A'(AB)$
6. $A'(AB) = (A'A)B$
7. $(A'A)B = 0 \cdot B$
8. $0 \cdot B = 0$
9. (a) Draw the circuits for each side in problems 3 and 6.
   (b) Construct tables that verify the equivalences in problems 3 and 6.
10. Draw circuits for $(B + C)A$ and $A(B + C)$. Does the commutative property justify their equivalence?

Find the simplest circuit that is equivalent to each of the following expressions, and then construct appropriate tables for each:

11. $A + B + A'$
12. $A(A' + A'B)$
13. $C + BC + ABC$
14. Construct a truth table for *P*.AND.*Q*.AND..NOT.*P* and find a simpler logical form. Compare with problem 6.
15. Find set equations similar to $A \cdot 1 = A$ and $A + 0 = A$.
16. Construct a truth table for *P*.OR.(*Q*.AND.*P*).OR.(*R*.AND.*Q*.AND.*P*) and find a simpler logical form. Compare with problem 13.

14.5 SIMPLIFICATIONS

BASIC PROPERTIES. The basic properties or rules that have been developed for Boolean algebra are collected in Figure 14.32 as a convenient reference. To emphasize that each equivalence is a claim that two circuits behave in the same manner, the circuits are also drawn.

It can be shown (Exercise 14.5, problem 17) that the Boolean properties in Figure 14.33 also hold.

In using these basic properties, it is important to distinguish between similarities and differences with ordinary algebra. Properties 1-5, 9, 10, and 12 are the same as encountered in basic algebra; Properties 6, 7, 8, 11, 13, and 14 are different. Concentration on those properties that are different will aid in using them at the appropriate time.

1. $A + B = B + A$
2. $AB = BA$
3. $(A + B) + C = A + (B + C)$
4. $(AB)C = A(BC)$
5. $A(B + C) = AB + AC$
6. $A + BC = (A + B)(A + C)$
7. $A + A' = 1$
8. $A \cdot A' = 0$
9. $A \cdot 1 = A$
10. $A + 0 = A$
11. $A + 1 = 1$
12. $A \cdot 0 = 0$

FIGURE 14.32

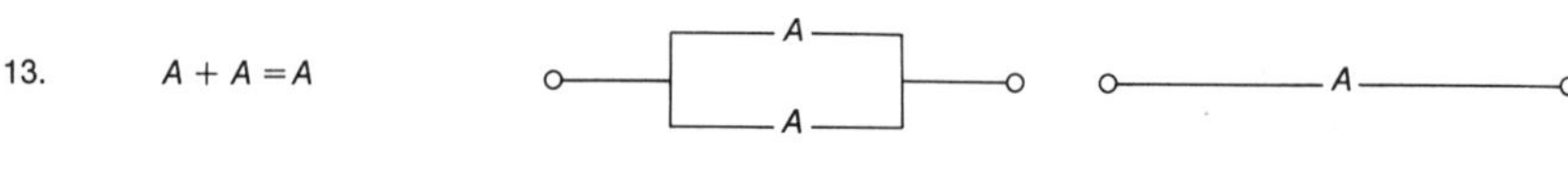

FIGURE 14.33

Several examples that demonstrate the use of these basic properties follow. Consider the circuit in Figure 14.34. Its representation is $AB + AB'$ and we simplify the circuit by working with the representation only.

$$\begin{aligned} AB + AB' &= A(B + B') && \text{(Property 5)} \\ &= A \cdot 1 && \text{(Property 7)} \\ &= A. && \text{(Property 9)} \end{aligned}$$

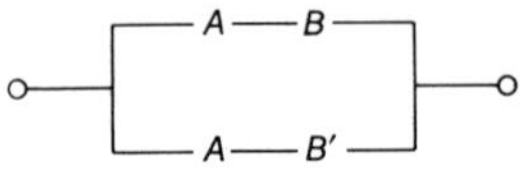

FIGURE 14.34

Hence the circuit in Figure 14.34 can be replaced by the single switch A. The reader is encouraged to check the original circuit to see that current does flow in the two circuits for the same switch settings.

A circuit that involves some of the less frequently used Boolean properties

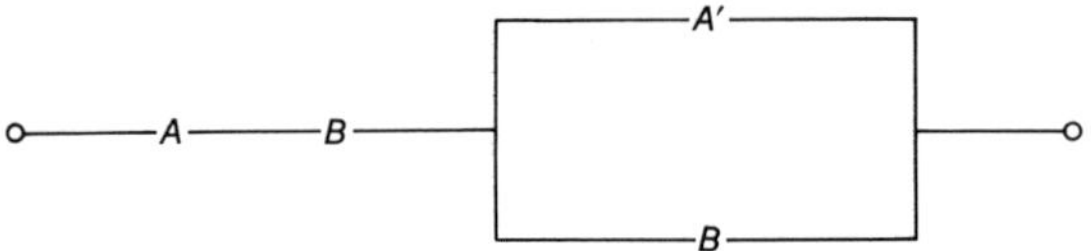

FIGURE 14.35

is shown in Figure 14.35. Starting with its representation, $(AB)(A' + B)$, we have

$$\begin{aligned} (AB)(A' + B) &= (AB)A' + (AB)B && \text{(Property 5)} \\ &= (BA)A' + (AB)B && \text{(Property 2)} \\ &= B(AA') + A(BB) && \text{(Property 4)} \\ &= B \cdot 0 + A \cdot B && \text{(Properties 8 and 14)} \\ &= 0 + AB && \text{(Property 12)} \\ &= AB. && \text{(Property 10)} \end{aligned}$$

Another example, $A + A'B$, shown in Figure 14.36, requires the second distributive property; that is,

$$\begin{aligned} A + A'B &= (A + A')(A + B) && \text{(Property 6)} \\ &= 1(A + B) && \text{(Property 7)} \\ &= A + B && \text{(Property 9)} \end{aligned}$$

Switch A' can be removed to produce an equivalent circuit.

As a final example, consider

$$ABC + AB'C' + A'B'C'.$$

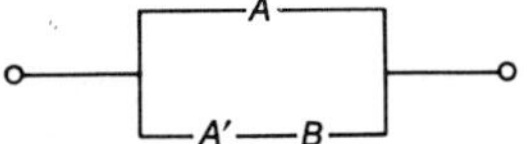

FIGURE 14.36

If we first consider the common factor A in the first two terms, we have

$$ABC + AB'C' + A'B'C' = A(BC + B'C') + A'B'C'.$$

However, no further simplifications can be performed with this grouping. Looking again at the original expression, the last two terms have $B'C'$ as a common factor. For this grouping, we are able to find a simpler Boolean expression.

$$\begin{aligned} ABC + AB'C' + A'B'C' &= ABC + (A + A')B'C' && \text{(Property 5)} \\ &= ABC + 1 \cdot B'C' && \text{(Property 7)} \\ &= ABC + B'C'. && \text{(Property 9)} \end{aligned}$$

The original circuit requires nine switches; the simplified form requires only five switches. Hence the simplification process reduces the number of switches by almost one half.

EXERCISE 14.5

Simplify the following Boolean expressions by using the basic rules in Figures 14.32 and 14.33:

1. $A(BA)$
2. $(B + A)B$
3. $A(B + A') + B$
4. $(C + D)(C + D')$
5. $A \cdot C + A'$
6. $BC + BC'$
7. $AB + A' + B$
8. $A + A'B + AC$
9. $(A + A'B + A)(A'B')$
10. $(AB + A'B')(A + B)$
11. $(A + B)(AB)$
12. $(A + B)(A'B')$
13. $(A + B) + A'B'$
14. $AB + (A' + B')$
15. $(A + A')(A + 0) + (A'B)A'$
16. $(A + AB)(B + CD)(A' + 1)$
17. Verify Properties 13 and 14 (Figure 14.33) by constructing the appropriate tables.
18. Draw the circuit for the simplified expressions and the original expressions in problems 8, 9, and 13.
19. Construct a table for the simplified circuit and the original circuit in problems 8, 9, and 13.
20. Find and simplify the Boolean expression representing each of the circuits in Figure 14.37.

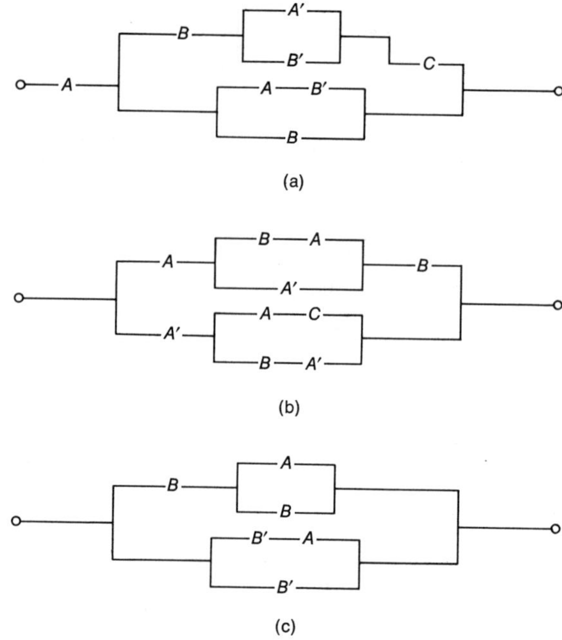

FIGURE 14.37

21. Simplify (P.AND.Q).OR..NOT.P by using the result of problem 5.
22. Simplify $(B \cap C) \cup (B \cap C')$ by using the result of problem 6.
23. Simplify $(A \cap B) \cup A' \cup B$ by using the result of problem 7.
24. Use Boolean properties to simplify each flowchart.

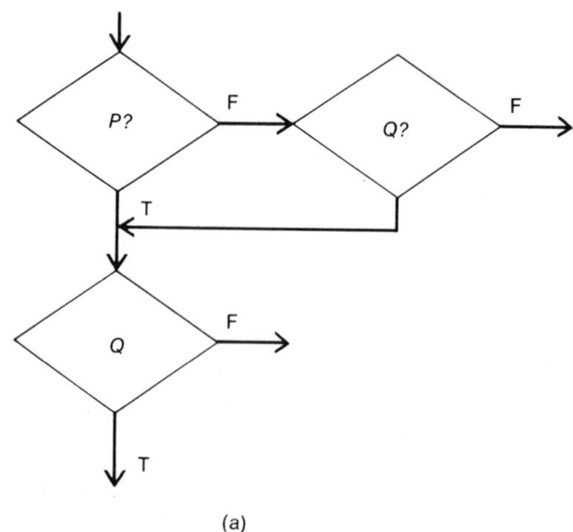

(a)

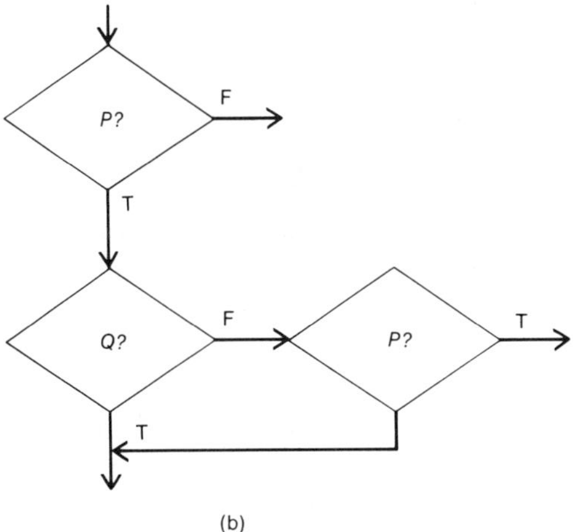

(b)

14.6 AN APPLICATION OF BOOLEAN ALGEBRA

We are now in a position to demonstrate the use of the binary number system in a basic arithmetic operation such as addition. Recall that the four basic addition combinations in base 2 are

$$\begin{array}{r} 00 \\ 0 \\ \hline 00 \end{array} \quad \begin{array}{r} 0 \\ 1 \\ \hline 01 \end{array} \quad \begin{array}{r} 1 \\ 0 \\ \hline 01 \end{array} \quad \begin{array}{r} 1 \\ 1 \\ \hline 10 \end{array}$$

These sums can be conveniently listed in table form (Table 14.8) by using the following designations:

$$\begin{array}{c} A \quad B \quad CS \\ 1 + 1 = 10 \end{array}$$

In other words, S represents the entry in the units column, and C the carry.

TABLE 14.8

A	B	C	S
0	0	0	0
0	1	0	1
1	0	0	1
1	1	1	0

A circuit that will indicate the correct sum for any combination can be constructed if its Boolean representation is first determined. This can be accomplished by assuming that A and B represent switches. The S column requires a circuit that allows current flow for the switch settings shown on lines 2 and 3. If we concentrate on line 2, observe that

$$A'B$$

representing

$$\circ\!\!-\!\!-\!\!-A'\!-\!\!-\!\!-B\!-\!\!-\!\!-\!\!\circ$$

will allow current to flow only when $A = 0$ and $B = 1$. Moreover, for line 3,

$$AB'$$

allows current to flow only when $A = 1$ and $B = 0$. Since we need current flow in both situations, $A'B$ and AB' are connected in parallel. Hence

$$A'B + AB'$$

is the Boolean representation of the desired circuit.

In a similar manner, the carry column (C) requires the single series connection

$$AB.$$

Current flow for each circuit is shown in Table 14.9.

TABLE 14.9

A	B	AB	$A'B + AB'$
0	0	0	0
0	1	0	1
1	0	0	1
1	1	1	0

In drawing these circuits, it is less confusing if the symbolic representations in Figure 14.38 are used. If A and B are inputs and K the output, the following holds:

1. In Figure 14.38(a), $K = 1$ when $A = 1$ AND $B = 1$, otherwise $K = 0$.
2. In Figure 14.38(b), $K = 1$ when $A = 1$ OR $B = 1$, otherwise $K = 0$.
3. In Figure 14.38(c), $K = 1$ when $A = 0$ AND $K = 0$ when $A = 1$.

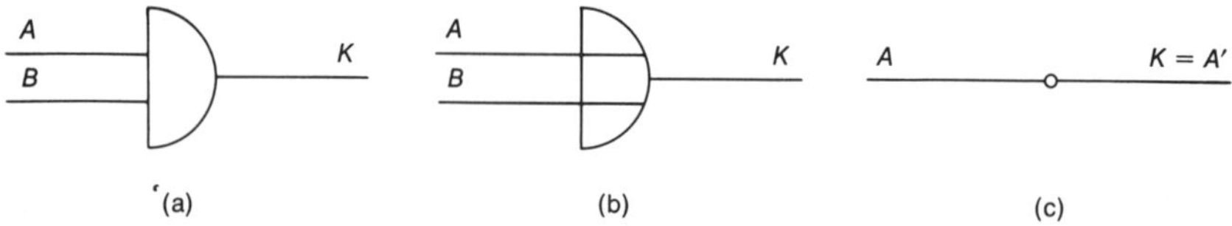

FIGURE 14.38 (a) AND gate; (b) OR gate; (c) NOT

The AND gate represents a series connection, the OR gate a parallel connection, and the NOT diagram relates to the complement of a switch.

The symbolic representations of $A'B$ and AB' are as shown in Figure 14.39. When $A = 0$ and $B = 1$, $A'B = 1$ [Figure 14.39(a)]; for $A = 1$ AND $B = 0$,

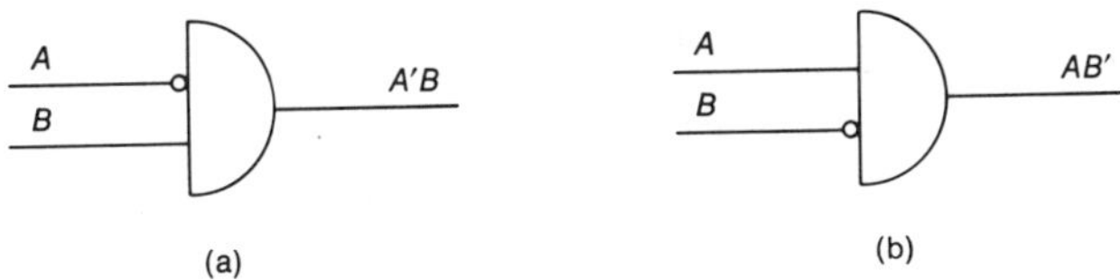

FIGURE 14.39 (a) $A'B$; (b) AB'

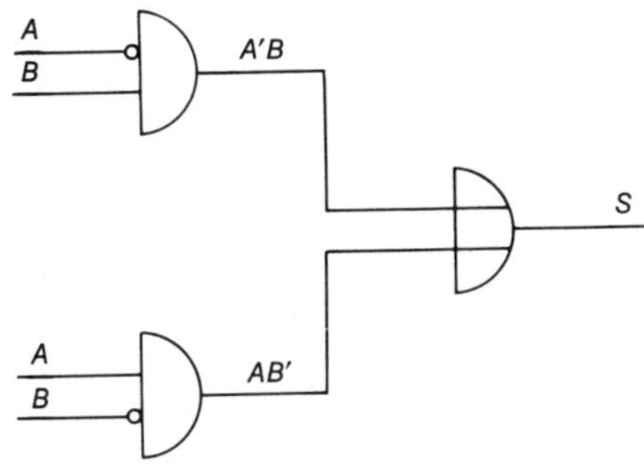

FIGURE 14.40

$AB' = 1$ [Figure 14.39(b)]. The parallel connection of $A'B$ and AB' requires the use of an OR gate, as shown in Figure 14.40.

This circuit should respond to values of A and B in accordance with the S column in Table 14.8. Two of the four possible input combinations are shown in Figure 14.41. The output of the OR gate in Figure 14.41(a) is 1, since the output of the lower AND gate is 1. In Figure 14.41(b), the output is 0 because each input to the OR gate is 0. Both results are as desired.

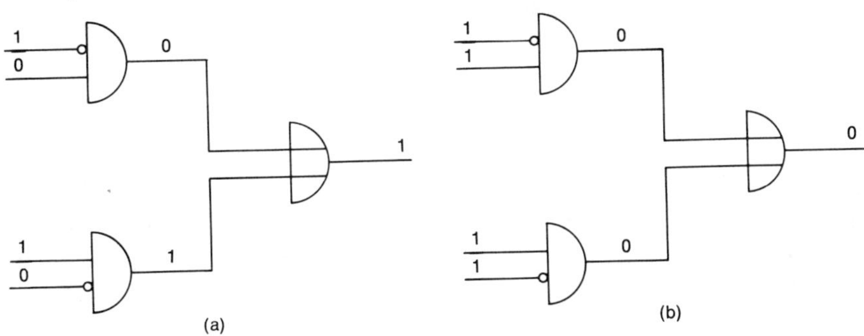

FIGURE 14.41 (a) $A = 1, B = O$; (b) $A = 1, B = 1$

The required circuits for S and C are combined in Figure 14.42. A complete check of the four possible input combinations indicates that S and C represent the sum of A and B. The name **half adder** is associated with this circuit because the possibility of a carry from another column is not included.

DeMORGAN'S RULES. A second representation of the S column in Table 14.8 involves the complement of AB, $(AB)'$. Since a complement repre-

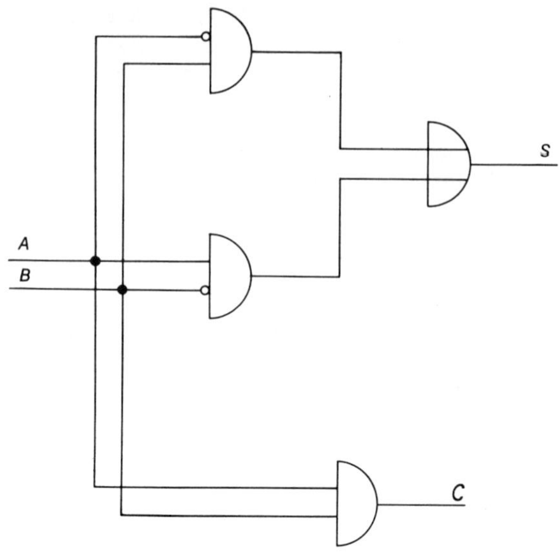

FIGURE 14.42

sents an opposite state, the values of $(AB)'$ can be determined easily. Table 14.10 also indicates the values of A', B', and $A' + B'$.

TABLE 14.10

A	B	AB	$(AB)'$	A'	B'	$A' + B'$
0	0	0	1	1	1	1
0	1	0	1	1	0	1
1	0	0	1	0	1	1
1	1	1	0	0	0	0

The columns for $(AB)'$ and $A' + B'$ are the same, indicating that these two circuits are equivalent. Since a similar result holds for $(A + B)'$ and $A'B'$ (Exercise 14.6, problem 1), we have

$$(AB)' = A' + B'$$
$$(A + B)' = A'B'.$$

These equivalencies, known as DeMorgan's rules, can be useful in changing the form of a Boolean expression and simplifying it. As an example, we will show that the S column in Table 14.8 can also be represented by

$$(A + B)\,(AB)'.$$

This can be verified by constructing a table or by showing that this expression is equivalent to $A'B + AB'$.

$$\begin{aligned}(A+B)(AB)' &= (A+B)(A'+B') && \text{(DeMorgan's rules)}\\ &= (A+B)A' + (A+B)B' && \text{(Property 5)}\\ &= AA' + BA' + AB' + BB' && \text{(Property 5)}\\ &= 0 + BA' + AB' + 0 && \text{(Property 8)}\\ &= BA' + AB' && \text{(Property 10)}\\ &= A'B + AB' && \text{(Property 2)}\end{aligned}$$

Hence, we conclude that the two expressions represent equivalent circuits that would perform the same function.

EXERCISE 14.6

1. Construct a table that verifies the equivalence of $(A+B)'$ and $A'B'$.
2. Find a Boolean expression for each column in the following table. Draw the corresponding circuit, using AND and OR gates.

A	*B*	*I*	*II*	*III*	*IV*	*V*	*VI*
0	0	1	0	1	0	1	1
0	1	1	0	0	1	0	0
1	0	0	1	1	1	1	0
1	1	1	0	1	1	0	1

3. Simplify, if possible, each Boolean expression in problem 2 and draw the simplified circuit.

Rewrite each of the following Boolean expressions, using DeMorgan's rules. Simplify when possible.

4. $(B'+C)'$
5. $(BC')'$
6. $A'+C$
7. BC'
8. $(A+B'+C)'$
9. $(AB'C)'$
10. $(A+B'C)'$
11. $(A+B)'(A'+B')$
12. Find a Boolean expression for each column in the following table:

A	*B*	*K*	*I*	*II*	*III*	*IV*	*V*	*VI*	*VII*	*VIII*
0	0	0	1	1	0	0	0	1	1	0
0	0	1	1	0	1	0	1	0	0	1
0	1	0	0	0	1	1	1	1	0	0
0	1	1	0	1	1	0	1	0	1	1
1	0	0	1	0	0	0	1	1	0	1
1	0	1	1	1	1	0	1	0	0	0
1	1	0	0	0	0	0	0	1	1	1
1	1	1	0	1	1	1	0	0	0	0

13. Simplify, if possible, each Boolean expression in problem 12 and draw the simplified circuit.

Simplify each of the following expressions:

14. $(A + BC' + C)'$
15. $(A'BC)'$
16. $(A'B + AB' + C)'$
17. $(A'B'C' + A'B'C')'$
18. (a) Extend Table 14.8 to include the possibility of a carry (1 or 0) being added to $A + B$.
    (b) Find a Boolean expression for the table in (a).
    (c) Draw the circuit for this expression using AND and OR gates. Use Figure 14.42 as a guide.
19. Construct a table and circuit that would compare two (one-digit) binary numbers, and determine if they are the same or not.
20. A room with two doors is to have a light switch at each door. The switches are to be connected so that either one can turn the light on or off. Construct the required circuit. (*Hint:* start with $A = 0$, $B = 0$, $L = 0$, and visualize what happens to A or B and L when either switch is flipped.)

REFERENCES

Allendoerfer, C. B. and Oakley, C. O., *Principles of Mathematics,* 3d ed. New York, McGraw-Hill, Inc., 1969.

Boole, G., *The Laws of Thought.* New York, Dover Publications, Inc., 1953.

Dinkines, F., *Elementary Concepts of Modern Mathematics.* New York, Appleton-Century-Crofts, 1964.

Hohn, F. E., *Applied Boolean Algebra,* 2d ed. New York, The Macmillan Company, 1966.

Nahikian, H. M., *Topics in Modern Mathematics.* New York, The Macmillan Company, 1966.

Siegel, P., *Understanding Digital Computers.* New York, John Wiley & Sons, 1961.

Whitesitt, J. E., *Boolean Algebra and Its Applications.* Reading, Mass., Addison-Wesley Publishing Company, Inc., 1961.

APPENDIX 1 HEXADECIMAL-DECIMAL CONVERSION TABLE

	0	1	2	3	4	5	6	7	8	9	A	B	C	D	E	F
00 _	0000	0001	0002	0003	0004	0005	0006	0007	0008	0009	0010	0011	0012	0013	0014	0015
01 _	0016	0017	0018	0019	0020	0021	0022	0023	0024	0025	0026	0027	0028	0029	0030	0031
02 _	0032	0033	0034	0035	0036	0037	0038	0039	0040	0041	0042	0043	0044	0045	0046	0047
03 _	0048	0049	0050	0051	0052	0053	0054	0055	0056	0057	0058	0059	0060	0061	0062	0063
04 _	0064	0065	0066	0067	0068	0069	0070	0071	0072	0073	0074	0075	0076	0077	0078	0079
05 _	0080	0081	0082	0083	0084	0085	0086	0087	0088	0089	0090	0091	0092	0093	0094	0095
06 _	0096	0097	0098	0099	0100	0101	0102	0103	0104	0105	0106	0107	0108	0109	0110	0111
07 _	0112	0113	0114	0115	0116	0117	0118	0119	0120	0121	0122	0123	0124	0125	0126	0127
08 _	0128	0129	0130	0131	0132	0133	0134	0135	0136	0137	0138	0139	0140	0141	0142	0143
09 _	0144	0145	0146	0147	0148	0149	0150	0151	0152	0153	0154	0155	0156	0157	0158	0159
0A _	0160	0161	0162	0163	0164	0165	0166	0167	0168	0169	0170	0171	0172	0173	0174	0175
0B _	0176	0177	0178	0179	0180	0181	0182	0183	0184	0185	0186	0187	0188	0189	0190	0191
0C _	0192	0193	0194	0195	0196	0197	0198	0199	0200	0201	0202	0203	0204	0205	0206	0207
0D _	0208	0209	0210	0211	0212	0213	0214	0215	0216	0217	0218	0219	0220	0221	0222	0223
0E _	0224	0225	0226	0227	0228	0229	0230	0231	0232	0233	0234	0235	0236	0237	0238	0239
0F _	0240	0241	0242	0243	0244	0245	0246	0247	0248	0249	0250	0251	0252	0253	0254	0255
10 _	0256	0257	0258	0259	0260	0261	0262	0263	0264	0265	0266	0267	0268	0269	0270	0271
11 _	0272	0273	0274	0275	0276	0277	0278	0279	0280	0281	0282	0283	0284	0285	0286	0287
12 _	0288	0289	0290	0291	0292	0293	0294	0295	0296	0297	0298	0299	0300	0301	0302	0303
13 _	0304	0305	0306	0307	0308	0309	0310	0311	0312	0313	0314	0315	0316	0317	0318	0319
14 _	0320	0321	0322	0323	0324	0325	0326	0327	0328	0329	0330	0331	0332	0333	0334	0335
15 _	0336	0337	0338	0339	0340	0341	0342	0343	0344	0345	0346	0347	0348	0349	0350	0351
16 _	0352	0353	0354	0355	0356	0357	0358	0359	0360	0361	0362	0363	0364	0365	0366	0367
17 _	0368	0369	0370	0371	0372	0373	0374	0375	0376	0377	0378	0379	0380	0381	0382	0383
18 _	0384	0385	0386	0387	0388	0389	0390	0391	0392	0393	0394	0395	0396	0397	0398	0399
19 _	0400	0401	0402	0403	0404	0405	0406	0407	0408	0409	0410	0411	0412	0413	0414	0415
1A _	0416	0417	0418	0419	0420	0421	0422	0423	0424	0425	0426	0427	0428	0429	0430	0431
1B _	0432	0433	0434	0435	0436	0437	0438	0439	0440	0441	0442	0443	0444	0445	0446	0447
1C _	0448	0449	0450	0451	0452	0453	0454	0455	0456	0457	0458	0459	0460	0461	0462	0463
1D _	0464	0465	0466	0467	0468	0469	0470	0471	0472	0473	0474	0475	0476	0477	0478	0479
1E _	0480	0481	0482	0483	0484	0485	0486	0487	0488	0489	0490	0491	0492	0493	0494	0495
1F _	0496	0497	0498	0499	0500	0501	0502	0503	0504	0505	0506	0507	0508	0509	0510	0511

11365A

	0	1	2	3	4	5	6	7	8	9	A	B	C	D	E	F
20 –	0512	0513	0514	0515	0516	0517	0518	0519	0520	0521	0522	0523	0524	0525	0526	0527
21 –	0528	0529	0530	0531	0532	0533	0534	0535	0536	0537	0538	0539	0540	0541	0542	0543
22 –	0544	0545	0546	0547	0548	0549	0550	0551	0552	0553	0554	0555	0556	0557	0558	0559
23 –	0560	0561	0562	0563	0564	0565	0566	0567	0568	0569	0570	0571	0572	0573	0574	0575
24 –	0576	0577	0578	0579	0580	0581	0582	0583	0584	0585	0586	0587	0588	0589	0590	0591
25 –	0592	0593	0594	0595	0596	0597	0598	0599	0600	0601	0602	0603	0604	0605	0606	0607
26 –	0608	0609	0610	0611	0612	0613	0614	0615	0616	0617	0618	0619	0620	0621	0622	0623
27 –	0624	0625	0626	0627	0628	0629	0630	0631	0632	0633	0634	0635	0636	0637	0638	0639
28 –	0640	0641	0642	0643	0644	0645	0646	0647	0648	0649	0650	0651	0652	0653	0654	0655
29 –	0656	0657	0658	0659	0660	0661	0662	0663	0664	0665	0666	0667	0668	0669	0670	0671
2A –	0672	0673	0674	0675	0676	0677	0678	0679	0680	0681	0682	0683	0684	0685	0686	0687
2B –	0688	0689	0690	0691	0692	0693	0694	0695	0696	0697	0698	0699	0700	0701	0702	0703
2C –	0704	0705	0706	0707	0708	0709	0710	0711	0712	0713	0714	0715	0716	0717	0718	0719
2D –	0720	0721	0722	0723	0724	0725	0726	0727	0728	0729	0730	0731	0732	0733	0734	0735
2E –	0736	0737	0738	0739	0740	0741	0742	0743	0744	0745	0746	0747	0748	0749	0750	0751
2F –	0752	0753	0754	0755	0756	0757	0758	0759	0760	0761	0762	0763	0764	0765	0766	0767
30 –	0768	0769	0770	0771	0772	0773	0774	0775	0776	0777	0778	0779	0780	0781	0782	0783
31 –	0784	0785	0786	0787	0788	0789	0790	0791	0792	0793	0794	0795	0796	0797	0798	0799
32 –	0800	0801	0802	0803	0804	0805	0806	0807	0808	0809	0810	0811	0812	0813	0814	0815
33 –	0816	0817	0818	0819	0820	0821	0822	0823	0824	0825	0826	0827	0828	0829	0830	0831
34 –	0832	0833	0834	0835	0836	0837	0838	0839	0840	0841	0842	0843	0844	0845	0846	0847
35 –	0848	0849	0850	0851	0852	0853	0854	0855	0856	0857	0858	0859	0860	0861	0862	0863
36 –	0864	0865	0866	0867	0868	0869	0870	0871	0872	0873	0874	0875	0876	0877	0878	0879
37 –	0880	0881	0882	0883	0884	0885	0886	0887	0888	0889	0890	0891	0892	0893	0894	0895
38 –	0896	0897	0898	0899	0900	0901	0902	0903	0904	0905	0906	0907	0908	0909	0910	0911
39 –	0912	0913	0914	0915	0916	0917	0918	0919	0920	0921	0922	0923	0924	0925	0926	0927
3A –	0928	0929	0930	0931	0932	0933	0934	0935	0936	0937	0938	0939	0940	0941	0942	0943
3B –	0944	0945	0946	0947	0948	0949	0950	0951	0952	0953	0954	0955	0956	0957	0958	0959
3C –	0960	0961	0962	0963	0964	0965	0966	0967	0968	0969	0970	0971	0972	0973	0974	0975
3D –	0976	0977	0978	0979	0980	0981	0982	0983	0984	0985	0986	0987	0988	0989	0990	0991
3E –	0992	0993	0994	0995	0996	0997	0998	0999	1000	1001	1002	1003	1004	1005	1006	1007
3F –	1008	1009	1010	1011	1012	1013	1014	1015	1016	1017	1018	1019	1020	1021	1022	1023

	0	1	2	3	4	5	6	7	8	9	A	B	C	D	E	F
40 _	1024	1025	1026	1027	1028	1029	1030	1031	1032	1033	1034	1035	1036	1037	1038	1039
41 _	1040	1041	1042	1043	1044	1045	1046	1047	1048	1049	1050	1051	1052	1053	1054	1055
42 _	1056	1057	1058	1059	1060	1061	1062	1063	1064	1065	1066	1067	1068	1069	1070	1071
43 _	1072	1073	1074	1075	1076	1077	1078	1079	1080	1081	1082	1083	1084	1085	1086	1087
44 _	1088	1089	1090	1091	1092	1093	1094	1095	1096	1097	1098	1099	1100	1101	1102	1103
45 _	1104	1105	1106	1107	1108	1109	1110	1111	1112	1113	1114	1115	1116	1117	1118	1119
46 _	1120	1121	1122	1123	1124	1125	1126	1127	1128	1129	1130	1131	1132	1133	1134	1135
47 _	1136	1137	1138	1139	1140	1141	1142	1143	1144	1145	1146	1147	1148	1149	1150	1151
48 _	1152	1153	1154	1155	1156	1157	1158	1159	1160	1161	1162	1163	1164	1165	1166	1167
49 _	1168	1169	1170	1171	1172	1173	1174	1175	1176	1177	1178	1179	1180	1181	1182	1183
4A _	1184	1185	1186	1187	1188	1189	1190	1191	1192	1193	1194	1195	1196	1197	1198	1199
4B _	1200	1201	1202	1203	1204	1205	1206	1207	1208	1209	1210	1211	1212	1213	1214	1215
4C _	1216	1217	1218	1219	1220	1221	1222	1223	1224	1225	1226	1227	1228	1229	1230	1231
4D _	1232	1233	1234	1235	1236	1237	1238	1239	1240	1241	1242	1243	1244	1245	1246	1247
4E _	1248	1249	1250	1251	1252	1253	1254	1255	1256	1257	1258	1259	1260	1261	1262	1263
4F _	1264	1265	1266	1267	1268	1269	1270	1271	1272	1273	1274	1275	1276	1277	1278	1279
50 _	1280	1281	1282	1283	1284	1285	1286	1287	1288	1289	1290	1291	1292	1293	1294	1295
51 _	1296	1297	1298	1299	1300	1301	1302	1303	1304	1305	1306	1307	1308	1309	1310	1311
52 _	1312	1313	1314	1315	1316	1317	1318	1319	1320	1321	1322	1323	1324	1325	1326	1327
53 _	1328	1329	1330	1331	1332	1333	1334	1335	1336	1337	1338	1339	1340	1341	1342	1343
54 _	1344	1345	1346	1347	1348	1349	1350	1351	1352	1353	1354	1355	1356	1357	1358	1359
55 _	1360	1361	1362	1363	1364	1365	1366	1367	1368	1369	1370	1371	1372	1373	1374	1375
56 _	1376	1377	1378	1379	1380	1381	1382	1383	1384	1385	1386	1387	1388	1389	1390	1391
57 _	1392	1393	1394	1395	1396	1397	1398	1399	1400	1401	1402	1403	1404	1405	1406	1407
58 _	1408	1409	1410	1411	1412	1413	1414	1415	1416	1417	1418	1419	1420	1421	1422	1423
59 _	1424	1425	1426	1427	1428	1429	1430	1431	1432	1433	1434	1435	1436	1437	1438	1439
5A _	1440	1441	1442	1443	1444	1445	1446	1447	1448	1449	1450	1451	1452	1453	1454	1455
5B _	1456	1457	1458	1459	1460	1461	1462	1463	1464	1465	1466	1467	1468	1469	1470	1471
5C _	1472	1473	1474	1475	1476	1477	1478	1479	1480	1481	1482	1483	1484	1485	1486	1487
5D _	1488	1489	1490	1491	1492	1493	1494	1495	1496	1497	1498	1499	1500	1501	1502	1503
5E _	1504	1505	1506	1507	1508	1509	1510	1511	1512	1513	1514	1515	1516	1517	1518	1519
5F _	1520	1521	1522	1523	1524	1525	1526	1527	1528	1529	1530	1531	1532	1533	1534	1535

11314

	0	1	2	3	4	5	6	7	8	9	A	B	C	D	E	F
60 _	1536	1537	1538	1539	1540	1541	1542	1543	1544	1545	1546	1547	1548	1549	1550	1551
61 _	1552	1553	1554	1555	1556	1557	1558	1559	1560	1561	1562	1563	1564	1565	1566	1567
62 _	1568	1569	1570	1571	1572	1573	1574	1575	1576	1577	1578	1579	1580	1581	1582	1583
63 _	1584	1585	1586	1587	1588	1589	1590	1591	1592	1593	1594	1595	1596	1597	1598	1599
64 _	1600	1601	1602	1603	1604	1605	1606	1607	1608	1609	1610	1611	1612	1613	1614	1615
65 _	1616	1617	1618	1619	1620	1621	1622	1623	1624	1625	1626	1627	1628	1629	1630	1631
66 _	1632	1633	1634	1635	1636	1637	1638	1639	1640	1641	1642	1643	1644	1645	1646	1647
67 _	1648	1649	1650	1651	1652	1653	1654	1655	1656	1657	1658	1659	1660	1661	1662	1663
68 _	1664	1665	1666	1667	1668	1669	1670	1671	1672	1673	1674	1675	1676	1677	1678	1679
69 _	1680	1681	1682	1683	1684	1685	1686	1687	1688	1689	1690	1691	1692	1693	1694	1695
6A _	1696	1697	1698	1699	1700	1701	1702	1703	1704	1705	1706	1707	1708	1709	1710	1711
6B _	1712	1713	1714	1715	1716	1717	1718	1719	1720	1721	1722	1723	1724	1725	1726	1727
6C _	1728	1729	1730	1731	1732	1733	1734	1735	1736	1737	1738	1739	1740	1741	1742	1743
6D _	1744	1745	1746	1747	1748	1749	1750	1751	1752	1753	1754	1755	1756	1757	1758	1759
6E _	1760	1761	1762	1763	1764	1765	1766	1767	1768	1769	1770	1771	1772	1773	1774	1775
6F _	1776	1777	1778	1779	1780	1781	1782	1783	1784	1785	1786	1787	1788	1789	1790	1791
70 _	1792	1793	1794	1795	1796	1797	1798	1799	1800	1801	1802	1803	1804	1805	1806	1807
71 _	1808	1809	1810	1811	1812	1813	1814	1815	1816	1817	1818	1819	1820	1821	1822	1823
72 _	1824	1825	1826	1827	1828	1829	1830	1831	1832	1833	1834	1835	1836	1837	1838	1839
73 _	1840	1841	1842	1843	1844	1845	1846	1847	1848	1849	1850	1851	1852	1853	1854	1855
74 _	1856	1857	1858	1859	1860	1861	1862	1863	1864	1865	1866	1867	1868	1869	1870	1871
75 _	1872	1873	1874	1875	1876	1877	1878	1879	1880	1881	1882	1883	1884	1885	1886	1887
76 _	1888	1889	1890	1891	1892	1893	1894	1895	1896	1897	1898	1899	1900	1901	1902	1903
77 _	1904	1905	1906	1907	1908	1909	1910	1911	1912	1913	1914	1915	1916	1917	1918	1919
78 _	1920	1921	1922	1923	1924	1925	1926	1927	1928	1929	1930	1931	1932	1933	1934	1935
79 _	1936	1937	1938	1939	1940	1941	1942	1943	1944	1945	1946	1947	1948	1949	1950	1951
7A _	1952	1953	1954	1955	1956	1957	1958	1959	1960	1961	1962	1963	1964	1965	1966	1967
7B _	1968	1969	1970	1971	1972	1973	1974	1975	1976	1977	1978	1979	1980	1981	1982	1983
7C _	1984	1985	1986	1987	1988	1989	1990	1991	1992	1993	1994	1995	1996	1997	1998	1999
7D _	2000	2001	2002	2003	2004	2005	2006	2007	2008	2009	2010	2011	2012	2013	2014	2015
7E _	2016	2017	2018	2019	2020	2021	2022	2023	2024	2025	2026	2027	2028	2029	2030	2031
7F _	2032	2033	2034	2035	2036	2037	2038	2039	2040	2041	2042	2043	2044	2045	2046	2047

	0	1	2	3	4	5	6	7	8	9	A	B	C	D	E	F
80 _	2048	2049	2050	2051	2052	2053	2054	2055	2056	2057	2058	2059	2060	2061	2062	2063
81 _	2064	2065	2066	2067	2068	2069	2070	2071	2072	2073	2074	2075	2076	2077	2078	2079
82 _	2080	2081	2082	2083	2084	2085	2086	2087	2088	2089	2090	2091	2092	2093	2094	2095
83 _	2096	2097	2098	2099	2100	2101	2102	2103	2104	2105	2106	2107	2108	2109	2110	2111
84 _	2112	2113	2114	2115	2116	2117	2118	2119	2120	2121	2122	2123	2124	2125	2126	2127
85 _	2128	2129	2130	2131	2132	2133	2134	2135	2136	2137	2138	2139	2140	2141	2142	2143
86 _	2144	2145	2146	2147	2148	2149	2150	2151	2152	2153	2154	2155	2156	2157	2158	2159
87 _	2160	2161	2162	2163	2164	2165	2166	2167	2168	2169	2170	2171	2172	2173	2174	2175
88 _	2176	2177	2178	2179	2180	2181	2182	2183	2184	2185	2186	2187	2188	2189	2190	2191
89 _	2192	2193	2194	2195	2196	2197	2198	2199	2200	2201	2202	2203	2204	2205	2206	2207
8A _	2208	2209	2210	2211	2212	2213	2214	2215	2216	2217	2218	2219	2220	2221	2222	2223
8B _	2224	2225	2226	2227	2228	2229	2230	2231	2232	2233	2234	2235	2236	2237	2238	2239
8C _	2240	2241	2242	2243	2244	2245	2246	2247	2248	2249	2250	2251	2252	2253	2254	2255
8D _	2256	2257	2258	2259	2260	2261	2262	2263	2264	2265	2266	2267	2268	2269	2270	2271
8E _	2272	2273	2274	2275	2276	2277	2278	2279	2280	2281	2282	2283	2284	2285	2286	2287
8F _	2288	2289	2290	2291	2292	2293	2294	2295	2296	2297	2298	2299	2300	2301	2302	2303
90 _	2304	2305	2306	2307	2308	2309	2310	2311	2312	2313	2314	2315	2316	2317	2318	2319
91 _	2320	2321	2322	2323	2324	2325	2326	2327	2328	2329	2330	2331	2332	2333	2334	2335
92 _	2336	2337	2338	2339	2340	2341	2342	2343	2344	2345	2346	2347	2348	2349	2350	2351
93 _	2352	2353	2354	2355	2356	2357	2358	2359	2360	2361	2362	2363	2364	2365	2366	2367
94 _	2368	2369	2370	2371	2372	2373	2374	2375	2376	2377	2378	2379	2380	2381	2382	2383
95 _	2384	2385	2386	2387	2388	2389	2390	2391	2392	2393	2394	2395	2396	2397	2398	2399
96 _	2400	2401	2402	2403	2404	2405	2406	2407	2408	2409	2410	2411	2412	2413	2414	2415
97 _	2416	2417	2418	2419	2420	2421	2422	2423	2424	2425	2426	2427	2428	2429	2430	2431
98 _	2432	2433	2434	2435	2436	2437	2438	2439	2440	2441	2442	2443	2444	2445	2446	2447
99 _	2448	2449	2450	2451	2452	2453	2454	2455	2456	2457	2458	2459	2460	2461	2462	2463
9A _	2464	2465	2466	2467	2468	2469	2470	2471	2472	2473	2474	2475	2476	2477	2478	2479
9B _	2480	2481	2482	2483	2484	2485	2486	2487	2488	2489	2490	2491	2492	2493	2494	2495
9C _	2496	2497	2498	2499	2500	2501	2502	2503	2504	2505	2506	2507	2508	2509	2510	2511
9D _	2512	2513	2514	2515	2516	2517	2518	2519	2520	2521	2522	2523	2524	2525	2526	2527
9E _	2528	2529	2530	2531	2532	2533	2534	2535	2536	2537	2538	2539	2540	2541	2542	2543
9F _	2544	2545	2546	2547	2548	2549	2550	2551	2552	2553	2554	2555	2556	2557	2558	2559

11315

	0	1	2	3	4	5	6	7	8	9	A	B	C	D	E	F
A0 _	2560	2561	2562	2563	2564	2565	2566	2567	2568	2569	2570	2571	2572	2573	2574	2575
A1 _	2576	2577	2578	2579	2580	2581	2582	2583	2584	2585	2586	2587	2588	2589	2590	2591
A2 _	2592	2593	2594	2595	2596	2597	2598	2599	2600	2601	2602	2603	2604	2605	2606	2607
A3 _	2608	2609	2610	2611	2612	2613	2614	2615	2616	2617	2618	2619	2620	2621	2622	2623
A4 _	2624	2625	2626	2627	2628	2629	2630	2631	2632	2633	2634	2635	2636	2637	2638	2639
A5 _	2640	2641	2642	2643	2644	2645	2646	2647	2648	2649	2650	2651	2652	2653	2654	2655
A6 _	2656	2657	2658	2659	2660	2661	2662	2663	2664	2665	2666	2667	2668	2669	2670	2671
A7 _	2672	2673	2674	2675	2676	2677	2678	2679	2680	2681	2682	2683	2684	2685	2686	2687
A8 _	2688	2689	2690	2691	2692	2693	2694	2695	2696	2697	2698	2699	2700	2701	2702	2703
A9 _	2704	2705	2706	2707	2708	2709	2710	2711	2712	2713	2714	2715	2716	2717	2718	2719
AA _	2720	2721	2722	2723	2724	2725	2726	2727	2728	2729	2730	2731	2732	2733	2734	2735
AB _	2736	2737	2738	2739	2740	2741	2742	2743	2744	2745	2746	2747	2748	2749	2750	2751
AC _	2752	2753	2754	2755	2756	2757	2758	2759	2760	2761	2762	2763	2764	2765	2766	2767
AD _	2768	2769	2770	2771	2772	2773	2774	2775	2776	2777	2778	2779	2780	2781	2782	2783
AE _	2784	2785	2786	2787	2788	2789	2790	2791	2792	2793	2794	2795	2796	2797	2798	2799
AF _	2800	2801	2802	2803	2804	2805	2806	2807	2808	2809	2810	2811	2812	2813	2814	2815
B0 _	2816	2817	2818	2819	2820	2821	2822	2823	2824	2825	2826	2827	2828	2829	2830	2831
B1 _	2832	2833	2834	2835	2836	2837	2838	2839	2840	2841	2842	2843	2844	2845	2846	2847
B2 _	2848	2849	2850	2851	2852	2853	2854	2855	2856	2857	2858	2859	2860	2861	2862	2863
B3 _	2864	2865	2866	2867	2868	2869	2870	2871	2872	2873	2874	2875	2876	2877	2878	2879
B4 _	2880	2881	2882	2883	2884	2885	2886	2887	2888	2889	2890	2891	2892	2893	2894	2895
B5 _	2896	2897	2898	2899	2900	2901	2902	2903	2904	2905	2906	2907	2908	2909	2910	2911
B6 _	2912	2913	2914	2915	2916	2917	2918	2919	2920	2921	2922	2923	2924	2925	2926	2927
B7 _	2928	2929	2930	2931	2932	2933	2934	2935	2936	2937	2938	2939	2940	2941	2942	2943
B8 _	2944	2945	2946	2947	2948	2949	2950	2951	2952	2953	2954	2955	2956	2957	2958	2959
B9 _	2960	2961	2962	2963	2964	2965	2966	2967	2968	2969	2970	2971	2972	2973	2974	2975
BA _	2976	2977	2978	2979	2980	2981	2982	2983	2984	2985	2986	2987	2988	2989	2990	2991
BB _	2992	2993	2994	2995	2996	2997	2998	2999	3000	3001	3002	3003	3004	3005	3006	3007
BC _	3008	3009	3010	3011	3012	3013	3014	3015	3016	3017	3018	3019	3020	3021	3022	3023
BD _	3024	3025	3026	3027	3028	3029	3030	3031	3032	3033	3034	3035	3036	3037	3038	3039
BE _	3040	3041	3042	3043	3044	3045	3046	3047	3048	3049	3050	3051	3052	3053	3054	3055
BF _	3056	3057	3058	3059	3060	3061	3062	3063	3064	3065	3066	3067	3068	3069	3070	3071

	0	1	2	3	4	5	6	7	8	9	A	B	C	D	E	F
C0 _	3072	3073	3074	3075	3076	3077	3078	3079	3080	3081	3082	3083	3084	3085	3086	3087
C1 _	3088	3089	3090	3091	3092	3093	3094	3095	3096	3097	3098	3099	3100	3101	3102	3103
C2 _	3104	3105	3106	3107	3108	3109	3110	3111	3112	3113	3114	3115	3116	3117	3118	3119
C3 _	3120	3121	3122	3123	3124	3125	3126	3127	3128	3129	3130	3131	3132	3133	3134	3135
C4 _	3136	3137	3138	3139	3140	3141	3142	3143	3144	3145	3146	3147	3148	3149	3150	3151
C5 _	3152	3153	3154	3155	3156	3157	3158	3159	3160	3161	3162	3163	3164	3165	3166	3167
C6 _	3168	3169	3170	3171	3172	3173	3174	3175	3176	3177	3178	3179	3180	3181	3182	3183
C7 _	3184	3185	3186	3187	3188	3189	3190	3191	3192	3193	3194	3195	3196	3197	3198	3199
C8 _	3200	3201	3202	3203	3204	3205	3206	3207	3208	3209	3210	3211	3212	3213	3214	3215
C9 _	3216	3217	3218	3219	3220	3221	3222	3223	3224	3225	3226	3227	3228	3229	3230	3231
CA _	3232	3233	3234	3235	3236	3237	3238	3239	3240	3241	3242	3243	3244	3245	3246	3247
CB _	3248	3249	3250	3251	3252	3253	3254	3255	3256	3257	3258	3259	3260	3261	3262	3263
CC _	3264	3265	3266	3267	3268	3269	3270	3271	3272	3273	3274	3275	3276	3277	3278	3279
CD _	3280	3281	3282	3283	3284	3285	3286	3287	3288	3289	3290	3291	3292	3293	3294	3295
CE _	3296	3297	3298	3299	3300	3301	3302	3303	3304	3305	3306	3307	3308	3309	3310	3311
CF _	3312	3313	3314	3315	3316	3317	3318	3319	3320	3321	3322	3323	3324	3325	3326	3327
D0 _	3328	3329	3330	3331	3332	3333	3334	3335	3336	3337	3338	3339	3340	3341	3342	3343
D1 _	3344	3345	3346	3347	3348	3349	3350	3351	3352	3353	3354	3355	3356	3357	3358	3359
D2 _	3360	3361	3362	3363	3364	3365	3366	3367	3368	3369	3370	3371	3372	3373	3374	3375
D3 _	3376	3377	3378	3379	3380	3381	3382	3383	3384	3385	3386	3387	3388	3389	3390	3391
D4 _	3392	3393	3394	3395	3396	3397	3398	3399	3400	3401	3402	3403	3404	3405	3406	3407
D5 _	3408	3409	3410	3411	3412	3413	3414	3415	3416	3417	3418	3419	3420	3421	3422	3423
D6 _	3424	3425	3426	3427	3428	3429	3430	3431	3432	3433	3434	3435	3436	3437	3438	3439
D7 _	3440	3441	3442	3443	3444	3445	3446	3447	3448	3449	3450	3451	3452	3453	3454	3455
D8 _	3456	3457	3458	3459	3460	3461	3462	3463	3464	3465	3466	3467	3468	3469	3470	3471
D9 _	3472	3473	3474	3475	3476	3477	3478	3479	3480	3481	3482	3483	3484	3485	3486	3487
DA _	3488	3489	3490	3491	3492	3493	3494	3495	3496	3497	3498	3499	3500	3501	3502	3503
DB _	3504	3505	3506	3507	3508	3509	3510	3511	3512	3513	3514	3515	3516	3517	3518	3519
DC _	3520	3521	3522	3523	3524	3525	3526	3527	3528	3529	3530	3531	3532	3533	3534	3535
DD _	3536	3537	3538	3539	3540	3541	3542	3543	3544	3545	3546	3547	3548	3549	3550	3551
DE _	3552	3553	3554	3555	3556	3557	3558	3559	3560	3561	3562	3563	3564	3565	3566	3567
DF _	3568	3569	3570	3571	3572	3573	3574	3575	3576	3577	3578	3579	3580	3581	3582	3583

11316

11317

	0	1	2	3	4	5	6	7	8	9	A	B	C	D	E	F
E0 _	3584	3585	3586	3587	3583	3589	3590	3591	3592	3593	3594	3595	3596	3597	3598	3599
E1 _	3600	3601	3602	3603	3604	3605	3606	3607	3608	3609	3610	3611	3612	3613	3614	3615
E2 _	3616	3617	3618	3619	3620	3621	3622	3623	3624	3625	3626	3627	3628	3629	3630	3631
E3 _	3632	3633	3634	3635	3636	3637	3638	3639	3640	3641	3642	3643	3644	3645	3646	3647
E4 _	3648	3649	3650	3651	3652	3653	3654	3655	3656	3657	3658	3659	3660	3661	3662	3663
E5 _	3664	3665	3666	3667	3668	3669	3670	3671	3672	3673	3674	3675	3676	3677	3678	3679
E6 _	3680	3681	3682	3683	3684	3685	3686	3687	3688	3689	3690	3691	3692	3693	3694	3695
E7 _	3696	3697	3698	3699	3700	3701	3702	3703	3704	3705	3706	3707	3708	3709	3710	3711
E8 _	3712	3713	3714	3715	3716	3717	3718	3719	3720	3721	3722	3723	3724	3725	3726	3727
E9 _	3728	3729	3730	3731	3732	3733	3734	3735	3736	3737	3738	3739	3740	3741	3742	3743
EA _	3744	3745	3746	3747	3748	3749	3750	3751	3752	3753	3754	3755	3756	3757	3758	3759
EB _	3760	3761	3762	3763	3764	3765	3766	3767	3768	3769	3770	3771	3772	3773	3774	3775
EC _	3776	3777	3778	3779	3780	3781	3782	3783	3784	3785	3786	3787	3788	3789	3790	3791
ED _	3792	3793	3794	3795	3796	3797	3798	3799	3800	3801	3802	3803	3804	3805	3806	3807
EE _	3808	3809	3810	3811	3812	3813	3814	3815	3816	3817	3818	3819	3820	3821	3822	3823
EF _	3824	3825	3826	3827	3828	3829	3830	3831	3832	3833	3834	3835	3836	3837	3838	3839
F0 _	3840	3841	3842	3843	3844	3845	3846	3847	3848	3849	3850	3851	3852	3853	3854	3855
F1 _	3856	3857	3858	3859	3860	3861	3862	3863	3864	3865	3866	3867	3868	3869	3870	3871
F2 _	3872	3873	3874	3875	3876	3877	3878	3879	3880	3881	3882	3883	3884	3885	3886	3887
F3 _	3888	3889	3890	3891	3892	3893	3894	3895	3896	3897	3898	3899	3900	3901	3902	3903
F4 _	3904	3905	3906	3907	3908	3909	3910	3911	3912	3913	3914	3915	3916	3917	3918	3919
F5 _	3920	3921	3922	3923	3924	3925	3926	3927	3928	3929	3930	3931	3932	3933	3934	3935
F6 _	3936	3937	3938	3939	3940	3941	3942	3943	3944	3945	3946	3947	3948	3949	3950	3951
F7 _	3952	3953	3954	3955	3956	3957	3958	3959	3960	3961	3962	3963	3964	3965	3966	3967
F8 _	3968	3969	3970	3971	3972	3973	3974	3975	3976	3977	3978	3979	3980	3981	3982	3983
F9 _	3984	3985	3986	3987	3988	3989	3990	3991	3992	3993	3994	3995	3996	3997	3998	3999
FA _	4000	4001	4002	4003	4004	4005	4006	4007	4008	4009	4010	4011	4012	4013	4014	4015
FB _	4016	4017	4018	4019	4020	4021	4022	4023	4024	4025	4026	4027	4028	4029	4030	4031
FC _	4032	4033	4034	4035	4036	4037	4038	4039	4040	4041	4042	4043	4044	4045	4046	4047
FD _	4048	4049	4050	4051	4052	4053	4054	4055	4056	4057	4058	4059	4060	4061	4062	4063
FE _	4064	4065	4066	4067	4068	4069	4070	4071	4072	4073	4074	4075	4076	4077	4078	4079
FF _	4080	4081	4082	4083	4084	4085	4086	4087	4088	4089	4090	4091	4092	4093	4094	4095

** From IBM Systems Reference Library manual, "IBM 1800 Data Acquisition and Control System Functional Characteristics," File No. 1800-01, Form A26-5918-3, pp. 88-92.*

APPENDIX 2 | FOUR PLACE LOGARITHM TABLE

Four Place Logarithm Table

n	0	1	2	3	4	5	6	7	8	9
10	0000	0043	0086	0128	0170	0212	0253	0294	0334	0374
11	0414	0453	0492	0531	0569	0607	0645	0682	0719	0755
12	0792	0828	0864	0899	0934	0969	1004	1038	1072	1106
13	1139	1173	1206	1239	1271	1303	1335	1367	1399	1430
14	1461	1492	1523	1553	1584	1614	1644	1673	1703	1732
15	1761	1790	1818	1847	1875	1903	1931	1959	1987	2014
16	2041	2068	2095	2122	2148	2175	2201	2227	2253	2279
17	2304	2330	2355	2380	2405	2430	2455	2480	2504	2529
18	2553	2577	2601	2625	2648	2672	2695	2718	2742	2765
19	2788	2810	2833	2856	2878	2900	2923	2945	2967	2989
20	3010	3032	3054	3075	3096	3118	3139	3160	3181	3201
21	3222	3243	3263	3284	3304	3324	3345	3365	3385	3404
22	3424	3444	3464	3483	3502	3522	3541	3560	3579	3598
23	3617	3636	3655	3674	3692	3711	3729	3747	3766	3784
24	3802	3820	3838	3856	3874	3892	3909	3927	3945	3962
25	3979	3997	4014	4031	4048	4065	4082	4099	4116	4133
26	4150	4166	4183	4200	4216	4232	4249	4265	4281	4298
27	4314	4330	4346	4362	4378	4393	4409	4425	4440	4456
28	4472	4487	4502	4518	4533	4548	4564	4579	4594	4609
29	4624	4639	4654	4669	4683	4698	4713	4728	4742	4757
30	4771	4786	4800	4814	4829	4843	4857	4871	4886	4900
31	4914	4928	4942	4955	4969	4983	4997	5011	5024	5038
32	5051	5065	5079	5092	5105	5119	5132	5145	5159	5172
33	5185	5198	5211	5224	5237	5250	5263	5276	5289	5302
34	5315	5328	5340	5353	5366	5378	5391	5403	5416	5428
35	5441	5453	5465	5478	5490	5502	5514	5527	5539	5551
36	5563	5575	5587	5599	5611	5623	5635	5647	5658	5670
37	5682	5694	5705	5717	5729	5740	5752	5763	5775	5786
38	5798	5809	5821	5832	5843	5855	5866	5877	5888	5899
39	5911	5922	5933	5944	5955	5966	5977	5988	5999	6010
40	6021	6031	6042	6053	6064	6075	6085	6096	6107	6117
41	6128	6138	6149	6160	6170	6180	6191	6201	6212	6222
42	6232	6243	6253	6263	6274	6284	6294	6304	6314	6325
43	6335	6345	6355	6365	6375	6385	6395	6405	6415	6425
44	6435	6444	6454	6464	6474	6484	6493	6503	6513	6522
45	6532	6542	6551	6561	6571	6580	6590	6599	6609	6618
46	6628	6637	6646	6656	6665	6675	6684	6693	6702	6712
47	6721	6730	6739	6749	6758	6767	6776	6785	6794	6803
48	6812	6821	6830	6839	6848	6857	6866	6875	6884	6893
49	6902	6911	6920	6928	6937	6946	6955	6964	6972	6981
50	6990	6998	7007	7016	7024	7033	7042	7050	7059	7067
51	7076	7084	7093	7101	7110	7118	7126	7135	7143	7152
52	7160	7168	7177	7185	7193	7202	7210	7218	7226	7235
53	7243	7251	7259	7267	7275	7284	7292	7300	7308	7316
54	7324	7332	7340	7348	7356	7364	7372	7380	7388	7396

Holt Master Tables

Four Place Logarithm Table (*Continued*)

n	0	1	2	3	4	5	6	7	8	9
55	7404	7412	7419	7427	7435	7443	7451	7459	7466	7474
56	7482	7490	7497	7505	7513	7520	7528	7536	7543	7551
57	7559	7566	7574	7582	7589	7597	7604	7612	7619	7627
58	7634	7642	7649	7657	7664	7672	7679	7686	7694	7701
59	7709	7716	7723	7731	7738	7745	7752	7760	7767	7774
60	7782	7789	7796	7803	7810	7818	7825	7832	7839	7846
61	7853	7860	7868	7875	7882	7889	7896	7903	7910	7917
62	7924	7931	7938	7945	7952	7959	7966	7973	7980	7987
63	7993	8000	8007	8014	8021	8028	8035	8041	8048	8055
64	8062	8069	8075	8082	8089	8096	8102	8109	8116	8122
65	8129	8136	8142	8149	8156	8162	8169	8176	8182	8189
66	8195	8202	8209	8215	8222	8228	8235	8241	8248	8254
67	8261	8267	8274	8280	8287	8293	8299	8306	8312	8319
68	8325	8331	8338	8344	8351	8357	8363	8370	8376	8382
69	8388	8395	8401	8407	8414	8420	8426	8432	8439	8445
70	8451	8457	8463	8470	8476	8482	8488	8494	8500	8506
71	8513	8519	8525	8531	8537	8543	8549	8555	8561	8567
72	8573	8579	8585	8591	8597	8603	8609	8615	8621	8627
73	8633	8639	8645	8651	8657	8663	8669	8675	8681	8686
74	8692	8698	8704	8710	8716	8722	8727	8733	8739	8745
75	8751	8756	8762	8768	8774	8779	8785	8791	8797	8802
76	8808	8814	8820	8825	8831	8837	8842	8848	8854	8859
77	8865	8871	8876	8882	8887	8893	8899	8904	8910	8915
78	8921	8927	8932	8938	8943	8949	8954	8960	8965	8971
79	8976	8982	8987	8993	8998	9004	9009	9015	9020	9025
80	9031	9036	9042	9047	9053	9058	9063	9069	9074	9079
81	9085	9090	9096	9101	9106	9112	9117	9122	9128	9133
82	9138	9143	9149	9154	9159	9165	9170	9175	9180	9186
83	9191	9196	9201	9206	9212	9217	9222	9227	9232	9238
84	9243	9248	9253	9258	9263	9269	9274	9279	9284	9289
85	9294	9299	9304	9309	9315	9320	9325	9330	9335	9340
86	9345	9350	9355	9360	9365	9370	9375	9380	9385	9390
87	9395	9400	9405	9410	9415	9420	9425	9430	9435	9440
88	9445	9450	9455	9460	9465	9469	9474	9479	9484	9489
89	9494	9499	9504	9509	9513	9518	9523	9528	9533	9538
90	9542	9547	9552	9557	9562	9566	9571	9576	9581	9586
91	9590	9595	9600	9605	9609	9614	9619	9624	9628	9633
92	9638	9643	9647	9652	9657	9661	9666	9671	9675	9680
93	9685	9689	9694	9699	9703	9708	9713	9717	9722	9727
94	9731	9736	9741	9745	9750	9754	9759	9763	9768	9773
95	9777	9782	9786	9791	9795	9800	9805	9809	9814	9818
96	9823	9827	9832	9836	9841	9845	9850	9854	9859	9863
97	9868	9872	9877	9881	9886	9890	9894	9899	9903	9908
98	9912	9917	9921	9926	9930	9934	9939	9943	9948	9952
99	9956	9961	9965	9969	9974	9978	9983	9987	9991	9996

Holt Master Tables

APPENDIX 3 ANSWERS TO SELECTED PROBLEMS

EXERCISE 1.1

1.

Card Number	Content
1	Place new cards in holes 13, 14 and 15
2	Erase slate, write number from card 13
3	Multiply by number on card 14
4	Round off answer to 2 decimal places
5	Replace number on card 13 with number from slate
6	Erase slate, write number from card 15
7	Subtract number on card 16
8	Write number from slate on card 15
9	If result is zero, go to card 11
10	Go back to card 2
11	Remove cards 13, 14 and 15 and save for boss
12	Go back to card 1
13	200
14	1.06
15	5
16	1

2. After executing the instruction on card 6, card 7 would be inspected as the next instruction. But since it contains data, confusion would result.

3. It *would* function properly since the program would branch around the data.

7. (a) Clear the card reader
   (b) Place the misread card in the card reader and start
   (c) If no further reader check then continue normal work routine
   (d) If second attempt to reread then continue to step (e), otherwise go to step (a)
   (e) Set the deck aside for later correction and go to next job.

11. (a) Get next student record
    (b) If senior skip to step (f)
    (c) If carrying more than 15 units skip to step (f)
    (d) If GPA 3.0 or better skip to step (f)
    (e) Skip to step (g)
    (f) Record name on scholarship list
    (g) If last student then stop
    (h) Otherwise return to step (a)

EXERCISE 1.2

1.

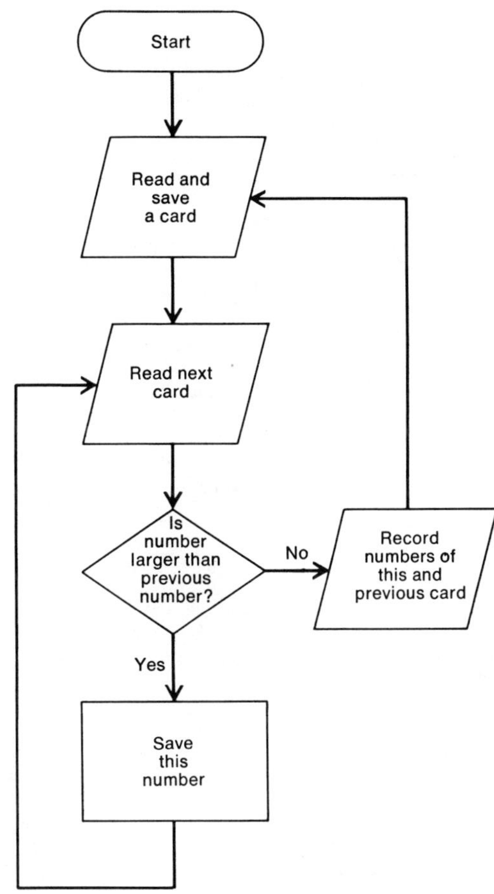

8.

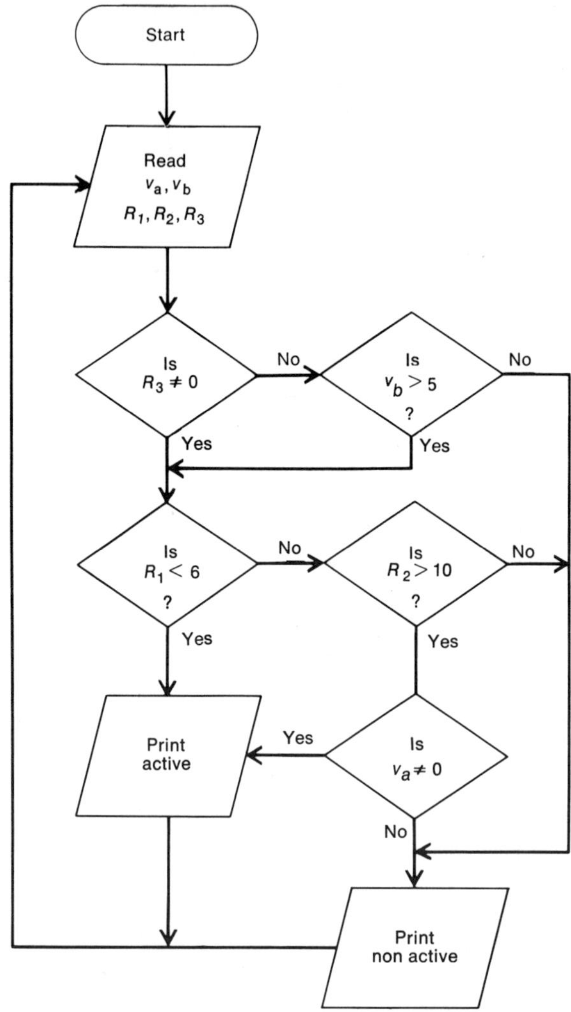

EXERCISE 1.3

1. (a) $H_2 = \{002,004,006,007,010,011,014,015,016,019\}$
   (b) $H_3 = \{001,003,005,008,009,012,013,017,018,020\}$
   (c) $P = \{004,008,010,019\}$
   (d) $F = \{002,003,005,006,007,009,011,013,014,015,016,017,020\}$
   (e) $V = \{001,012,018\}$

2. (a) $\{002,006,007,011,014,015,016\}$
   (b) $\{001,012,018\}$
   (c) $\{002,004,006,007,008,010,011,014,015,016,019\}$
   (d) All members of the universal set except 004, 010, 019.
   (e) The complement of $\{1,12,18\}$
   (f) The complement of $\{004,008,010,019\}$
   (g) ϕ
   (h) $\{008\}$

3. (a) $P \cup V$ (b) $P \cup F$ (c) V (d) F (e) P

4. (a) T (b) T (c) F (d) T (e) T (f) T

5. (a) $H_2 \cap F$ (b) $P \cap H_3$ (c) $P' \cap H_2$ (d) $V' \cap H_2$ (e) F

6. (a) ϕ (b) U (c) ϕ (d) ϕ (e) U (f) ϕ
(g) ϕ (h) U (i) U (j) ϕ

7. (a) {2} (b) U (c) {3,5} (d) {1,4} (e) ϕ (f) {1,3,4,5}
(g) {1,3,4,5} (h) ϕ

9. (a) $\{b\}$ (b) $\{a,b,c,d,f\}$ (c) $\{c,e,f\}$ (d) $\{a.d.e\}$ (e) $\{e\}$
(f) $\{a,c,d,e,f\}$ (g) $\{a,c,d,e,f\}$ (h) $\{e\}$

10. (a) U (b) ϕ (c) B (d) A (e) ϕ

11. (a) male employees under 21
(b) women employees over 60
(c) women employees in data processing
(d) women employees under 21
(e) employees between 21 and 60, inclusive
(f) employees in data processing under 21
(g) employees not in data processing or 21 or over
(h) male employees 21 or over

13. (a) {2,6} (b) {4} (c) {4,5} (d) {1,5} (e) {1}
(f) {2,6} (g) {1,3,5} (h) {1,3,5} (i) {2,3,6} (j) ϕ

15. (a) 9 (b) 24 (c) 9 (d) 18 (e) T

17. (a) T (b) T (c) F (d) F (e) T
(f) F (g) F (h) T (i) F

18. (a) {1,2} (c) {3,5} (g) $A \cap B', B \cap A', A', \cdots$

EXERCISE 1.4

1. (a) (b) (c) (d)

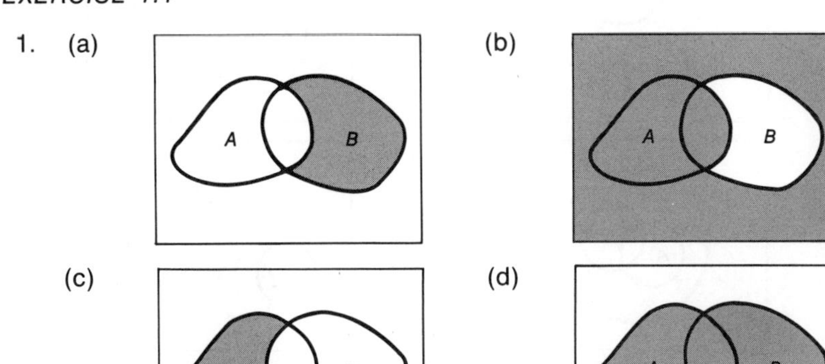

(e)

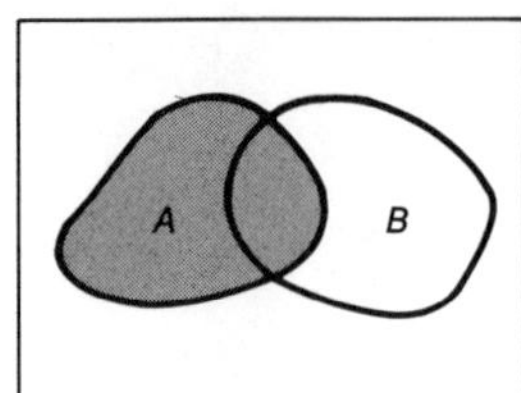

(f)

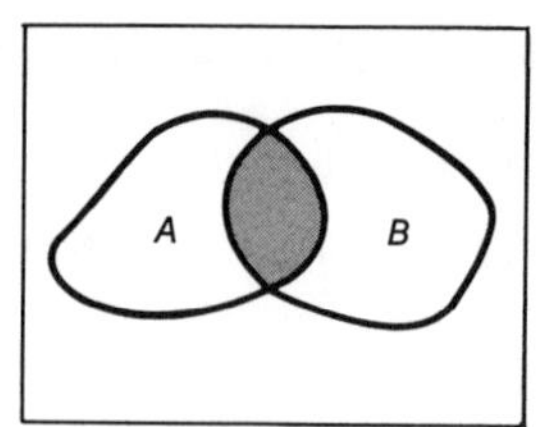

3. $1 - A \cap B$, $2 - A \cap B'$, $3 - B \cap A'$, $4 - A' \cap B'$

5. (a) $A \cup B'$ (b) $A' \cap B$ (c) $A' \cap B'$ (d) $A \cap B' \cap C$
(e) $A \cap (B \cup C)'$ (f) $B' \cap (A \cap C)$

7. $(A \cap B') \cup (B \cap A')$

8. (a) $\{3\}$ (b) $\{3,4\}$ (c) $\{4\}$ (d) $\{3\}$ (e) $\{2\}$ (f) $\{2,4\}$
(g) $\{3\}$ (h) $\{1,3,4\}$ (i) $\{3,4\}$

9. (a) $M \cap F$ (b) $(M' \cap T) \cap A$ (c) $M \cap (T \cup F)'$ or $M \cap (T' \cap F')$
(d) $M' \cap (A' \cap B')$ (e) $(M' \cap F) \cap B$ (f) $M \cap (T' \cap F') \cap (A' \cap B')$

10. $1 - A \cap B \cap C$, $2 - A \cap C \cap B'$, $5 - C \cap A' \cap B'$, $8 - A' \cap B' \cap C'$

11. (a) $(A \cap C \cap B') \cup (B \cap C \cap A')$ (b) $B \cap A'$ (c) $(A \cup C) \cap B'$
(d) B (e) $A \cap (B \cup C)$ (f) $(C \cap A' \cap B') \cup (A \cap B' \cap C')$

13. (a)

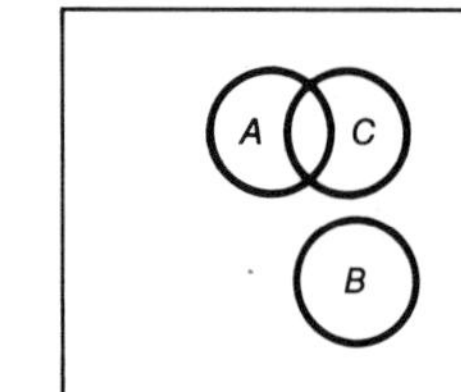

(b)

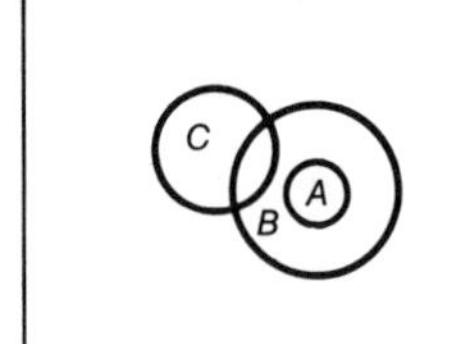

(c)

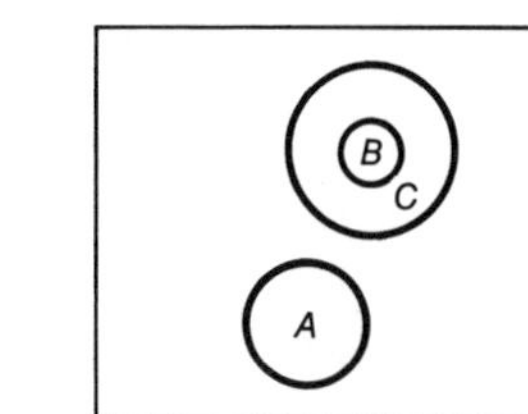

(d)

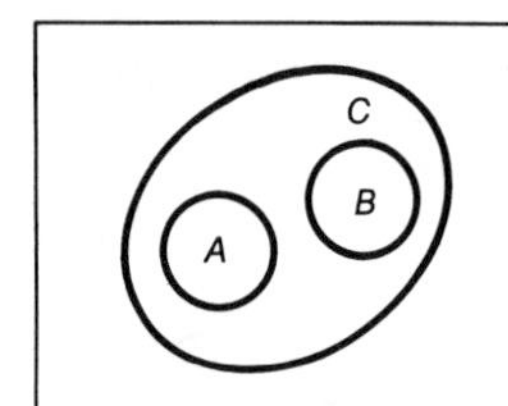

(e)

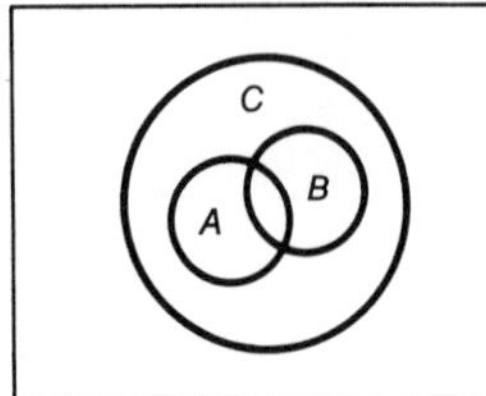

(f)

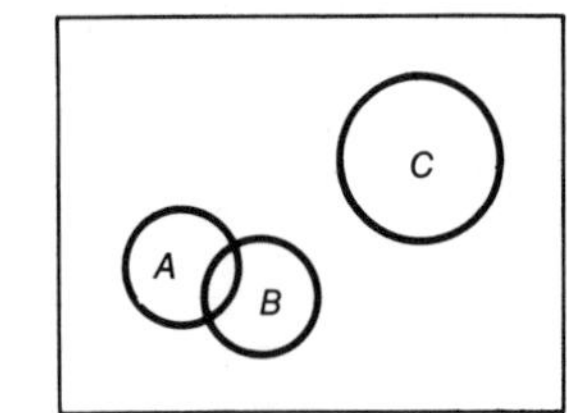

15. (a) $I - C - U$
(b) $U - C - C - I$
(c) $U - U - I$
(d) $C - C - U - C$
(e) $C - I - C - I - U$
(f) $C - U - C - I - C$

EXERCISE 1.5

1.

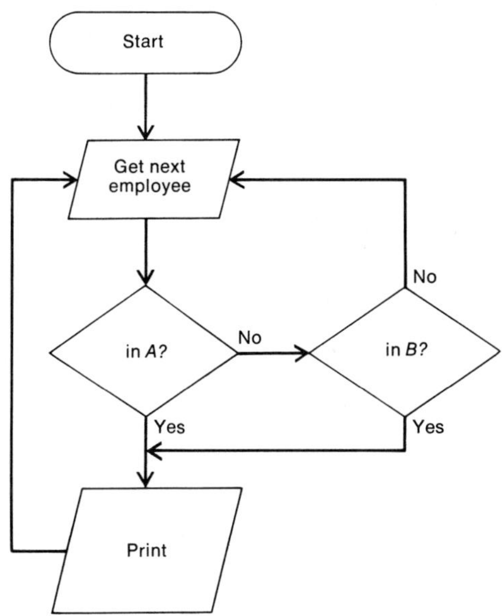

9.

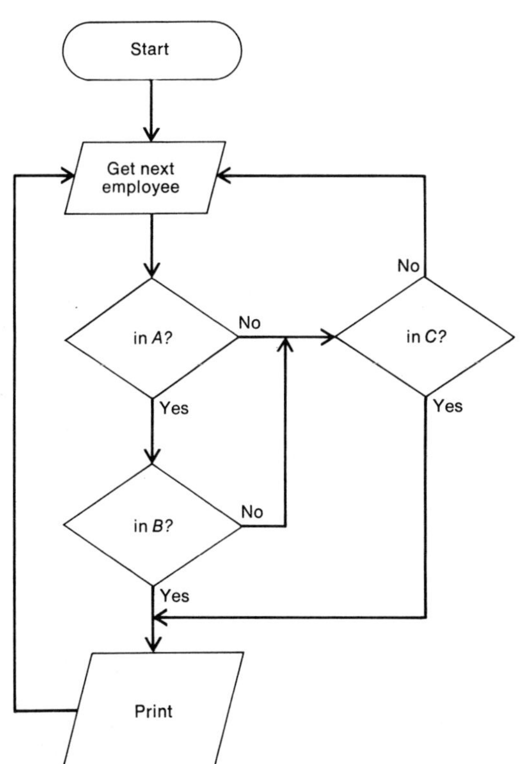

17.

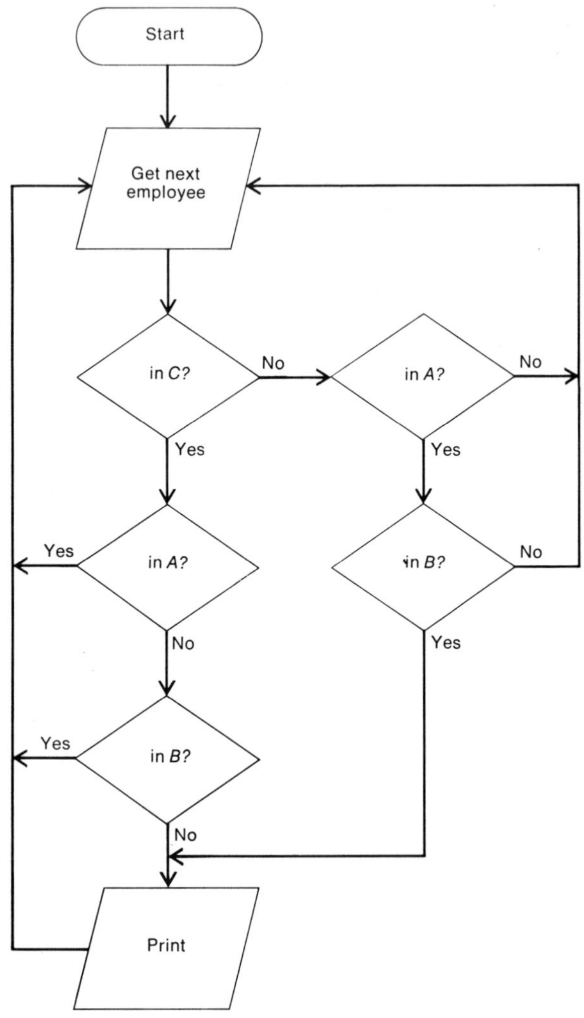
Start
Get next employee
in C?
No
in A?
No
Yes
Yes
Yes
in A?
in B?
No
No
Yes
Yes
in B?
No
Print

25.

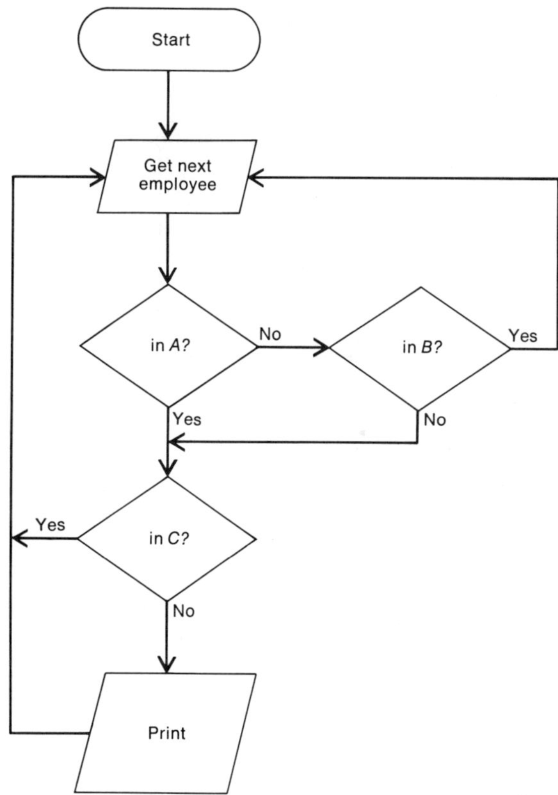

EXERCISE 1.6

1. (a) Figure 1.25(a)
   (b) For Figure 1.25(a)−314; for 1.25(b)−337
   (c) 66; 43

3. (a) $A \cap B'$ provides best flowchart. (b) $B \cap A'$

5. (a) Use $(B \cap A) \cap C$ (b) Use $(C \cup A) \cup B$
   (c) The number of checks for $A' \cap (B \cup C)$ is 626; $A' \cap (C \cup B)$−602; $(B \cup C) \cap A'$−640; $(C \cup B) \cap A'$−600. See problems 9 and 10.
   (d) Number of checks, $(A \cup C) \cap B'$−628; $(C \cup A) \cap B'$−603; $B' \cap (A \cup C)$−641; $B' \cap (C \cup A)$−620.

7. (a) {1,2,4} (b) {1,2,4} (c) {1,2,3,4} (d) {1,2,3,4}

9. (b) 189, 95, 36, 153, 53, 100 (c) $284 + 189 + 153 = 626$

10. (b) $B, B', B' \cap C, B' \cap C'$, $(B \cap A') \cup (B' \cap C \cap A')$,
    80, 204, 72, 132 $36 + 53$,
    $(B \cap A) \cup (B' \cap C \cap A)$
    $44 + 19$

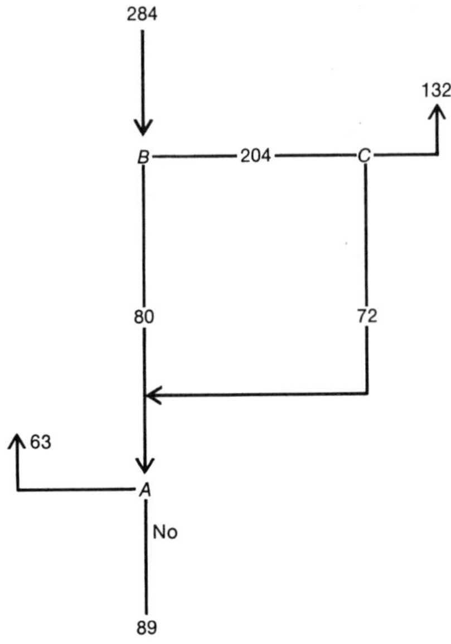

11. (b) $C, C', C' \cap A, C' \cap A'$, $(C \cap B') \cup (C' \cap A \cap B')$,
120, 164 35 129 $72 + 32$
$(C \cap B) \cup (C' \cap A \cap B)$
$48 + 3$

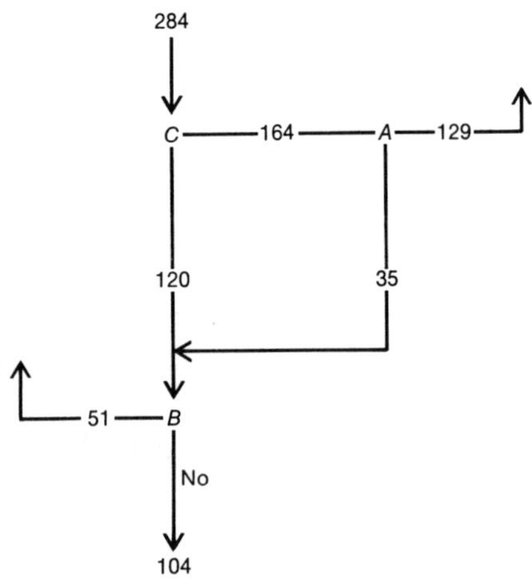

EXERCISE 2.1

1. (a) Yes (b) Yes 3. (a) Yes (b) Yes 5. (a) No (b) Yes
7. If m and n are natural numbers then x and y are also natural numbers where $x = m + n$, $y = m \times n$

EXERCISE 2.2

1. (a) $1 < 2$ (b) $3 < 27$ (c) $100 > 87$ (d) $-5 < 12$
(e) $4 > -4$ (f) $-4 > -5$ (g) $9 > 1$ (h) $-27 < 0$

(i) $0 = 0$ (j) $1/5 < 1/4$ (k) $1/6 > 7/48$ (l) $-1/100 < 0$
(m) $a + b > a$ (n) $a + b = b + a$ (o) $a + b + c = c + a + b$
(p) $0 > -1/b$

3. (a) $16 > 8 > 0 > -15/16 > -1 > -5$ (b) $6c > 9c/5 > 0 > 4c/5 > -6c$
5. (a) Yes (b) No $(-1)(-2) = +2$
7. $\sqrt{2} \times \sqrt{2} = 2$

EXERCISE 2.3

1. $(8773 + 5891) + (-4992) \rightarrow 4664 + (-4992) = -328$
   $8773 + [5891 + (-4992)] \rightarrow 8773 + (899) = 9672$
5.

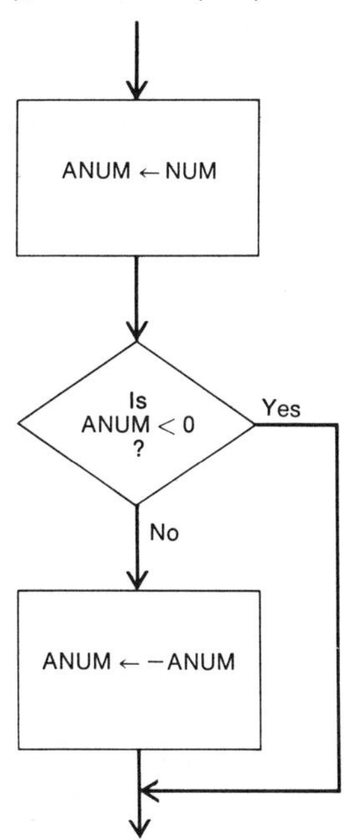

7. (a) The product is negative (b) The product is positive
9. (a) Either *a* or *b* (but not both) is negative
   (b) Either *a* or *b* or both are zero.
11. The result 996847 is the complement (ignoring sign) of the correct result.
15. (a) 1 (c) 0 (e) 0 (f) 3 (h) 21

EXERCISE 2.4

1. $6 \times 10^1 + 2 \times 10^0$ 3. $4 \times 10^2 + 2 \times 10^1 + 1 \times 10^0$
5. $9 \times 10^4 + 2 \times 10^3 + 1 \times 10^2 + 0 \times 10^1 + 5 \times 10^0$
7. $2 \times 10^4 + 2 \times 10^3 + 1 \times 10^2 + 0 \times 10^1 + 1 \times 10^0$
9. $9 \times 10^1 + 2 \times 10^0 + 1 \times 10^{-1} + 0 \times 10^{-2}$
11. $2 \times 10^2 + 0 \times 10^1 + 0 \times 10^0 + 0 \times 10^{-1} + 0 \times 10^{-2} + 1 \times 10^{-3}$
13. $5 \times 10^2 + 0 \times 10^1 + 1 \times 10^0 + 6 \times 10^{-1} + 8 \times 10^{-2} + 7 \times 10^{-3}$

15. $1 \times 10^{-1} + 5 \times 10^{-2} + 6 \times 10^{-3}$

17. $9 \times 10^{-1} + 0 \times 10^{-2} + 0 \times 10^{-3} + 6 \times 10^{-4}$

19. $2 \times 10^{-2} + 0 \times 10^{-3} + 3 \times 10^{-4}$

21. $9 \times 10^{-3} + 0 \times 10^{-4} + 9 \times 10^{-5}$

23. 2.2706×10^{4} 25. 4.801×10^{3} 27. 5.962×10

29. 6.439×10^{2} 31. 6.53×10^{-1} 33. 4.93×10^{-3}

35. 8.02×10^{-6} 37. 9.752×10^{6} 39. 8.01×10^{-2}

41. 8.27×10^{-5} 43. 4.3×10^{-3} 45. 9.01×10^{-9}

47. 7.2×10^{13} 49. 3.0 51. 1.2×10^{-7}

53. 1.2×10^{5} 55. 3.5×10^{-13} 57. 8.0×10^{2}

59. 5.0×10

61. 3 63. 4 65. 2 67. 5 69. 5 71. 3 73. 2

EXERCISE 3.1

2.

Decimal Digit	Excess-three Code			
	8	4	2	1
0	0	0	1	1
1	0	1	0	0
2	0	1	0	1
3	0	1	1	0
4	0	1	1	1
5	1	0	0	0
6	1	0	0	1
7	1	0	1	0
8	1	0	1	1
9	1	1	0	0

3.

Decimal Digit	Two-of-five code				
	6	3	2	1	0
1	0	0	0	1	1
2	0	0	1	0	1
3	0	1	0	0	1
4	0	1	0	1	0
5	0	1	1	0	0
6	1	0	0	0	1
7	1	0	0	1	0
8	1	0	1	0	0
9	1	1	0	0	0

However, this is not the only possibility, consider $a = 4$; then what is b?

EXERCISE 3.3

1. 0, 100, 1001, 11011, 111111, 1000000, 1011011, 1001010001
3. 1, 2, 5, 15, 16, 17, 90
7. 1, 1000, 1101, 100100, 1111111, 10000000, 10000001, 110101011
9. 0, 3, 7, 29, 32, 31, 33, 85
11. 0.75, 0.59375, 0.65625, 0.80859375, 0.11328125
13. 3.25, 5.625, 28.21875, 1.40625, 109.703125
15. 0.1, 0.0101, 0.101010010111, 0.001100110011, 0.000111111011, 0.000000011101
17. 11000.011, 1000111101.1000001, 1010101010.001000110101, 111111.001110101110
19. 0.011, 101, 1111.0100001, 0.111001100110, 10110.100110011001, 10001.010100000110, 1001.011000010100

EXERCISE 3.4

1.	111	3.	10010	5.	100000	7.	100000
9.	1001010	11.	1001	13.	101	15.	1100
17.	1001	19.	10000111	21.	10111101	23.	10010001001
25.	10011	27.	1101	29.	11001.1	31.	1011.01001···

EXERCISE 3.5

1. 00100110, 11011010, 01111111, 10000000, 11111111, 10110111, exceeds 8 bits, exceeds 8 bits

3. 01111 5. 01 7. 01101

		Carry in	Carry out	Overflow
9.	01100011	No	No	No
11.	11111011	No	No	No
13.	00000101	Yes	Yes	No
15.	11111111	No	No	No
17.	00101110	No	Yes	Yes
19.	11100110	No	No	No
21.	00000000	Yes	Yes	No
23.	11110011	No	No	No

24. (a) largest positive number 31
smallest positive number 0
largest negative number −32 (in magnitude)
smallest negative number −1 (in magnitude)

EXERCISE 3.6

1. 7, 111 3. 113, 001110001 5. 63, 111111
7. 3706, 111001111010 9. 3662, 111001001110
11. 19.21875, 10011.00111 13. 9, 1001 15. 29, 11101
17. 409, 110011001 19. 511, 111111111
21. 512, 1000000000 23. 48.53125, 110000.10001
25. 17, F 27. 21, 11 29. 100, 40 31. 417, 10F
33. 751, 1E9 35. 421, 111 37. 4368 39. 8656
41. 40986 43. 1001 45. 10C8 47. 206C

51.

Base 6	Base 10	Base 1	Base 10
a	0	*ba*	6
b	1	*bb*	7
c	2	*bc*	8
d	3	*bd*	9
e	4	*be*	10
f	5	*bf*	11
		ca	12

EXERCISE 3.7

3. Multiplication Table for Base 8

×	0	1	2	3	4	5	6	7
0	0	0	0	0	0	0	0	0
1	0	1	2	3	4	5	6	7
2	0	2	4	6	$^{1}0$	$^{1}2$	$^{1}4$	$^{1}6$
3	0	3	6	$^{1}1$	$^{1}4$	$^{1}7$	$^{2}2$	$^{2}5$
4	0	4	$^{1}0$	$^{1}4$	$^{2}0$	$^{2}4$	$^{3}0$	$^{3}4$
5	0	5	$^{1}2$	$^{1}7$	$^{2}4$	$^{3}1$	$^{3}6$	$^{4}3$
6	0	6	$^{1}4$	$^{2}2$	$^{3}0$	$^{3}6$	$^{4}4$	$^{5}2$
7	0	7	$^{1}6$	$^{2}5$	$^{3}4$	$^{4}3$	$^{5}2$	$^{6}1$

5. 7773 7. 10711 9. 100111 11. 6552.22
13. 518C 15. C290 17. 10008 19. CB.2B1
21. 4441 23. 12362 25. 10427 27. 30712
29. 4577.1 31. 2572 33. F70 35. 4BD1A
37. 65.E93 39. 52 41. 1606 43. 136573
45. 2A 47. A0 49. 284C0A 51. Yes to all questions
53. Yes; 1, 4, 6

54.

+	a	b	c	d	e	f
a	a	b	c	d	e	f
b	b	c	d	e	f	ba
c	c	d	e	f	ba	bb
d	d	e	f	ba	bb	bc
e	e	f	ba	bb	bc	bd
f	f	ba	bb	bc	bd	be

EXERCISE 4.1

1.

P	Q	P.AND..NOT.Q
T	T	F
T	F	T
F	T	F
F	F	F

The truth values for *P*.AND.NOT.*Q* can be indicated in row form as FTFF. This form will be used in all answers in this chapter.

3. FFFT 5. (*G*.AND.*D*).OR.(*S*.AND.*R*) 7. *P*.OR.(*B*.AND.*S*)

9. (*U*.AND.*O*).AND.*W* 11. *L*.AND.(*G*.OR.*E*)

13. *M*.OR.(*C*.AND.*I*).OR.(*G*.AND.*F*).OR.(*F*.AND.*B*) 15. (*I*.OR.*S*).OR.*L*

17. *R*.OR.(*P*.AND.*S*.AND.*W*) 19. T 21. F

23.

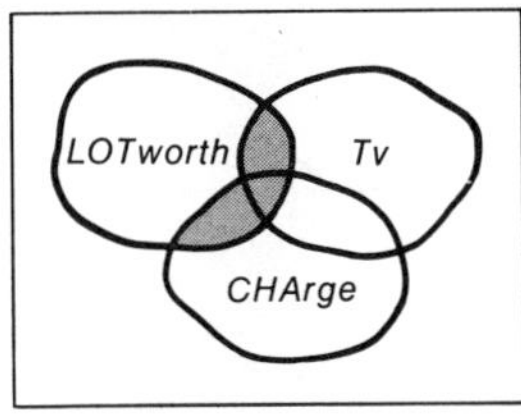

25.

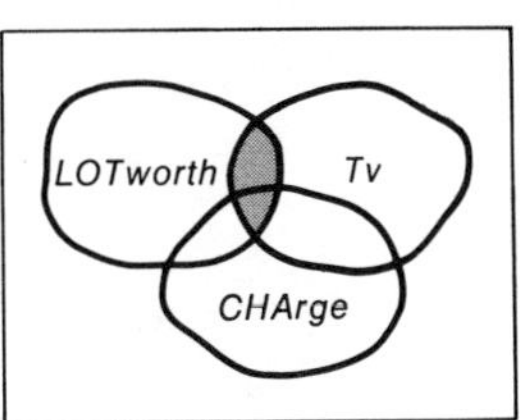

27. T 29. F 31. T 33. F 35. FTTF

37. TFFT 39. .NOT.*P*.AND.*Q* 41. *P*.AND..NOT.*Q*

43. The employee is 50 or under or has worked 5 years or less

EXERCISE 4.2

1. (*P*.AND..NOT.*Q*).OR.*R* 3. (*P*.OR.*Q*).OR.*R*

5. (*P*.AND.*Q*).OR.(*R*.AND.*S*) 7. (*P*.OR.*Q*).OR.(*R*.AND.*S*)

9. Using the truth values as shown in Table 4.5, the condensed form is TFTT TTTT

11. TFTT FFFF 13. TFFF 15. TTFF 16. Yes 17. No
18. Yes 19. No 21. F 23. T 25. T 27. F
29. FTTF FFFF

EXERCISE 4.3

13.

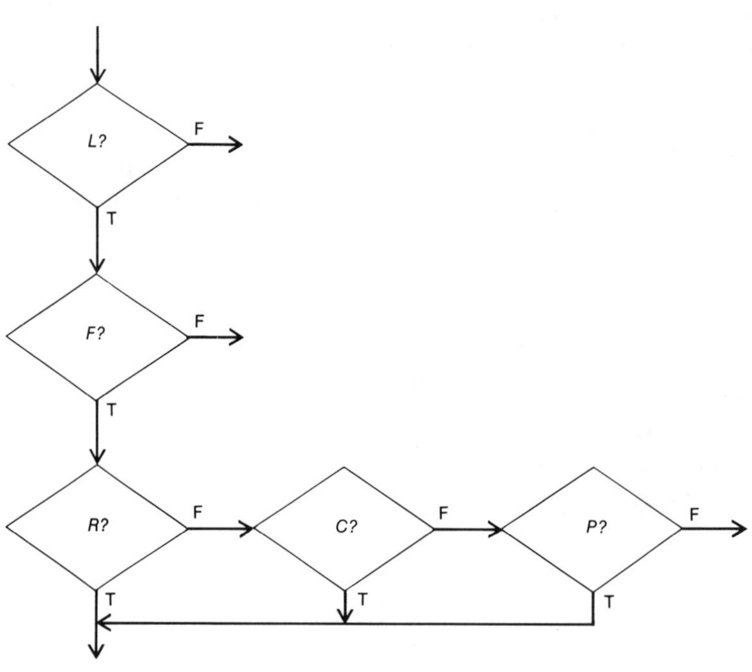

17.

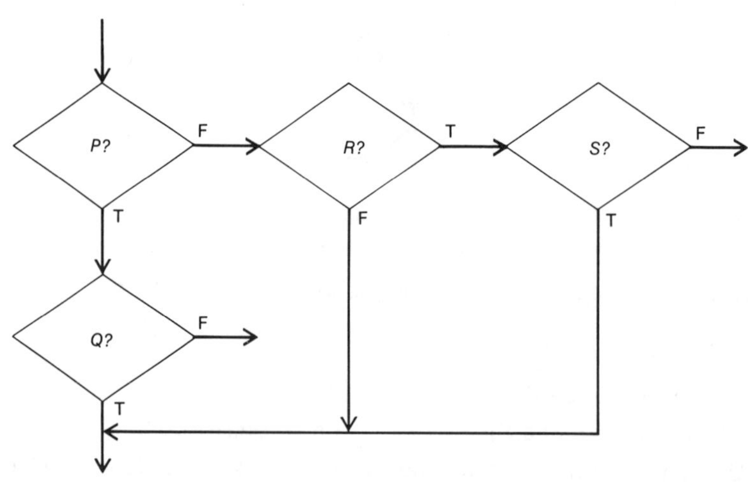

19.

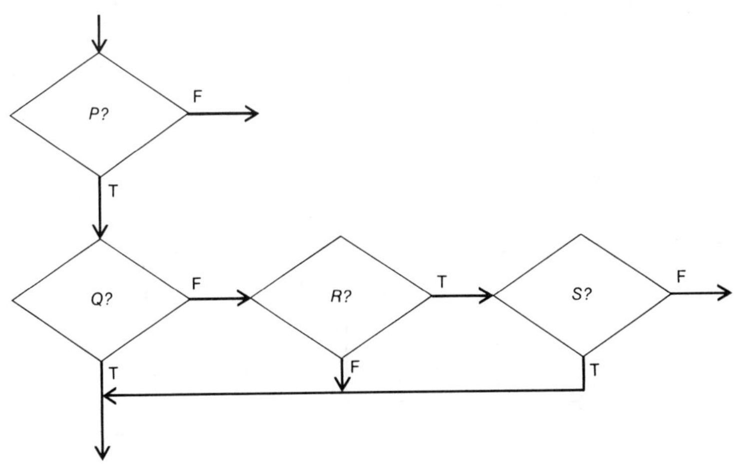

21.

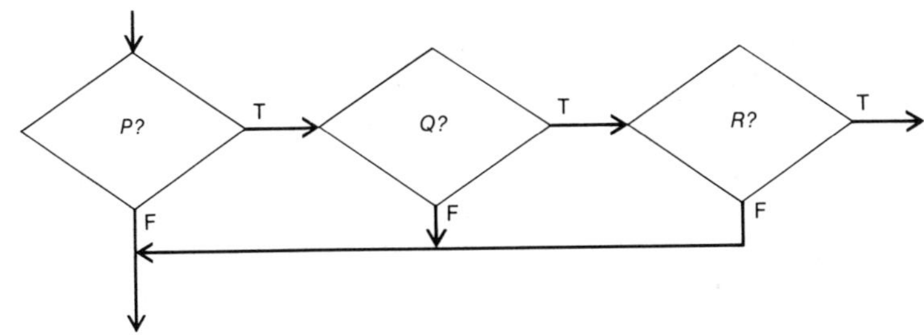

23. FTTT TTTT

25. FFFT TTTT

EXERCISE 4.4

1. If W, then P; $W \to P$
3. If W, then H; $W \to H$
5. If O.OR.(W.AND..NOT.O), then B; (O.OR.(W.AND..NOT.O))$\to B$
7. If (U.AND.B).AND.I, then E; ((U.AND.B).AND.I)$\to E$
9. If C, then (E.AND..NOT.L); $C \to$ (E.AND..NOT.L)
11. (M.OR.L) $\to$ (R.OR.P)

23.

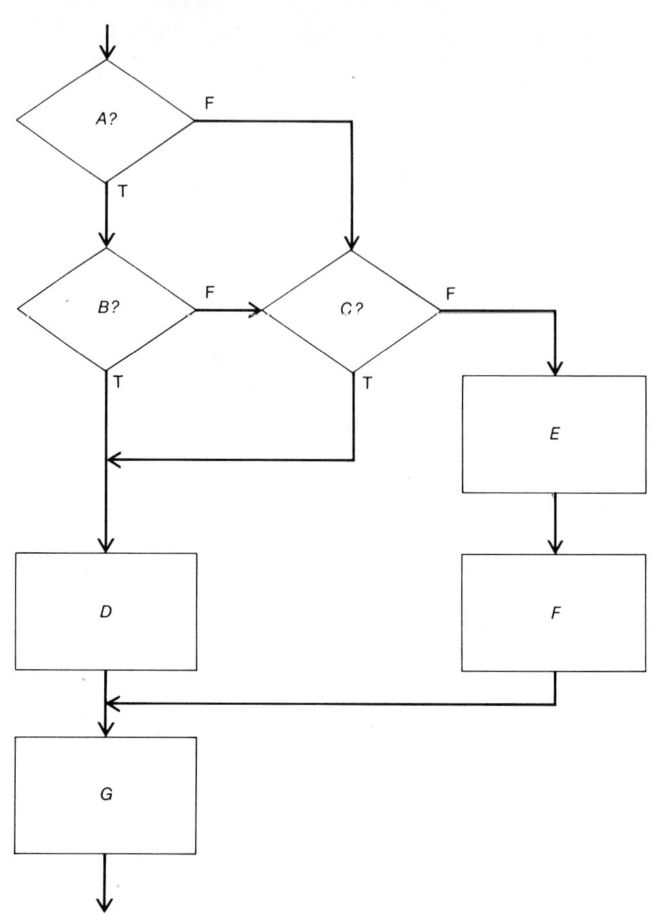

25.

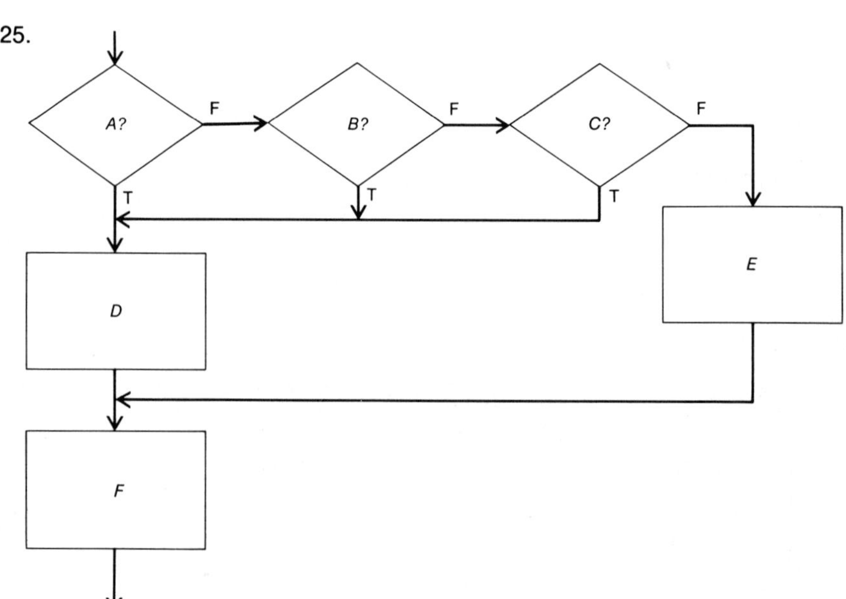

29. (U.OR.C) → (R.AND.P)

30. (N.AND.(W.OR.P).AND.O) → ((S.AND.I).OR.(C → A))

33.

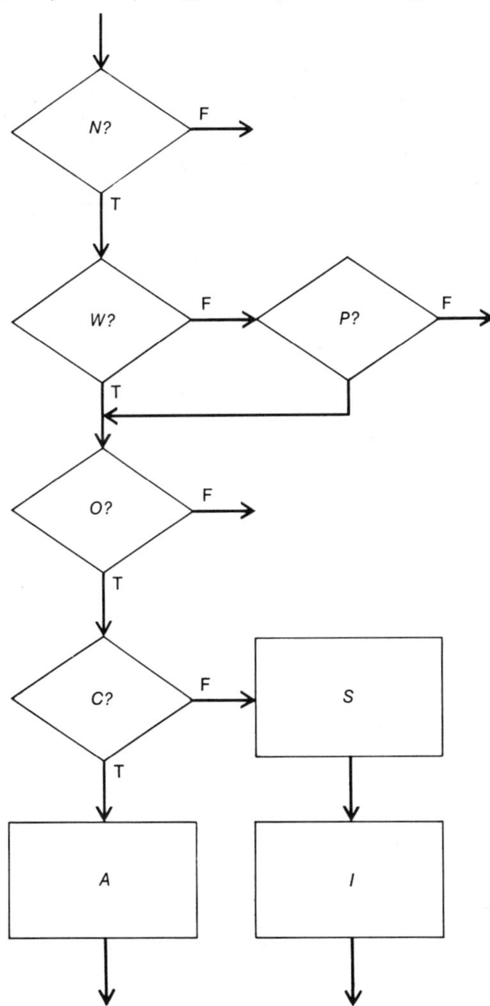

EXERCISE 4.5

1. FTTT 3. TTFT 5. FTTT 7. .NOT.P.OR.Q 9. Q.OR.P

11. (a) TFTT (b) TTFF (c) line 1 only (d) Yes (e) Yes

13. $P \rightarrow Q$ TFTT; .NOT.Q FTFT: Both true—line 4 only; .NOT.P is true on line 4

15. (a) TTTF—line 1,2,3; (b) FTFT—line 2,4;
(c) Both true—line 2, on line 2—P is true (Note .NOT.(.NOT.P) is logically equivalent to P)

17. (a) TTFT—lines 1,2,4 (b) TTFF—lines 1,2
(c) No; .NOT.Q is false on line 1.

19. FTTF 21. FTTF 23. (a) TTFF (b) TFTT

EXERCISE 4.6

1. (a) 0000, 1011, 1011 (b) 1011, 1111, 0100 (c) 1011, 1011, 0000
   (d) 0000, 1111, 1111

3. (a) *D*—now true (b) *F*—now true
   (c) *B*—now false and *E*—now false
   (d) *B*—now false and *E*—now false
   (e) *B*—now false and *F*—now true (f) *B*—now false

5. (a) Employee—now over 65 with special instead of extended insurance coverage
   (b) Employee—now on hourly pay with no special insurance coverage
   (c) Employee—now 21 and on salary
   (d) Same as (c)

7. Employee is between 21 and 65.

9. Use EOR with a mask of 11000000. Then use TEST UNDER MASK instruction with a mask of 11000000. A condition code equal to 0 indicates under 21 and female.

11.

1011	1011	1101	1101
	→	→	→
1101	0110	0110	1011
	(a)	(b)	(c)

A and *B* have been interchanged.

13.

1100	1100	0011	0011
	→	→	→
0011	1111	1111	1100
	(a)	(b)	(c)

EXERCISE 4.7

1. .NOT.*P*—FFTT: *P*.OR.*Q*—TTTF; .NOT.*P*.AND.(*P*.OR.*Q*)—FFTF
   .NOT.*P*.AND.*Q*—FFTF

3. Antecedents simplify to: (a) *P* (b) *P* (c) *P*.OR.*Q* (d) *P*.AND.*Q*

4. (a) *P* (b) *Q* (c) True (d) False
   (e) *P*.OR.*R* (f) *P*.AND.*R*

7. (a)

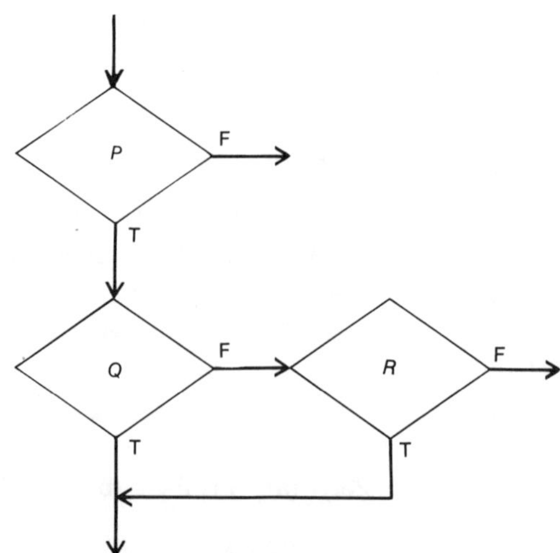

(c)

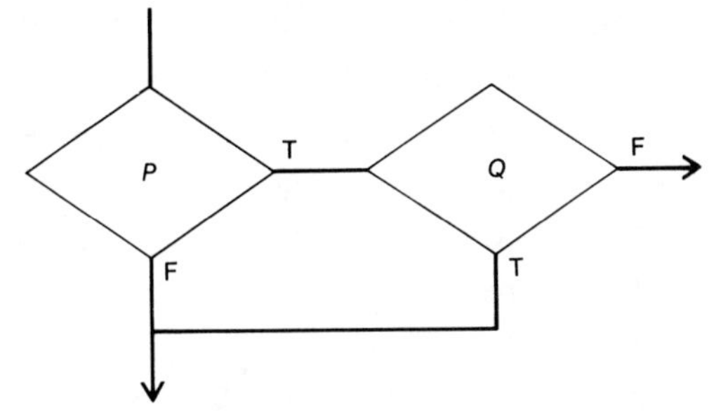

(e)

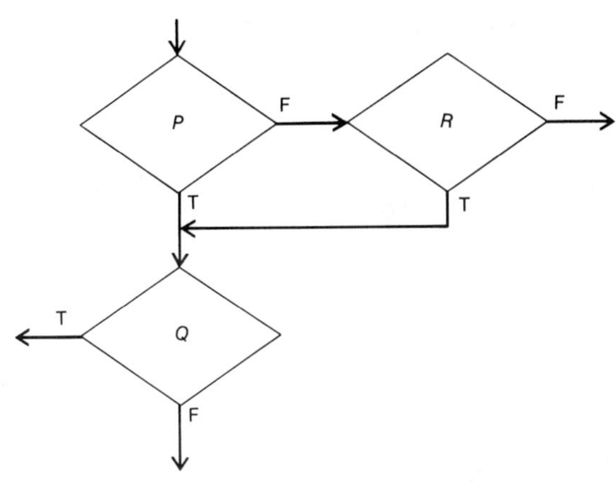

9. (a) .NOT.*P*.OR..NOT.*Q* (b) .NOT.*P*.AND..NOT.*Q*
(c) $(A \cap B)' = A' \cup B'$; $(A \cup B)' = A' \cap B'$

11. Antecedents simplify to: (a) .NOT.*P*.OR.*Q* (b) .NOT.*P*.AND..NOT.*Q*
(c) *Q*.AND.*R* (d) *Q*.AND.(*P*.AND..NOT.*R*) (e) (.NOT.*P*.OR.*R*).OR.*Q*
(f) *P* (g) *P*

EXERCISE 5.1

1. 5, 3, −5 3. 4, 3, 0 5. 14, −4, 9 7. −1, 5, −1
9. −5, −3, 5 11. −6, 4, −6 13. −1, −17, −11 15. 1, −63, 31
17. 59; A**2 + 5.0*(B+4.0*C) 19. 0; C*(B+2.0) − (A+1.0)
21. −8; A*B*(C−1.0)**2
23. 198; 2.0*(C**2 + 1.0)**2 − A*B**2
25. −34; 2.0*((A+B)** 2 − 3.0*(A*C))
27. −52; 2.0*A*(A−3.0*(C+2.0))
29. 42; −(3.0*A+2.0*(B**2−(C+A)**2))

EXERCISE 5.2

1. $4x$ 3. $5x$ 5. $3xy^2$ 7. $-5xyz$ 9. 0 11. 0
13. $n-m$ 15. $38x-4$ 17. $2a$ 19. m^2+2
21. $y-3q-2p$ 23. $8m-8n$ 25. -6 27. $2m-2n$
29. $2x+2y$ 31. $21m-14n$ 33. $-5a-15b$ 35. $27x+18y$

37. $6ax - 4az$ 39. $-3mx - 3my + 3mz$ 41. $6m^4 - 12mn^2$
43. $6x^3 - 3x^2 + 15x$ 45. $3a^4 - 3a^3 + 9a^2 - 3a$ 47. $x^2 + 3x + 2$
49. $y^2 + 7y + 12$ 51. $z^2 + 6z + 9$ 53. $2a^2 + 6a - 20$
55. $-10a^2 + 11a - 3$ 57. $10x^2 - 7xy - 6y^2$ 59. $18x^2 + 12x + 2$
61. $-10x^2 + 8xy + 2y^2$ 63. $a^3 - 2a^2 - 3a$ 65. $a^6 - 9$
67. $-x^2y^2 - axy^2 + bx^2y + abxy$ 69. $-z^4a + z^3a - 3za + 3a$
71. $-9a^2 + 6ab - b^2$ 73. $y^2 + 2ay + a^2$
75. $abn^3 - abn + ab - an^3 + an - a$ 77. $3a + 3b^3 - 3ab$ 79. $-a^2$
81. $-ab^2$ 83. $6ax - 4x + 4x^3$ 85. $-4my + 8mz - 8myz$

89.

Expression	Mult.	Add/Subt.
(X**2 + 3.0*X−1.0) + (X**2 − 3.0*X−1.0)	4	5
2.0*X**2 − 2.0	2	1
3.0*M*(2.0*M**3 − 4.0*N**2)	7	1
6.0*M**4 − 12.0*M*N**2	7	1
−3.0*A*(4.0*B + A − (2.0*(B−5.0*A) − 3.0*A**2))	7	5
−9.0*A**3 − 33.0*A**2 − 6.0*A*B	7	3

EXERCISE 5.3

1. $2(3x+2)$ 3. $x(x-1)$ 5. $3xy(2y^2-1)$ 7. $x(a-c-dx)$
9. $y(4x - 9y + 3)$ 11. $xy(x^{m-1} - y^{n-1})$ 13. $(x + 5)(x + 1)$
15. $(x - 1)(x + 1)$ 17. $2(2x - 1)(2x + 1)$ 19. $(x - 4)(x + 3)$
21. $(x + 4)(x - 3)$ 23. $z(z + 1)^2$ 25. $4(x - 2)^2$
27. $(z^2 + 1)(z + 1)(z - 1)$ 29. $3(x - 2)(x - 7)$ 31. $z^3(z - 3)^2$
33. $2a(x - 4)^2$ 35. $(x - 3)^2$ 37. $x^n(x - 3)(x + 2)$
39. $(p + 1)^2(p - 1)$ 41. $-8(2x - 3)(2x + 3)$
45. (A+3.0)**3; 2 multiplications, 1 addition
(A**2 + 6.0*A + 9.0)*(A+3.0); 3 multiplications, 3 additions

EXERCISE 5.4

1. Unequal 3. Unequal 5. Equal 7. Equal 9. Unequal
11. Equal 13. Unequal 15. Unequal 17. a) 60/120 b)40/120
c) 50/120 d) 135/120 e) 104/120 f) 25/120 g) 8/120
19. (a) $\dfrac{a}{a^2 - ab}$ (b) $-\dfrac{-a}{a^2 - ab}$ (c) $\dfrac{a^2}{a^2 - ab}$ (d) $\dfrac{-a^4(a - b)}{a^2 - ab}$
21. $-\dfrac{-1}{b}; -\dfrac{1}{-b}; \dfrac{-1}{-b}$ 23. $-(-b)$
25. $-\dfrac{-a}{b - a}; \dfrac{a}{-(b - a)}; -\dfrac{-a}{-(b - a)}; -\dfrac{a}{a - b}; \dfrac{-a}{a - b}$

27. $-\frac{-(a+1)}{a}$; $\frac{-(a+1)}{-a}$; $-\frac{a+1}{-a}$

29. $\frac{a^2c}{bd}$ 31. $-\frac{a}{b}$ 33. $-\frac{x^2-1}{y(y^2+2)}$ 35. $\frac{-3x^3z^2}{6y(x+y)}$

37. $-\frac{5(a^3-b^3)}{a^2}$ 39. $\frac{a}{a}=1$ 41. $\frac{5x^2+4x+1}{x(x+1)}$

43. $\frac{1-b}{a-b}$ 45. $-\frac{1}{x^2-1}$

47.

	Expression	Mult/Div.	Add/Subt.
#31	A*(−1.0/B)	2	1
	−A/B	1	1
#33	((X+1.0)/(−Y))*((X−1.0)/(Y**2 + 2.0))	4	4
	−(X**2 − 1.0)/(Y*(Y**2 + 2.0))	4	3
#35	((3.0*X**3)/(2.0*Y))*(−Z**2/(3.0*(X+Y)))	9	2
	(−3.0*X**3*Z**2)/(6.0*Y*(X+Y))	8	2
#39	((2.0+A)/A)+(−2.0/A)	2	3
	1.0	0	0
#41	(2.0*X/(X+1.0))+((3.0*X+1.0)/X)	4	3
	(5.0*X**2 + 4.0*X + 1.0)/(X*(X+1.0))	5	3
#43	1.0/(A−B) + B/(B−A)	2	3
	(1.0−B)/(A−B)	1	2

EXERCISE 5.5

1. $\frac{5x^2}{2}$ 3. $\frac{4x^2y}{-z}$ 5. $\frac{1}{abc}$ 7. $x-2$ 9. $2xz^2-4z$

11. $-\frac{a+b}{b}$ 13. $\frac{1-x}{x}$ 15. $\frac{-(x+4)}{x}$ 17. $\frac{x^2}{2}$ 19. $\frac{-2b^3}{3a^2}$

21. $\frac{5y}{y+4}$ 23. $\frac{x^2}{2(x+1)}$

25.

	Expression	Mult/Div.	Add/Subt.
#7	(2.0*X−4.0)/2.0	2	1
	X−2.0	0	1
#11	(A**2 − B**2)/(B**2 − A*B)	5	2
	−(A+B)/B	1	2
#17	(X*Y)/(4.0*Z)*(2.0*X*Z/Y)	7	0
	X**2/2.0	2	0
#21	(Y−4.0)*(5.0*Y/(Y**2−16.0))	4	2
	5.0*Y/(Y+4.0)	2	1

EXERCISE 5.6

1. $\frac{45}{49}$ 3. $\frac{2x^2y}{z^2}$ 5. $-\frac{r^2s^3}{9t^6}$ 7. 1 9. $\frac{a^2+b^2}{a+b}$ 11. $\frac{x}{x+1}$ 13. $\frac{2xyz}{x-1}$

15.

	Expression	Mult/Div	Add/Subt
#5	(4.0*R*S/(6.0*T**3))/(6.0*T**3/(−R*S**2))	13	1
	−R**2*S**3/(9.0*T**6)	11	1
#13	X*Y/(5.0*Z**2)/(1.0/(X + 1.0))*(10.0*Z**3/(X**2 − 1.0))	12	2
	2.0*X*Y*Z/(X − 1.0)	4	1

EXERCISE 5.7

1. $\frac{13}{6}$ 3. $\frac{3}{10}$ 5. $\frac{15}{44}$ 7. $\frac{3x-ab^2}{abx}$ 9. $\frac{a^2-b^2}{ab}$ 11. $\frac{11x+13}{12}$

13. $-\frac{2x}{3}$ 15. $\frac{2x-a+b}{(x-a)(x+b)}$ 17. $\frac{2}{z-1}$ 19. $\frac{6x+3}{x^2+x-2}$

21. $\frac{(x-y)}{y(x-z)}$ 23. $\frac{a}{b+a}$

EXERCISE 5.8

1. 2^2 3. 10^1 5. a^3 7. z^4 9. $1/2^3$

11. $1/10^2$ 13. $1/2^3$ 15. $1/2^5$ 17. $1/x^3$ 19. 2^4

21. 2^5 23. 10^9 25. 10^0 27. x^5 29. a^{13}

31. b^2 33. x^0 35. 10^{-8} 37. 10^{-3} 39. x^{-1}

41. x^{-5} 43. 10^1 45. 2^{-1} 47. 2^0 49. x^1

51. x^0 53. z^{-2} 55. $(x^a \cdot x^b)x^c = x^a(x^b \cdot x^c)$

$x^{a+b} \cdot x^c = x^a \cdot x^{b+c}$

$x^{a+b+c} = x^{a+b+c}$

59. $4a^2x^4$ 61. $25a^2$ 63. $-a^8b^{12}c^4$ 65. $b^{2m}x^{2mn}$

67. $2^{xz}x^{2z}y^z$ 69. $(-a)^mx^{2m}$ 71. $(-x)^ay^{2a}$

73. $\frac{y^4}{x^4}$ 75. $\frac{y^{10}}{x^{5a}z^{5b}}$ 77. $\frac{4^na^{2n}}{x^{2n}}$ 79. $\frac{a^{12}}{b^6}$ 81. $\frac{y^{2a}}{x^{2a}}$

83. $\frac{(-1)^m}{2^mx^{3m}}$ 85. $\frac{a^nx^{2mn}y^{3n}}{3^n}$ 87. $\frac{108}{y^4}$ 89. $\frac{a^2}{b}$ 91. $\frac{9}{4y^3}$

93. $\frac{36a^2b^4}{c^2}$

95. #65 (B*X**IN)**(2*IM)
B**(2*IM)*X**(2*IM*IN)

#69 (−1.0*A*X**2)**IM
−A**IM*X**(2*IM)

#93 (3.0*A*B**2/C**3)**2*(1.0/(2.0*C**2))**(−2)
36.0*A**2*B**4/C**2

EXERCISE 5.9

1. 4 3. −2 5. 4 7. 25 9. 1/2 11. 2/3

13. 5 15. 8 17. $3\sqrt{2}$ 19. $2\sqrt[3]{3}$ 21. $x\sqrt{5ax}$

23. $\sqrt{3}/2$ 25. $\sqrt[3]{3}/2$ 27. $1/a$ 29. $\sqrt[3]{4x^2}/2yz$

31. $2\sqrt{15}$ 33. $5\sqrt[3]{10}$ 35. $xy^2\sqrt{2}$ 37. $\sqrt{yz}/z$

39. #29 (4.0*X**2/(8.0*Y**3*Z**3))**0.3333
(4.0*X**2)**0.3333/(2.0*Y*Z)

#30 ((A + B)/(A**2*B**2))**0.5
(A + B)**0.5/(A*B)

#36 (3.0*X*Y**2)**0.3333*(X**2*Y**5)**0.3333
X*Y**2*(3.0*Y)**0.3333

EXERCISE 6.1

1. Conditional 3. Identity 5. Identity 7. Identity

9. Identity 11. −7, $x = 1$ 13. −3/4, $x = 1$ 15. −10, $x = 4$

17. 1/4, $x = 2$ 19. 3, $x = 15$ 21. 1/15, $x = 14$ 23. No

25. No 27. No 29. Yes

31. A = P*(1.0 + R/N)**(N*T)

33. Z = 2.0*V**5 − 5.0*V**2 + 8.0*V − 9.0

35. D = (S − P)/(S*T)

EXERCISE 6.2

1. $x = 30/7$ 3. $y = 2$ 5. Contradiction 7. $x = 7/6$

9. $x = 42$ 11. $z = 61/2$ 13. Identity 15. $x = -21/2$

17. $y = -17$ 19. $y = -7/5$ 21. $x = 9$ 23. $x = 26/9$

25. 4000 27. 400, 800, 1000 29. 80 minutes 33. 23

35. 23 hrs.; 26 hrs.

EXERCISE 6.3

1. $d = \dfrac{s-p}{sn}$ 3. $c = \dfrac{AB-D}{A}$ 5. $y = -\dfrac{3}{z-5}$ 7. $c = \dfrac{ak-b}{a-1}$

9. $F = \dfrac{9C}{5} + 32$ 11. $R_2 = \dfrac{RR_1R_3}{R_1R_3 - RR_3 - RR_1}$ 13. $d = \dfrac{i}{1+it}$

EXERCISE 6.4

1. $x > 3/2$ 3. $y > -3$ 5. $x \geq 1$ 7. $x > 0$ 9. $x \leq -7$

11. $x \leq -7/3$ 13. $2 < x < 3$ 15. $-8/25 < w < 0$ 17. $x > 3$ and $x < -1$

EXERCISE 6.5

1. −1, −2 3. −1/2, −1 5. 0, −5/3 7. 1, −1

9. -2 11. $3, -2$ 13. $1, -5$ 15. $4, 5$ 17. $4, -5$
19. $2, -8$ 21. -2 23. $4, -4$ 25. $6, -6$ 27. 5

EXERCISE 6.6

1. 4 3. $1/16$ 5. b^2 7. $0, 6$ 9. $-1/3, 1$
11. $(-a \pm \sqrt{a^2-4b})/2$ 15. $5, -3, 1$ 17. $3, 9r, -5s$ 19. $m, -n, -1$
21. $\dfrac{-5 \pm \sqrt{13}}{6}$ 23. $\dfrac{1 \pm \sqrt{29}}{2}$ 25. $\dfrac{-5 \pm 2\sqrt{5}}{2}$ 27. $-\dfrac{2}{5}, \dfrac{4}{3}$
29. 8 31. $12, 13, 14$ 33. 20
35. First run; 60 values for A, 60 values for B
Second run; 75 values for A, 72 values for B

EXERCISE 7.2

1. Function 3. Function (not defined at $x = 0$) 5. Function
7. Not a function 9. Function 11. $1, 2, 6$ 13. $y \leq 1$
15. $y \leq -1$ 17. $y \geq 0$ 19. $y > 0$ 21. $1 \leq y < 0$
23. $y = \frac{1}{2}(x-3)$
25. $y = \dfrac{z}{4} - 1$ 27. $y = z^2 - 2$ 29. $y = \dfrac{1}{1-z}$ 31. (a) 2
(b) 0 (c) $a^2 - 3a + 2$ (d) $b^2 - 3b + 2$ (e) $(b^2 - a^2) - 3(b - a)$
33.
$$y = \begin{cases} 120x & \text{if } x \leq 5 \\ 50 + 110x & \text{if } 5 < x \leq 20 \\ 250 + 100x & \text{if } x > 20 \end{cases}$$
$3050
35.
```
   Y = X**2
   IF(X) 5,5,4
 4 Y = -Y
 5 ....
```

EXERCISE 7.3

1. $y = 2x + 1$ 3. $y = -x + 1$ 5. $y = -2x + 1$ 7. $y = -2$
9. $y = 4$ 11. $y = -3$ 13. $y = -\dfrac{x}{4} + 1$ 15. $3x + y - 2 = 0$
17. $4x - y + 3 = 0$ 19. $3x - y + 2 = 0$ 21. $2x - y = 0$
23. $y - 5 = 0$ 25. $2y - 1 = 0$ 27. $x - 2y + 4 = 0$
29. $8x - 6y + 3 = 0$

EXERCISE 7.4

1.

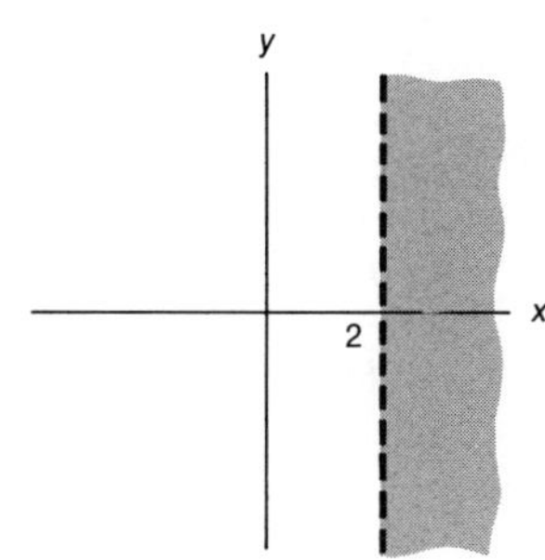

EXERCISE 8.1

1. min $(-3/2, -1/4)$; $x = -1, -2$
3. min $(0, 0)$; $x = 0$
5. max $(-1/2, 5/4)$; $x = (-1 \pm \sqrt{5})/2$
7. min $(-3/4, -1/8)$; $z = -1/2, -1$
9. max $(0, 1)$; $x = 1, -1$
11. max $(1/2, -3/4)$; $x = (1 \pm \sqrt{-3})/2$
13. min $(1/2, -81/4)$; $x = 5, -4$
15. max $(1/4, 825/64)$, min $(3, -4)$; $x = -1, 2, 4$
17. min $[(-2/3)(2 - \sqrt{7}), (-16/27)(10 + 7\sqrt{7})]$,
max $[(-2/3)(2 + \sqrt{7}), (16/27)\ (7\sqrt{7} - 10)]$; $x = -2, -4, 2$

EXERCISE 8.2

3. $3/5, 4/5, 3/4$
5. $3/5, -4/5, -3/4$
7. $-\sqrt{2}/2, \sqrt{2}/2, -1$
9. 1, 0, undefined
11. $1/2, \sqrt{3}/2, \sqrt{3}/3$
13. $-\sqrt{15}/4, 1/4, -\sqrt{15}$
15. first: $\sin\theta = \sqrt{2}/2$, $\cos\theta = \sqrt{2}/2$; third: $\sin\theta = -\sqrt{2}/2$, $\cos\theta = -\sqrt{2}/2$
17. first: $\cos\theta = 4/5$, $\tan\theta = 3/4$; second: $\cos\theta = -4/5$, $\tan\theta = -3/4$
19. third: $\cos\theta = -\sqrt{6}/3$, $\tan\theta = \sqrt{2}/2$; fourth: $\cos\theta = \sqrt{6}/3$, $\tan\theta = -\sqrt{2}/2$
21. first: $\cos\theta = 12/13$, $\tan\theta = 5/12$; second: $\cos\theta = -12/13$, $\tan\theta = -5/12$

23.

Angle	Sine	Cosine	Tangent
120°	0.8660	−0.5	−1.7321
150°	0.5	−0.8660	−0.5774
180°	0.	−1.0	0.
210°	−0.5	−0.8660	0.5774
240°	−0.8660	−0.5	1.7321
270°	−1.0	0.	undefined
300°	−0.8660	0.5	−1.7321
330°	−0.5	0.8660	−0.5774
360°	0.	1.0	0.
390°	0.5	0.8660	0.5774
420°	0.8660	0.5	1.7321
450°	1.0	0.	undefined

25. .7660, .7660, −.3640, −.1737, −.1737, .2588, .9659

27. 80°, 20°, 40°, 75°, 55°

29. $\sin\theta = \frac{y}{r}$; $\cos\theta = \frac{x}{r}$; $\tan\theta = \frac{x}{y}$

$$\frac{\sin\theta}{\cos\theta} = \frac{y/r}{x/r} = \frac{y}{r}\cdot\frac{r}{x} = \frac{y}{x} = \tan\theta$$

31. C = SQRT(A**2 + B**2 + 2.0*A*B*COS(AC/57.296))

EXERCISE 8.3

7. $3 = \log_2 8$ 9. $0 = \log_2 1$ 11. $-2 = \log_{10} 0.01$
13. $3/4 = \log_{16} 8$ 15. $-2 = \log_{1/4} 16$ 17. $3 = \log_a b$
19. $2^4 = 16$ 21. $8^{-1} = 1/8$ 23. $10^1 = 10$ 25. $5^{-3} = 1/125$
27. $1/8^{-1} = 8$ 29. $e^x = 5$ 31. 2 33. 3
35. Not defined 37. 10,000 39. 1 41. 1/512
43. 2 45. 8

EXERCISE 8.4

1. .30 3. 1.48 5. 3.95 7. 3.90 9. 2 11. 60
13. 100,000 15. 300 17. $10^{0.90} = 8$ 19. $10^{0.96} = 9$
21. $10^{1.60} = 40$ 23. $10^{4.48} = 30{,}000$ 25. $10^{4.95} = 90{,}000$
27. $10^{0.6} = 4$ 29. $10^{2.0} = 100$ 31. $10^{1.60} = 40$ 33. $10^{1.60} = 40$

EXERCISE 8.5

1. 0.50 3. 1.55 5. 3.43 7. 4.66 9. 9.4
11. 370 13. 220 15. 61,000 17. 160 19. 490
21. 217 23. 450 25. 270 27. 31,000

EXERCISE 8.6

1. 0.3927 3. 4.3927 5. 6.3927 − 10 7. 1.9945
9. 2.7173 11. 9.6007 − 10 13. 1.9403 15. 0.4343
17. 99.5 19. 0.348 21. 98.54 23. 296600
25. 4.569×10^{-7} 27. 9.753×10^{6} 29. 381.1 31. 210.6
33. 19,900 35. 0.05783 37. 8.815 39. 1536
41. 1.770 43. 1.984 45. (a) \$1629 (b) \$131,800
47. 2.132

EXERCISE 9.1–1

1. (1, 2) 3. (−2, 3) 5. (−2, −5) 7. (−4, 0)

EXERCISE 9.1–2

1. (1, −4) 3. (1, 2) 5. (3, 3) 7. −38/11, 73/11
9. (5, 2) 11. (14, 6)

EXERCISE 9.1–3

1. 5, 2 3. −5/9, 5/9 5. −7/4, −2 7. 54/35, 39/35
9. (5, 2) 11. 19/9, 4/9
13. $\frac{b+c}{2}, \frac{b-c}{2}$ 15. $\frac{b_2c_1 - b_1c_2}{a_1b_2 - a_2b_1}, \frac{a_1c_2 - a_2c_1}{a_1b_2 - a_2b_1}$

EXERCISE 9.2

1. (0, −2) 3. Inconsistent 5. (81, 9) 7. Inconsistent
9. $\left(\frac{a(1-b^2)}{a^2-b^2}, \frac{b(a^2-1)}{a^2-b^2}\right)$

EXERCISE 9.3

1. (1, 1, −1) 3. (3/5, −12/5, −3/5) 5. Inconsistent
7. (51/28, 17/28, 5/28) 9. Inconsistent 11. (1/2, 1, 1/3)
13. (−1/7, 4/7, 0) 15. (1, 1, −1) 17. (1/2, 0, −1)
19. Inconsistent 21. (1/2, 1, 1/3) 23. Inconsistent

EXERCISE 9.4

1. $74\frac{1}{2}, 24\frac{1}{2}$ 3. 33/7 5. No 7. $y = 2x^2 - 5x - 4$

EXERCISE 9.5–1

1. (2, 3) 3. (−1, 1) 5. 61/13, −36/13 7. −5/58, −123/58
9. Inconsistent 11. Inconsistent

EXERCISE 9.5–2

1. (0, 1/2, −1) 3. (1, 1, −1)

EXERCISE 10.1

1. 2×1 3. 2×3, 1 5. 3×3, 5 7. 2×4, 5 9. 2×1
11. $a = 1, b = 5, c = 3, d = -1$ 13. $x_1 = 2, x_2 = 1$ 15. $x_1 = 4, x_2 = 3$
17. $\begin{bmatrix} -4 & 6 \\ 10 & 0 \\ 7 & 3 \end{bmatrix}$ 19. $\begin{bmatrix} -5 & -6 & -2 \\ -9 & 0 & -2 \\ 0 & -1 & 0 \end{bmatrix}$ 21. $\begin{bmatrix} a_1+4 & b_1+3 \\ a_2+2 & b_2-1 \end{bmatrix}$
23. $\begin{bmatrix} 0 & 0 \\ 0 & 0 \\ 0 & 0 \end{bmatrix}$ 25. $\begin{bmatrix} 1 & 3 \\ 5 & 9 \end{bmatrix}$ 27. $\begin{bmatrix} 1 & -2 & -3 \\ 3 & -2 & 1 \\ 5 & 2 & 7 \end{bmatrix}$

EXERCISE 10.2

1. $\begin{bmatrix} -3 & 11 \\ -10 & -11 \\ 4 & -6 \end{bmatrix}$ 3. $\begin{bmatrix} 9 \\ -23 \\ 17 \end{bmatrix}$ 5. $\begin{bmatrix} 24 & -17 \\ 4 & -1 \end{bmatrix}$ 7. $\begin{bmatrix} 2 & 27 \\ 12 & 10 \end{bmatrix}$

9. $\begin{bmatrix} 8 & 1 \\ 2 & -2 \end{bmatrix}$ 11. $\begin{bmatrix} 6m + 8z \\ -2m + 3z \end{bmatrix}$ 13. $\begin{bmatrix} 8x + 6z \\ x + 5z \end{bmatrix}$

15. $\begin{bmatrix} -10 & 9 & -56 \\ 15 & 45 & 51 \\ 21 & 101 & 21 \end{bmatrix}$ 17. $\begin{bmatrix} 1 & 2 & 5 \\ 5 & 10 & -15 \\ 8 & -9 & 1 \end{bmatrix}$ 19. $\begin{bmatrix} 2x + 5z \\ 3x - y \\ 5x + 2y - z \end{bmatrix}$

21. $\begin{bmatrix} 1 & 2 & 4 \\ 15 & -10 & 25 \\ 4 & 7 & -5 \end{bmatrix}$ 23. $\begin{bmatrix} 1 & 2 & 4 \\ 3 & -2 & 5 \\ -4 & -7 & 5 \end{bmatrix}$ 25. $\begin{bmatrix} 4 & 7 & -5 \\ 3 & -2 & 5 \\ 1 & 2 & 4 \end{bmatrix}$

27. $\begin{bmatrix} 1 & 2 & 4 \\ 3 & 0 & 9 \\ 4 & 7 & -5 \end{bmatrix}$ 29. $\begin{bmatrix} a_{11} & a_{13} & a_{12} \\ a_{21} & a_{23} & a_{22} \\ a_{31} & a_{33} & a_{32} \end{bmatrix}$ 31. $\begin{bmatrix} a_{11} + a_{12} & a_{12} & a_{13} \\ a_{21} + a_{22} & a_{22} & a_{23} \\ a_{31} + a_{32} & a_{32} & a_{33} \end{bmatrix}$

33. $a = -5$, $b = -7$, $c = 2$, $d = 3$ (both cases)

35. $\begin{bmatrix} 0 & 0 \\ 0 & 0 \end{bmatrix}$ 37. $\begin{bmatrix} 10 & 1 & -1 \\ -1 & 22 & -39 \\ 15 & 21 & -36 \end{bmatrix}$ (both cases)

EXERCISE 10.3

1. $B = A^{-1}$ 3. $B = A^{-1}$ 5. $B = A^{-1}$ 7. $B \neq A^{-1}$ 9. $B = A^{-1}$

11. $B \neq A^{-1}$ (but $B \simeq A^{-1}$) 13. Exercise 8, (1, 8)

15. Exercise 1, (8, –7) 17. Exercise 9, (12, 5)

19. Exercise 12, (3.289, 0.143) 21. Exercise 10, (–3, –4)

23. $B = A^{-1}$ 25. $B = A^{-1}$ 27. $B = A^{-1}$ 29. $B \simeq A^{-1}$

31. Exercise 22, (1, –1, 1) 33. Exercise 23, (1, 5, –2)

35. Exercise 30, (–2.9998, 2.9984, 1.0002) 37. (1, –1, 1, –1)

EXERCISE 10.4

1. $\begin{bmatrix} -4 & 7 \\ 3 & -5 \end{bmatrix}$; $x = 2$, $y = -1$ 3. $\begin{bmatrix} -2 & -3 \\ -1 & -2 \end{bmatrix}$; $x = 8$, $y = 9$

5. $\begin{bmatrix} 3/11 & -4/11 \\ 2/11 & 1/11 \end{bmatrix}$; $x = 1$, $y = -1$ 7. $\begin{bmatrix} 2/11 & 1/11 \\ -3/11 & 4/11 \end{bmatrix}$; $x = 5$, $y = -3$

9. $\begin{bmatrix} -1 & -3 & 1 \\ 16 & 45 & -14 \\ 10 & 29 & -9 \end{bmatrix}$; $x = 1$, $y = 2$, $z = 3$

11. $\begin{bmatrix} -.156 & .031 & .406 \\ -.375 & -.125 & .375 \\ .156 & -.031 & -.073 \end{bmatrix}$; $x \simeq 0.99$, $y \simeq -0.99$, $z \simeq 0.99$

13. $\begin{bmatrix} -.075 & .015 & -.209 \\ .157 & .269 & .239 \\ .201 & .060 & .164 \end{bmatrix}$; $x \simeq 5.002$, $y \simeq 4.008$, $z \simeq -0.988$

15. $\begin{bmatrix} 0 & -2 & 1 & 1 \\ -3 & -2 & 2 & 1 \\ 1 & 1 & -1 & 0 \\ 3 & 3 & -2 & -2 \end{bmatrix}$; $w = 2$, $x = -1$, $y = -2$, $z = 3$

17. No solution; inconsistent system

EXERCISE 11.1

1. yes, no, no, yes, yes, no, no 3. (a) 270, 290 (b) 0, 68 (c) 0, 120 (d) 74, 0 (e) 100, 0 (f) 4, 6 5. (a) $370 (b) $510 7. (a) $530 (b) $540 9. $203 11. $18

EXERCISE 11.2

1. (a) (4/5, 24/5) (b) $(15, 10)\begin{bmatrix} 4/5 \\ 24/5 \end{bmatrix} = 60$, $(20, 5)\begin{bmatrix} 4/5 \\ 24/5 \end{bmatrix} = 40$ (c) 57 (d) $.41 3. (a) 25/17 36/17 (b) deficiency (c) $.28

5. $.25 7. $.24 9. 29 6/7 11. 52 2/7 13. 11

15. (a) 68¢ (b) 23¢ (c) 39¢ 17. 20

EXERCISE 11.3

1. 24 3. 43 5. 33 7. 81 13. 14 15. 23

EXERCISE 11.4

1. 18 3. 45 5. 10 7. 26 9. 23

EXERCISE 11.5

3. $220 5. 25 7. 11 9. 19 11. 22 13. 19
15. 42 17. 71 19. 77 4/7 21. 68

EXERCISE 12.1

1. 2, 4, 6, 8, 10 3. 1, 4, 9, 16, 25 5. 0, 3, 8, 15, 24

7. −1, 1, −1, 1, −1 9. 0, 2, 0, 2, 0 11. 2, 9/4, 64/27, 625/256, 7776/3125

13. −1, −4, −9, −16, −25 15. 2, 4, 6, 8, 10 17. −1/3, 0, 1/3, 2/3, 1

19. −1, −3, −9, −27, −81 21. 1, 2, 6, 24, 120

23. 1, 1, 1/2, 1/6, 1/24   25. 1, 8.5, 5.2, 4.1, 4.0, 4.0, 4.0, 4.0
27. $a_{20} = 4.0$; $4.0 = \sqrt{16}$ 29. $a_n = 2n$ 31. $a_n = 2n - 1$
33. $a_{n+2} = a_{n+1} + a_n$, $a_1 = a_2 = 1$ 35. $a_n = (-1)^{n+1}$
37. $a_n = (-1)^n \dfrac{n+1}{n}$ 39. $a_n = 1 - (0.1)^n$
41. 39.0625; 9.08×10^7; 7.94×10^{397}

EXERCISE 12.2

1. Arithmetic; $a_n = 2n$; $a_9 = 18$ 3. Arithmetic; $a_n = 3n - 2$; $a_7 = 19$
5. Geometric; $a_n = -(81/2)(-2/3)^n$; $a_6 = -32/9$ 7. Geometric; $a_n = (1/2)^n$; $a_9 = 1/512$ 9. Arithmetic; $a_n = 13n - 26$; $a_7 = 65$
11. Arithmetic progressions: 1, 2; Geometric progressions: 7, 8, 10
13. $a_n = 60n - 104$; $a_n = 4(2^n)$ 15. Arithmetic progression: $A_n = 5000 + 600(n - 1)$; $10,400   17. Fifth year

EXERCISE 12.3

1. Diverges   3. Diverges   5. Diverges   7. Diverges; $|a_m| \leq 1$
9. Diverges; $|a_n| \leq 2$ 11. Converges; $a_i \to e$ as $i \to \infty$; $|a_i| \leq e$
13. (a) 1, 1/2, 1/4, 1/8   (b) 1, 1/4, 1/16, 1/64   (c) 1, 1/8, 1/64, 1/512

EXERCISE 12.4

1. 15   3. 55   5. −39   7. 73/12   9. −3   11. 126
13. 84   15. −31/12

EXERCISE 12.5

1. Arithmetic progression; 252   3. Geometric progression; 3280
5. Geometric progression; 0   7. Arithmetic progression; 175
9. Geometric progression; 255/64   13. 32,767; 1,073,741,823; 1,073,741,824
15. $3.68, $69.83   17. 42.9 ft

EXERCISE 12.6

1. 2   3. 4/3   5. Diverges   7. 256/3   9. $1/(1-x)$ 11. 2/3
13. 5/9   15. 9/111   17. 14/110   19. 1/10   21. $|r| \nless 1$
23. 72.90, 65.61, 59.05

EXERCISE 13.1

1. $\sqrt{7}$

n	x_n
0	1
1	4
2	2
3	2

$\sqrt{8}$

n	x_n
0	1
1	4
2	3
3	2
4	3
5	2
.	.
.	.
.	.
.	.

$\sqrt{9}$

n	x_n
0	1
1	5
2	3
3	3

$\sqrt{10}$

n	x_n
0	1
1	5
2	3
3	3

3.

n	x_n
0	150
1	153
2	153

EXERCISE 13.2

1. (a)

n	x_n
1	5.50
2	5.25
3	5.38
4	5.44
5	5.47
6	5.46
7	5.47
8	5.47

Root = 5.47

(b)

n	x_n
1	−1.50
2	−1.25
3	−1.38
4	−1.44
5	−1.47
6	−1.46
7	−1.47
8	−1.47

Root = −1.47

3. (a)

n	x_n
1	4.50
2	4.25
3	4.13
4	4.19
5	4.16

Root = 4.16

3. (b)

n	x_n
1	−2.50
2	−2.25
3	−2.13
4	−2.19
5	−2.16

Root = −2.16

5.

n	x_n
1	−1.50
2	−1.25
3	−1.38
4	−1.32
5	−1.35
6	−1.34
7	−1.33
8	−1.33

Root = −1.33

7.

n	x_n
1	5.50
2	5.25
3	5.38
4	5.44
5	5.41
6	5.40
7	5.39
8	5.39

Root = 5.39

EXERCISE 13.3

1. (a)

n	x_n
1	5.43
2	5.46
3	5.46

Root = 5.46

(b)

n	x_n
1	−1.43
2	−1.46
3	−1.46

Root = −1.46

3. (a)

n	x_n
1	4.14
2	4.16
3	4.16

Root = 4.16

3. (b)

n	x_n
1	−2.14
2	−2.16
3	−2.16

Root = −2.16

5.

n	x_n
1	−1.17
2	−1.25
3	−1.29
4	−1.31
5	−1.32
6	−1.32

Root = −1.32

7.

n	x_n
1	2.89
2	2.92
3	2.92

Root = 2.92

EXERCISE 13.4

1.

x	y
0.00	0.00
−4.33	−3.40
−3.20	−4.27
−2.91	−4.04
−2.99	−3.98
−3.01	−4.00
−3.00	−4.00

3.

x	y
0.00	0.00
1.20	0.25
1.15	1.15
0.97	1.11
0.98	0.98
1.00	0.99
1.00	1.00

5. Does not converge

7.

x	y
0.00	0.00
−0.25	3.50
0.63	3.13
0.53	4.45
0.86	4.30
0.83	4.79
0.95	4.75
0.94	4.93
0.98	4.91
0.98	4.97
0.99	4.97
0.99	4.99
1.00	4.99
1.00	5.00

9.

x	y
0.00	0.00
0.50	−5.67
−2.34	−5.83
−2.42	−4.89
−1.95	−4.86
−1.93	−5.02
−2.01	−5.02
−2.01	−5.00
−2.00	−5.00

EXERCISE 13.5

1.

x	y
−4.33	−4.27
−2.91	−3.98
−3.01	−4.00
−3.00	−4.00

3.

x	y
1.20	1.15
0.97	0.98
1.00	1.00

5. Does not converge

7.

x	y
−0.25	3.13
0.53	4.30
0.83	4.75
0.94	4.91
0.98	4.97
0.99	4.99
1.00	5.00

9.

x	y
0.50	−5.83
−2.42	−4.86
−1.93	−5.02
−2.01	−5.00
−2.00	−5.00

11.

x	y	z
0.14	−1.17	2.33
1.48	−1.84	2.83
1.88	−1.96	2.96
1.97	−1.99	2.99
1.99	−2.00	3.00
2.00	−2.00	3.00

13.

w	x	y	z
6.00	3.50	5.17	1.43
3.32	1.81	2.82	3.54
2.70	3.19	2.86	3.02
3.03	3.06	3.04	2.95
3.02	2.97	3.00	3.01
2.99	3.00	2.99	3.00
3.00	3.01	3.00	2.99
3.01	3.00	3.01	3.00
3.00	3.00	3.00	3.00

EXERCISE 14.1

1. No 3. No 5. Yes 7. Yes

9.

11.

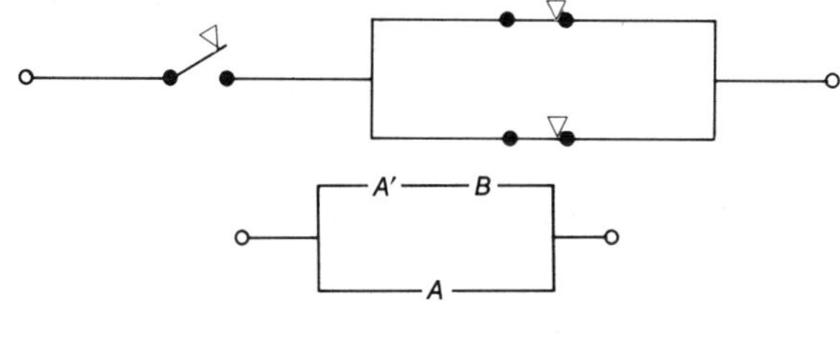

0111

13. TTTF—If 0 replaces F, 1 replaces T and 0111 is reversed, they are the same. A truth table begins with TT and ends with FF while a table showing current flow is reversed.

15. $A \cup A' = U, A \cap A' = \phi$

EXERCISE 14.2

1. (a) $(A+B)C$ (c) $A(B+C)D$ (e) $(B+C)+D$
   (g) $AB(C+A)+A'D$ (i) $A(BC+A)A'$

3. Current flows: a, d, e, g, h

4. (a) A — B — C / D

(g)

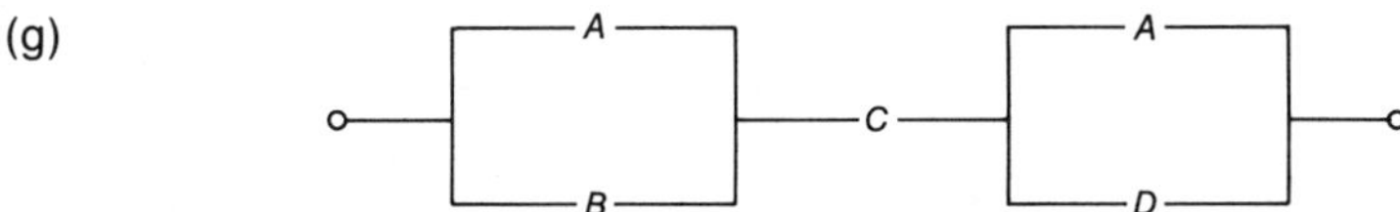

5. ABD and AC 6. (g) ABC, ABA, $A'D$

7. (a) P.AND.(Q.OR.R); $A \cap (B \cup C)$
   (c) A.AND.B.OR.B.AND.C; $(A \cap B) \cup (B \cap C)$

9. (b) 011, 101, 110, 111 (c) 101, 110

EXERCISE 14.3

1. 00011111—using the values of A, B, and C as shown in Table 13.7

3. (a) $BC+1=1$ (b) $A+B\cdot 1=A+B$ (c) $A+B+1=1$

(d) $(A + B) \cdot 1 = A + B$; (a) and (c) are equivalent, (b) and (d) are equivalent. 5. (a) $B + BC = B(1 + C) = B \times 1 = B$
(b) $AB + AB' = A(B + B') = A \times 1 = A$
(c) $A + AB + AC = A(1 + B + C) = A \times 1 = A$
(d) $A'A + A' = A'(A + 1) = A' \times 1 = A'$ 7. $(B + A)(B + C)$

9. $1 \times A + 1 \times C$ 11. $C + AD$ 13. $A'A + A'A'$

15. (a) $A \cup (A \cap B)$ (c) A 17. (a) $A \cup (A \cap B) \cup (A \cap C)$ (c) A

EXERCISE 14.4

1. Commutative 3. $A + A' = 1$ 5. Commutative 7. $A'A = 0$

9.

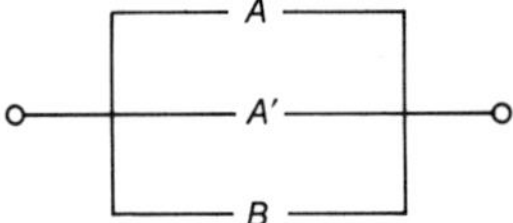

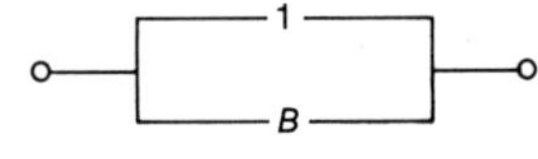

o—A'———A—B—o   o—A'—A———B—o

11. B 0101 13. C 01010101 15. $A \cup U = A$, $A \cup \phi = A$

EXERCISE 14.5

1. AB 3. B 5. $A' + C$ 7. $A' + B$

9. 0 11. AB 13. 1 15. $A + B$

17. A 10; $A + A$ 10; $A \cdot A$ 10

20. (b) $A(BA + A')B + A'(AC + BA') = B$

21. Q.OR..NOT.P 23. $A' \cup B$

EXERCISE 14.6

1. 1000

3. I $A' + B$; II AB'; III $B' + A$; IV $B + A$; V B'; VI $A'B' + AB$

5. $B' + C$ 7. $(B' + C)'$ 9. $A' + B + C'$

11. $A'B'(A' + B') = A'B'$ 15. $A + B' + C'$ 17. $A + B + C$

INDEX

Abacus, 67
Abscissa, 248
Absolute value, 53, 54
Achilles and the tortoise, 412, 413
Adding machine, Pascal, 68
Addition, of algebraic quantities, 160-162
of binary numbers, 85-87
of decimal numbers, 44, 45, 53, 54, 55
of hexadecimal numbers, 105-107
of matrices, 324
of octal numbers, 105-107
of rational numbers, 174
Additive identity, 53
Additive inverse, 56, 202, 203
Address, computer, 99
Algebraic expression, 154
Algebraic fraction, 170
Algol, 130, 140, 156
Algorithm, 2-6
Alternating sequence, 398
AND gate, 472
AND statements, 114
Antilogarithm, 280
Approximation, in Fortran, 171
Arbitrary constant, 166, 225
Argument, 245
Fortran, 235
Arithmetic, integer, 56, 57
Arithmetic operations, decimal numbers, 44-64
hexadecimal numbers, 105-108
octal numbers, 105-108
Arithmetic progression, 395-397, 408-410
Arithmetic statement, Fortran, 199, 200
Array, 322
Artificial variable, 383
Assignment statement, Fortran, 199, 200
Associative property, 52
union, 35
intersection, 35
Boolean algebra, 464, 467
Axis, 237

Balgol, 131, 140
Bar graph, 237
BCD, 70, 71
Binary arithmetic, 85-94
addition, 85-87
multiplication, 88
subtraction, 87-88, 91-94
using complements, 91-94
Binary coded decimal format, 70
Binary complement, 91-93
Binary device, 69
Binary digit, 73
Binary fraction, 74
conversion of, 81-85
Binary numbers, 71-95
Binary point, 74
Binomial, 162, 178
Bit, 73

Block diagram, 6
Bolzano method, 421-426
Boolean algebra, 39, 446-475
Boolean operators, 446
Boolean properties, 32, 457-469
Boolean variables, 446
Byte, 98, 99

Carry, binary, 86
Cartesian coordinates, 248, 249
Characteristic, 281
Cipherization, 60
Circuit, 447
Closure, 45, 46
Cobol, 130, 140, 156, 198
Coding, computer storage,
binary, 71, 72, 89-92
binary coded decimal, 70, 71
excess-three, 71
two-out-of-five, 71
Coefficient, 154
Coefficient of restitution, 395
Column matrix, 323
Column vector, 323
Common logarithm, 278
Common ratio, in geometric progression, 397
Commutative property, 45, 46
union of sets, 33
intersection of sets, 33
Boolean algebra, 463, 467
Complement, 89-93
sets, 17
circuits, 451
Completing the square, 223
Complex root, 227
Computer, Honeywell Series 15, 94, 99
IBM 650, 71
IBM 1130, 94
IBM System/360, 85
Condition code, 145
Conditional equation, 196-198
Conditionals, logic, 129
Connector symbol, 8
Constraint, 354
Continuity, 238
Continuous function, 239
Convergence, 400-407
of progressions, 401, 402
test, 432, 433
Convergence conditions, 441
of Jacobi method, 432, 433
of Newton's method, 418, 420
Conversion, number base, 74-85
Conversion tables, 101-103
Coordinate, 248, 249
Cosine, 269-271
Cost function, 364
Cramer's rule, 314, 318, 319

Decimal numbers, 44-64
Decision symbol, 8
Degree, 268
DeMorgan's Laws, 119, 473-475
Denominator, 170
Dependent linear system, 302
Dependent variable, 234, 243
Determinant, 312-319
Diagonal, of matrix, 330
Digit, 44
significant, 63, 64
Dimension, of matrix, 323
DIMENSION statement, Fortran, 236
Discontinuous function, 239
Discriminant, 227
Distributive property, 52, 162
sets, 38
Boolean algebra, 459-462, 467
Divergent solutions, Gauss-Seidel method, 439-441
Jacobi method, 431
Diverging sequence, 401
Division, in Fortran, 171, 175
of hexadecimal numbers, 108
of octal numbers, 108
of rational numbers, 179-181
with exponents, 185
Domain, 234, 237, 243
Double-and-add number conversion, 79

Element, of determinant, 313
Elementary transformation matrix, 331, 332
End, logical, 11
Equality, of rational numbers, 171
Equation, 196

Equation (*continued*)
quadratic, 221
Equations, simultaneous, 295-320
Equivalence, 140
Equivalent circuits, 458
Equivalent equations, 200
Equivalent linear system, 299
Excess-three coding, 71
Exclusive OR (EOR) 115
Expansion, of polynomial, 163
Exponent, 59, 60, 184-192, 274
division, 185
Fortran, 188, 189
fractional, 191, 192
multiplication, 184, 185
negative, 185, 186
zero, 185, 186
Exponential function, 273-274
Expression, 154-156
algebraic, 154
Fortran, 155, 156

Factor, 155, 165
prime, 167
Factoring, 165-169
Fixed point arithmetic, 57
Floating point, 104
Flow direction, 7
Flowchart, 6-12
logic, 123
sets, 28
symbols, connector, 8
decision, 8
flow direction, 7
input/output, 7
processing, 7
termination, 8
Formula, recursive, 393, 396, 397
Fortran, 130, 140, 155-159, 171, 175, 199, 200, 228, 229, 350, 405, 406
arithmetic statement, 199, 200
assignment statement, 199, 200
DIMENSION statement, 236
exponent, 188, 189
expression, 155, 156, 162, 163
function, 245
subscript, 325
subscripted variable, 236
variable, 155
Fraction, algebraic, 170
Function, 234-259, 289
continuous, 239
discontinuous, 239
domain of, 234, 237
exponential, 273, 274
in Fortran, 235, 245, 246
linear, 249-254
logarithmic, 274, 275
multivalued, 245
quadratic, 262-267
sequence, 389
trigonometric, 267-272
Functional notation, 244-245
Fundamental principle of fractions, 172

Gauss method, 306-308
Gauss-Jordan method, 308
Gauss-Seidel method, 437-443
Geometric progression, 397, 398, 410, 411
Geometric series, 412-414
Graph, 237-239

Half adder, 473
Half life, 408
Halfword, 94
Hexadecimal arithmetic, addition, 105-107
division, 108
multiplication, 108
subtraction, 107, 108
Hexadecimal number, 95-108
Hexadecimal-decimal conversion table, 103
Hierarchy of operations, 157-159
in Fortran, 189
Hindu-Arabic numeral, 44
Histogram, 237
Honeywell Series 15 computer, 94, 99

IBM 650 computer, 71
IBM 1130 computer, 94
IBM card, 13-15
IBM System/360 computer, 85, 145
Identical equation, 196, 197

Identity, 196
Identity element, Boolean algebra, 465, 467
Identity matrix, 330
Imaginary number, 227
Implication, 140
Inclusive OR, 115
Inconsistent linear system, 302, 303
Independent variable, 234, 243
Indeterminate form, 172
Index, of radical, 191
Inequality, 216-220, 256-259
  addition, 217
  multiplication, 218
Infinite loop, 11
Infinite series, 412-414
Input/output symbol, 7
Integer, 47, 48
Integer arithmetic, 56, 57
Interest, computation of, 390, 391
Interpolation, 282-288
Intersection of sets, 16
Inverse, 48, 56
  additive, 202, 203
  matrix, 334, 335, 342-344
  multiplicative, 203, 204
Irrational number, 49
Iteration, 9
Iterative method, 417

Jacobi method 429-434

Key column, 377
Key variable, 369

Least common denominator (LCD), 182, 183, 206, 207
Library, computer, 245
Linear combination, 299, 300
Linear equations, simultaneous, 295-320
Linear function, 249-254
  slope, 251, 253
  y intercept, 251, 252
Linear inequality, 257-259
Linear interpolation, 282-288
Linear programming, 353-387
Linear system, solution of, determinants, 312-319
  Gauss-Seidel method, 437-443
  Jacobi method, 429-434
  linear combination, 299, 300
  substitution, 296, 297
Logarithm, 275-291
  characteristic, 281
  common, 278
  mantissa, 281
  natural, 278
  table, 287
Logarithmic function, 274, 275
Logical end, 11
Logical equivalence, 117
Loop, 9
  infinite, 11

Mad (Michigan Algorithm Decoder), 130, 140
Magnitude, 53
Mantissa, 281
Mathematical model, 208
Matrix, 322
  addition, 324
  inverse of, 334, 335, 342-344
  multiplication, 327-329
  multiplication by computer, 348-350
Minor, 316
Mixed mode, in Fortran, 188
Model, mathematical, 208, 209
Monomial, 162
Multiplication, binary, 88
  of algebraic quantities, 162, 163
  of hexadecimal numbers, 108
  of matrices, 327-329
  of octal numbers, 108
  of rational numbers, 173, 174, 179-181
  with exponents, 184, 185
Multiplicative inverse, 56, 203, 204

Natural logarithm, 278
Natural number, 44
Negation, 116
Newton's method for square root, 417-420
Number systems, binary, 72-74
  decimal, 44-94
  hexadecimal, 95-98

Number systems (*continued*)
octal, 95-98
Numerals, 44
Numerator, 170

Objective function, 354
Octal numbers, 95-108
Octal arithmetic, addition, 105-107
division, 108
multiplication, 108
subtraction, 107, 108
Octal-decimal conversion table, 102
OR gate, 472
OR statements, 115
Order, of radical, 191
Ordinate, 248
Origin, 248

Parabola, 263
Parallel connection, 447
Pascal adding machine, 68
Pascal, Baise, 67, 68
Periodic function, 271
Pi, calculation of, 402-404
Pigeonhole example, 4, 5
Pivot, 378
Pivotal equation, 370
Pivotal row, 377
Place value, 60, 61
PL/1, 156
Polar coordinate system, 269
Polynomial, 154, 163, 220, 266, 267
Premultiplication, of matrices, 331
Prime factor, 167
Principal root, 191, 192
Processing symbol, 7
Profit lines, 357
Progression, arithmetic, 395-397, 408-410
geometric, 397, 398, 410, 411
Properties of real numbers, 50-54, 56
absolute value, 53, 54
associative law, 52
closure, 45, 46
commutative law, 45, 46
distributive law, 52
inverse, 56
magnitude, 53

Quadrant, 248
Quadratic, 166-168
Quadratic equation, 221
Quadratic formula, 225-227
Quadratic function, 262-267

Radian, 272
Radical, 191-193
Radicand, 191
Radius vector, 267
Range, 234, 237
Ratio, common, 397
Rational number, 48, 49, 170-183, 413, 414
Real axis, 50
Real number, 50
Rectangular cartesian coordinates, 248, 249
Recursive formula, 393
Regula falsi, 427
Repeating decimal, 62
Root, 191, 196, 220, 221
complex, 227
principal, 191, 192
Roots, calculation of, Bolzano method, 421-426
false position method, 426-429

Scalar product, 329
Scientific notation, 61, 62
Second order determinant, 313
Sector, 99
Sequence, 389-411
alternating, 398
arithmetic, 395-397
bounded, 400
geometric, 397, 398
unbounded, 400
Sequence function, 389
Series, 404-414
Series connection, 447
Sets, 13-41
Sign, of fraction, 172, 173
Significance, 60
Significant digit, 63, 64
Simplex method, 380
Simplex tableau, 381-385
Simultaneous equations, 295-320
Sine, 269-271

Slack variable, 368
Slope, 251, 253
Sorting, 50, 51
Square root computation, 417-420
Storage address, 99
Storage register, 46
Subprogram, 245
Subroutine, 245
Subscript, 325
Subscripted variable, Fortran, 236
Subscripting, Fortran, 325
Subsets, 16
Subtraction, binary, 87-88, 91-94
using complements, 91-94
of algebraic quantities, 160-162
of hexadecimal numbers, 107, 108
of octal numbers, 107, 108
Summation, 347, 405
Summation notation, 347
Symbolic logic, 114

Tangent, 269-271
Term, 155
Termination symbol, 8
TEST UNDER MASK, 145
Transformation matrix, 331, 332
Trigonometry, 267-273
Truth tables, 120
Truth values, 113
Two's complement, 91

Unbounded sequence, 400
Union of sets, 17
Universal set, 16
USA Standards Institute, 6
USASI, 6

Variable, 154, 155
algebra, 154
dependent, 234, 243
Fortran, 15
independent, 234, 243
Vector, column, 323
Venn diagram, 21
Vertex, 262

Word, 94
Word problem, 208-210

y intercept, 251, 252

Zero, of a function, 264